HISTORY OF

THE THIRTIETH REGIMENT

Dedicated

TO THE MEMORY

OF THE OFFICERS AND MEN

OF THE EAST LANCASHIRE REGIMENT

WHO GAVE THEIR LIVES FOR THEIR
COUNTRY IN THE GREAT WAR
AND PROVED THEMSELVES
WORTHY SUCCESSORS OF
THOSE WHOSE DEEDS
ARE RECORDED IN
THIS BOOK.

HISTORY OF
THE THIRTIETH REGIMENT
NOW THE
FIRST BATTALION EAST LANCASHIRE REGIMENT
1689–1881

BY
LIEUT.-COLONEL NEIL BANNATYNE

PREFACE

FORTY years ago when the 30th became the First Battalion of the East Lancashire Regiment, some of the officers felt that it was desirable to have a historical record of the old corps to that date. They had many difficulties to overcome. One great obstacle was that as they were serving in India they could not pursue their inquiries in person. Another was that the officers of the Public Record Office had not had time to arrange fully and to index the mass of documents handed over to them from the public departments, particularly by the Admiralty and War Office. In spite of all difficulties a most interesting and, under the circumstances, a wonderfully correct "Historical Record of the 30th Regiment" was produced in 1887.

When, through the indefatigable labours of the officers of the Public Record Office, and of other great libraries, the sources of information were increased and rendered more easily accessible, it was felt that the "Historical Record" should be amplified and in some places revised, and the present history is the result.

The author desires to express his gratitude to all who have helped him in collecting material, especially to the staffs of the British Museum, of the Public Record Office and of the Royal United Service Institution. He particularly wishes to record his indebtedness to the late Major Wylly when librarian of the last named institution for his ever ready help and sympathy and to Miss E. M. Fairbrother, student at the British Museum and Public Record Office.

He trusts that he is not too sanguine in cherishing a hope that, even at his advanced age, he may live to see a history of the heroic deeds, for so they must be described, of all the battalions which form the East Lancashire Regiment.

AUGUST, 1923.

CONTENTS

PAGE

CHAPTER I

Raising of the Regiment. Defence of the Castle of Namur. Forcing of the French Lines between the Lys and the Scheldt. Siege and Assault of Namur. Disbandment. 1688–1698 1

CHAPTER II

Regiment raised again as Marines. Capture of Gibraltar. Battle of Malaga. Defence of Gibraltar. Capture of Barcelona. Victory of S. Estevan. Relief of Barcelona. Capture of Alicante. Defence of Lerida. Capture of Sardinia and Minorca. Defeat of French Fleet in Firth of Forth. Defence of Denia. Lincolnshire assigned as Recruiting Ground. Disbanded. 1702–1714 23

CHAPTER III

Reinstated in former Seniority in Army. Placed on Irish Establishment. Second Defence of Gibraltar. Expedition to Lorient. Lord Anson's Victory off Finisterre. 1715–1754 89

CHAPTER IV

Transferred to the British Establishment. Expeditions to Rochefort, St. Malo, Cherbourg, St. Malo a Second Time, and Action at St. Cast. Capture of Belleisle. Gibraltar, Britain, Ireland. 1755–1781 . 136

CHAPTER V

War of American Independence. Battle of Eutaw Springs. Created the Cambridgeshire Regiment. Suppression of Two Risings of Negroes in Dominica. 1781–1791 158

CHAPTER VI

French Revolutionary War. Defence of Toulon. Capture of S. Fiorenzo, Bastia and Calvi. Hotham's Naval Victory off Hyères. Four Companies Drafted and Sent Home to Recruit. 1791–1798 176

CHAPTER VII

Occupation of Messina. Siege and Capture of Valetta. Expedition to Egypt. 1799–1802 202

PAGE

CHAPTER VIII

The Regiment divided into Two Battalions and serves for two years in Ireland, after which the Battalions separate. First Battalion: Expedition to the Elbe. Expeditions to the Coast of Java and Macao. Service in Southern India. Second Battalion: Irish Service. Portugal, Gibraltar, Cadiz. Return to Portugal. The Battalion joins Wellington in the Lines of Torres Vedras. 1803–Oct., 1810 . . . 225

CHAPTER IX

Second Battalion: Peninsula. Torres Vedras. Fuentes Onoro, Sabugal, Ciudad Rodrigo, Badajoz, Salamanca, Villa Muriel. 1810–1813 . . 246

CHAPTER X

Second Battalion: Siege of Antwerp. Regimental Depot re-formed at Colchester. Return of Prisoners of War 296

CHAPTER XI

Quatre Bras and Waterloo. The End of the Second Battalion. 1815–1818 309

CHAPTER XII

India. The Mahratta War. The Second Battalion Absorbed by the First. Asseerghur. Seven Years in the Nizam's Territory. Return to the Coast and Embarkation for Home. Peace Service at Home, in the Colonies, and in the Ionian Islands up to the Outbreak of the Crimean War. 1811–1853 359

CHAPTER XIII

The Crimean War. 1854–1856 391

CHAPTER XIV

Gibraltar. Home. Canada. Home. India. The End. 1856–1881 . . 445

Index 469

ILLUSTRATIONS

FACING PAGE

Drummer and Musketeer, 1689 1
Three representative Commanding Officers: General Sir Chas. Wills, K.B.; Colonel Jas. Thos. Mauleverer, C.B.; Colonel Alexander Hamilton, C.B. . . 22
Officer, 1736 117
Officoer, 1742 122
Private, 1745 123
Grenadier, 1751 133
Officer and Drummer, 1755 136
Officer, 1772 152
Grenadier, 1772 153
Officer and Private, 1792 176
Officer, 1806 232
Private, 1806 233
Company Officer and Private, 1815 308
Fifer, 1816 355
Colonel and Officer, 1818 358

MAPS

Flanders . . . .
Namur . . . .
Spain and Portugal . . . .
Toulon and District . . . .
Lower Egypt . . . .
Actions of March 13 and 21 (Battle of Alexandria) and Battalion action at the Green Hill, August 17, 1801 . . . .
Spanish and Portuguese Frontier—Ciudad Rodrigo . . . .
Badajoz and its defences . . . .
Quatre Bras: Gitaut's Cuirassiers' advance . . . .
Quatre Bras: Position of Halkett's Brigade and Picton's Division . . .
Waterloo: Position at 11 a.m. . . . .
Waterloo: Wellington's dispositions to meet the attack of the Imperial Guard .
South-West Crimea . . . .
Inkerman, Shell Hill: First phase of the battle . . . . . .
Sebastopol: Relative position of principal forts . . . . . .

CONTENTS

PAGE

CHAPTER I

Raising of the Regiment. Defence of the Castle of Namur. Forcing of the French Lines between the Lys and the Scheldt. Siege and Assault of Namur. Disbandment. 1688–1698 1

CHAPTER II

Regiment raised again as Marines. Capture of Gibraltar. Battle of Malaga. Defence of Gibraltar. Capture of Barcelona. Victory of S. Estevan. Relief of Barcelona. Capture of Alicante. Defence of Lerida. Capture of Sardinia and Minorca. Defeat of French Fleet in Firth of Forth. Defence of Denia. Lincolnshire assigned as Recruiting Ground. Disbanded. 1702–1714 23

CHAPTER III

Reinstated in former Seniority in Army. Placed on Irish Establishment. Second Defence of Gibraltar. Expedition to Lorient. Lord Anson's Victory off Finisterre. 1715–1754 89

CHAPTER IV

Transferred to the British Establishment. Expeditions to Rochefort, St. Malo, Cherbourg, St. Malo a Second Time, and Action at St. Cast. Capture of Belleisle. Gibraltar, Britain, Ireland. 1755–1781 . 136

CHAPTER V

War of American Independence. Battle of Eutaw Springs. Created the Cambridgeshire Regiment. Suppression of Two Risings of Negroes in Dominica. 1781–1791 158

CHAPTER VI

French Revolutionary War. Defence of Toulon. Capture of S. Fiorenzo, Bastia and Calvi. Hotham's Naval Victory off Hyères. Four Companies Drafted and Sent Home to Recruit. 1791–1798 176

CHAPTER VII

Occupation of Messina. Siege and Capture of Valetta. Expedition to Egypt. 1799–1802 202

PAGE

CHAPTER VIII

The Regiment divided into Two Battalions and serves for two years in Ireland, after which the Battalions separate. First Battalion: Expedition to the Elbe. Expeditions to the Coast of Java and Macao. Service in Southern India. Second Battalion: Irish Service. Portugal, Gibraltar, Cadiz. Return to Portugal. The Battalion joins Wellington in the Lines of Torres Vedras. 1803–Oct., 1810 . . 225

CHAPTER IX

Second Battalion: Peninsula. Torres Vedras. Fuentes Onoro, Sabugal, Ciudad Rodrigo, Badajoz, Salamanca, Villa Muriel. 1810–1813 . . 246

CHAPTER X

Second Battalion: Siege of Antwerp. Regimental Depot re-formed at Colchester. Return of Prisoners of War 296

CHAPTER XI

Quatre Bras and Waterloo. The End of the Second Battalion. 1815–1818 309

CHAPTER XII

India. The Mahratta War. The Second Battalion Absorbed by the First. Asseerghur. Seven Years in the Nizam's Territory. Return to the Coast and Embarkation for Home. Peace Service at Home, in the Colonies, and in the Ionian Islands up to the Outbreak of the Crimean War. 1811–1853 359

CHAPTER XIII

The Crimean War. 1854–1856 391

CHAPTER XIV

Gibraltar. Home. Canada. Home. India. The End. 1856–1881 . . 445

Index 469

ILLUSTRATIONS

FACING PAGE

Drummer and Musketeer, 1689 1
Three representative Commanding Officers: General Sir Chas. Wills, K.B.; Colonel Jas. Thos. Mauleverer, C.B.; Colonel Alexander Hamilton, C.B. . . 22
Officer, 1736 117
Officer, 1742 122
Private, 1745 123
Grenadier, 1751 133
Officer and Drummer, 1755 136
Officer, 1772 152
Grenadier, 1772 153
Officer and Private, 1792 176
Officer, 1806 232
Private, 1806 233
Company Officer and Private, 1815 308
Fifer, 1816 355
Colonel and Officer, 1818 358

MAPS

Flanders . . . . [illegible]
Namur . . . . [illegible]
Spain and Portugal . . . . [illegible]
Toulon and District . . . . [illegible]
Lower Egypt . . . . [illegible]
Actions of March 13 and 21 (Battle of Alexandria) and Battalion action at the Green Hill, August 17, 1801 . . . . [illegible]
Spanish and Portuguese Frontier—Ciudad Rodrigo . . . . [illegible]
Badajoz and its defences . . . . [illegible]
Quatre Bras: Gitaut's Cuirassiers' advance . . . . [illegible]
Quatre Bras: Position of Halkett's Brigade and Picton's Division . . . . [illegible]
Waterloo: Position at 11 a.m. . . . . [illegible]
Waterloo: Wellington's dispositions to meet the attack of the Imperial Guard . [illegible]
South-West Crimea . . . . [illegible]
Inkerman, Shell Hill: First phase of the battle . . . . [illegible]
Sebastopol: Relative position of principal forts . . . . [illegible]

DRUMMER AND MUSKETEER, 1689.

OFFICER, 1736.

OFFICER, 1742.

PRIVATE, 1745.

GRENADIER, 1751.

OFFICER AND DRUMMER, 1755.

OFFICER, 1772

GRENADIER, 1772.

OFFICER AND PRIVATE, 1792.

OFFICER, 1806.

PRIVATE, 1806.

COMPANY OFFICER AND PRIVATE, 1815.
The Field Officers and Regimental Staff still wore the cocked hat.

FIFER, 1816.

COLONEL AND OFFICER, 1818.

CHAPTER I

RAISING OF THE REGIMENT. DEFENCE OF THE CASTLE OF NAMUR. FORCING OF THE FRENCH LINES BETWEEN THE LYS AND THE SCHELDT. SIEGE AND ASSAULT OF NAMUR. DISBANDMENT.

1688–1698

ON November 5th, 1688, William, Prince of Orange, at the invitation of the leaders of the Whig Party, landed at Torbay with a force of English, Scottish and Dutch troops and commenced his march to London, gathering strength as he advanced.

On December 18th he entered the capital, which his father-in-law, King James II, quitted for Rochester. Four days later James fled from his kingdom, and Barillon, the French ambassador and the representative of French domination in England, was expelled the kingdom by Prince William's orders.

On February 13th, 1689, the Convention Parliament offered the crown of England to Prince William and his wife Mary, daughter of James II, and they were proclaimed on that day King and Queen of England, France and Ireland.

The Catholics in Ireland, however, who adhered to the cause of James II, were supreme in Leinster, Munster and Connaught and, on March 12th, James landed at Kinsale with some French support and the promise of more.

On May 11th the Commissioners of the Parliament of Scotland offered the crown of that Kingdom to William and Mary. It was accepted, and two days later war was declared against France in the name of the three kingdoms.

Louis XIV had made war against the United Provinces as soon as he heard of William's expedition to England.

As it was evident that the success of the revolution would be followed by a French war as well as a civil war in Ireland, and perhaps in Scotland, William had from the first been in communication with the leading men in their respective counties with a view to raising soldiers.

Among other noblemen and landed proprietors, Sir George Saunderson, Baronet, of Saxby, Lincolnshire, and fifth Viscount Castleton, in the Kingdom of Ireland, received a commission to raise a regiment of foot, of which he was to be colonel. He had vast estates in Lincolnshire and Yorkshire, and Saxby is a few miles to the east of the City of Lincoln. The House of Commons votes show that the regiment was primarily intended for service in Ireland. Lord Castleton's commission is dated the eighth of March, 1689, which may be considered the Regimental Birthday.

Lord Castleton had played a prominent part in public life. As a young man he had shared in the rising in 1659 against the tyranny of the Army and the Rump, and had been sent to the Tower. At the Restoration he had become the senior Knight of the Shire for Lincolnshire, a position he held till 1698. In 1667 he had been commissioned to raise one of the many troops of horse authorized for defence at the time. In 1669 he was one of the founders of the Lincoln race meeting.

The lieut.-colonel was Thomas Fairfax, a cadet of the great Yorkshire house. He is described by his friend Evelyn in his diary as "a soldier, a traveller, an excellent musician, a good-natured, well-bred gentleman." He had been a captain in the Guards, and in 1678 had served a campaign in Flanders, as major in Sir Harry Goodrickes' foot. Under King James II he had commanded a regiment of foot in Ireland, until the corps was remodelled by Tyrconnel. He brought with him to his new regiment, Matthew Des Vaux, who had been captain of grenadiers, and George Burston, who had been an ensign under him in Ireland.

The appointment of a major was not published at this time.

The regiment was to consist of 12 battalion companies and 1 of grenadiers, and the beating orders issued to captains on March 16th, fix the establishment of each company at 69 privates, 3 sergeants, 2 corporals, and 1 drummer. The levy money was 20 shillings. York was named as the rendezvous to which each company was to march as soon as complete, or nearly complete.

In a few days after the issue of beating orders, passes were granted for Lord Castleton and his lieut.-colonel to proceed to York, for Captain Saunderson and Lieutenant Bedford to Sheffield, and for Captain Saunderson and Lieutenant Godfrey to Lincoln. They are the first company officers we hear of as appointed to the regiment. The captain who went to Yorkshire to recruit was probably the Hon. Charles Saunderson, the eldest son and heir to the title. When a boy he had been appointed to the Navy and had joined his ship, but nothing further is known of his sea service. Captain the Hon. Thomas Saunderson was the second son; he always recruited his company in Lincolnshire. Bedford retired in 1708; to the end he recruited his company in Yorkshire.

By the beginning of June the regiment was united at York, clothed and armed. The uniform was a hat with a broad brim, a grey coat, waistcoat and breeches with shoes and stockings; the facings were purple and the drummers' coats and waistcoats were also of that colour with grey facings. The grenadiers wore fur caps instead of hats, to enable them to sling firelocks easily before handling their grenades. An overcoat or surtout was supplied on active service.

A regiment of that day presented a brilliant appearance. The officers were resplendent with laced and plumed hats, powdered perukes, cravats of rich lace, coats with knots of ribbon at the shoulders and laced with gold at the seams. A broad embroidered baldrick over the right shoulder was crossed with the silk sash worn over the left. The mounted officers wore the cuirass, the captain's gorget was gilt, that of the lieutenant of blued steel, and the ensign's bright steel with silver studs. The captain's colour waved at the head of every company except the grenadiers. The pikemen, selected for their size and strength, wore the headpiece of iron and the corselet, over which was a broad shoulder belt to sustain the sword. The drummers wore hanging sleeves and the royal cipher on back and breast.

The clothing was free, supplied biennially after the first year. There is no record of the arms supplied to Castleton's regiment, but from the numbers supplied to other regiments at this time, we may assume that the number of pikes was 168 and that 74 firelocks with slings and bayonets were issued to the grenadiers, and that the musketeers of 5 battalion companies also were armed with firelocks, but without slings or bayonets. The remaining 7 companies would receive matchlocks, without rests, slings or bayonets. The slow match used in discharging the matchlock was worn round the hat, tied in a loose knot in front and, in the field, lighted at both ends. The pike was 11 feet long in the shaft. The grenadiers carried a hammer hatchet on the right side, and a broad belt over the left shoulder to sustain a grenade pouch. All ranks wore swords. The sergeants carried the halberd, 6 feet long in the shaft, and the officers carried the half-pike or spontoon, except the ensigns carrying the colours and the grenadier officers; the latter carried a light fusil.

The pay was issued under two heads—

	Subsistence.		Off-reckonings.		Total.	
	s.	*d.*	*s.*	*d.*	£	*s.*
Colonel and captain	15	6	4	6	1	0
Lieut.-colonel and captain	12	0	3	0	0	15
Major and captain	10	9	2	3	0	13
Captain	5	0	3	0	0	8
Lieutenant	3	0	1	0	0	4
Ensign	2	6	0	6	0	3
Staff—						
Adjutant	3	6	0	6	0	4
Quartermaster	3	6	0	6	0	4
Chirurgeon	3	6	0	6	0	4
Chirurgeon's Mate	1	6	0	6	0	2
Chaplain	5	0	1	0	0	6

The gross pay of the 3 field officers includes 8 shillings a day, their pay as captains of companies. The captain-lieutenant who commanded the colonel's company had the rank and pay of a lieutenant only. In the 30th he was always the senior lieutenant.

	s.	*d.*	*s.*	*d.*	*s.*	*d.*
Sergeant	1	0	0	6	1	6
Corporal	0	9	0	3	1	0
Drummer	0	9	0	3	1	0
Private	0	6	0	2	0	8

No rations were issued and there were no barracks. The soldier paid his billetmaster $2\frac{1}{2}$*d.* a day for board and lodging.

The officers' off-reckonings were retained to satisfy any charges which Government had against them, and the balance was paid to them whenever there was a general settlement with the regiment. The off-reckonings of the other ranks formed a fund at the disposal of the colonel, from which he paid for regimental clothing, side arms and accoutrements, the firearms being supplied by Government. The drill of the period is interesting. In the first place great precision was insisted on in the handling of the cumbrous weapons with which the men were armed. The reason of this is obvious when we remember that cartridges were not in general use, that both ramrod and primer had to be used and that half the regiment carried a lighted match. Although the companies fell in by files, 6 deep, line was formed for action and reliance seems to have been placed more upon fire than on shock tactics. In line the pikemen, 6 deep, were massed in the centre, the musketeers, 3 deep, formed the right and left wings. The captains and lieutenants were in front of their musketeers. The 12 ensigns, each with his company's colour in hand, were in front of their pikes. The commanding officer was in front of the ensigns. The captain of grenadiers with half his company was in front of the commanding officer and a section of grenadiers was on each flank under a lieutenant. When preparing for action each ensign handed his colour to the right-hand front rank man of his division of pikes and, taking the man's spear, fell in in his place. The colours were borne by the pikemen, whose privilege it was to carry them, into the midst of the pikes. Firing was usually by the second and third ranks alternately. The front rank knelt for them to fire over it, and its fire was always held in reserve and only given by word of the commanding officer. This was to guard against the danger of being caught unloaded, which was a cause for anxiety as long as the muzzle-loader was in use. In a few years an improved system of firing by platoons instead of ranks was introduced, and when all regiments were armed with the firelock and bayonet, the pikemen were no longer of use and disappeared.

While the regiment was in process of formation the following events of military importance had taken place:—

On March 8th, 1689, Lord Marlborough had taken 12 regiments to Flanders to assist the Dutch, hard pressed by the French.

On March 28th the first Mutiny Act was passed by Parliament.

In Ireland King James II had raised an army and, assisted by French troops, had occupied the country, with the exception of Enniskillen and Londonderry, the latter of which was closely besieged. In Scotland King William's Government was busy raising troops to support the 3 Scottish regiments which accompanied him from Holland in opposing a threatened rising of some of the Highland clans. In England William had nearly completed arrangements for the relief of Londonderry, and for the despatch of an army under the Duke of Schomberg to assert his authority in Ireland. Apparently the intention was to employ Castleton's on Irish service, and on June 16th it was ordered to march from York to Preston. The regiment was now completed to 13 companies, 60 strong, and a committee of the House of Commons had visited it to see that the men had received all they were entitled to.

The following were the original officers, as far as is known:—

COLONEL—

George Viscount Castleton . . . . appointed 8th March, 1689.

LIEUT.-COLONEL—

Thomas Fairfax . . . . . . „ „ „ „

MAJOR—

Thomas Barrington . . . . . appointed 8th March, 1689.

CAPTAINS—

Charles Saunderson	,,	16th	,,	,,
Thomas Saunderson	,,	,,	,,	,,
Edward Turney (left the Regiment 18th July, 1689)	,,	,,	,,	,,
John Symons	,,	,,	,,	,,
Matthew Desvaux (Grenadiers)	,,	,,	,,	,,
John Nash (not mentioned after July, 1689)	,,	,,	,,	,,
Richard Beaumont	,,	,,	,,	,,
Edward Phillips	,,	,,	,,	,,
George Whichcott	,,	,,	,,	,,
Francis Coney (not mentioned again).				

LIEUTENANTS—

Isaack Knight	,,	,,	,,	,,
Thomas Bedford	,,	,,	,,	,,
Gabriel Mustemberger	,,	18th	June	,,
Charles Steigar	,,	30th	,,	,,

ENSIGNS—

William Bishop	,,	15th	,,	,,
Peter de St. Just	,,	1st	Nov.	,,

Of the above officers, Captain Beaumont was of Lascelles Hall, Yorkshire. George Whichcott was of a well-known Lincolnshire family. Whether he is identical with the Knight of the Shire in 1698 and 1705 is uncertain. A George Whichcott was executor of the will of Colonel Pownall of the 30th. There were many families of the name of Coney in Lincolnshire, one connected through the Suttons with Lord Castleton. It is not known to which Capt. Francis Coney belonged. Lieut. Isaack Knight was of the family of Kiddington in Lincolnshire. Charles Steigar was a Huguenot nobleman of Berne, Switzerland.

The movement to Ireland never took place, for on July 27th the Highland clans under Viscount Dundee routed the Scottish Army under General Mackay at Killiecrankie, and all available troops were moved to Mackay's support. On August 3rd an order was issued for Castleton's to march from Preston, 6 companies to Kingston-on-Hull, 1 company to Tynemouth, and 6 companies to Berwick-on-Tweed.

Dundee had been mortally wounded at Killiecrankie, and after the bloody repulse of the Highlanders by the Cameronians at Dunkeld on August 21st, the rebellion began to die out; the regiment therefore did not cross the border at this time.

Schomberg had landed in Ireland in the beginning of August, and the men of his green regiments died like flies from bad weather, bad clothing, want of food and privation of every sort, due to the dishonesty of the public departments backed by their own inexperience and want of discipline. Schomberg soon called for the more seasoned regiments which had served under Mackay, and thus caused the move of Castleton's into Scotland in the middle of November to replace them. Four companies were left to hold Hull, 1 at Tynemouth and 1 at Berwick, and only 7 entered Scotland.

The following are the promotions and appointments for 1690 :—

CAPTAINS—

Howard Phipps, before April, probably *vice* Turney.
Vincent Grantham, 1st May, *vice* Nash.
George Brathwayt, 4th May, *vice* Desvaux. Captain Brathwayt took over the Grenadiers.
George Burston, 6th August, probably *vice* Francis Coney.

LIEUTENANTS—

James Livesay	30th Sept.
John Saunders	3rd Nov.
John Green	,, ,,

ENSIGNS—

Richard Bolton	1st March.
John Alured	,, April.
Richard Sutton	,, ,,
John Green	,, ,,
Abraham Vanbelle	,, ,,
Richard Middlemore	,, July.
John Hicks	3rd Nov.

Of the above, Captain Phipps belonged to a distinguished Lincolnshire family which afterwards held the titles of Mulgrave and Normanby.

The Granthams were also a well-known county family; the first Lord Castleton's mother was a daughter of the Vincent Grantham of that day.

George Burston was the son of an Irish clergyman. He had been dismissed from the Irish Army for being a Protestant, and had served as a volunteer in Lord Tollemache's regiment at the Revolution.

John Saunders, captain of the grenadiers for 15 years, always recruited his company in Lincolnshire, but it is not known to which of the many families of Saunders he belonged. He had been appointed to the Marines in October, but had not joined them.

Richard Sutton, connected with the Saundersons, sat as one of the members of Parliament for Newark in 1702, the other member being James Saunderson, youngest son of Lord Castleton.

Richard Middlemore belonged to a distinguished family of Barton-on-Humber, Lincolnshire.

On January 1st a royal order was issued to recruit Castleton's to 100 private soldiers in each company.

Recruiting must have been brisk. On February 27th an order was issued to pay the £40 levy money for each company to Colonel Lord Castleton as soon as the commissaries had held their musters. But the regiment did not retain the recruits long. Schomberg in his despatches strongly urged the Government to send no young regiments to Ireland, but to supply drafts to fill up the older regiments which had an established discipline, and in consequence Lord Castleton was ordered on March 13th to send the recruits under his command in charge of conducting officers to Belfast, 220 to be delivered to the officers of Colonel Hastings' regiment, which had fought at Killiecrankie, and 300 to the officers of Sir Henry Belasys' regiment, the conducting officers to return. This draft swept away the forty men per company just raised.

Towards the end of the year the course of events in Ireland enabled King William to leave the command in that country to Baron Ginkell, and he hoped that when he took the field in Flanders at the head of the allied army in 1691, he would have with him a strong force of English and Scottish troops. Castleton's regiment was therefore assembled at Hull to march south. A royal order of December 6th directs that upon the arrival of seven companies of the regiment "from our Kingdom of Scotland," several companies should march to Hull.

The following are the promotions and appointments for

1691

CAPTAINS—

William Godfrey, before March.
Abraham Rogers, 3rd October.

LIEUTENANTS—

Edward Coney, 6th October to Capt. Charles Saunderson.
Richard Bolton, 19th November to Capt. Charles Saunderson.

ENSIGNS—

Richard Dealtry, 3rd March to Capt. Godfrey.
George Saunderson, 3rd March to Capt. Phillips.

George Saunderson was probably of the Saundersons of Thoresby Abbey, Lincolnshire, and great-grandson of the first Lord Castleton.

The regiment quitted Hull on January 21st and was assembled at Portsmouth about February 20th.

On March 19th the regiment was almost completely rearmed, all musketeers receiving the firelock and bayonet. Whether the number of pikes was reduced we do not know, but there is no trace of new pikes being issued, and if the number in possession still stood at about 170 the proportion to the number of firelocks would be much smaller, the strength of all ranks having increased during two years by a third. The probability is, however, that the actual number of pikes was reduced as the colours of the nine junior companies disappeared about this time, only those of the colonel, lieut.-colonel and major remaining. It is possible that the new bayonets were of the socket and ring pattern which enabled the soldier to fire with the bayonet fixed. Most likely the blade was sword shaped; two or three years later the blades were certainly of that pattern.

The firelock was certainly a great improvement on the old matchlock, but it shared with it one great defect in the wooden ramrod or "scowrer" supplied at 7s. a hundred and certain sooner or later to warp and break. This was one of the reasons why the first fire given by a regiment in action was more effective than those later in the day, and why commanding officers tried to reserve it as long as they could.

On December 14th the order came at last to proceed on active service. The regiment now numbered 925 effectives and embarked, 625 men on board Dutch men-of-war and 300 on hired transports, at Portsmouth. It landed at Willemstad and went into winter quarters at Bergen op Zoom at the beginning of January, 1692. Bergen is sixteen miles south of Willemstad.

There is no return of officers who landed with the regiment in Holland. Probably they were very few. The British armies in Flanders were in winter quarters where one officer a company was held to be enough; the others had leave to return home to recruit, attend Parliament, look after their affairs, or amuse themselves. The King resided near London in the winter, and rejoined his army in the spring, usually waiting till Parliament had risen. He expected all officers to be present and the army to be ready to take the field on his arrival. Hence there was a rush of officers of all ranks from London to join before him, and every conveyance and sailing packet was crowded.

1692

The establishment of the regiment for the year was: Companies 13, officers 44, non-com. officers 104, and private soldiers 780, including 69 officers' servants. The drummers are included in the 104 non-commissioned officers.

The following are the promotions and appointments for the year:—

CAPTAINS—

Henry Frankland for Capt. Edward Phillips . . 1st April
Ric. Beaumont, junr. *vice* Ric. Beaumont, senr. (Here father is succeeded by son) . . . . . . ,, Aug.

LIEUTENANTS—

William Lascelles Capt. Lieut. for Frankland . . 1st April
Isaac Knight Capt. Lieut. for Lascelles . . . . ,, Aug.
—— Wolstenholme to Capt. Charles Saunderson . . 7th March
William Leach to Capt. George Burston . . . 1st April
John Sye to Major Thomas Barrington . . . . ,, ,,
William Whichcott to Capt. Thomas Saunderson . . ,, ,,
Thomas Broxholme to Capt. Vincent Grantham . . ,, ,,
Benjamin Marolfe to Capt. Henry Frankland . . 28th Oct.
Marc Antony Bernard to Capt. Abraham Rogers . 26th Nov.

ENSIGNS—

Robert Jackson to Capt. Thomas Saunderson . . 1st April
John Bernard to Capt. George Whichcott . . . ,, ,,
Thomas Broxholme to Capt. Vincent Grantham . . ,, ,,
Francis Filbridge to Capt. Will. Godfrey . . . 12th June
George Hazard to Capt. Vincent Grantham. . . 1st Aug.
George Terwhit to Capt. Thomas Saunderson . . 30th Nov.

Of the above, the Syes were a Lincolnshire family closely related to the Saundersons.

The Broxholmes were well known in Lincolnshire. A Broxholme was

Willemstadt
HOLLAND
Zeebrugge
Ostend
NORTH
SEA
Bergen-op-Zoom
R·Scheldt
Putten
Fort Lillo
Brasschaet
Antwerp
Bruges
Ghent
Oudenarde
B E L G I U M
Ypres
Courtrai
Brussels
Armentieres
Waterloo
Hal
Wavre
Lille
Genappe
R·Lys
Nivelles
Tournai
Soignies
R·Meuse
F
Namur
Quatre-Bras
R·Scheldt
Mons
Charleroi
R
Bavai
Ligny
Cambrai
Maubeuge
Dinant
A
Le Cateau
N
C
E
Scale of Miles.
0 10 20 30 40 50 60 70 80 90 100

among the twelve Knights of the Royal Oak created in the county by King Charles II. John Broxholme was Member of Parliament for Lincoln City 1640–47 and William Broxholme for Grimsby 1678–80.

The Bernards most likely came from Epworth, Lincolnshire. The Terwhits were also a Lincolnshire family and nearly related to the Saundersons. The name is also spelt Tirwhit or Tyrwhit.

The landing at Willemstad had an unfortunate effect on the regiment. The ill-health contracted at Bergen op Zoom seriously impaired the efficiency of Castleton's in the campaign of 1692. Two-thirds of the regiment, however, took the field in spring, when Lord Castleton received £200 to provide transport wagons, and £10 to buy a horse for the ammunition tumbril. Sixty horses or mules were allowed for regimental transport.

The French Government had two objectives in 1692; one the capture of Namur by an army commanded by the King in person, and the second an invasion of England by troops assembled near La Hogue, and protected by the fleet under Tourville.

In the beginning of May the French Army of the Netherlands directed by one strong Government was concentrated about Mons, and King William was at Brussels trying to hasten the concentration of the scattered forces of the Allies.

On the 18th he was at Louvain. From here he detached Castleton's and the Royal Welch Fusiliers to strengthen the garrison of Namur. The 30th was commanded by Lieut.-Colonel Fairfax. Castleton, who was an old man, did not take the field.

The French King invested Namur on the 23rd with 40,000 men, having a covering army of 60,000 under Luxembourg between him and the Allies. On the same day the Allies heard of the defeat of the French fleet off La Hogue by Admiral Russell. All thought of the invasion of England was now at an end.

On May 29th the concentration of King William's army having been completed, he thought himself strong enough to attempt the relief of Namur, but torrents of rain made movement impossible. On June 9th the city of Namur surrendered, and the French were able to contract their lines round the castle and to reinforce their covering army. The castle and a new work called Fort William made a gallant defence after the city had surrendered. Fort William fell on June 13th, but the castle held out for a week longer and then capitulated. The garrison, of which the 30th and the Royal Welch Fusiliers formed part, marched out with the honours of war, drums beating and colours flying, and joined King William at Louvain. Louis XIV returned to Paris in triumph after his capture of Namur.

On June 29th, King William reviewed the English army at Genappes. He had on parade 15 battalions, of which the 30th was one. Two days later he reviewed 10 regiments of the Scottish army. Genappes is of interest to us, for on the night before Waterloo, 123 years later, the 30th again passed through the place in the Duke of Wellington's rearguard.

On July 21st, King William left Genappes and marching by Nivelles reached Hal on the 22nd. Luxembourg, whose movements had been roughly parallel to those of the Allies, now made a false move. He encamped with his main body at Steenkirk, while Boufflers, his ablest lieutenant, lay seven miles off at Manny St. Jean. William determined to overwhelm Luxembourg on the 23rd before Boufflers could come to the aid of his chief. The scheme failed through faulty execution. As the 30th were scarcely engaged, it is sufficient to say that the allied vanguard was opposed without support to the French army and after fighting heroically was driven back. King William retired to the position at Hal from which he had advanced, covering his retreat with the grenadier companies of his army. This last movement is probably the only one in which men of the 30th took part. The regimental loss was 1 non-com. officer killed, 1 non-com. officer and 1 private wounded.

The army was rather infuriated than depressed by its defeat, and the King by the reckless gallantry he showed in the retreat had quite regained the confidence of his men who blamed the Dutch generals for the mismanagement of the battle.

In the spring of 1815 the 30th again marched from Hal to Steenkirk, when Sir Charles Alten's division was assembling for the campaign of Waterloo.

On August 1st the French decamped without beat of drum, leaving behind the wounded prisoners they had taken in the battle. They thus forfeited any moral superiority they may have gained from their victory.

On August 16th the allied army passed the Scheldt, and on the 21st Brigadier Ramsay taking the 30th and 5 other battalions marched towards Bruges where he joined General Talmash, who had 9 battalions with him. The object of this detachment from the main army was to hold out a hand to the Duke of Leinster who was expected to land at Ostend with 15 battalions from England. Leinster landed on August 22nd and being joined by Talmash occupied Dixmude and Furnes without opposition. On September 30th his army went into winter quarters, the 30th and 4 other battalions forming the garrison of Bruges. They along with the garrisons of Ostend and Ghent were called into the field in December owing to a move on the part of the French, but rain rendered the roads impassable and the troops on both sides returned to quarters. Winter quarters were estimated to last about 220 days.

1693

The establishment of the regiment remained the same in 1693 as in former years.

The following promotions and appointments were made :—

MAJOR—

Captain Charles Saunderson *vice* Barrington	4th March

CAPTAINS—

William Marshall *vice* Barrington	4th ,,
Isaac Knight *vice* Grantham	10th ,,
Charles Stygar *vice* G. Whichcott	1st May
Richard Sutton *vice* Marshall	1st June

LIEUTENANTS—

(Richard) Middlemore to Capt. Ric. Beaumont	10th March
Henry Fookes (Foulkes) to Capt. Marshall	1st May
William Smith to Capt. Frankland	1st ,,

SECOND LIEUTENANTS—

James Brathwayt to Capt. George Brathwayt (Grenadiers)	1st March
John Compton to Capt. Knight	1st July
Peter Minshall to Capt. Stygar	1st May

ENSIGNS—

—— Smith to Capt. Burston	1st ,,
Philip Brunskill to Lt.-Col. Fairfax	1st ,,
Alexander Swayn to Capt. Knight	1st ,,
—— Davison to the Colonel	1st ,,
William Whitoft to Major Saunderson	1st July
William Singleton to Capt. Beaumont.	1st ,,

Charles Saunderson on promotion kept his own company and Marshall took over Major Barrington's.

When the campaign opened in 1693 Castleton's did not take the field, but remained in garrison at Bruges.

Early in May, King William was at Brussels facing a French army, much stronger than his own, commanded by Luxembourg. The latter however was ordered to detach strong forces to the Rhine which placed the two armies on an equality. They remained facing each other till June 28th, when want of supplies forced Luxembourg to move towards the Meuse. William, then, to cause a diversion and also to feed his troops at his enemies' expense, detached the Duke of Wirtemburg to make an inroad into the parts of Flanders which the French had seized at the beginning of the war and which had not yet been visited by a hostile army. The French had covered the "Conquered Country," as it was called, by fortified lines which were strongest between the Lys and Scheldt where it was proposed that Wirtemburg should attack. The Duke, taking with him 40 squadrons, marched rapidly to Oudenarde, followed by General Ellenburg with 13 battalions including Argyle's and Bath's regiments.

At Oudenarde on July 4th the Duke picked up Castleton's which had marched from Bruges with 9 other battalions, and pushed on with them and his cavalry till he came in touch with the enemy near d'Ottignies. On the evening of the 6th, Ellenburg, with the remaining infantry and 12 guns and some pontoons, arrived within a league of Wirtemburg's force and halted to rest his men, sending on his quartermasters and markers to take up his ground. Wirtemburg at the same time closed to his right so that Ellenburg should come into line on his left.

The rain, which had been incessant for a week, now became worse, and except Count Horn's regiment, none of Ellenburg's force could reach Wirtemburg, indeed several men were drowned in attempting to cross the torrents in the darkness. To make the enemy think that the whole force was following him, Count Horn caused his drummers to play English, Scottish, and German marches, so that it might be believed that Bath's, Argyle's and the allied troops were coming into line. Fires were lit along the line and the markers were placed on sentry and visited with as much noise as possible by the quartermasters.

On the morning of the 7th, Ellenburg marched into position, but Wirtemburg judged it better to put off the attack for another day. To cheer the men after their hardships he ordered six gallons of corn brandy to be served out to each company.

The lines about to be attacked, consisting of a rampart and ditch with bastions at intervals, followed the line of the Espierre, a stream which, rising near Courtrai on the Lys, runs across the country between the two rivers and falls into the Scheldt at Pont D'Espierre. In some places the line is behind the river and in some the river is made to run through the ditch.

Wirtemburg had 23 battalions with 12 guns, of which 2 were 12 pounders, and 40 squadrons. The lines were held by 12 battalions of French and 24 squadrons under the Marquis de la Valette. He appears to have used no artillery, though some old guns were found in the lines.

The Duke of Wirtemburg had been busy preparing fascines to fill the ditch from the time of his arrival, and on the morning of July 8th he attacked in three columns, each led by a British regiment, Castleton's (30th) on the right against the redoubt of Beau Verde, Argyle's in the centre, and Bath's (15th) on the left. Each attack was supported by a detachment of guns, while the fire of the 2 12-pounders was directed on the French reserves and cavalry. The order of attack was the same in each column.

First an advanced party of grenadiers under a captain, then a detachment of 30 pikemen from each battalion in that column, carrying fascines and their pikes tied in bundles four by four to lay under the fascines to form a bridge, then the remainder of the grenadiers of the column, then the leading battalion.

The pikemen advanced to the edge of the ditch, covering their bodies with the fascines which they threw into the ditch and then made way for the grenadiers who charged. A good many fascines were carried away by the current and the water was up to the men's necks, but they got across somehow, and a better bridge must have been made, for we have a " draught," said to be very correct, showing Castleton's crossing in column, the pikes massed in the centre about the colours, and musketeers in several ranks round them.[1]

The French cavalry and reserves who ought to have charged would not face the fire of the guns, and on the allied cavalry crossing in the centre, where Argyle's regiment had captured a bridge, the whole French force quitted the field.

It is impossible to say what the losses of the regiment or of the army were. The opposition is said to have been stiff, and Mr. d'Auvergne mentions casually in his history that the two subalterns of Argyle's leading company were shot, but he does not pretend to give a full statement of the day's losses.

A disgraceful scene of plunder and indiscipline took place that night, 2 churches and 12 villages being burned. Wirtemburg thereupon published an order that men plundering would be put to death, and those setting fire to anything would be burned alive ; this was no idle threat, men were still burned by order of court martial. At the same time he ordered the inhabitants to supply *gratis* to officers and men, rations of meat and corn brandy.

On July 12th the Duke marched through the pass of St. Leger to near

[1] Only two colours are shown. We do not know when the major's colour was done away with.

Courtrai, and on the following day 1,200 horse and 600 foot were sent to force the *Pont à Tresein,* which was defended by 2,000 men; this is a bridge two miles from Tournay on the causeway between that fortress and Lille. The attack was made on the bridge by 200 grenadiers, supported by the fire of the artillery and of infantry from the houses near it and succeeded with small loss, only 14 or 15 men being killed and 1 English sergeant.

Mr. d'Auvergne, the chaplain of Lord Bath's regiment, says in his account of this action: "Our English and Scottish grenadiers who had the van, behaved themselves very well and got the approbation and applause of their commanding officer who was an eye-witness of their bravery."

On July 17th the Duke of Wirtemburg moved to Annapes for facility in foraging the Lille district, and as the march lay open to attack from Douay, he formed a rear-guard of Castleton's, Argyle's, and Bath's regiments. At Annapes the Duke distributed a ducat a man to the grenadiers and pikemen who had led the attack on the French lines on the 8th, and to the grenadiers who had led at the Pont à Tresein. Wirtemburg now began to withdraw from the conquered country, and his movements were hastened by a letter from King William telling of the defeat of our main army at Landen and ordering him to rejoin. On the 21st a march of 6 leagues brought him to Espiers, where he passed the French lines. On the 22nd he repassed the Scheldt at Oudenarde and by the King's order halted at Alost.

In July, after Wirtemburg had quitted the main army, the French under Luxembourg had captured Huy and threatened Liége. King William, alarmed for the safety of the Meuse fortresses, had marched to Landen, within 20 miles of Luxembourg's quarters, at the same time detaching from his army, which was already too weak, 8,000 men to Liége and Maestricht.

On July 19th, Luxembourg attacked and defeated the allied army after a day's desperate fighting in which the loss of the English and Scottish regiments had been especially severe. The King, who was in the thick of the fight, had one shot through his periwig, another through his sleeve, another carried off the knot of his scarf, slightly bruising his side. He fell back to Louvain and on the 21st encamped at Eppeghem. Finding his army in good heart, and the French too severely punished to think of pursuing, he stood his ground there and halted Wirtemburg at Alost. On August 2nd he concentrated his army at Wemmel and reviewed his troops.

Luxembourg's despatch announcing the victory of Landen was carried to Paris by M. d'Artagnan, who is an old friend to most of us, for he is *the* d'Artagnan of *The Three Musketeers.*

The French after a fortnight's inactivity began to move to the west. On the 8th, Luxembourg was at Nivelles, faced by the Allies at Hal.

During its halt at Hal the allied army was rejoined by the prisoners taken by the French at Landen, and here occurred a scene which was common enough in those days. Several British fugitives from the battle had been captured making for the sea-coast and brought back as deserters. Six guardsmen and 30 others were condemned for this crime, but the King pardoned half the guardsmen and 24 of the others, so after drawing lots for life and death the nine losers were hanged at the head of their respective regiments.

Early in October both armies retired to winter quarters. The 30th with 8 other regiments formed the garrison of Bruges.

1694

The establishment was the same as in previous years. The appointments and promotions were as follows :—

Thomas Saunderson to be Colonel *vice* Lord Castleton who retired.

Lieut.-Colonel John Ward to be Lieut.-Colonel *vice* Thomas Fairfax promoted Colonel of the Northumberland Fusiliers.

Captain Charles Wills from Erle's regiment, to be Major.

All three promotions dated November 6th.

Lieutenant Thomas Bedford to be Captain *vice* Rogers, 20th August.

Lieutenant Fookes to be Capt.-Lieutenant *vice* Bedford.

This name is spelt also Fowke or Foulkes.

William Saunderson to be Ensign to Captain Saunderson .	16th March
William Bernard „ „ „ Captain Rogers. .	16th March
William Wedall „ „ „ Captain Sutton. .	20th August
Lieutenant Richard Bolton to be Adjutant . . .	20th August
Henry Clarke to be Quartermaster	1st Jan.

When Lord Castleton prepared to retire from the service he naturally wished to leave the regiment he had raised to his heir. The death of Major Charles Saunderson interfered with this purpose for a time, but in November Lieut.-Colonel Fairfax was appointed colonel of the Northumberland Fusiliers, Lord Castleton retired and Thomas Saunderson, now heir to the title, was promoted colonel of the regiment. The promotion is apparently from senior captain ; we have no record of the majority made vacant by the death of Charles Saunderson having been filled up. Two outsiders, Ward and Wills, were brought in as Lieut.-Colonel and Major ; this is perhaps the price Lord Castleton had to pay for permission to keep the regiment in his own family. Colonel Thomas Saunderson led the regiment in the field except when employed as brigadier.

There is little of regimental interest in the campaign of 1694. Both sides were exhausted and could do little more than watch each other. The 30th were in the brigade of Sir David Collier, an officer of the Scots army. Thirty years later the regiment served under him again in the defence of Gibraltar.

On March 8th, King William arrived and ordered his army to concentrate at Louvain. Collier with his brigade joined the King there on June 4th. After some indecisive manœuvring the armies returned to winter quarters early in October, the 30th forming, as usual, part of the garrison of Bruges.

1695

The establishment for the year was as before.

The promotions and appointments were as follows :—

CAPTAINS—

Henry Fowke *vice* Beaumont	10th June
Clifford Whichcott *vice* Godfrey	10th June
John Saunders	10th to 20th June

CAPT.-LIEUTENANT—

Richard Bolton	10th June

LIEUTENANTS—

John Bernard to Capt. Jas. Braithwait	9th to 19th June
Will. Whitoft to Capt. George Burston	10th to 20th June
Alexander Swain to Capt. Jas. Braithwait	10th to 20th June
Will. Singleton to Lieut.-Colonel John Ward	10th to 20th June
John Davison to Captain John Saunders	15th July
Richard Place to Captain Isaac Knight	15th July

ENSIGNS—

Thomas Sedgwick to John Saunders	10th to 20th June
Theophilus de Vauclen to Charles Steigar	10th to 20th June
Robert Hume to Henry Fowke	10th to 20th June
John Sharpe to Isaac Knight	15th July
Stanhope Yarborough to The Colonel	15th July

The Lieutenant Davison mentioned above was probably a son of Davison of Elvet, Durham. William Davison had six sons whose names appear in the following order in the Davison Estate Act for working coal mines in the county of Durham: Alexander, Thomas, William, Charles, James, and Joseph.

The first Davison to appear in the regiment was —— Davison appointed Ensign to Lord Castleton in 1693. In 1695 he was promoted as above and his name is given by Mr. Dalton as John. He does not appear again. The second was Alexander, of whose appointment there is no trace, but who was placed on half-pay as a lieutenant in 1698. He did not rejoin. The third was Charles, appointed Lieutenant in 1702, the fourth William, Second-Lieutenant in the same year, the fifth James, Second-Lieutenant in 1704. There was also another Alexander Davison of a younger generation who was Adjutant of the regiment in 1720. John may be a clerical error for Thomas; the abbreviations constantly used by our ancestors such as Thos. and Jho. were a fruitful source of confusion. In any case four and perhaps five brothers served in the regiment.

The appointments were all signed by the King in the field, those of June at Becelaere when threatening the city of Ypres.

On May 14th the King arrived from England and formed two armies, one under himself with Headquarters at Arseel near Ghent and the other on his left under the Elector of Bavaria with Headquarters at Brussels. Collier's brigade joined the King's army and on parade stood on the left of the half brigade of Guards in the second line. Both armies were about 50,000 and in addition there was an allied force of 20,000 at Liége, formed of Brandenburgers and troops of the Bishopric.

The French army was not much inferior in numbers, but it had lost Luxembourg, who had died after the last campaign. Villeroi, a much weaker man, had succeeded. He was fortunate in having de Boufflers as his second, whose enterprise and resolution might make good his own defects.

The object of the Allies was the capture of Namur and, to draw the French away from that city, the King began a series of manœuvres to alarm Villeroi for his left. On June 3rd, William's Headquarters were at Becelaere, threatening Ypres and the line of the Lys. Villeroi at once moved to his left into the lines between the Lys and the Scheldt. For a fortnight by threatening first one fortress, then another, the King induced Villeroi to move more and more troops to his left until there were no French field troops east of the Scheldt except a small column under the Marquis d'Harcourt watching the Meuse. On the 17th the King saw that the time had come, and leaving his own army under de Vaudemont he marched on Namur with that commanded by the Elector of Bavaria.

The Elector marching by Ninove and Genappes (and Quatre Bras) converged on Namur with the Brandenburgers on the 23rd, Athlone's cavalry covering the movement and seizing the boats on the Sambre and Meuse. The investment had been accomplished ; everything had run as designed with one exception. De Boufflers, becoming uneasy, had paid a visit to Charleroi ; there the truth was at once evident to him. Taking some artillery and engineers he hurried east and picking up 7 regiments of dragoons from d'Harcourt's force threw himself into Namur just before the girdle round it was complete.

De Vaudemont had during this time placed his army as a covering force in front of Brussels.

King William soon satisfied himself that no dangerous counter stroke was to be feared from Villeroi, so the siege artillery was ordered up from Maestricht and Vaudemont was called upon for every man he could spare from the covering army ; thus the 30th came to be employed in the siege.

On June 28th, Major-General Ramsay reached Temploux, a league and a half from Namur, with Castleton's and 9 other British battalions and 4 battalions of allied troops.

Before attacking the city it was necessary to take possession of the heights of Bouge, which cover Namur on the west and were fortified with redoubts connected by a covered way. On the day the 30th reached Temploux, Dutch troops had opened trenches on the heights and by July 6th had carried them close up to the covered way connecting the redoubts of Coquelet and Balart. British troops now took over the trenches and on the 8th orders were given to storm the two redoubts and the covered way. Saunderson's and 5 other British battalions were brought up from Temploux, and towards evening two columns of attack were formed.

The 4 battalions of English and Scots Guards supported by 2 English line battalions formed the left column of attack, and the 2 battalions of Dutch Blue Guards, the 1st Battalion Royal Scots, the Fusiliers, Saunderson's, the present 23rd and 25th and Lauder's regiments formed the right.

The order of attack was nearly the same as that used by the Duke of Wirtemburg in forcing the lines on the Lys in 1693. First a body of fusiliers in three ranks carrying fascines, then the grenadiers, with three grenades apiece, then the pioneers with woolsacks and gabions, and then the battalions. The left attack was entirely successful ; after the pioneers had deposited their packs under cover of the fire of the fusiliers and grenadiers, the British Guards advanced without firing to the palisades and drove the enemy from them with one volley. The covered way was occupied and the enemy pursued down the hill close to the town.

The Dutch Blue Guards met with more resistance. Their right flank was confined by a steep ravine and they came under fire from the redoubt of St. Fiacre (C), and the enemy on the opposite bank of the ravine. It was not until they were reinforced first by the Royal Scots and Fusiliers and then by

Saunderson's and the 23rd and 25th, while Lauder's crossed the ravine to keep down the flanking fire, that the attack succeeded. In the assault Saunderson's had Lieutenant Swain killed, Captain Knight and Lieutenant John Bernard wounded.

All the advanced works from the Coquelet to the river Meuse were now in the possession of the besiegers and the approaches were pushed forward to the main works of the city, especially towards the gate of St. Nicolas (O) near the river, for here the batteries of the Brandenburgers on the other side of the Meuse could assist in the attack.

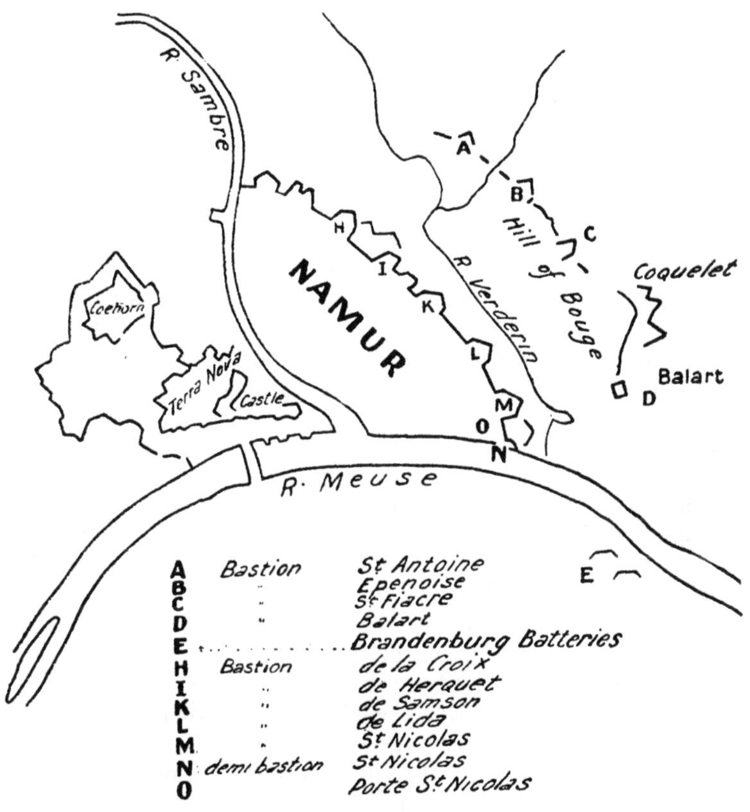

NAMUR

On July 16th all was ready for the assault on the defences of the gate of St. Nicolas

Colonel Clifford Walton's account of the attack is so spirited and clear that permission has been asked and obtained to make use of it :

" The Porte St. Nicolas was defended by the bastion St. Nicolas (M) on the left of the gate and the demi-bastion St. Roch (N) on the Meuse with a curtain connecting them ; in front of the curtain a ravelin ; and in front of the ravelin a counterguard. Along the whole front of the glacis flowed the Verderin.

" Towards five o'clock in the afternoon the attack commenced by the advance of 500 grenadiers from all the regiments in camp, except the Guards supported by 2 brigades under Brigadiers Selwyn and Lord George Hamilton.

"Selwyn's brigade consisted of the 23rd and 25th, Saunderson's and Lauder's. Major-General Ramsay commanded the attack.

"The grenadiers marched straight up to the palisades of the covered way and discharged their grenades over them. The 23rd and Saunderson's were the next to come up, the enemy's fire from the covered way was terrific and while the batteries of the Allies galled the French in their works the French redoubts Epinoise and St. Fiacre fired with fatal effect upon the English regiments as they marched up the glacis, but the assailants could not be driven back. Then the French spring four fougasses on the glacis. The English fled backwards as the earth opened and belched forth its deadly load; no man knew whether his next step might not place him again on the very nest of one of those fearful messengers of death, yet even this did not deter the British troops from again advancing. Again they reached the top of the glacis and then began to lodge woolpacks and gabions on the palisades over against the Bastion St. Nicolas (M); the work progressed rapidly until the enemy managed to set fire to the woolsacks; the assailants were thus again exposed for a long time to the full fire of the besieged, but the men stuck to their ground obstinately, refusing to give way again; some of the grenadiers even leaped over the palisades and fell fighting in the very thick of the foe.

"Meanwhile a body of Dutch troops crept along the bank of the river close under the covered way of the ravelin, the whole right face of the counterguard having been knocked to pieces. Three times were they repulsed with carnage but ultimately they also effected a lodgement and thus an effectual footing from the Meuse to beyond the Bastion St. Nicolas."

The British loss was from seven to eight hundred men, Ingoldsby's (23rd) and Saunderson's (30th) being the principal sufferers. No officers were killed in Saunderson's, but Captain Fowke (Foulke), Lieutenant Hazard, Ensigns Partridge and Pazaster were wounded. Pazaster is perhaps a mistake for Pallister, and the officer wounded may have been Hugh Palliser or Pallister who served as lieutenant in the following year though there is no record of his appointment as ensign. There was no person of the name of Pazaster in the regiment, nor as far as can be known in the army.

The town of Namur capitulated on July 24th, one of the conditions being a truce for forty-eight hours to enable the defenders to withdraw to the castle.

On the 28th the British troops crossed the Sambre above the town to take part in the attack on the castle and the works known as Coehorn, the name of the famous engineer who designed them.

On August 8th, Captain Richard Sutton greatly distinguished himself when in command of a covering party of 40 men of the regiment. Lord Cutts (Salamander) had in person placed the parties to protect the men working in the trenches. Captain Sutton with his party was on the left of the line. The men had just taken up their ground when a sudden sortie was made from the castle by 200 mounted dragoons and 500 grenadiers. The attack on our right was weak, but 100 of the dragoons boldly charged Captain Sutton's party. His men had been strongly warned against firing without word of command and he held his fire till the dragoons were almost on him, when at his word half the detachment fired. In the confusion caused by such a deadly volley he fell back and had nearly gained the trenches when the dragoons, encouraged by the thought that his men were now unloaded and defenceless, came on again. A volley from the other half company checked them and now the Spanish and Bavarian horse which were at hand charged the French and pursued them up to the palisades of their works. His behaviour on this occasion gained Captain Sutton the friendship of Lord Cutts, the idol of the English army. The incident speaks well not only for Captain Sutton but for the discipline of the regiment.

By August 11th the Allies had 136 guns and 50 mortars playing on the French works, and on the 19th the breaches being practicable a general assault was ordered for next day.

The British attack, which alone concerns us, was on the left, on the Terra Nova, a line of works which touches the Sambre with its right.

The attack was headed by a forlorn hope of 60 men followed by the grenadiers of the Guards regiments; they were followed by the grenadiers of all British regiments, the 17th and 18th, Buchan's and Mackay's regiments, were in support.

The grenadiers attacked with their accustomed gallantry but not with their usual success. When the grenadiers had lost half their number and the 4 regiments which reinforced them were little better off, Lord Cutts, who was wounded himself, hearing that the Bavarians in the neighbouring breach on his right were equally unsuccessful, was on the point of withdrawing his men when it occurred to him there was one chance left. He saw or guessed that the enemy, busy at the breaches, had neglected the salient between them. His call for volunteers was answered at once, Ensign Cockle of Mackay's was put at their head, and the colours of that regiment were ordered to follow closely on the volunteers. The salient was carried sword in hand and the colours planted, Mackay's rushed after while Ensign Cockle slewed round one or two guns and swept the place with their fire.

In this attack the regiment lost Lieutenant Middlemore of the grenadier company, killed.

Two days afterwards de Boufflers offered to surrender if not relieved by the 26th. He was granted the honours of war and marched out on that date with colours flying, drums beating, arms carried, bullet in mouth, matches lighted, and accompanied by 6 guns. His garrison of 13,000 men had fallen to 5,000.

We have no returns of the regimental losses from day to day and are indebted to the Rev. E. d'Auvergne for the names of the officers killed and wounded. There are two returns in "King William's Chest" which give the total loss during the siege. They are neither dated nor signed and they do not agree with d'Auvergne's, but they were evidently prepared for the King's information and probably at a few days' interval.

The following we take to have been made out immediately on the fall of the city and is obviously incomplete:—

"A list of what officers and soldiers is kill'd and wounded during the siege of Namur:—

		Killed.	*Wounded.*
"Saunderson.	Captains	—	3
	Lieutenants	—	3
	Ensigns	—	2
	Sergeants	5	3
	Private Men	56	121

"The number of private men killed in this regiment is only exceeded by the first regiment of Guards which is 65, and the number of men wounded is only exceeded by the second regiment of Blew Guards which is 136." (The Dutch Blue Guards were on the English establishment.)

The second return is as follows:

" Liste des Soldats morts et blessés devant Namur des Regiments Anglais depuis le commencement jusques à la fin du siège.

	Morts.	*Blessés.*
Saunderson	80	112"

We know that this is the later and the correct statement because 80 recruits were called for to replace the dead; some of the wounded had evidently died between the dates of those returns. The 80 recruits were furnished by a draft from a regiment in Ireland. It is the only occasion on which the regiment did not find its own recruits.

In addition to the officers already named it is certain that Lieutenant William Smith died of his wounds at Namur, because his widow received a pension on that account. No doubt there were other casualties the record of which has been lost.

As far as we know the total loss stands:—

		Officers.	*N.C.O.s and Men.*
Killed or died of wounds.—	Lieut.	Swain	
	,,	Middlemore	
	,,	Smith	80
Wounded.—	Captain	Fowke	
	,,	Knight	

(Another captain was reported to the King as wounded, but Mr. d'Auvergne does not give the name.)

	Lieut.	John Bernard	
	,,	Hazard	
	Ensign	Partridge	
	,,	Pazaster (Pallister)	112

There were hospitals in existence but they were not yet fit to cope with great operations of war, and the King probably did all he could for the wounded when he gave the Colonel a pound for each man. He also gave Colonel Saunderson 500 guilders (about £45) to be distributed among the wounded officers.

1696

The establishment was the same as in previous years.

The promotions and appointments were as follows:—

Isaac Tousain, Surgeon	1st March
(Isaac) Duplex, Ensign to Captain Bedford	1st ,,
Blaise Gervaisot, Lieut. to Captain Brathwayt . .	1st ,,
Hugh Palliser, Lieut. to Captain Foulkes	30th May
Andrew Day, Ensign to Major Wills	30th ,,
Ric. Bolton, Captain *vice* Foulkes	20th Sept.
James Harris, Captain-Lieut. *vice* Bolton	20th ,,

The campaign of this year possesses not the least interest regimentally. The French were in superior numbers to the Allies, who entrenched themselves to defend Brussels and the line of the Bruges canal. The King joined his army in May, but on August 16th "there being no appearance of any motion of consequence in the army," he left the field. Saunderson's which had been, during the summer, in Tiffin's brigade at Marykirk, went into winter quarters at Ostend.

1697

The establishment remained the same.

The following promotions and appointments were made:—

Charles Wills to be Lieut.-Colonel for Ward . . .	1st Aug.
Richard Sutton to be Major for Wills	1st ,,
Uriah Brereton to be Captain of Ward's company . .	1st ,,
Walter Palliser to be Captain for Steigar . . .	5th May
Bernard Pinnock to be Lieutenant to Lieut.-Colonel Ward.	31st ,,
Francis Filbridge, Lieutenant to Knight. . . .	10th July
Richard Tirwhit, Ensign to Whichcott	10th ,,
John Dalton, Ensign to Brathwayt	31st May
William Singleton to be Adjutant	31st ,,

The change among the field officers was caused by Ward being cashiered by a Council of War "for speaking things not for His Majesty's service." The King did not take a very serious view of the matter, for Ward was employed two years later at Nevis in the West Indies.

The regiment saw no fighting in 1697; the war was dying out. On September 20th the Peace of Ryswick was signed.

Soon after the Peace an order was issued for the reduction of two companies, but it does not seem to have been carried out. Later in the year the 30th, which was among the many corps to be disbanded, was ordered home. It landed in the Thames about November 21st, and was quartered at Gravesend, Dartford, Crayford, Erith and Woolwich, for a week and then was assembled at Southwark.

About the middle of January, 1698, the regiment started on its march to Lincolnshire, 1 company for Brigg, 2 for Gainsborough, 2 for Newark, 2 for Grantham, and the remaining 6 for the City of Lincoln. On February 16th an order was issued to Brigadier Selwyn to proceed to Lincoln to muster and disband the Regiment, which was ordered to assemble there. On March 4th, General Selwyn carried out this order in presence of the Mayor of the City. In marching the regiment, for disbandment, to the centre of its own recruiting ground and in other matters King William showed a fairness unusual in that age. Each soldier was allowed to take with him his clothes, belt and knapsack, he was paid three shillings for his sword and was given fourteen days' subsistence money and a pass allowing him a convenient time to reach his home. This just treatment must have been one of the reasons why the ranks were filled at once when four years later Colonel Saunderson and his officers were ordered to raise a regiment on the same ground.

By an order of March 16th, the officers of Saunderson's and their servants were placed on the half-pay of the regiment. The numbers were 1 colonel,

1 lieut.-colonel, 1 major, 9 captains, 14 lieutenants, 12 ensigns, 1 chaplain, 1 quartermaster, and 65 servants at fourpence a day. Only 12 captains are accounted for; Clifford Whichcott died about the time when the reduction of the regiment was decided upon and the vacancy had not been filled. In July, Phillip Griffin was appointed a captain on half-pay in Saunderson's *vice* Whichcott. With this exception, there were no changes in the regiment till 1701, when the country began to make some slight preparation for war. In that year Lieut.-Colonel Wills and Captains Frankland and Brathwayt were transferred to the full pay of Lord Charlemont's newly raised regiment. It is not necessary to say anything about Wills, who returned to the 30th, but Frankland rose to be lieut.-colonel of his regiment and was taken prisoner when commanding it at Almanza. Brathwayt retired or was killed early. Major Richard Sutton in the same year was transferred to Webb's, now "The King's Regiment," served throughout Marlborough's campaigns and died a lieut.-general. Captain Bolton was transferred to Howe's (later the 15th) regiment and was wounded at Schellenburg.

As soon as the Peace of Ryswick was signed, Louis XIV began to prepare for fresh hostilities. In 1700, Charles II, the last King of Spain of the Austrian line, died without issue. There were two claimants to the throne, Philip, Duke of Anjou, grandson of Louis XIV, supported by France and known to his adherents as Philip V, and the Archduke Charles, second son of the Emperor Leopold, known to his party as Charles III of Spain. The latter was supported by the Empire and the United Provinces. Louis was too quick for his opponents. On the death of Charles II, the Duke of Anjou had hurried to Madrid and had been received as King. Britain was somewhat apathetic about the balance of power and the succession to the Spanish throne, but Louis XIV took a step which roused her effectually. On September 20th, 1701, on the death of James II, the exiled king, he recognized his son, the "pretended Prince of Wales," as King of England, Scotland and Ireland.

This insult to the nation gave William his opportunity. Parliament authorized the raising of 40,000 soldiers and 6 regiments of marines. Of these 6 regiments the senior was that originally raised by Lord Castleton in 1689 and now known as the First Battalion, East Lancashire Regiment.

THREE REPRESENTATIVE COMMANDING OFFICERS

WARS OF KING WILLIAM III
AND QUEEN ANNE.

GENERAL SIR CHARLES WILLS, K.B.
1694–1701, 1705–16.

NAMUR, S. ESTEVAN, BARCELONA, LERIDA, SARDINIA, ALMENARA, SARAGOSSA, BRIHUEGA, PRESTON.

REVOLUTIONARY
AND NAPOLEONIC WARS.

COLONEL ALEXANDER HAMILTON, C.B.
1787–1829.

TOULON (W), BASTIA, HYÈRES, EGYPT, CADIZ, FUENTES ONORO (W), SALAMANCA,, VILLA MURIEL, QUATRE BRAS (W).

THE CRIMEAN WAR.

COLONEL JAMES THOMAS MAULEVERER, C.B.
1844–62.

ALMA
INKERMAN (W)
REDAN (W)
CRIMEA

CHAPTER II

REGIMENT RAISED AGAIN AS MARINES. CAPTURE OF GIBRALTAR. BATTLE OF MALAGA. DEFENCE OF GIBRALTAR. CAPTURE OF BARCELONA. VICTORY OF S. ESTEVAN. RELIEF OF BARCELONA. CAPTURE OF ALICANTE. DEFENCE OF LERIDA. CAPTURE OF SARDINIA AND MINORCA. DEFEAT OF FRENCH FLEET IN FIRTH OF FORTH. DEFENCE OF DENIA. LINCOLNSHIRE ASSIGNED AS RECRUITING GROUND. DISBANDED.

1702–1714

A COMMISSION was issued on February 12th, 1702, to Colonel the Hon. Thomas Saunderson to raise a regiment of marines of 12 companies each of 66 non-commissioned officers and men, and Colonel Thomas Pownall and William Dornell were appointed his lieut.-colonel and major. The commissions of the field officers were among the last signed by King William. He died on March 8th and was succeeded by Queen Anne, only surviving child of the exiled king, James II, by his first marriage. Her half-brother, the "Pretended Prince of Wales," was excluded from the succession at the Revolution. The Queen signed the commissions of the company officers on March 10th.

The officers of Saunderson's who had been placed on the half-pay of the regiment in 1698 were called up, and those available, 22 in number, joined. Colonel Wills, who was now serving in Charlemont's, rejoined three years later. The 22 officers would bring with them their soldier servants from half-pay. The following are lists of (1) the officers placed on half-pay in 1698 and (2) the officers of Saunderson's Marines in 1702. Those who served in both years are numbered.

SAUNDERSON'S REGIMENT

		1698		1702
Colonel	(1)	Thomas Saunderson	(1)	Thomas Saunderson
Lieut.-Colonel		Charles Wills		Thomas Pownall
Major		Richard Sutton		William Dornell
Captains	(2)	George Burston	(2)	George Burston
		Henry Frankland		Andrew Abington
	(3)	Isaac Knight	(3)	Isaac Knight
	(4)	Thomas Bedford	(4)	Thomas Bedford
		Richard Bolton	(5)	Walter Palliser
		James Brathwayt	(6)	Uriah Brereton
	(5)	John Saunders		John Thompson
	(6)	Walter Palliser		John Casewell
	(7)	Urian (*sic*) Brereton	(7)	John Saunders
Capt.-Lieutenants	(8)	James Harris	(8)	James Harris
Lieutenants	(9)	William Singleton	(9)	William Singleton
		Robert Jackson	(10)	Bernard Pinnock

SAUNDERSON'S REGIMENT

		1698		1702
Lieutenants		William Whitehof	(11)	Peter de St. Just
		George Hassard	(12)	Francis Filbridge
	(10)	Marc Antonio Bernard	(13)	Marc Antony Bernard
	(11)	Francis Filbridge	(14)	Nathaniel Potter
	(12)	Natheniel Potter	(15)	Hugh Palliser
	(13)	Peter de Singest		William Forbes
		Alexander Davison		Edmund Harris
	(14)	John Bernard	(16)	John Bernard
		Blaise Gervaisot		Charles Davison
	(15)	Hugh Pallister		
	(16)	Bernard Pinnock		Second Lieutenants:
Ensigns		Stanhope Yarborough		George Ord
	(17)	Andrew Day		William Nichol
		William Weddell	(17)	George Saunderson
		John Partridge		Charles Christian
	(18)	George Saunderson		Gilburn Scroope
	(19)	John Sharpe		Thomas Burke
		Isaac Duplex	(18)	Theophilus de Vauclen
		Nicholas Tyrwhit		William Gardner
		Robert Hume		John Vangensimmer
		John Dalton	(19)	John Sharpe
	(20)	Theo Vauclen	(20)	Andrew Day
		William Saunderson		William Davison
				Edmund Thwaites
Chaplain	(21)	John Walley	(21)	John Whalley
Quart. Master	(22)	Henry Clarke	(22)	Henry Clarke

Of the officers appointed in 1702, Lieut.-Colonel Dornell was from the half-pay of Lord Seymour's marines, disbanded in 1698. Capt. Abington was from the half-pay of Erle's Foot. Capt. Casewell, Lieutenants George Ord and John Vangensimmer were from the half-pay of Dutton Colt's marines, disbanded in 1698. Gilburn Scroope was formerly an ensign in the Duke of Bolton's regiment. Vangensimmer was a foreigner, but naturalised. He had served on Admiral Aylmer's fleet.

On May 30th, Capt. David Ward was appointed to Casewell's company from the 1st Guards.

On September 24th, Andrew Day was appointed quartermaster *vice* Henry Clarke, deceased. Lord Castleton and Colonel Saunderson joined in petitioning the Treasury for a pension for Clarke's widow. They stated that the late quartermaster had fought in every action in which the 30th had been engaged in the late war.

Of the newly appointed officers, Colonel Pownall strengthened the connection with Lincolnshire, of which he was High Sheriff in 1693 and 1703. His military life included service in the Royal Dragoons, the Queen's Dragoons and Arran's Horse, of which he was lieut.-colonel in 1694.

The regiment was raised upon the same ground as when first formed in 1689, and was as much as ever a Castleton corps; many of the former soldiers must have rejoined. The pay and off reckonings of all ranks were exactly the same. The second lieutenants drew ensigns' pay.

The Queen directed that the establishment of six regiments of marines "do commence and take place from the respective times of training." She thus fixed the rank and seniority of each regiment. The 30th was the senior of the six.

In April war had been declared by Britain, the Empire and the United Provinces against France.

At the end of that month the regiment began its march to Salisbury, which had been named as the rendezvous to which the companies should move as soon as complete. Most of the routes have been lost, but we have four dated April 30th, one from Sheffield, near which were Lord Castleton's seat, Sandbeck, and the family estates where his son, Capt. Saunderson, had recruited in 1689; another from Wakefield—this was almost certainly Capt. Bedford's company. The other two are for Lincolnshire companies, one from Grantham, the other from Lincoln.

On May 16th a fresh order was issued directing the regiment to concentrate at Portsmouth. Five companies were to march from Banbury and seven from Salisbury. The five which the order was intended to intercept at Banbury must have been fresh Lincolnshire companies, for those which marched under the routes of April 30th must have passed Banbury, which is only five marches from Grantham, long before. The Yorkshire companies passed to the west of Banbury on their march to Salisbury. Among the seven companies ordered to march from Salisbury to Portsmouth were the two from Yorkshire and the two from Lincolnshire already mentioned, one from Chelsea and two whose origin is unknown. The two latter were no doubt from the north, but the one from Chelsea is hard to account for; neither before this date nor for many years afterwards did the regiment recruit in London. The company may have been brought to the Thames from Hull or from Newcastle on the men-of-war escorting the convoys of colliers to London, as was done with recruits in the following years.

The regiment must have been united at Portsmouth soon after May 20th.

Colonel Saunderson had now formed and assembled his regiment, and it will be as well to take a look at it and see what it is like. What strikes one first is that the Castleton grey and purple have gone and the regiment is clothed in red with yellow facings. More than that, it is a modern regiment. The flags, the smoking match, the tall pikes and corselets which gave a medieval touch to the picture have vanished, and we see a regiment with a pair of colours, and armed throughout with flintlock and bayonet of a pattern which was to last with little alteration for more than 140 years. The broad-leaved hat had given way to the marine cap. Here there is some uncertainty, for there is no authentic picture or record of what the marine cap really was. We know that it was similar to that of the grenadiers, but, curiously enough, we have no pattern or exact description of the grenadier cap of this period.

Immediately after the concentration at Portsmouth, the Duke of Ormonde was ordered to make a draft from the regiment to complete those in the Isle of Wight. The regiments in the island were those for the West Indies, which were kept there to prevent desertion before embarkation. An order to proceed to the West Indies was considered equal to a death warrant. Three hundred men were drafted into Columbine's regiment. This was very hard upon Saunderson's, which was one of the few which had found no difficulty in filling its ranks. To complete the regiment again to its establishment, the four companies of Colonel Saunderson, Lieut.-Colonel Pownall, Captain Knight and Captain Saunders (grenadiers) were sent back to Lincolnshire to recruit. The companies had left parties in the county when they marched to Portsmouth. The four companies after marching through Lincolnshire were to cross the Humber from Barton to Hull, where three were to remain and the fourth was to continue the march to Berwick-on-Tweed. In September a second company marched to Berwick.

On December 17th the regiment had three companies in the Isle of Wight, four at Portsmouth, two at Hull and two at Berwick.

About the time the regiment arrived at Portsmouth the colonels and field officers of marines had met in consultation about the payment of their corps and had addressed a memorial to the Lord High Treasurer, humbly desiring him to appoint Mr. Walter Whitfield their particular paymaster, so that the Paymaster-General of the Land Forces might have no claim to meddle with the marines. The memorial, which is in Colonel Saunderson's handwriting, was signed by Thomas Saunderson, George Villiers, Alexander Lutterell, Lord Shannon and Thomas Holt.

Mr. Whitfield was appointed and started well, but the marines suffered a good deal from his conduct in later years. It required a very strong man to resist the temptations put in his way by the want of method on the part of the Treasury and the Admiralty. By the end of the year the subsistence of the marines was four months in arrear, and Mr. Whitfield pointed out to the Admiralty that disturbances in quarters and desertions were frequent, owing to the men being in want. In the beginning of the following year he was able to pay off arrears, but neither the 30th nor any other marine regiment was ever regularly paid while under the Admiralty. The officers suffered with the men, but the service would often have come to a standstill if the officers had not paid for the men's subsistence out of their own pockets.

1703

The establishment was the same as in the previous year.

The following promotions and appointments were made:—

Geo. Ord to be Captain *vice* Andrew Abington, deceased		25th Feb.
Sir William Mansell, Bart., to be Capt. *vice* John Thompson		10th March
William Gardner to be First Lieut. *vice*	Forbes	5th April
George Saunderson ,, ,, ,,	Singleton, deceased	25th ,,
Edmund Thwaites ,, ,, ,,	Pinnock	6th Dec.
John Sharpe ,, ,, ,,	John Bernard posted to a company of Invalids	,, ,,
Andrew Day ,, ,, ,,	Peter de St. Just	,, ,,
Rothwell Stow to be Second Lieut.		11th Feb.
Daniel de Vauclen ,, ,,		25th ,,
Thomas Brooke ,, ,,		,, ,,
Joseph Mason ,, ,,		,, ,,
William Pownall ,, ,,		25th July
Richard Turner ,, ,,		1st Nov.

William Dawes to be Second Lieut.	. . . . . .	1st Nov.
George Dobson ,, ,,	*vice* Nicholls, deceased .	6th Dec.
John Brown ,, ,,	,, Thwaites. . . .	,, ,,
Sam. Elford ,, ,,	,, Sharpe	,, ,,
Thomas Newdigate ,, ,,	,, Day	,, ,,
M. A. Bernard to be Adjutant	*vice* Singleton, deceased . .	24th July
Edward Claringburn to be Chaplain	*vice* Whalley . . .	7th ,,

By the new year Colonel Saunderson had again completed his regiment to the establishment, and in January and February some of the companies in the south were ordered on board ship. Six companies served in a squadron which convoyed " the trade " as far as Lisbon and returned to join Sir George Rooke's fleet at Portsmouth. On March 27th the four companies in the north were ordered to the Thames. The Colonel's and Captain Knight's marched to Tilbury, Lieut.-Colonel Pownall's and Captain Saunders' companies were brought south on H.M. ships *Lynn* and *Reserve,* two men-of-war acting as convoy to the colliers from Newcastle to London.

Captain Knight's company embarked and served on H.M.S. *Tryton* in the home fleet for the remainder of the year, the grenadiers remained on the *Reserve,* the Lieut.-Colonel's was turned over from the *Lynn* to H.M.S. *Panther,* and the Colonel's embarked on H.M.S. *Romney.* The following statistics are given as a fair specimen of the strength usually embarked and the effect of a short cruise.

H.M.S. *Romney,* Colonel's company; strength, June 11th, 1704: Capt.-Lieut. Harris, Lieut. Brooke; other ranks, 63. The ship served in the Nore fleet. Men were discharged from time to time at the Channel ports and the company landed and marched to Canterbury in August, eight deaths having occurred.

H.M.S. *Panther,* Lieut.-Colonel's company; strength, July 1st, Lieuts. Pinnock and Nichols; other ranks, 54. After cruising in the Channel and North Sea, the company was landed at Portsmouth in December, Lieut. Nichols and five privates having died. Lieut. Pinnock had ceased to belong to the regiment on December 6th, reason not stated.

In June, 1703, some companies in the Isle of Wight were called upon to embark under rather unusual circumstances.

Admiral Sir George Rooke was, from his services, skill and character, the recognized head of his profession. He had, however, long been a martyr to gout and, what still more interfered with his usefulness, he was resolutely opposed, on grounds both of strategy and seamanship, to the " forward " policy in naval matters adopted by Marlborough and the Government and supported by the younger admirals. His idea of strategy was confined to concentrating in the narrow seas, and he held, justly as a seaman, that the " capital " ships of that day should be in port before the autumn gales came on. He sailed in spring, threatened Rochefort, and returned to Spithead in June. His marines were landed in the Isle of Wight and he himself went on leave to Bath for a month's cure of the gout.

The Government seized the opportunity. Sir Cloudesley Shovel, Rooke's second in command, was ordered to re-embark forthwith all the marines he could lay his hands on and sail for " The Straights." This time five companies of the regiment were embarked.

The fleet was got to sea with all speed and, accompanied by a Dutch squadron, made a triumphal progress up the Mediterranean. It protected "the trade" to Smyrna and proclaimed the Archduke Charles, the chosen of the Allies, King Charles III of Spain in the Italian and Sicilian ports.

The fleet met with no opposition, but this new manifestation of the power of the maritime nations startled Europe and confirmed Marlborough in his forward policy.

The justness of Sir George Rooke's judgment as a seaman remained to be tested.

The fleet reached Portsmouth in safety about the beginning of November and the Chatham ships sailed for their own port. When off the Kentish coast the "Great Storm" of November 23rd—"The storm which late o'er pale Britannia passed"—nearly overwhelmed them. Many ships foundered and most of the others were dismasted or otherwise disabled. The companies of Captains George Ord and David Ward had a wonderful escape. Throughout the cruise they had served on H.M.S. *Stirling Castle*, but on the 22nd an order met them to transfer at once to H.M.S. *Royal Oak*. A few hours after the transfer was made the storm broke and the *Stirling Castle* was cast away, very few being saved. The *Royal Oak* weathered the storm and landed the two companies at Chatham on December 11th. During the cruise fifteen men had died.

The other ships of the fleet carrying men of the 30th were equally fortunate.

On the east coast, however, the regiment was not so happy. The *Reserve*, on which Captain Saunders and the grenadier company had served since they left Newcastle had, during the summer, been employed as convoy to 100 merchant vessels to Archangel and back. On their return both the *Reserve* and the ships under convoy were, as was usual in those days, short of water and provisions. Captain Anderson of the *Reserve* appointed Yarmouth as the rendezvous for the convoy and took his own ship to Tynemouth, where he sent a boat ashore for Shields pilots to take him up the river. At the same time he allowed Captain Saunders, his subalterns, Lieutenants John Bernard and Andrew Day, and their servants, to go ashore with a master's mate and midshipman, no doubt for supplies for the different messes. The pilots, on coming off, refused to incur the risk of taking the *Reserve* over the bar with the wind in the quarter in which it was. Captain Anderson therefore at once made for Yarmouth to meet his convoy and to water and provision his ship, leaving Captain Saunders and the other officers on shore at Tynemouth. On the morning of the 23rd, Captain Anderson, accompanied by the surgeon, landed with two boats for supplies, leaving his ship at anchor in Yarmouth Roads. While he was on shore the storm broke and the *Reserve* foundered at her anchors, not a man on board being saved. The whole grenadier company perished except the officers and their servants and some sick landed before sailing to Archangel. The officers were sent to Lincolnshire to raise a new company.

By Christmas the whole regiment was in quarters in Kent, except the Colonel's company on the *Romney*, Captain Brereton's on the *Eagle*, and the grenadiers reforming in Lincolnshire.

We have no returns from some ships, but, judging from those we have, we are safe in estimating the deaths in the regiment from February to December at over 130 (including those lost on the *Reserve*), being about one-sixth of the strength. The ships were overcrowded, the main deck ports were so near the water as to be seldom open, but, above all, bad water and bad beer account for the dreadful mortality on shipboard. Iron tanks were unknown, and the water quickly became putrid in the wooden barrels, but it is doubtful if the water or the contractor's beer had most to answer for.

On November 23rd, Lieut.-Colonel Pownall, who had been pricked for Sheriff of Lincolnshire, was granted twelve months' "furlow."

1704

The establishment was raised in this year and fixed at 12 companies of 100 privates, 4 sergeants, 2 drummers, 4 corporals, a total of 1,320 non.-com. officers, rank and file. The officers were 1 colonel, 1 lieut.-colonel, 1 major, 9 captains, 12 first lieutenants, 12 second lieutenants, 1 adjutant, 1 quartermaster, 1 chirurgeon and 1 chirurgeon's mate, 1 chaplain, a total of 41.

The following appointments and promotions were made :—

Thomas Pownall to be Colonel, *vice* Saunderson, who retired 15th Dec.
William Dornell to be Lieut.-Colonel, *vice* Pownall ,, ,, ,,
George Burston to be Major, *vice* Dornell ,, ,, ,,
James Harris to be Captain, *vice* Saunderson ,, ,, ,,
M. A. Bernard to be Captain-Lieutenant, *vice* Harris ,, ,, ,,
Rothwell Stow to be Lieutenant to the grenadiers, *vice* Sharpe, 29th Sept.
William Davison ,, ,, Capt. Burston, *vice* Day, deceased, 29th Sept.
Thomas Brookes ,, ,, Capt. Ward, 29th Sept.
Joseph Mason ,, ,, Capt. Bedford, 15th Dec.
Dan. de Vauclen ,, ,, Capt. Ord, *vice* Filbridge, transferred 1st Dec.
Richard Leathat to be Second Lieutenant to Capt. Knight, 29th Sept.
Thomas Burston ,, ,, ,, ,, Capt. Bedford, 29th Sept.
William Townrow ,, ,, ,, ,, the Colonel, 29th Sept.
James Davison ,, ,, ,, ,, Capt. W. Palliser, 18th Dec.
John Thompson to be Quartermaster *vice* Mitford, deceased, 5th Oct.

Although placed definitely at the beginning of the year under the Lord High Admiral, the marine regiments had up to the end of 1703 worked without any special regulations, being treated, when on board ship, as soldiers embarked for the cruise for a special purpose. On January 6th, 1703–4, Prince George, the Lord High Admiral, ordered Admiral Sir Cloudesley Shovel " to call as many of the Flaggs and Captains that had been with him in the streights as could be conveniently got together, to consider and offer their thoughts in relation to the Marines." The recommendations of this council, which was followed by one composed of colonels of marines, were approved of, in great measure, by Prince George. They were marked by some jealousy of military authority on board ship, the number of marine officers embarked was to be limited to one of each company, and that of subordinate rank, and, as in the marines disbanded in 1698, the men were to be trained for the navy and when efficient were to be encouraged to become blue-jackets. This was not what Government wanted, and the Prince's decision was over-ruled. An efficient landing force was what Marlborough was aiming at, so the military side of the marine's training was rather emphasized. On board ship he was part of the ship's complement for victuals only, not less than fifteen marines were to embark on any one ship, and they were always to be accompanied by an officer of their own corps. Practically, unless there was a prospect of campaigning, the subalterns only embarked.

On February 16th permanent Headquarters were assigned to each marine regiment. To Colonel Saunderson's, Canterbury; to Lord Shannon's, Maidstone; Colonel Holt's, Chichester; Colonel Fox's, Southampton; Brigadier Seymour's, Exeter; and Colonel Lutterel's, Plymouth.

The increase of the establishment for the marines was caused by the resolve of the British Government to make a great effort in the Mediterranean. Each regiment was called on to enlist in two months 500 fresh men in addition to replacing the heavy losses of the previous year. After discussion with the Lord High Admiral, the colonels of marines presented a memorial, which is in Colonel Saunderson's handwriting, urging that the bounty should be raised from £2 to £3. The Treasury agreed and recruiting was started at once. Government was so much in earnest that by March 13th bounty money for the required number of recruits was advanced, with an intimation that at the end of two months the captain's pay would be charged with £3 for every recruit he was short of his complement.

GIBRALTAR

In February embarking orders had been issued, and by April 3rd the following were on board Sir George Rooke's fleet bound for the Mediterranean:—

	Strength.		*Guns.*	*Crew.*
Lieut.-Col. Pownal's Company under Lieut. Thwaites	53,	on *Cambridge*	80	500
Capt. Thomas Bedford's Company under Lieut. W. Davison	56	,, ,,		
Capt. Thomas Burston's Company under Lieut. Andrew Day	50	,, *Boyne*	80	500
Capt. Walter Palliser's Company under Lieut. Dan. de Vauclen	57	,, ,,		
Capt. Uriah Brereton's Company under Lieut. Jos. Mason	66	,, *Eagle*	70	440
Capt. George Ord's Company under Lieut. Thos. Newdigate	43	,, *Royal Oak*	76	500
Capt. David Ward's Company under Lieut. Edmund Harris	48	,, *Firm Prize*	70	440
Capt. Sir Wm. Mansel's Company under Lieut. Charles Davison	47	,, ,, ,,		
	420			
In May twenty-one men from the companies of Dornell and Knight embarked on H.M.S. *Lennox* under Lieut. George Saunderson at Deal and joined Capt. Ord's company in the Mediterranean	22			
	442			

In all cases the officers are included in the strength.

FRANCE
GALICIA
Corunna
ASTURIAS
BISCAY
NAVARRE
LEON
Burgos
Villamuriel
Palencia
Duenas
Cabecon
R. Douro
Tordesillas
OLD CASTILE
R. Ebro
S. Estevan
Balaguer
CATALONIA
Gerona
Saragossa
Lerida
Igualada
Barcelona
Tarragona
Tortosa
ARAGON
Oporto
R. Douro
Almeida
R. Tormes
Salamanca
Ciudad-Rodrigo
Celorica
Coimbra
R. Mondego
Figueras
Castel Branco
MADRID
SPAIN
ESTREMADURA
PORTUGAL
Villa Velha
R. Tagus
Torres Vedras
Sobral
Portalegre
NEW CASTILE
Valencia
VALENCIA
LISBON
Elvas
R. Guadiana
Badajoz
Almanza
Denia
Alicante
ALTEA BAY
MURCIA
R. Guadalquivir
Seville
ANDALUSIA
Carthagena
Malaga
Cadiz
C. Trafalgar
Gibraltar
Scale of Miles
50 0 20 40 60 80 100

From the fact that no field officers or captains embarked, it appears that the eight companies were intended for service on board ship only. Dornell's and Knight's companies were left on duty guarding prisoners of war at Dover, the Colonel's company was in the Home Fleet, and the grenadiers reforming in Lincolnshire.

On May 8th the fleet under Sir George Rooke sailed from Lisbon for Nice. There was a rumour that Nice was attacked, which was found to be baseless.

Prince George of Hesse Darmstadt, who was well known in the English army for courage and skill as a soldier, had embarked on Sir George Rooke's fleet as representative of Charles III. In 1697 he had conducted the famous defence of Barcelona, and he still possessed great influence in Catalonia. At his suggestion Rooke agreed to stay for twenty-four hours before Barcelona, hoping that a rising of the Catalans might deliver the city to the Allies. These hopes were disappointed; the Catalans were not organized and did not rise, and Rooke, who had disembarked his marines, took them on board again and continued his course towards Nice.

Off Hyêres, however, he heard that the French fleet had sailed from Brest under the Count of Toulouse with the intention of entering the Mediterranean and uniting with the fleet from Toulon. Rooke thereupon made for Gibraltar and Lisbon to meet Sir Cloudesley Shovel, who had sailed from England with reinforcements. On his way he met Toulouse, but could not bring him to action or prevent his entering Toulon. Off Lagos, Rooke met Sir Cloudesley Shovel and the Allied fleet now numbered seventy-two ships.

The Kings of Spain and Portugal urged from Lisbon that this great strength should be used for the capture of Cadiz. Sir George Rooke, on the other hand, held that, strong as he was, he could not attempt Cadiz without more troops. After weeks of indecision it was determined, on the suggestion of Prince George of Hesse, to attack Gibraltar, which was known to be weakly garrisoned, although strongly armed.

The plan of attack was settled while the fleet watered in Tetuan Bay.

For the navy it was arranged that Admiral Byng, having first transferred the marines of his squadron to other ships, should take the principal part in the naval attack. Admiral Byng's orders were to close, as near as the depth of water would allow, with the Spanish works between the North and South Mole and bombard with seventeen sail of the line and three bomb vessels. The remainder of the fleet were to support Byng and to cover the landing of the marines under the Prince of Hesse, to the north of Gibraltar.

A council of marine officers was held on board the *Boyne* to arrange the details of this operation. It was ordered that every marine on his landing should have eighteen charges of powder and ball, each grenadier two grenades with match in proportion, that the officer commanding each company should have one of the largest cartridge cases filled up with musket cartridges of powder and shot, and that there should be a sufficient quantity of shovels and light crows and men with hatchets to cut fascines.

Apparently escalade was not contemplated and, as no guns were to be landed, it is evident that the Prince of Hesse could only hope to make a demonstration against the north front of the fortress On June 21st the fleet, headed by Byng's squadron, entered Gibraltar Bay. The marines, under the

Prince of Hesse, were landed at the mouth of the Guadarama. They were threatened by a small party of Spanish horse, which was quickly driven off by the fire of the covering ships, and the Prince took up a position from sea to sea across the isthmus.

His landing strength was about 1,800 English and 400 Dutch.

That of the 30th (Saunderson's) was, as nearly as we can judge :—

	Officers included.
Lieut.-Colonel Pownall's Company, under Lieut. Thwaites with Cadet M. Brown attached	46
Capt. Bedford's Company, under Capt Bedford, Lieut. Wm. Davison, Cadet Robert Bedford. (Capt. Bedford had joined off Gibraltar from home.)	50
Capt. Brereton's Company, under Lieut. Mason	59
Capt. Burston's Company, under Lieut. Andrew Day	43
Capt. Palliser's Company, under Lieut. Daniel de Vauclen	55
Capt. Ord's Company, under Lieuts. Newdigate and Saunderson	49
Capt. Ward's Company. No musters or pay books. Lieut. Edmund Harris	46
Capt. Sir Wm. Mansell's Company. No musters or pay books. Lieut. C. Davison	45
Total	393

There had been fourteen deaths on the voyage and some men had been landed at Channel ports. Capt. Ord's company had detached sixteen privates under Sergt. Pearce for duty on the *Roebuck* cruiser, but had received a party of twenty-one under Lieut. Geo. Saunderson.

Admiral Byng had great difficulty in warping his ships into position and the attack was delayed till the morning of the 23rd, when the Spaniards on the sea front were driven from their guns after a gallant defence of five hours, during which 15,000 shot or shell were " hove into " the town by the fleet. Capts. Jumper of the *Lennox* and Hicks of the *Yarmouth*, seeing the defence slacken, sent armed parties ashore from their ships and Sir George Rooke ordered Capt. Whitaker to support with the boats of the fleet. Owing to the explosion of a magazine, by which two lieutenants and forty seamen were killed and sixty wounded, the advanced parties fell back on their boats, but on Capt. Whitaker's arrival the fight was restored and the southern defences of Gibraltar firmly occupied. The Prince of Hesse had done what he could on the north side, but without any real progress. He and Admiral Rooke now joined in a summons to the Spanish Governor to surrender ; the summons was accompanied by a threat to put the garrison to the sword in case of non-compliance.

The Governor, who had no choice, and had fully done his duty, consented to surrender at 4 a.m. on the morning of the 24th, and sent hostages. His garrison of regulars and militia was about 500 strong, and had caused about 330 casualties to the assailants.

The Spanish flag was again hoisted in Gibraltar, but now in the name of our ally, Charles III. Unfortunately the sailors, enraged by the Prince of Hesse forbidding plunder, broke loose from all restraint and nearly wrecked the town, an act which not only had a very bad effect on the allied cause, but inflicted increased sufferings on the marines who garrisoned the place for the following twelve months.

Some of the marines also misbehaved, and one was hanged. The proceedings were characteristic of the times. An English and a Dutch marine were condemned to death, but allowed to settle by a cast of the dice which should suffer. The Dutchman "hove" ten and the Englishman nine, and was hanged accordingly.

After the capture of Gibraltar the allied admirals submitted to the Kings of Spain and Portugal at Lisbon a proposal to attack Cadiz if sufficient troops were sent to them. Pending a reply, Admiral Rooke did what he could to put Gibraltar in a state of defence, the marines being left on shore as a garrison. The fleet meanwhile covered Gibraltar and watered by squadrons on the Barbary coast. All doubts in the minds of the Allies as to their future operations were settled by the news that the French fleet from Toulon under the Count of Toulouse was in the neighbourhood of Malaga, and on August 5th Sir George Rooke's scouts viewed the enemy ten leagues to the eastward (windward). A battle appeared imminent, but the Count of Toulouse manœuvred to join his galleys which he had left at Malaga, and Sir George Rooke took advantage of the delay to send a few ships in to Gibraltar for 500 marines, but without their officers. He was weak-handed and no doubt wished to distribute them among his gun crews. The Prince of Hesse, like the fine soldier he was, sent over 1,000 men, only keeping 500 in garrison.

The men were brought out and distributed through the fleet. It is impossible to trace how many of the 30th (Saunderson's) were embarked at this time or in what ships they served, but it is natural to suppose that two-thirds of the men were taken and that they were placed in their old ships. Although the logs of all the ships with which the 30th was connected mention the landing after the battle of those marines brought from Gibraltar, only that of the *Cambridge* records their coming on board. The entry is: "10 Aug. Shipped 39 Marines brought in a yacht from Gibraltar. Sent ashore on 22nd." Probably these were Bedford's men, forty-seven of whom had landed from the same ship for the capture of the fortress on June 21st. The case of the *Firm Prize*, on which the companies of Capts. Ward and Sir William Mansel had come from England, was quite different.

The Master's Log of that ship records:

Gibraltar Bay. 21 July. Landed all our marines.
31 July. This day we sent our boats and brought all our marines off.
1 Augt. Our two lieutenants of marines remained behind in the garrison.

It is evident therefore that, before the general re-embarkation of marines in anticipation of an action near Malaga, the *Firm Prize* had re-embarked the two companies she had brought out, but without their officers, Charles Davison and Edmund Harris, who remained and shared in the defence of Gibraltar. The reason was that the crew was too weak to work the ship without the marines. It is as well to finish the story of the *Firm Prize* and its two companies here as far as we know it.

The ship took her fair share in the battle of August 13th. The log records :—

" Off Malaga, 13th August. We were very much disabled in our masts and rigging, had 25 men killed outright and 50 wounded." As there are no pay lists, the regimental loss is unknown.

The *Firm Prize* did not land her marines again during the cruise, but when the fleet returned to Portsmouth her log shows that Capt. Ward's and Capt. Sir Wm. Mansell's companies were disembarked on October 16th. The route for those companies to march from Portsmouth to Canterbury is dated October 10th.

To return to the movements of the fleets. Sir George Rooke having strengthened his crews with the marines and the Count of Toulouse having been rejoined by his galleys, both sought an engagement, but passed each other in the night. In the forenoon of the 12th, however, they met, the wind being east and Rooke to windward, but the wind was so light that it was the morning of the 13th before he could close with the enemy.

The fighting was very severe, and on the whole the English and Dutch had the best of it. Had it not been for the expenditure of ammunition at Gibraltar, Rooke might have gained a great victory, but nine English ships had to haul out of action for want of powder and shot.

At nightfall the fleets were close together, but neither commander was anxious to push matters to a decision. The French, who had suffered severely, were unaware that the English had expended nearly all their ammunition. By distributing what remained, Rooke was able to furnish each ship with eight to twelve rounds per gun.

The following day was almost a calm and the fleets remained in presence of each other, repairing damages. At a council of war held in the evening on Rooke's flagship, it was decided that the fleet had done all that honour required and that it should return at once to Gibraltar, but it was further agreed that, if the French fleet were met with, it should not be avoided. On the following morning the French fleet was seen a few leagues to windward and on the 16th it was out of sight. Both sides claimed the victory, but the Allies reaped the advantage. Toulouse had failed to drive their fleet off and recover Gibraltar.

We have no means of knowing the regimental loss, but the *Lennox*, which brought out Lieut. Saunderson's party, lost one-fourth of her complement, 23 killed and 78 wounded ; the *Boyne*, on which were Capt. Burston's and Capt. Palliser's companies, had 14 killed and 52 wounded ; the *Cambridge*, on which were Col. Pownall's and Capt. Bedford's companies, had 11 killed and 27 wounded ; the *Eagle*, which had Capt. Brereton's company, had 7 killed and 57 wounded ; the *Firm Prize*, as already mentioned, had 25 killed and 50 wounded. The logs of the *Boyne*, *Cambridge*, *Royal Oak* and *Eagle* show that the marines who had been embarked for the battle of Malaga were relanded at Gibraltar on August 21st and 23rd. They had not been entered in the ships' books.

A congratulatory address was voted by Parliament to Sir George Rooke and his fleet for their victory off Malaga and commemorative medals were struck.

Sir George sailed for England with the main body of his fleet on August 25th, leaving a squadron of light ships under Sir John Leake at Lisbon. A Spanish army under the Marquis of Villadarias was already preparing to attempt the recapture of the fortress, and his advanced guard was taking up its ground, to the north of Gibraltar, as Sir George Rooke's fleet left the bay. On October 4th a French fleet under Admiral de Pointis arrived from Toulon, bringing 3,000 French marines to strengthen Villadarias, who had under him 8,000 Spaniards and a siege train of seventy guns and a dozen mortars.

On October 21st trenches were opened, the French fleet making a diversion by opening fire from the bay. On the 23rd the batteries were commenced. The work was carried on with great spirit, many Spanish grandees serving as volunteers in the trenches. The front attacked was very short, being little more than the space between the northern point of the rock and the bay, and consisted of a curtain, pierced by the North Gate and flanked on the right by a bastion and tower thrown well forward and a little way up the rock. On the left the front was flanked by works on the New Mole which raked the enemy's trenches; thirty-four guns were mounted on the front and thirty on the Mole. On October 10th one of his captains reported to Sir John Leake at Lisbon that the garrison feared nothing the enemy could do unless they were attacked at the same time by sea and land and so forced to divide their men so much that they would never be off duty. Except when pressed, half the garrison rested while the other half was on duty.

On November 8th the Spanish breaching batteries opened, and the superiority of their fire was at once manifest. The rampart of the bastion to the right of the North Gate consisted merely of a bank of sand faced with light stones. The Spanish guns carried everything before them, dismounted the guns and overwhelmed the defenders with showers of stones and sand. Colonel Nugent, the Governor, and Colonel Fox were killed. The command of the marines fell to Lieut.-Colonel Jacob Bor of the 32nd, who served with great courage and ability throughout the siege.

November 10th was fixed by the Marquis of Villadarias for what he intended to be the final attack on Gibraltar. Five hundred volunteers, supported by a strong column, were to climb the rock from the east side, guided by a goatherd, and descend upon the weakest part of the town. A detachment was to be conveyed in boats to Europa Point and to attack from the south, while the main body under Villadarias was to assault the north front. The attack by sea and land which the garrison had feared was now to take place. The effective strength of the defenders had fallen to 1,000 men, and the fall of Gibraltar seemed certain.

Help, however, was at hand. On November 5th, Sir John Leake sailed from Lisbon with eighteen English and Dutch men-of-war, and on the 9th entered Gibraltar Bay. A great part of the French fleet was watering at Cadiz, but those ships left in the bay were captured, and the boats which were to have landed the attacking party at Europa Point were crowded into the Guadarama for safety.

Although Leake's appearance had deprived Villadarias of the support of the French ships and rendered the attack from the south impossible, the Marquis did not at once abandon the two parts remaining of his original plan. At nightfall on the 11th the 500 volunteers climbed the eastern cliff and, crossing the summit to the south of Middle Hill, reached St. Michael's Cave, where they waited for their supports and for the attack from the north to commence. The supports never appeared, nor was any attack delivered from the north. At the last moment Villadarias must have realized that under the changed circumstances his plan was hopeless.

The unlucky Spanish volunteers in St. Michael's Cave were discovered at daybreak and attacked by Prince Charles, younger brother of Prince George of Hesse. The Prince was wounded, but 500 marines, headed by Colonel Bor, charged and overthrew the Spaniards, of whom only 100 escaped by the way they came. The Spanish leader was wounded and taken and his brother killed, both fighting bravely at the head of their men, spontoon in hand.

In a despatch of the 13th, Sir John Leake informed the Admiralty of his success, and added, "I doubt the garrison will not have the same success in ye defence of the town. It is not provisions and powder but officers and men that are wanted." A week later he writes that the garrison of Gibraltar is still in our possession but that he cannot guarantee for how long; the enemy had raised another battery and dismounted most of our guns, but the principal thing wanted by the defence was men.

This closes the first part of the defence, in which the marines alone had for four months held their own, but with great difficulty and heavy loss from constant fatigue and exposure as well as from the enemy's fire.

On December 5th reinforcements of 1,970 officers and men of the Guards, Donegal's and Barrymore's and of Waese's Dutch reached the Prince of Hesse, and he took full advantage of his increased strength in the brilliant sorties of December 11th and 20th.

1705

The establishment was the same as in the previous year.

The following appointments and promotions took place :—

Charles Wills to be Colonel by purchase, *vice* Pownall, who retired 13th October.
Andrew Corbett, sub-brigadier of the first troop of the Horse Guards, to be Captain, *vice* Mansell, who exchanged 5th March.
Hugh Palliser to be Capt.-Lieut. *vice* Bernard, deceased 13th October.
Will. Pownall to be First Lieut. to Capt. Ord, *vice* Dan. de Vauclen, deceased, 24th May.
Thomas Newdigate ,, ,, ,, Capt. Harris, *vice* Thwaites, deceased 25th November.
George Dobson ,, ,, ,, Capt. Ward, *vice* Brooke, deceased 25th December.
James Saunders, Second Lieut. to Capt. Brereton, *vice* Mason, promoted 13th January.
Henry Long ,, ,, ,, Capt. Harris, *vice* Townrow, 6th April.
Charles Raynsford ,, ,, ,, Major Burston, *vice* Pownall, promoted 24th May.
George Dobson to the Grenadiers, *vice* Elford, deceased 15th December.
James Saunders to the Grenadiers, *vice* Dobson, promoted 25th December.
Arthur Storey to be Adjutant, *vice* M. A. Barnard, promoted 13th January.

The appointment of Colonel Wills was published at home on December 15th, but he had taken command of the regiment on October 13th, and that is the true date of his appointment. Hugh Palliser in like manner became capt.-lieut. on that date; his predecessor had died on October 1st.

In January, 1705, 4,000 French troops joined the besiegers at Gibraltar, and encouraged them to try their fate in another grand assault, but they were decidedly repulsed with a loss to the garrison of only 147 killed and wounded. The marines again distinguished themselves on this occasion.

On January 29th the French Marshal Tessé took command of the besiegers, but could make no progress. The heavy rain filled his trenches and, thanks to Admiral Leake's assistance, the defenders had now the superiority in guns and ammunition.

On February 4th and 6th reinforcements of 700 men, Guards, Donegal's, and Dutch, joined the Prince, and on the 18th he completed a new battery behind the breach in the north curtain. This put the seal to the safety of Gibraltar, and to celebrate the occasion he caused fifty gallons of punch to be brewed, named the new battery "The Queen's," fired a royal salute from it at the enemy and drank Her Majesty's health, the troops giving three cheers. On the 24th the Prince made another successful sally, killing and wounding sixty-five of the enemy.

On March 9th, Admiral Leake, who had been collecting troops and supplies at Lisbon, returned to the Bay of Gibraltar, where he found the French squadron under de Pointis. Leake had thirty-five sail, English and Dutch, and de Pointis' squadron was destroyed at once and 1,200 men of Mountjoy's and a Dutch regiment were landed.

At the end of the month Marshal Tessé began to withdraw his guns, and on April 8th he retreated.

Competent judges put the loss of the French and Spaniards at 10,000 or 12,000 men.

There is no separate record of the part the regiment played in the defence. The marines were usually spoken of by the name of the field officer commanding as in the War in Catalonia, 1705–07, when all the marines were usually described as Wills' Marines, though he had men from several regiments under him besides our own.

The strength of the six companies of the regiment fell from 302 to 175 during the operations from first landing at Gibraltar to the termination of the siege. How many of the 175 were effective we do not know.

Lieutenant Andrew Day died on board ship two days before the Battle of Malaga, whether wounded at the capture of Gibraltar or not we cannot say. He had been received into the regiment by Colonel Saunderson at the request of the Duke of Marlborough.

Lieut. Dan. de Vauclen died on shore about the time of the termination of the siege, from what cause we know not. We have no accurate knowledge even of the date of his death, but we know that his successor was appointed on May 24th. The date of appointment, however, was often weeks after the vacancy occurred. We know of Lieut. Mason having been more than once wounded and losing an arm, from his petition and the warrant granting him a year's pay (£85) as compensation for mutilation. He was made a brevet-captain in 1706 for his service in the defence of Gibraltar. Lieut. Edmond

Harris ceased to belong to the regiment in October, 1704, or earlier. Nothing is known of the cause. He landed at Gibraltar.

Lieut. Thwaites was not among the effectives embarked at the termination of the siege, and as he does not appear to have served again and died in the following winter, he was probably invalided from Gibraltar.

Nothing is known of the place or cause of death of Mitford, the quartermaster, in October, 1704. It was probably at Gibraltar, for his successor, Thompson, was sent out to the companies there.

Cadet M. Brown landed at Gibraltar in June, but his name does not appear again.

Of the officers who landed in June, 1704, for the capture of Gibraltar, Captain Bedford, Lieutenants Charles Davison, George Saunderson, Thomas Newdigate, William Davison and Cadet Robert Bedford re-embarked in July, 1705, along with the following, who were sent out from home from time to time to fill up vacancies, Lieutenants James Davison, James Saunders and Bedford (Burston), Surgeon M. Poumies and Quartermaster J. Thompson. There was no Lieut. Bedford ; the purser's clerk who entered the name probably meant Thomas Burston, who was appointed to Bedford's company in September, 1704, and must have been sent out to fill the place of William Davison, promoted into Burston's company.

We cannot do better than close the account of the defence of Gibraltar with the following quotation from Boyer's *Annals of Queen Anne* : " During Leake's absence the Prince of Hesse redoubled his diligence for preventing the designs of the enemy and spent all the days in the works and most part of the night in the covered way. This example had so good an effect that the garrison did more than could be humanly expected and the English Marines gained immortal honour."

ENGLAND

While the older soldiers were gaining honour at Gibraltar a new regiment had sprung up at home. In May, 1704, fresh arms for 592 men were supplied by the Ordnance, 527 on augmentation and sixty-five to replace those lost with the grenadier company in the *Reserve*. In June the new grenadier company was ordered to march from Lincoln and Grantham, where it had been formed, to the regimental Headquarters at Canterbury. On July 13th, Brigadier Holt was ordered to review at Canterbury " Colonel Saunderson's regiment which had been recently ordered to make up its men to 100 in each company." It was now considered fit for service. Throughout 1704 two companies served on the Home Fleet and were relieved from time to time.

Towards the end of the year several important orders were issued by the Lord High Admiral. On November 11th he directs that 6*d*. a month be deducted from the pay of each non-com. officer and man of the marines while aboard ship, and that the money be paid into the chest at Chatham. The marines if hurt, maimed or disabled on service on board ship, to be entitled to the same benefits from the chest as the seamen. A further order directs the 6*d*. a month to be deducted whether the men are on shipboard or not. Another order extends the benefits from the chest to the garrison defending Gibraltar. The first list of men relieved under this order contains the names of seven men of the regiment sent home from Gibraltar. The records are incomplete. Another order directs that 6*d.* a month be deducted from the marines' pay on board ship, of which the ship's chaplain is to receive 4*d.* and the surgeon 2*d.* A further order of December 8th authorizes another stoppage of 6*d.* a month while at sea to be applied to the use of Greenwich Hospital, to the benefits of which the marines were to have the same rights as seamen.

On December 20th, Colonel Pownall had succeeded Colonel Saunderson in command. It is usually said that Colonel Saunderson died then, but this is very doubtful. It is certain that his will is dated December 1st, 1704, and it was proved on July 18th, 1707, by his younger brother, the Hon. James Saunderson. Colonel Pownall bought the regiment from Colonel Saunderson or his heir.

In January, 1705, Captain Ord and Chirurgeon Pawncy were ordered to Gibraltar for duty; probably Matthew Poumies is meant. The date of his appointment is uncertain, but he was serving in March, 1703, as was his mate, John Watts. A Matthew Poumies was naturalized in May, 1702; a certificate states that he had taken the Sacrament at St. Martin's, Westminster, in 1701. The yearly issue of clothing for the men at Gibraltar was to have been in Captain Ord's charge, but it was not ready and he did not go, but continued to live at Berwick-on-Tweed, as he did throughout his service in the regiment. The surgeon was ordered to take with him "a half chest of good wholesome medicine as well internall as externall supplied by the Apothecaries' Company of London."

In the winter of 1704–5 Captain Ord's company, under Second Lieutenant Brown, sailed on H.M.S. *Kingston,* convoying the trade, to New York and Virginia. A new Governor for the latter colony went out on the *Kingston,* which brought home the Governor relieved. Seventeen men of the company died during the cruise.

BARCELONA

The spring of 1705 found the British Government and the Duke of Marlborough firm in their Mediterranean policy, but the Great Fleet was this year entrusted to Sir Cloudesley Shovel. Under convoy of the fleet a number of transports conveyed an army of 6,500 men under Lord Peterborough, with Colonel Hans Hamilton as quartermaster-general and Colonel Charles Wills as adjutant-general. The capture of Barcelona and the occupation of Catalonia, which was believed to be favourable to Charles III, was the first object of the expedition.

From the orders issued to Colonel Pownall it is evident that a battalion of eight companies of the 30th was intended to act on land under his command. He and his adjutant and all the company officers were to embark with their men. The orders for embarkation were upset from various causes, one of which was a secret expedition or raid directed by the Admiralty on the coast towns from "Romansgate" to Hythe to capture sailors for service on the fleet; on which three companies were engaged. The following actually embarked:—

H.M.S. *Lennox,* Colonel's company, Capt.-Lieut. M. A. Bernard, Second Lieut. Sam. Elford.

H.M.S. *Grafton*, Major Burston.

H.M.S. *Berwick*, Capt. W. Palliser, Lieut. Hugh Palliser.

H.M.S. *Association*, Capt. Isaac Knight, Lieut. Nathan Potter, Second Lieut. Ric. Leathat.

H.M.S. *Royal Sovereign*, Capt. Ward, Lieut. Brooke, Second Lieut. John Brown.

The strength of all ranks was 443; fifty men of Bedford's company without an officer were on the *Royal Sovereign*. Colonel Pownall with Lieut. Storey, his adjutant, embarked at Portsmouth on H.M.S. *Albemarle*, on May 13th, and was transferred to the *Royal George*, the flagship, on the 30th, and there we lose sight of him. It is unlikely that he sailed. He must have been already negotiating for the sale of the regiment, which took place in a few months. He died in 1707. Storey joined at Barcelona and served in Catalonia.

The fleet reached Lisbon in June and sailed for Barcelona in July. When passing Gibraltar two young regiments were landed and the veteran garrison was embarked. H.M.Ss. *Yarmouth* and *Nottingham* took on board nine officers, one cadet, and 175 other ranks of the 30th, who had survived the siege, and also Sergeant Pearce and ten men, the remains of the party which had served for a year on H.M.S. *Roebuck* in Sir John Leake's fleet. Lieut. Charles Davison had gone home to fetch some men of Corbet's company. He and the detachment from Gibraltar joined the regiment in Barcelona Bay.

Corbet's company, under Lieut. Turner, had embarked in January on H.M.S. *Dorset*, and had landed in June at Portsmouth, ten men having died. Half the company re-embarked and were landed at Torbay, and these are the men whom Charles Davison brought out to Barcelona.

On August 11th, Sir Cloudesley Shovel's fleet reached Barcelona, and the landing of the troops commenced on the following day; 1,100 marines disembarked, but as far as we know, none of the 30th, though they may have landed as a covering party and re-embarked. The notices in the logs are ambiguous. The captain of H.M.S. *Grafton* (Burston's company) records: "12 August Signal for our soldiers to embark in order to land, we sent away our boats with our Marines. At 9 ye soldiers began to land, we having several small frigates to cover the descent." By soldiers the troops on the transports are meant. Some if not all the companies of the 30th were on board three weeks later. On August 18th, King Charles landed.

Our army disembarked to the north of the town; the coast to the south is too precipitous and is under the fortress of Montjuich and the fort of St. Bertran, which protects the short mile of road between that fortress and Barcelona. For three weeks after landing nothing was done and discontent rose high.

On September 2nd Lord Peterborough consented to carry out a plan suggested by the Prince of Hesse, Sir Cloudesley Shovel readily offering to co-operate. On that evening the Prince and Peterborough, from their camp north of the town, led a body of 1,400 men due west, with the intention of bending to the south during the night and so marching round Barcelona. Major-General Stanhope followed with 1,300 men to protect the left and rear. Our leaders hoped that Montjuich would be surprised at daybreak and captured by escalade.

To carry out his share of the undertaking, Sir Cloudesley Shovel had caused a body of marines to be transferred at nightfall from their ships to the *Antelope* and six other light vessels, with a view to landing south of Barcelona and assisting to isolate Montjuich; two companies of the 30th, Walter Palliser's and Ward's, certainly formed part of this detachment, and others may have done so. Some men of Burston's were there. In 1713 a claim was made for the subsistence of the marines at Montjuich on this occasion, and the Commissioners appointed to pay off and disband the six regiments, ordered that the amount should be divided equally between all six, so all the marine regiments must have served in this action. A battalion of twelve companies would muster about 600 men. The Admiral would scarcely employ seven vessels for less.

The Prince of Hesse did not reach Montjuich till after daybreak, but attacked without hesitation; he was repulsed and when, changing his object, he endeavoured to isolate the fortress by the capture of St. Bertran, he was mortally wounded by a shot from Montjuich to which, in his eagerness, he had passed too close. There was some confusion after his fall, but the position was restored by the capture of St. Bertran by a party of Miquelets (Catalan irregulars). It is at this stage that the companies of the 30th and other marine regiments came into action; at 1 a.m. on the 3rd by Sir Cloudesley's order more marines had been placed on the light craft to strengthen those already embarked. Soon after daybreak the sound of guns and the rattle of musketry round Montjuich could be heard plainly on the fleet, and when it was known that St. Bertran had been taken the *Antelope* and her consorts ran in and landed the marine battalion. From the ships it was seen by the puffs of smoke and reports that it was engaged in a smart skirmish as soon as landed. Montjuich was now cut off effectually from Barcelona and guns, stores and entrenching tools were landed to the south of the town. A lucky shot blew up the magazine of Montjuich on the 7th and its surrender followed. The part which the marines, directed by Sir Cloudesley Shovel, played in the capture of Montjuich has been rather overlooked by historians. It is given in the ships' logs.

After the fall of Montjuich the siege of Barcelona was commenced; the army remained in its camp to the north-east of the town and had charge of operations on that side; to the south the Admiral landed over 3,000 of his seamen and heavy guns and artificers. With the assistance of the Catalans the sailors formed batteries to breach the south-western face of the defences. We do not know with which division of our forces the 30th acted.

By September 14th the breaching batteries had opened fire, and in ten days the town was summoned. The Governor, Velasco, did not dare to face an assault with his half-hearted men and capitulated on condition that he and his garrison should march out with the honours of war on October 1st. The townspeople, however, seized the gates and admitted the Miquelets. Lord Peterborough, to save life, thereupon embarked Velasco and as many of his troops as would go with him, about a third, and sent them to Malaga.

Sir Cloudesley Shovel sailed for England with the allied fleets on October 13th. He had re-embarked a large number of marines, but no effectives of the 30th except a small party of grenadiers.

The losses of the regiment at Barcelona are hard to ascertain, but they appear to have been about forty-six non-com. officers and men from leaving England to the commencement of the siege and ninety-five from all causes during the siege. Second Lieut. Samuel Elford died before the fleet reached Barcelona Bay, and Capt.-Lieutenant M. A. Bernard died towards the termination of the siege, whether from wounds or sickness is unknown. He was one of the three Bernards who served in the regiment under Colonel Saunderson in Flanders.

On October 13th, Colonel Charles Wills succeeded Colonel Pownall as colonel of the 30th, having purchased the regiment. After being transferred from the 30th to Charlemont's Foot in 1701, Wills had commanded that regiment with great distinction at Cadiz, Vigo, and in the West Indies, and had joined the expedition to Barcelona as adjutant-general. He was now placed in command as brigadier of all the marines left in Lord Peterborough's army. They were formed in two battalions, to one of which the 30th contributed the staff and three-fourths of the officers and men. Each battalion was 750 strong.

Lord Peterborough was anxious to send British troops as soon as possible to occupy Lerida, which has been called the key of Catalonia, but the want of money stood in his way. He was unable to repair the breaches he had made in the walls of Barcelona, much less equip a force for the field. The marines were the worst off, as no one had contemplated their taking the field at a distance from the coast. Their clothes too were very bad. His Lordship was relieved from some of his anxieties by the public spirit shown by Colonel Wills and the marine officers, who volunteered to purchase and maintain a sufficient number of mules for themselves and their men. As regards the clothing, nothing could be done and the men would have to make it last till a fresh supply came out with the spring fleet.

The number of mules bought gives us a state of the brigade; the following was the distribution :—

Col. Wills	8	Mules
Two Lieut.-Colonels	8	,,
Three Majors	9	,,
Seven Captains	14	,,
One Major of Brigade (Captain)	4	,,
Twenty-four Lieutenants	24	,,
One Adjutant and One Quartermaster, one each	2	,,
Two Chirurgeons	2	,,
Two Mates	2	,,
Fifteen hundred men, one to fifty	30	,,
Commissary to the Marines	2	,,
	105	,,
For Adjutant General	12	,,
Total	117	,,

At 80 dollars (£19) apiece for purchase and maintenance for one year, the expense was £2,223. The last charge seems to be for staff mules for Wills, first as adjutant-general and afterwards as brigadier.

The 30th found the brigadier, 1 major, 4 captains, 10 lieutenants, the adjutant, the quartermaster, 1 surgeon and 1 mate, that is 20 officers out of 45. It gave 500 men, or two-thirds of a battalion, but as it had always six companies in Catalonia, its contribution to the strength must on an average have been three-fourths of a battalion of eight companies.

The figure 500 is arrived at in this way. The strength of the six companies on leaving England, including 33 of Corbet's company attached, was 476; to this must be added 186 from Gibraltar and 12 of Dornell's company from H.M.S. *Grafton,* making a total of 674. The casualties up to the time the fleet sailed for home were 141, perhaps more, so 500 is probably near the mark.

The officers were Colonel Charles Wills, Capt.-Lieut. Hugh Palliser, the Adjutant, Arthur Storey, Quartermaster John Thompson, Surgeon Matthew Poumies, Surgeon's Mate John Watts; Major George Burston and Lieut. William Davison; Capt. Isaac Knight, Lieut. Nathan Potter and Sec. Lieut. R. Leathat; Capt. Thomas Bedford and Sec. Lieut. Thomas Burston; Capt. Walter Palliser and Sec. Lieut. James Davison; Capt. David Ward, Sec. Lieuts. John Brown and Thomas Brooke. In addition to the foregoing, who were serving with their companies, Lieut. Charles Davison had brought a detachment of Corbet's company, Sec. Lieut. Newdigate one of Ord's, and Sec. Lieut. James Saunders one of Brereton's. It is remarkable as showing the continuity of regimental life that, of the officers named, Burston, Bedford, Knight, the two Pallisers and Potter had all served with Col. Wills in Flanders when the regiment was known as Saunderson's Foot. M. A. Bernard too, who died a few days before Wills returned to the regiment, was an old comrade of King William's wars. Of the younger officers, Thomas Burston was the son and the three Davisons were the brothers of old comrades, and James Saunders is believed to have been a brother of John Saunders commanding the grenadiers in Kent, who also had served at Namur.

Lord Peterborough assured Col. Wills and his officers that he would repay them at the earliest opportunity, and there is no reason to doubt his sincerity, but the opportunity never came, for the army in Spain was always in want, and it was not till sixteen years afterwards that the officers or their representatives received the amount due for the mules and their upkeep.

By November 12th the field force was equipped, and one marine battalion had marched for Lerida. The other followed on the 13th. Lord Rivers' regiment started on the 17th and Palmes's Dutch marines on the 19th. The march could be safely made in detachments, for it was covered by a body of Catalans and Neapolitans who had occupied Lerida under Prince Henry of Hesse. Major-General Conyngham was in command and his own dragoons formed part of the force. While this small army was collecting for the defence of Catalonia, Lord Peterborough had entered and overrun Valencia and already dreamt of an advance on Madrid. Between December 14th and 25th he ordered General Conyngham to attack Mequinenca, near the junction of the Ebro, Segre and Cinca, "in order that upon the success thereof he might be in a better position to detach troops to Lord Peterborough in Valencia." This order shows an almost inconceivable ignorance of the state of affairs on the part of Lord Peterborough. As was to be expected, the French and the adherents of the Duke of Anjou were preparing to make a great effort to retake

Barcelona. Their design was favoured by the wide dispersion of Peterborough's troops and still more by the absence of the British fleet.

Marshal Tessé had massed in Aragon all his troops available in Spain while an army was collected in Roussillon and a fleet fitted out at Toulon. The plan was that Tessé should cross the lower Ebro and passing between Peterborough and Conyngham, march straight on Barcelona, there to be joined by the army from Roussillon and the fleet from Toulon. He would give up his direct communication with Aragon and Madrid and base his army on the fleet and on Roussillon.

Lieut. Brooks, of Ward's company, died or was killed towards the end of the year. His successor, Dobson, was sent out in the spring fleet of 1706.

1706

The establishment was the same as in previous years.

The following promotions and appointments were made :—

Joseph Mason to be Brevet Captain in the Regiment .	1st	April
Hugh Palliser to be Capt., *vice* Harris, deceased . .	13th	,,
Nathan Potter to be Capt.-Lieut., *vice* Palliser . . .	13th	,,
Charles Davison to be Capt.-Lieut., *vice* Potter, deceased .	1st	Sept.
Richard Turner to be First Lieut. to Knight, *vice* Potter .	13th	April
John Thompson from Quartermaster First Lieut. to Dornell, *vice* George Saunderson, deceased	13th	May
James Davison, First Lieut. to Corbet, *vice* C. Davison .	1st	Sept.
William Dawes, First Lieut. to Walter Palliser, *vice* Hugh Palliser	15th	,,
Thomas Burston, First Lieut. to Knight, *vice* Turner, deceased	10th	Dec.
Edward Hutchinson, Sec. Lieut. to Burston, *vice* Saunders .	12th	Jan.
Geo. Sherrard, Sec. Lieut. to Dornell, *vice* Dawes . .	1st	March
James Littleton, Sec. Lieut. to Ord, *vice* Newdigate . .	6th	June
Thomas Dawes, Sec. Lieut. to Bedford, *vice* Burston . .	10th	Dec.
Edward Aylmer, Sec. Lieut. to Colonel, *vice* Dobson . .	15th	,,
Lieut. C. Davison to be Adjutant, *vice* Storey, deceased .	2nd	Nov.
William Norman to be Quartermaster, *vice* Thompson . .	13th	May
The Rev. William Merridew to be Chaplain, *vice* Claringburn .	12th	,,

Charles Davison was adjutant and captain-lieutenant, an unusual combination. His predecessor, Arthur Storey, a Northumberland man, died at sea, perhaps invalided from Catalonia, and Colonel Wills had to fill up the vacancy on the spot. The date of the appointments of Dawes and Sherrard cannot be right. The vacancy was caused by the promotion of Hugh Palliser to capt.-lieut. on October 13th, 1705.

Edward Aylmer was the first of three of his name to join the regiment. From their Christian names they were probably nearly related to the distinguished admiral. The family interest must have been strong, for in 1698, at the age of thirteen, Edward was appointed a muster-master of marines ; a year later when the marines were disbanded he was placed on half-pay.

In 1706 the regiment lost two veterans who had served in it in Flanders under Colonel Saunderson, Harris and Potter. Harris appears to have died at home. Potter served at the capture of Barcelona and no trace of his going home has been found. He probably died or was killed on service in Catalonia.

The French opened the campaign of 1706 early in January. To divert attention from the Ebro their left was the first to move and it attacked high up on the river Cinca opposite General Conyngham's right. On the 23rd of the month General Conyngham sent Colonel Wills with 400 British and 200 Dutch marines to strengthen a post of our dragoons at a place called Tamarite. As Major Burston was with the detachment we may conclude that the 400 British represented the battalion of which the 30th formed the greater part. Four hundred may well have been the battalion's effective strength, for it was sickly in Barcelona.

On the 24th, Wills moved forward to S. Estevan, where he occupied the village with his own men, having the Dutch marines two miles in rear as a support; twenty-five of Conyngham's dragoons under Capt. Matthew accompanied Wills. On the same day General Conyngham, in case further assistance was required, moved up to Tamarite with his own regiment and 600 foot and summoned all the Miquelets to join him.

Major Burston's account of the action fought on the next day is so good and it is so impossible to alter a word without spoiling the effect that it is here given in full:

" On the next day, which was the 25th, about noon, while the officers at St. Estevan were all at dinner, nine Squadrons of the enemy appeared in a little plain not above half a mile from the village among the mountains, having a narrow road leading from it to the village and passing between two hills, so that it was no easy thing to attack us that way.

" Our troops took the alarm, and all our men were under arms in an instant, and Mr. Wills who commanded, ordered Captain Matthew with the dragoons to enter the lane, and discover the enemy, a hundred foot being detached to sustain them, with orders to possess themselves of a small cottage and garden-plot lying on the side of the way, and post themselves there to defend the passage, and prevent the enemy advancing to the village.

" The rest of the foot were ordered to march up the hill on the right side of the land, which lay directly between the town and the plain where the enemy were. This body marched in six columns, keeping as near to one another as they could; with orders that when any of the columns should discover the enemy, they should beat the Grenadiers' March as a signal to the rest, which was to be answered by the rest, as well those on the hills as those in the lane.

" These dispositions were so well made, and the orders so punctually executed, that at the very same moment that our dragoons enter'd the plain, killing the advanced or videt Centinel of the enemy, that very moment our infantry shew'd themselves, and beat the Grenadiers' March upon the hills, and as all the Marines wore Grenadier Caps, the several Divisions looked like the several companies of Grenadiers belonging to particular regiments, and the enemy not doubting but there was a regiment of foot to every company of Grenadiers, believed we had six regiments of foot there, whereas we had but six divisions of fifty men in a division.

" This stratagem had its full effect, for the enemy immediately remounted and march'd off in a great hurry, leaving most of their pickets and some provisions behind them; and which was still more, their infantry, who were in full march to join them, seeing the retreat of their horse, halted, and afterwards marched back also; we followed and kept pace with them on the top of the hill, for near two mile, frequently sending down a platoon or two to fire at them; but seeing no greater matter to be done that way, gave it over, and returned to the village.

"Lieut.-Genl. Cunningham having notice of what had happened, march'd with all speed and joyn'd us about midnight, so that we were then in all 1,200 foot and 250 Dragoons, with a confus'd rabble of about 800 Miquelets, good men enough and well enough armed, but under no command and observing no manner of order.

"Early the next morning the enemy allarm'd us again; their body consisted of 4,000 men, regular troops, commanded by Monsieur D'Hasfield, an experienced officer, besides a great number of Miquelets.

"Upon this alarm Mr. Wills ordered Major Burston to advance with 400 Miquelets only, with directions to go up the hill, and take possession of their former posts till the rest of their troops got to their arms. He march'd, indeed, and had 400 Miquelets with him to the said posts: But when Lieut.-Genl. Cunningham and the rest of the troops came up, they found him there all alone: for as soon as the enemy appeared, the Miquelets run all away from him to a man, and made no stop till they came to another hill about a mile off.

"It is to be observed here, that as our men marched on the top of the mountains on one side of a narrow valley, so the enemy appeared on the mountain on the other side: the valley lying between both bodies, with this difference only, that as the enemy on their side of the hill were naked and expos'd to us even to the ground they trod on: so on our side there run along a rising ground parallel to the said valley, which was a kind of natural breast-work to our men: so that the enemy could not discover them, except only from the shoulders upward. It was upon this breastwork that Lieut.-Genl. Cunningham, stepping up afterwards to observe the motion of the enemy, received a musket shot in the groyne, of which wound, to the great regret of the whole Army, he died the next day.

"The platoons on either side pelted one another for some time: but the situation prevented the enemy attacking us, the ascent on our side being very steep. And as our men stood in cover, their men must have been expos'd to our fire, both in descending on their side and in advancing on ours: so they contented themselves with drawing up their whole body in view that we might judge of their strength, which they hoped would terrify us, but they were mistaken in that also.

"During this affair the Miquelets on both sides made a great fire after their manner, running and scampering about the field, now this and then that party alternately flying and pursuing, but to what purpose we knew not: for we never found any of them kill'd or hurt either on one side or other: and indeed it looked more like a parcel of fellows playing at Prison Bars, than like a pitched battle between soldiers that were in earnest with one another.

"In the meantime the enemy had like to have surprised us all: For Count D'Hasfield had detached 600 Grenadiers to attack us another way, and these men taking a bye-way among the mountains which we knew nothing of, were upon our right flank before we had the least notice of them: and advancing directly upon that place where Mr. Webb with a party of Lord Shannon's Regiment was posted, they drove him down the hill to Cunningham's dragoons: and had not 2 of Mr. Webb's men who had straggled from him come running to Mr. Burston, and given him notice that the enemy was just upon him, the fate of our little Army had been the same.

"Upon this notice, Major Burston advanced without waiting for orders, and with 25 files of Marines clamber'd up the hill, which was very steep and high on our side, and met the enemy, who were doing the like on the other side, and came up to them at the distance of only 20 paces.

"The enemy came on with great fury, giving the word to one another aloud, *Sans Quartier, Sans Quartier,* upon which seeing there was nothing to do but to fall on, he call'd to his men, and said, *you must expect no quarter, lads, so a God's name fire at 'em:* and to give his men their due, they did their duty, and behav'd bravely enough.

"The Generals, Cunningham, Wills and Palms were at this instant all together on the left, consulting upon the further dispositions of the field, and seeing their right thus suddenly and furiously attacked, they immediately sent supplies from several quarters after one another: and the two first came in person with what speed they could, so that the enemy, after about an hour's obstinate fight was obliged to retire, leaving behind them about 350 of their men kill'd upon the spot, or wounded, among whom was a French Colonel who commanded them.

"But this could not be done so soon, but that our men who were at first outnumbered by about five to one, suffer'd some loss, and Major Burston himself was left for dead among the slain, having received no less than 13 wounds, most of them being stabbs and cuts with the bayonets of the enemy, and had their main body advanced in time and supported their grenadiers, who, to do them justice, behaved very well, our whole body might have been in great danger.

"But to the great wonder of our Generals, they continued drawn up on the other side of the valley aforesaid, firing pretty briskly by platoons, but never advanced to engage our whole body, and at length seeing their detachment driven back, they gave it over, and march'd off, carrying abundance of wounded men off with them: their own accounts publish'd at Saragossa, acknowledging that they lost 1,000 men in this action, leaving the victory to our men who were not half their number. All that we could suppose to be the reason of this was, that they believ'd our strength to be much greater than it was."

Besides giving a clear account of the action of S. Estevan, Burston's narrative contains a good deal of interesting information.

In the first place, he tells us that the marine cap closely resembled that of the grenadier, but we are still in the dark as to the pattern of the grenadiers'.

He shows us also that a battalion of marines in Catalonia put in the field eight companies of twenty-five files. The strength of the six companies of the 30th would therefore be a trifle over 300 effectives. Now one of the few Admiralty returns we have gives the total strength of the regiment at this time "In Spain and on the Descent" as 383, a striking testimony to the correctness of Burston's figures, for eighty-three sick and on command is quite a natural proportion. The "two of Mr. Webb's men" who gave Burston the alarm of the flank attack were undoubtedly two marines belonging to the grenadier company of Shannon's regiment under Captain Webb, who was recommended for a majority by Wills in the following year. There are many reasons for believing that the regiment was armed with the ring and socket bayonet in 1702, but Burston's narrative puts the matter beyond a doubt. When, at the moment of closing with the French, he gave the order to fire it is impossible to suppose that bayonets were not already fixed. The French account published in the *Paris Gazette* confirms Burston's. It says that the leading of the main attack was entrusted to M. de Polastron, who charged at the head of six companies of grenadiers supported by six companies each sixty strong. The grenadiers advanced with great resolution with their bayonets in the muzzles of their muskets, and in spite of the sustained fire of the British and Dutch came to close quarters and overthrew the first line. It is evident that the British could fire with fixed bayonets and the French could not.

The enemy claimed to have annihilated our grenadiers, meaning the marines, and to have killed a colonel; this was our friend Burston, but he was tougher than they thought. Besides the twelve companies which delivered this most gallant and dangerous assault they had in action nine battalions belonging to the regiments Isle de France, de la Couronne and Sillery, and if Count d'Asfeld had shown in the management of his supports ability equal to that of Conyngham and Wills we might have had the worst of it.

When to the great regret of the army General Conyngham fell mortally wounded, Brigadier-General Wills took command. About three o'clock the enemy retreated, and after remaining some hours on the battlefield Wills fell back on Balaguer and not on Lerida, showing that he was uneasy about his right. He of course could have no idea of the great French combination which was so nearly to drive us from Catalonia.

The British and Dutch acknowledged to 300 casualties in the action, the French to 400, but letters published in Brussels from French sources say that one captain of grenadiers was killed, the lieut.-colonel of the regiment of Sillery and five captains and some subalterns wounded and about seventy soldiers killed and wounded. Probably seventy soldiers killed is meant. All accounts agree that the fighting was very fierce and frequently muzzle to muzzle.

A priest took Burston into his house, where his wounds were sewn up by a Spanish barber, under whose care he made a quick recovery. During his convalescence Colonel Palmes, at the head of the Dutch marine officers, all in full uniform, paid a visit to Burston to congratulate him on his gallant conduct.

On March 17th an end was put to all ideas of an advance to Madrid on Peterborough's part, for on that day Marshal Tessé at the head of 12,000 men crossed the Ebro, and leaving Lerida on his left marched through the mountains on Barcelona. There was nothing left for the British and their allies but to follow him with what forces they could collect, harass his rear and interfere as far as possible with his siege of the capital of Catalonia.

This duty was vigorously performed by Prince Henry of Hesse with his Catalan and Neapolitan troops from Lerida. He was assisted by a general rising of the Catalans in the rear of the French under the Conde de Cifuentes.

Lord Peterborough sent orders to Brig.-General Wills on March 23rd and 25th to bring all troops under his command down the Ebro to Tarragona on the coast south of Barcelona.

By the end of March there were assembled there Lord Rivers' regiment, the two battalions of British marines and the two battalions of Dutch under Palmes; Lord Peterborough joined them from Valencia with two regiments of dragoons about April 20th.

Marshal Tessé had reached Barcelona on April 3rd and found there the fleet which had arrived the day before from Toulon and 9,000 French troops under Lègal who had marched from Roussillon. The siege of Barcelona was commenced at once.

Nothing could have saved the town and King Charles (Archduke of Austria) from capture but that Tessé was too late: the British fleet was already at sea under Sir John Leake, and if a bold front was shown by the besieged and strenuous efforts made by their friends outside to hamper the besiegers there was good reason to hope that Sir John would arrive in time.

Strong appeals were made by King Charles and Prince Henry of Hesse to Lord Peterborough to join the Prince and Cifuentes in an attack on the French rear, but nothing would persuade him to move; the utmost he did was to send fifty grenadiers to the Prince under Captain Scott. Capt. Scott, who belonged to Holt's regiment, was given a company in the 30th in the following spring.

Peterborough has been much blamed for his inactivity, but Wills and Burston both thought he was right. It was easy enough for the Catalan irregulars to make an onfall on the French lines and if beaten to regain the shelter of the hills, not much the worse, but Peterborough had no right to engage his troops in such warfare without absolute necessity.

Sir John Leake's fleet was drawing nearer every day, and had on board it five fresh British battalions and strong drafts for the regiments in Catalonia. Among them were four companies of the 30th.

On April 29th the fleet was heard of at Altea Bay, and Peterborough prepared to join it by collecting a flotilla of small craft. He sent Lord Rivers' regiment back to Lerida and on May 5th ordered Brigadier Wills to embark the marines. On the 8th inst., Sir John Leake being two leagues south by west of Barcelona, Peterborough boarded the flagship and the marines under Wills in their small vessels sailed along the coast abreast of the fleet.

The French Admiral when Sir John Leake was within a day's sail of him had abandoned the army on shore and made for Toulon, leaving his colleague Marshal Tessé to get out of the scrape as best he could.

As soon as the British fleet had anchored in the Bay the marines and troops were disembarked to meet an expected assault, for the breaches were reported practicable. The French fire continued all that night and the day following, but on the night of May 9th loud explosions from their lines showed that Marshal Tessé was destroying his ammunition, and on the 10th, as the captain of H.M.S. *Essex* records in his log, "they raised their siege and scoured off." Prince Henry and the Catalans pursued, the British forces did not.

The news of the relief of Barcelona was received with great joy in England, and a vote of thanks was passed by Parliament and a commemorative medal struck.

The four companies of the regiment which had arrived on the fleet re-embarked and sailed with Lord Peterborough's expedition for the recovery of Valencia and the Spanish coast towns.

The two battalions under Wills were sent into garrison in Catalonia.

ALICANTE

The four companies of Corbet, Brereton, Ord and Saunders (grenadiers) after landing for the relief of Barcelona, re-embarked on Sir John Leake's fleet and sailed on May 15th for Valencia, where Lord Peterborough landed with his army. The fleet then turned to the reduction of the coast towns of the province. On June 13th, Sir John Leake was at Carthagena. The town surrendered without fighting and he occupied it with a garrison of 600 marines, including two companies of the 30th. On the 16th, Sir John sailed for Alicante. Colonel Mahoni, a brave Irish officer, who was in command of the town, refused to surrender, and some marines were landed to strengthen the Miquelets who were blockading the town on the land side. By the 22nd, Brigadier Gorges with a detachment from Peterborough's army arrived to join in the attack and 500 seamen were landed to help him. All the marines on board were now landed, each man carrying twenty-five rounds of ammunition. On the 23rd, Sir George Byng commenced to cannonade the enemy's defences. On the 25th, Sir John Jennings arrived from Carthagena with the two companies of the 30th, and the other marines who had been left in garrison. The *Fubs* yacht was employed to land them before Alicante. *Fubs* was King Charles the Second's name for the Duchess of Portsmouth, and he had named the yacht after her. The fire of Sir George Byng's squadron and the batteries on shore was kept up till June 28th. On that day the town was stormed by the seamen, marines and land forces. After entering, the attacking party found itself under fire from the windows of the houses, but soon the defenders were driven into the castle and the town occupied and plundered. The marines embarked three days afterwards.

The little we know of the share which the 30th took in the capture of Alicante is derived from the pay-lists of the ships on which different parties of the regiment re-embarked. While on shore during the siege their doings and their losses cannot be traced. Many parties may have re-embarked of which the record has not been found.

H.M.S. *Essex* re-embarked Lieutenant Rothwell Stow, and thirty-four of the grenadier company, and eight men of Brereton's, under Sergeant Pickering. After the embarkation the *Essex* was in action with the castle, where the gallant Mahoni still held out for a time, and two of Rothwell Stow's grenadiers were killed. Another man of the 30th was killed by the fire of the castle on H.M.S. *Grafton*. Another party of the grenadier company was engaged in the siege under Lieutenant James Saunders, who after serving at the defence of Gibraltar, the capture of Barcelona, action of S. Estevan and relief of Barcelona, with a part of Brereton's company, had joined the grenadiers at Barcelona. The ship on which this party re-embarked did not return home till the spring of 1707. Saunders was landed at Lisbon in April of that year with a number of sick, and there we lose sight of him : he was dead by November, 1707.

In addition to Sergeant Pickering's party on the *Essex*, H.M.S. *Rupert* embarked nineteen of Brereton's company at Alicante. The company had left England with a strength of forty-three under Lieutenant Gardner. The officer and fifteen men cannot be traced. Probably they were dead or on a hospital ship. Gardner was dead before April 10th, 1707.

Lieutenant James Littleton and forty-seven of Ord's company re-embarked on H.M.S. *Cambridge* after the storming of Alicante. Of these, five died off Alicante, four on passage home and three were left in Lisbon hospital.

Captain Corbet's company was divided in 1706, a strong party in Catalonia, another in the Home Fleet and another in the Mediterranean. The probate of his will shows that he died "On the High Seas" on February 10th, 1707. A part of his company must have served at Alicante.

While the greater part of the regiment was serving in Spain or the Mediterranean, the companies of Dornell and Hugh Palliser (formerly Harris) were guarding prisoners of war in Kent, or employed on the Channel Fleet. On August 11th, 1706, Sergeant Hoyes, Corporal Birch, and thirty-eight men of Dornell's under Lieutenant Richard Turner embarked on H.M.S. *Crowne* for the West Indies. The officer, sergeant, corporal, and thirteen men had died by Christmas; the survivors of the detachment returned to England in 1707.

On August 24th, 1706, a detachment of Hugh Palliser's company was transferred from the *Albemarle* to the *Cruiser*. Palliser, who had been recruiting in Lincolnshire, joined it and on February 2nd, 1707, he and twenty-four men landed at Alicante. His company was in garrison there for a year.

CATALONIA AND ALMANZA

We must return to the six companies left in Catalonia. According to Burston they had been left in garrison because "they were so tattered and ragged and indeed by long service and want of clothing were almost naked." The new clothing for the 30th was somewhere on the fleet but apparently could not be got at in time for the Valencia expedition.

On May 29th, 1706, the battalion in which the 30th were serving along with the fusiliers and 1,000 Neapolitans marched to Gerona under Colonel Wills, now a Major-General in the Spanish service. The other battalion of marines garrisoned Tortosa, but Wills was still charged with its comfort and discipline. Both battalions were 600 strong. The marines passed the remainder of 1706 tranquilly in garrison, troubled only by want of supplies and money. Wills complains in official letters of having to live from hand to mouth and being unable to form a magazine. The officers had still to maintain the mules they had provided at Barcelona for the brigade. They had also to pay for the care of their sick and wounded men. The old practice, when there was no hospital, was for the colonel to receive 20*s*. a man for his wounded with which he did his best for them, but in Spain the regiment received nothing.

Towards the end of the year the officers of marines petitioned Lord Galway, now commanding our troops in Spain, to be allowed to serve under his lordship in the field. The course of events leading up to this petition was as follows.

Lord Peterborough after the reduction of Alicante and the coast towns had in August effected a junction at Madrid with King Charles III and a force from Barcelona, and with an Anglo-Portuguese army under Lord Galway and

Das Minas which had advanced from Lisbon. Peterborough now left the Peninsula, and the French under the Duke of Berwick received such heavy reinforcements that Galway was obliged in September to retire to Valencia. Here he placed his army in winter quarters to await the arrival of reinforcements from England and to prepare for a fresh advance on Madrid in spring.

It was this prospect of a campaign on a great scale, deciding the fate of Spain, which moved the marine officers to present their petition to him.

The memorial represents that the marine officers are mightily ambitious of joining the army in Valencia, the cause which occasioned their being flung into garrison being removed, that their men are well inured to the country and desirous of service, that without help from public funds the officers had equipped their brigade with everything but tents and that the memorial was presented with the approbation of their Brigadier, Wills, who is satisfied that it is purely owing to their hearty desire of serving Her Majesty.

The memorial was signed for the 30th by Lieut.-Colonel George Burston, Brevet-Major David Ward, Lieut.-Colonel Walter Palliser, Lieutenants Charles Davison (Adjutant), William Davison, Edward Hutchinson and George Dobson.

For Holt's regiment, by Lieutenants William Scott and Hugh Wilson.

Bor's regiment, Lieut.-Colonel Cobham, Captains Cory and Stuart, and Lieutenant Sullivant.

Shannon's regiment, Lieut.-Colonel St. John Webb and Lieutenant Powell.

Lutterel's regiment, Lieutenants Poole and Paslett.

Seymour's regiment, Captain Walker, Capt.-Lieutenant Lethule, Lieutenants Bush, English and Davis.

Burston, who gives us this list, gives a brevet to Walter Palliser which was not bestowed by Lord Galway till shortly afterwards, but it is believed that at this time both Ward and Palliser were brevet-majors. For some reason not known Ward was now senior in the regiment to Palliser, though he got his company in it some months later.

The number of officers present when this memorial was drawn up by Major Burston was small, as the marines were in winter quarters and many were at home recruiting.

At the request of his brother officers Major Palliser waited upon Lord Galway at Valencia and presented their petition. His lordship was much pleased, but was forced to decline the proffered services of the marines on account of the King's anxiety as to the safety of Barcelona. His Majesty was timid about being left without British troops, and not all Wills' efforts could persuade him that his fears were groundless. The three battalions were kept at Gerona till the campaign of 1707 turned against us, when General Stanhope moved the fusiliers to Barcelona and the marines to Martorel to cover the southern outlets of the passes through the hills to the north of Barcelona.

In 1706 recruiting had been carried on in the old districts, Captain Ord's company in his own county of Northumberland, Bedford's near Wakefield, where he had recruited for Lord Castleton in 1698, Knight in Lincolnshire, Lieutenant James Davison in his native county of Durham, William Davison in Lincoln, and Hugh Palliser in the City of Lincoln. In the Public Record Office there is an autograph letter from Palliser to the Admiralty Secretary saying that he expects shortly to have thirty to forty recruits and asking permission to send them south by sea from Hull.

1707

The following promotions and appointments were made:

Charles Wills, Esq., to be a Brigadier	1st Jan.
Lieut.-Colonel G. Burston to be Colonel in the army . .	14th July
David Ward to be Lieut.-Colonel of Foot	24th Aug.
Major G. Burston to be Lieut.-Colonel, *vice* Dornell, deceased	14th Nov.
Major Charles Williams from Shannon's regiment to be Major	,, ,,
Captain John Saunders to be Major of marines from 15-8-05	20th Sept.
Lieutenant John Irwyn from Holt's regiment to be Captain, *vice* Knight, deceased	1st Jan.
John Gardner to be Captain, *vice* Brereton	1st Feb.
Charles Davison to be Captain, *vice* Corbet, deceased .	9th Feb.
Captain Michael Medford from Lord Paston's regiment to be Captain, *vice* Dornell, deceased	24th Dec.
William Davison to be Capt.-Lieutenant, *vice* Charles Davison	9th Feb.
Richard Leathat First-Lieutenant to Burston, *vice* W. Davison	,, ,,
James Davison ,, ,, ,, Charles Davison . .	22nd Feb.
Henry Long ,, ,, ,, J. Gardner, *vice* W. Gardner, deceased .	10th April
John Roper Second-Lieutenant to Corbet	9th Feb.
Fairfax Burston ,, ,, ,, W. Palliser, *vice* J. Davison	22nd Feb.
John Walker ,, ,, ,, Ward, *vice* Brown, retired	24th Feb.
Edward Martyn ,, ,, ,, H. Palliser, *vice* Long .	10th April
Edward Quarles ,, ,, ,, Burston, *vice* Raynsford, transferred to grenadiers, *vice* Saunders, deceased . . .	21st Nov.
Lieutenant Joseph Mason to be Adjutant, *vice* C. Davison, prom.	9th Feb.

Lieut.-Colonel Dornell died at Canterbury, but when his will was proved, his domicile was stated to be St. James's, Westminster, so he seems to have contented himself with an occasional visit to his command. Burston succeeded him at Canterbury.

On January 1st the regiment lost Captain Knight, who died at Lisbon. He had been recruiting in Lincolnshire during the season of winter quarters and was probably on his way to join with the reinforcements from England. He had been appointed lieutenant when Lord Castleton first raised the regiment in 1689; he was capt.-lieutenant to his lordship in 1692, and was wounded at the assault of Namur as a captain under Colonel Saunderson in 1695. He had served at the capture of Barcelona in 1705, but returned to England to recruit immediately afterwards.

It is impossible to say where John Gardner came from. Only one of that name appears in Mr. Dalton's army lists and he is an ensign.

Raynsford's appointment to the grenadiers is rather a comical one, for he was only ten years old, but the object evidently was to put him in a home company and to give an effective subaltern to Burston's company in Catalonia where Major Francis Raynsford, father of the grenadier, was fighting alongside the 30th. The boy had leave of absence till he was fifteen years old to complete his education.

The principal event in the year 1707 was the union of England and Scotland which the Commissioners of both countries had approved of in the previous year. On January 16th, 1707, the Act of Union passed the Scottish Parliament and was touched with the sceptre, and on March 6th, Queen Anne, seated on the throne in the House of Lords, gave her assent to the English Act. The Union of the crosses of St. George and St. Andrew formed the national flag of Great Britain from April 1st.

On February, 8th Sir Cloudesley Shovel arrived at Alicante with the fleet from England. He brought Lord Rivers and 8,000 British and Dutch troops. The Duke of Marlborough hoped that with such a reinforcement King Charles, Galway and Das Minas might advance again and recover Madrid. The King, however, although he had taken part in the concentration at Valencia, insisted on returning to Barcelona and taking with him the Austro-Spanish troops. By so doing he practically destroyed all hope of success.

ALMANZA

On April 10th, Galway and Das Minas advanced with the object of destroying the French magazines in Murcia. On the 22nd, they heard that Berwick lay at Almanza awaiting the arrival of the Duke of Orleans with reinforcements. Berwick had 25,000 against the 15,500 which was all the Allies could muster, but in spite of the disparity of force, Galway and Das Minas determined to make an attempt to defeat Berwick before Orleans could arrive.

On the 26th, they attacked and, chiefly through the giving way of the Portuguese, who formed half the army, were totally defeated. Galway, who was badly wounded, fell back to Alcira and strove with scanty means to put Valencia in a state of defence. In the previous year his hand had been carried off by a round shot, necessitating the amputation of the arm, and he had only recovered sufficiently to take the field again in November.

It is constantly said that no part of the 30th was engaged at Almanza, and Colonel Parnell in his *War of the Spanish Succession* hazards a conjecture that the marine battalion in Lord Galway's army was probably Bor's. His opinion is entitled to the highest respect, but Bor himself was not there, and his Lieut.-Colonel, Cobham, was in Catalonia, and had a detachment of his regiment with him. The fact is that we know very little about this marine battalion at Almanza, not even the name of its commander. The marine officers who had been recruiting at home came out in the winter and spring and brought with them a large number of marines, said to amount to two battalions, but for the most part probably drafts for the battalions in Catalonia and the coast garrisons. We know that when Lord Galway was preparing to advance, these garrisons were whipped up for seasoned soldiers. Now among the officers lately returned from England were Captain Hugh Palliser and his brother Walter's subaltern, James Davison. Hugh Palliser's company was at

Alicante, close in rear of Lord Galway's army, and he and Davison were in the battle, where he was badly wounded, shot through both cheeks, and Davison was taken prisoner. It is impossible not to believe that he had with him his company of veterans from Alicante. Perhaps Walter Palliser, who had presented the marine officers' petition to Lord Galway, was also in the action. Nothing was more likely than that Lord Galway, when obliged to decline the offered services of the marines, should invite their representative to accompany him in the campaign. Palliser appears soon afterwards in the musters as lieut.-colonel, but the date of the brevet and the action for which it was bestowed are unknown. Probably the marine battalion at Almanza like those under Wills in Catalonia was a composite one, but it is all conjecture.

James Davison was exchanged and rejoined in 1709.

On May 19th, Lord Galway retired with his cavalry across the Ebro, and on June 14th the Dukes of Orleans and Berwick had united their armies on the left bank of that river and were ready for the invasion of Catalonia.

On retiring across the Ebro one of Lord Galway's first acts had been to call for the garrison of Gerona. General Stanhope had anticipated his wishes, and on May 14th he wrote to his lordship from Barcelona: "The fusiliers and marines will march on Friday towards Tarragona (half-way to Galway's quarters at Tortosa), the first from here, where they have received their clothing, the other from Martorel."

The annual clothing for the marines had not arrived and Stanhope proposed to clothe them for the campaign in the clothes sent out for some of our regiments lost at Almanza. It is difficult to get at the strength of the companies of the 30th in Catalonia in 1707.

The marine paymaster, two years later, presented to the Duke of Ormonde a summary of musters, but it shows the strength of the marines as far in excess of what was possible and cannot be trusted. A muster-master was sent out at the end of 1707 to muster the marines in Spain and in the Mediterranean and the figures in the musters for 1708 seem to be more in accordance with what we know to be facts.

LERIDA

In the absence of proper musters we may accept General Wills' figures as to the number of marines at the beginning of the Lerida campaign. He puts each of his battalions at 600. The six companies of the 30th which had landed in 1705 were still there, and to those was added Corbet's old company, now commanded by Charles Davison.

In April, Lord Galway had appointed Burston lieut.-colonel of Holt's regiment, and had made David Ward major of a marine battalion. He had also given the company vacant through Knight's death to Lieutenant William Scott, the brigade-major of marines. All Lord Galway's promotions were cancelled by the Admiralty, but this was not known in Catalonia till after the fall of Lerida.

The campaign opened, as far as the 30th were concerned, with their advance to the frontier, which by the middle of June was threatened by the armies of the Dukes of Orleans and Berwick. To their numerous forces Lord Galway could only oppose his 4,000 cavalry and the three British battalions from Gerona, backed by a swarm of Catalan irregulars, full of zeal but quite undisciplined. The King did not assist him, but still pursued his suicidal policy of keeping his own troops in Barcelona, or between that city and Roussillon.

By his skilful dispositions Lord Galway was able for some time to maintain the line of the Ebro and the Cinca. Meanwhile he worked night and day in forming a new army, and especially in collecting the stragglers from Almanza. He records the valuable assistance given in this by Captain Scott, the brigade-major of marines, whom he had appointed to the 30th. On August 27th his lordship was able to write to the Duke of Marlborough that he hoped to have a dozen good battalions "*besides the marines and fusiliers.*"

The 30th must have found some opportunities of distinguishing itself during this frontier war, for on July 14th Lieut.-Colonel Burston was given the brevet of colonel, and on August 24th David Ward that of lieut.-colonel.

At this time also Lord Galway recommended Colonel Wills for the rank of brigadier in the British service. Wills, unknown to his lordship, was already a brigadier.

On August 24th the French crossed the Segre at Balaguer, and Lord Galway fell back to Igualada at the northern end of the passes leading through the mountains to Barcelona. Lerida was now open to attack by the enemy and Lord Galway had selected for its garrison his firmest battalions, the fusiliers and marines, together with two Dutch and one Portuguese corps. Prince Henry of Hesse Darmstadt was governor of the place and Major-General Wills commanded the British troops under him.

Colonel Parnell puts the strength of the garrison at 1,800 regulars and 800 Miquelets. If we accept this, then it is probable that each battalion of marines now was about 400 to 450 strong, the Dutch and Portuguese about 200 each and the fusiliers about 300 to 350. A return of the Allies' troops in Spain in 1706 gives the two battalions of marines as 1,000 men, the Dutch in the Lerida district as 600. But 500 Dutch and Portuguese were captured in Mequinenca, when the French advanced. The strength of the British line regiments is not given in the return, but a note says "Their battalions are hardly to be mentioned at 400 each, one with another."

According to Colonel Parnell, the 30th formed one complete battalion, the marines of different regiments forming the other, but it appears most probable that there were two mixed battalions to one of which the 30th furnished the staff and three-fourths of the officers and men, or perhaps more, as it had now seven companies in Catalonia.

Lerida lies on a slope on the right or northerly bank of the river Segre. Balaguer is nearly 20 miles up the river to the north-east. Tarraga is 25 miles slightly to the north of east of Lerida. The mountains which cover Barcelona are about twenty miles south-east of Lerida—they run, like the river Segre, in a general line from north-east to south-west. The distance from Lerida to Barcelona is a trifle over 80 miles.

Lerida had stood many sieges and was well fortified. In the midst of the town is a rocky hill crowned by the castle, which showed itself capable of defence long after the town had fallen. There is a detached work called Fort Gardin outside the walls of the town, but it played little or no part in the defence, being small and distant from the French point of attack.

On September 10th the Duke of Orleans appeared before Lerida and invested it, but delayed his attack, partly to await his siege train and partly because he had been forced to detach 5,000 men to assist in the defence of Toulon, which was besieged by Prince Eugene and Sir Cloudesley Shovel. The latter had 150 men of the regiment serving on his fleet, and they probably little knew when they landed and cleared the French from the line of the Var, how much they were helping their comrades in Catalonia. Early in October the French troops, freed by the failure of the Allies in front of Toulon, returned; the siege train had arrived; and a Spanish force of six battalions had joined the Duke of Orleans, who now could muster 22,000 men. Lord Galway with from 14,000 to 15,000 troops, many of indifferent quality, had advanced to Tarraga.

On October 2nd the Duke of Orleans attacked Lerida on the eastern side, and, although on the 8th the defenders made a spirited sortie and did a good deal of damage, he was able to open fire from nineteen guns and four mortars on the 9th. The guns were employed to breach the walls of the town and the mortars to keep down the fire from the castle, which raked some of the besiegers' batteries. The town wall was badly built and fell fast, and on the 11th two practicable breaches were opened.

The assault was given on the night of the 12th by a brigade headed by six companies of grenadiers and accompanied by a working party of 800 men. The rest of the army was in support. The French were twice driven back by the garrison led by Wills and the Prince of Hesse, but they made good a lodgment in the breaches and held them in spite of a counter-attack by 1,100 men, some of whom, issuing from the town, fell upon the flank of the attacking column. The French were now able to sap up the breaches and the town was untenable. In a letter to Lord Galway, Major-General Wills gives the following account of the defence up to this point.

"LERIDA CASTLE,
"18th *October*, 1707.

"The enemy opened the trenches before this town on the 1st instant, and the 9th they battered the town wall with twenty pieces of canon, and the 10th they had made two large breaches fit to storm but continued battering till the 12th at 6 in the evening, when they made two attempts but were repulsed; but they made lodgment under both breach and sapt through the stone bastion and through the middle wall and carried their entrenchments to the innermost breach which would have covered their second attempt and flanked our new work. We having defended the town to the last extremity, I spoke to the Prince of Hesse to demand a capitulation for the Burgers, and told him I would stand in the breach with the troops till they had terms, but he said it was no matter for the Burgers. Then we called a Council of War and it was agreed to retire the troops to the Castle, after having carried off our canon and ammunition to the castle, and then the troops marcht to the upper part of the town where we now are. You may depend upon it we will dispute every foot of ground with them for I would rather see the soldiers dead than prisoners."

On the 14th, the defenders withdrew from the city to the castle. On the following day the Duke of Orleans occupied the town, and after placing strong guards to prevent any interference on the part of the garrison of the castle, allowed two men per squad to take off their accoutrements and pillage for eight hours. Fifty of the inhabitants who tried to save their belongings were killed. It is possible that this plunder might have been prevented and the citizens' lives saved if Prince Henry had listened to the proposal of Wills to defend the breach till Orleans gave the inhabitants terms.

On the 16th, trenches were opened against the castle, but progress was slow and the loss heavy.

On the 21st, mortars were placed in position and caused heavy loss to the besieged, who had no bomb-proof protection except a strongly built church, which served as a store-house and magazine.

On the 25th, a message reached Tarraga from Lerida that the provisions for the castle were still untouched and that the great church stood proof to the enemy's bombs.

On that day, however, the breaching batteries opened fire, and, either from a mine being sprung, or from a shell igniting the ammunition in it, the church was blown up, and the thirteen survivors of its garrison of fifty were taken prisoners. Apparently the food supplies were destroyed as well.

By the 29th the French sap reached the covered way and the guns of the defenders were dismounted, except one large and two small pieces.

On this day Lord Galway advanced from Tarraga to Las Borjas, three leagues from Lerida, but south of the Segre.

The Duke of Orleans met this move by destroying the bridges on the Segre and posting the Duke of Berwick on the north bank with 14,000 men to defend the passage of the river. It was now evident to all that Galway was too weak to attempt the relief of the castle. On November 6th he advanced to the left bank of the Segre opposite Lerida with fourteen squadrons, and communicated by signals and letter with Prince Henry. In a letter to Lord Sunderland he reports: "Prince Henry writes ye 6th inst. provisions of flesh, wine and brandy were out. Living on bread and water. Less than 800 duty men."

Intimating to the Prince that he was helpless to save him, Lord Galway retired to Las Borjas.

On the 9th two breaches were opened, by which the enemy could enter twenty abreast, and the Duke of Orleans gave orders for the assault on the following night.

The garrison had for some time been living on a small daily allowance of oatmeal and rice, and the end of their supply even of that was in sight. The Prince of Hesse therefore decided not to sacrifice their lives uselessly, and beat the chamade at 6 p.m. on the 10th, the assaulting columns being then formed, and lying down in the trenches, and the match lit to explode the mines in the *fausse braie.*

Colonel Burston was empowered to treat with the Duke of Orleans, and for the rest of his life took more pride in his conduct of the negotiations than in his exploits in the field. It is safe to say that the Duke of Orleans had strong reasons to be on good terms with the officers of the British army. His colleague, the Duke of Berwick, would in the case of a counter-revolution take his uncle Marlborough's place at the head of that army, and preparations for a counter-revolution were already being made. Whether from that cause, or from the diplomacy on which Colonel Burston prided himself, or from a generous admiration for a gallant defence, the Duke at once gave honourable terms on condition that Fort Gardin was included in the capitulation. He even permitted the deserters from his army to march out in the British ranks, saying, " They have been rogues to me and will be rogues to you."

The principal terms were that the garrison should march out by the breach with two guns and a few rounds of ball, that the French should furnish 150 wagons for sick, wounded and stores, and that a French escort should conduct all to Lord Galway's Headquarters at Las Borjas, after which the garrison was free to serve again. The inhabitants of Lerida were referred to the mercy of Philip V.

On the 14th the garrison marched out, headed by the Prince of Hesse and General Wills. Six hundred men followed them under arms and a like number in wagons. The emaciated state of the whole excited the compassion of the French. The Prince led his men past the Duke of Orleans, saluting with his sword, and was received with great politeness and complimented heartily on his defence, but the Duke of Berwick informed General Wills that he was under arrest in retaliation for the conduct of Lord Galway to a French officer in somewhat similar circumstances. Wills was at once released on his application to the Duke of Orleans.

In forwarding a copy of the capitulation and General Wills' letter, already quoted, to Lord Sunderland, Lord Galway speaks in high terms of General Wills, " who has done very great service in defending ye place so long and so well. I hope Her Majesty has been pleased to make him a Brigadier in her service. He's a brave and good officer yt deserves to be encouraged."

General Wills had already been made a brigadier, and later was made a major-general in the British service for his conduct at Lerida.

In a further report to Lord Sunderland on December 29th, Lord Galway says, " there is in Tarragona ye Regiment of Fusiliers yat came from Lerida, it is reduced to 150 men. There are also 500 Marines from Lerida besides 100 in Majorca and 130 in Alicante." From this it is evident that all the marines in Catalonia had joined Wills in Lerida and shared in its defence.

The losses of the French during the campaign and siege were admittedly great, but have never been divulged.

When invested in Lerida, Colonel Wills had managed to pass out his horses, but the regimental transport mules which the officers had provided and maintained at their own expense had to be destroyed, as they were starving.

On the termination of the siege the troops on both sides retired into winter quarters. The French held the line of the Segre and Lord Galway occupied Tarragona and Reuss, which is about 10 miles to the north of it.

About the end of December, Major-General Wills and Colonel Burston sailed for home.

Captain Scott, who had been town major of Lerida, was taken prisoner after its fall.

Lord Galway's promotions of several marine officers having been disallowed at home, Colonel Burston was entrusted by the officers concerned with a memorial to H.R.H. the Lord High Admiral, and Lord Galway wrote a letter in support of their petition, of which the following is a summary.

He begins by stating that he had filled vacancies in the marines on the recommendation of General Wills, and goes on to say, "It would be a discouragement to all officers, particularly the Marines who ever since their coming to this country have been always upon action, if those that are younger and stay at home get commissions over their heads. I cannot but in general give a great character to the whole body of Marines." He especially recommends Colonel Burston and Major Scott, the brigade-major of marines and town major of Lerida.

Burston presented the memorial and Lord Galway's letter without any great success, and also presented a petition on his own behalf to the Queen, praying for some consideration for his long and faithful services to the Crown, particularly for his service and suffering in Spain. In this application he was strongly supported by General Wills. His petition called forth the following report from Lord Pembroke, who had become Lord High Admiral on the death of the Queen's husband, Prince George :—

"It appears from a certificate of General Wills that Colonel Burston behaved as a gallant and true officer and had no less than thirteen cuts and stabs at the battle of St. Estevan in Catalonia ; and that he has the character not only of an experienced but of an active and stout officer. But as to the reward for which he petitions, the Lord High Admiral, although of opinion that his merits entitle him to the Queen's bounty, is not able to inform her out of what moneys the petitioner can be gratified."

Upon this the Queen referred the petition to the Lord High Treasurer, Lord Godolphin, but he refused to advise Her Majesty to bestow her bounty on Colonel Burston on the ground that he was not permanently disabled, though Burston alleged that he had *nearly* lost the use of his left hand.

Only one notice of the recruiting parties in 1707 has come to light. It is a letter from the Admiralty Secretary to the commanders of H.M. ships of the Newcastle convoy, ordering them to receive on board all such marine recruits as Captain Irwyn sends and to land them at Sheerness ; so Irwyn did not serve in the Lerida campaign. He was a Northumberland man.

TOULON

Earlier in the war the Duke of Marlborough had pointed out that the speediest way to force the French to make peace was to attack on two fronts. He proposed that he and Prince Eugene, each taking 20,000 troops, should be transported to the Mediterranean ; they would take Toulon and raise again the rebellion in the south of France, and he had no doubt that the French would sue for peace. In 1707 his plan was partially adopted by the Allies, but instead of 40,000 good troops under Marlborough and Eugene, or under Prince Eugene alone, less than 25,000 were sent under Eugene and his cousin, the Prince of Savoy, although it was well known that the two could not act in harmony. Only the British and Dutch fleets under Sir Cloudesley Shovel were thoroughly well commanded and adequate to their task.

In June 150 men of the regiment marched from Canterbury to Portsmouth and were distributed among the ships in the fleet. As they were not embarked by companies, it is hopeless to attempt to trace them. The fleet sailed for the Mediterranean, and by June 30th was off Hyères. In passing Alicante, Sir Cloudesley had carried off more than half the marines in garrison there, including a party of twenty-six non-com. officers and men and two women belonging to six different companies of the 30th under Lieut. John Roper, of Charles Davison's company. There always were a number of casuals at Alicante, perhaps recovered men from hospitals or hospital ships. Roper's party served throughout the operations at Toulon and, except four who died, were landed in England before the end of the year.

To oppose the advance of Prince Eugene and the Duke of Savoy against Toulon, the French had occupied the line of the river Var, and the first service Sir Cloudesley Shovel performed for the Allies was to turn the French out of their position. The marines were ordered into the boats and half a dozen ships ran in and opened fire, raking the French lines while the boats pulled for the shore. The French left their lines in confusion and the marines, leaping ashore, rushed into them. The Duke of Savoy and Prince Eugene then crossed the river and invested Toulon on July 26th. At first the Allies had some success, but it was soon evident that their means were inadequate, although the British admiral landed heavy guns, with gunners to work them, and did everything in his power to make good deficiencies.

On August 20th the Duke of Savoy and Prince Eugene commenced to retreat, and at their request the allied fleets accompanied the movement, so as to threaten the French right if they pursued.

The fleet returned to English waters in October, and on the night of the 24th of that month H.M.S. *Association*, Sir Cloudesley's flagship, struck the Gilstone Rock off Scilly and sank in two minutes; not a man was saved. The *Eagle* and *Romney*, which followed the admiral, also went down, only three of their crews being picked up.

The ships' books were lost, and it is impossible to be certain, but it is believed that it was on this occasion that Second Lieut. Aylmer was drowned and that he had a party of the regiment with him. A week later Henry Aylmer was appointed second lieutenant in his brother's place.

In this year there were held two out of the three courts martial on men of the regiment under Queen Anne of which any record has been found, and for convenience sake an account of all three is given here. They were held while the Queen's husband, Prince George of Denmark, was Lord High Admiral. All the courts assembled in the great hall of the Horse Guards.

The first court was held on March 21st, 1703, when Uriah Williams, John Bolton and Francis Griffen, private soldiers in Captain Ward's company, were brought before it for mutiny and found guilty. The sentence was "that Uriah Williams be shot to death at the head of the companies of the regiment at Canterbury within 20 days and not before 14 from the date of the President's signature to the proceedings of the Court.

"That John Bolton do run the Gauntlet 4 several times through such part of the regiment as the Commanding Officer may direct, and that Francis Griffen do run the Gauntlet three."

A memo. attached states that the Queen upon the application of the prisoners was graciously pleased to pardon the life of Uriah Williams and to remit the punishment ordered to be inflicted on the others.

The men's misconduct had been provoked by their being called upon to re-embark before being paid by the Victualling Department the sums due for savings on ship's rations during a former cruise. The Prince expressed his serious displeasure with the Department and ordered a victualling officer to go at once to the quarters of the company wherever they might be, and to pay the men.

In 1706 a warrant was issued for a court to assemble at Portsmouth for the trial of some men of the company commanded by Captain Harris, but the court did not sit.

The second court martial of which we have knowledge was held on May 31st, 1707, for the trial of six privates for desertion. The court was of opinion "that the several prisoners were in a high manner guilty of the mutiny laid to their charge, for which the Court has adjudged that the five first of them be severally shot."

There was some doubt about the terms of enlistment of the sixth man and the court declined to sentence him.

Apparently desertion was considered a form of mutiny. The usual limit of the time during which the sentence could be carried out, namely, not sooner than the fourteenth or later than the twentieth day from the signing of the sentence, does not seem to have been fixed by the court, for the five prisoners were held over till a more serious case had been tried.

In this case thirty-six privates were charged with mutiny before a court martial held on July 18th, 1707. The court acquitted one and found twenty of the accused guilty of the charge. The sentence was as usual that they be shot to death not sooner than the fourteenth or later than the twentieth day after the signing of the sentence. With regard to the fifteen other men the court was not satisfied about the terms of their enlistment, "as the Act of Parliament requires to make them obnoxious to punishment," and sent them back to prison to await Her Majesty's pleasure.

The Queen confirmed the finding and sentence of both courts martial on July 21st, but directed that the twenty-five condemned men should cast lots for their lives and only one be put to death, " Her Majesty being graciously pleased to pardon the others who are after the execution to return to their Colours."

With regard to the men remanded by the two courts, Her Majesty was pleased to pass by their offence in regard they had been but lately taken into her service and through ignorance have transgressed their duty.

The prisoner who was unlucky in casting lots and had been reserved to die now had a piece of good fortune. Through some oversight he was not sent down to Canterbury for execution until more than twenty days had elapsed, and the officer in command there, presumably Lieut.-Colonel Dornell, refused to act until the matter had been referred to Her Majesty. The Judge Advocate's Department, after consulting counsel, advised Her Majesty that the man could not lawfully be put to death. The result was that on September 19th an Admiralty order directed that the man who was to suffer death, along with the rest of the mutineers of that regiment, be transferred to Colonel Handasyde's regiment, serving at that time in Jamaica.

If he had been executed as intended before August 8th, the other men would have been released to join their companies, and so would have escaped Jamaica. It looks as if some official had been able to take advantage of a quibble and thwart Her Majesty's kind intention, but on the whole the story of these three courts martial gives a more favourable idea of the humanity of that age than is generally entertained.

The Prince, of course, knew well that the chronic discontent in the marines was due to irregular payments on the part of the Treasury.

1708

The following promotions and appointments were made :—

Bt.-Major Charles Williams from Shannon's Regiment, to be Major, 12th August, to rank from	14th Nov., 1707
Captain William Scott from Holt's Marines, to be Captain, *vice* Gardner	24th December
James Littleton to be First Lieutenant to Captain Ward, *vice* Dobson, deceased	25th March
Charles Wills to be Second Lieutenant to Captain Gardner, *vice* Hutchinson, deceased	2nd ,,
Joseph Clayton to be Second Lieutenant to Captain Ord, *vice* Littleton, promoted	25th ,, aged 19
Lancelot Daws to be Second Lieutenant to Captain Irwyn, *vice* George Firebrace	5th Nov.
Patrick Aylmer to be Second Lieutenant to Lieut.-Colonel Burston	5th ,,
Lieutenant William Dawes to be Adjutant, *vice* Mason, resigned	25th Dec.

There are some fine examples here of how promotion was worked under Queen Anne, and we will take the shortest first.

The Charles Wills appointed to a commission without purchase was a son of the General, and he was not yet 10 years old. Gardner, who sold his company to Captain Scott, had bought it from Brereton the year before, but could not find the money to pay for it. Brereton then applied to the Admiralty to have his company back, a Major Churchill came forward as security for Gardner, but the latter being still impecunious Brereton came down on Churchill, upon which the Admiralty, who were watching matters, called upon Gardner to see that no harm came to Churchill. Wills now intervened, telling the Admiralty that Gardner was a man of very scandalous character, unfit to be in Her Majesty's service, and strongly recommending Captain William Scott of Holt's regiment, who had been his brigade major in Catalonia and had greatly distinguished himself while serving in the 30th under a commission as captain given him by Lord Galway. Scott therefore was allowed to purchase the company for £400, the price which Gardner had promised to pay Brereton.

The case of Major Williams is more complicated. In 1707 Lord Galway had given Capt. Charles Williams, of Shannon's regiment, a brevet majority. David Ward, of the 30th, was at that time acting as substantive major of a marine battalion under a commission also from Lord Galway. Galway's commissions were cancelled but, when a vacancy was made by Burston's promotion, Williams was brought into the 30th as major. David Ward, who was now a brevet lieut.-colonel, protested against a man being brought into the regiment over his head after his two years' service in the field as major. It was decided that Williams should keep the majority, but that he and Ward should both draw major's pay. Williams was now worse off than when he was a captain, for there was no company vacant for him, and he only drew 5s. a day, his major's pay. To retrieve his mistakes the Admiralty Secretary tried to get a company for Williams. In 1707, Ralph Ord, a boy of eighteen, had been allowed to succeed to the company of his father, George Ord. The Secretary now wrote in the name of Prince George, who was in his last illness, to George Ord to say that he might return to command his company if he liked, but that the Prince could not suffer a youth such as his son to command a company. Ord fought the question and won, and Ralph Ord kept the company. Williams went on service and was taken prisoner, and it was not till 1710 that he got a company through the death of young Ord in the West Indies. To complete the story, when the regiment was broke in 1713–14, Ward and Williams were put on half-pay as majors, but when it was re-instated by George I, Ward was the major and Williams a captain.

The commissions of Lancelot Daws and Patrick Aylmer were not signed as usual by the Lord High Admiral. Prince George was on his deathbed, and the Queen signed for him.

In this year Lieut. John Bernard, who had been wounded at Namur when serving in the regiment under Colonel Saunderson, was appointed to an invalid company.

General Wills reached home in the beginning of 1708 at a time of great public anxiety. A reaction had taken place in popular feeling after the Union: there had not been time for its good results to develop or for old animosities to die down. The Jacobites and the High Church Party, who were shocked by the terms of the provisions for the security of the Presbyterian Church in Scotland, were active against the Government.

Louis XIV thought that an attempt at a counter-revolution had a chance of success, and assembled 4,000 men at Dunkirk and half a dozen men-of-war under the Chevalier de Fourbin. The design was that when de Fourbin was ready to sail, the Chevalier de St. George, son of James II and styled "Pretended Prince of Wales" by the British Government, should join the expedition and be landed in Scotland with the 4,000 French troops and as many of his exiled adherents as would follow him.

On February 18th all marine regiments were ordered to assemble at the nearest ports and wait for orders. Gen. Wills was directed to strengthen Rochester and Chatham, and Adam Cardonell, Marlborough's secretary, wrote to inquire how many men he could muster. He had only 150 of the 30th, for 450 were already on board ship and eight companies were in Spain. A month later the 150 helped to man the ships sent over to bring regiments home from Flanders.

On the first alarm, Admiral Byng's squadron had been stationed off Dunkirk. Two ships of the squadron are of interest to us. Lieut. James Littleton with forty other ranks of Ord's company were on H.M.S. *Salisbury*; the party was commanded by William Davison, Wills's capt.-lieutenant. H.M.S. *Ludlow Castle* also carried forty men of the regiment, but we have not the detail.

The sailing of de Fourbin's expedition was delayed by the illness of the Chevalier de St. George, who had an attack of measles, but it got to sea on March 10th, when Byng had been blown off his station by a gale. Sir George pursued with sixteen sail and came up with de Fourbin's small squadron in the Firth of Forth on April 12th. Our fleet came in at night and anchored off the Bass Rock in thirty-five fathoms to wait for day, but de Fourbin, who was higher up the Firth, heard the words of command and the plunge of the anchors, and before it was light had cut his cables and made for the open sea. At daybreak Sir George Byng sighted the enemy to the north and ordered a general chase, which began from the Bass at 6 a.m. on the 13th and ended off Buchanness on the morning of the 14th. The *Ludlow Castle* and *Dover* were the first to come up with the French and engaged their larger ships, trying to hold them till the remainder of our fleet came into action, but night fell and the wind dropped. Capt. Hosier had pressed the *Salisbury* during the day and now carried on through the darkness without shortening sail. At 2 a.m. a couple of shots were fired close to him. They were from the *Ludlow Castle*, which had cut off a Frenchman and stuck to him in spite of the dark night. The French now hailed that they surrendered and almost immediately their ship fell aboard the *Salisbury*, a hundred of whose men sprang on board and took possession. The ships remained locked together till morning, when the prize turned out to be the old *Salisbury*, captured by the French years before and now recaptured by her namesake and successor. She carried numerous noblemen, gentry and officers to head the rebellion which they had hoped would take place. De Fourbin with the remainder of his force reached Dunkirk in safety. The *Salisbury* made some more good prizes during her cruise, and Capt.-Lieutenant Davison applied for a captain's share of prize money, but the Admiralty would not grant it as his pay was only that of a lieutenant.

After the danger of an invasion and rebellion had passed, General Wills returned to Catalonia to take command of the British infantry in the allied army which was now under Count Guido von Staremberg, a distinguished Austrian officer. General Stanhope commanded all the troops in British pay. Mr. Dalton thinks that the 30th were with Wills in an unsuccessful attempt to relieve Tortosa, to which the French laid siege in June. After the failure of this attempt the allies were so reduced in numbers that they could only watch the French and await the arrival of the British fleet which was bringing a reinforcement of 8,000 Imperial troops from Italy. The fleet arrived at the end of July.

SARDINIA AND MINORCA

The allied army was now able to hold its own, but suffered from want of supplies, chiefly of breadstuffs and horses. The advisers of Charles III had long had their eye on Sardinia as a source of supply, and the Duke of Marlborough was equally anxious to secure the great harbour of Port Mahon in Minorca for the British fleet.

Admiral Sir John Leake undertook the reduction of both islands. The expedition to Sardinia was first taken in hand. Major-General Wills was nominated to command the troops, which consisted of 1,000 Spaniards under Cifuentes, and 500 British marines. The fleet assembled at Barcelona, and on July 23rd there is an entry in the log of H.M.S. *Elizabeth*—" Three men-of-war arrived from Tarragona with marines." Those were doubtless the 500 veterans of the defence of Lerida, whom Lord Galway mentioned in his despatch of December 29th, 1707, as being at Tarragona. The muster of June 25th, which is the latest we have, shows 339 of the regiment at Tarragona on that date. This number includes a few at Alicante, probably men on their way home and not yet struck off by their companies.

On August 11th, 1708, the expedition arrived off Cagliari, the capital of Sardinia, a city of 40,000 inhabitants, well fortified and armed.

After summoning the Governor to surrender, Sir John Leake threw a few shells into the town and disembarked the marines and Spanish troops, and a battalion of 900 seamen, the whole under Major-General Wills. The inhabitants now forced the Governor, who was perhaps not very reluctant, to ask for terms. He and his small garrison were granted the honours of war and the town and island taken possession of in the name of Charles III. Having installed Cifuentes as Governor, Sir John Leake re-embarked the marines and sailed for Minorca. Major-General Wills returned to England by land, as it was termed. That is, he landed at Genoa and visited Florence, where he was received with the highest honour by the Archduke and his Court; he then travelled post to Holland, whence he crossed to England.

The attack on Minorca required a larger force, and General Stanhope assembled on transports 800 Spaniards, 600 Portuguese and Southwell's regiment—600 strong. With these he joined Leake off Port Mahon on September 13th. Next day the town was surrendered to him, but Fort St. Philip, situated on a neck of land to the west of the harbour, held out. On the 16th, 500 marines were landed. Mahon is to the south-east of the island, and the two other fortified places, Fornella and Citadella, are in the north-east and south-west respectively; Citadella surrendered at once to a small detachment, and Fornella was reduced by H.M.Ss. *Dunkirk* and *Centurion*, after a severe engagement of four hours.

General Stanhope was now able to devote his whole strength to the siege of Fort St. Philip. The fort was square in shape with regular bastions and ravelins, and had as an advanced work a dry stone wall terminating in the sea at each end, and flanked by four towers. The garrison consisted of 500 French marines and as many Spanish soldiers.

The place was so inaccessible that nine days were consumed in dragging into position the train brought from Barcelona and the heavy main deck guns landed from the men-of-war by Sir John Leake. The latter, having done all he could to further the objects of the expedition, sailed for home on the 19th with the ships which he judged could not safely face a winter at sea. He left Sir Edward Whittaker with seventeen sail to support General Stanhope, and every marine and all the provisions which could be spared with safety.

On the night of the 23rd the batteries were commenced, and on the 28th fire was opened. The dry stone wall was soon breached in two places and the flanking towers silenced. Two storming parties were formed. At one breach Wade led Southwell's regiment and at the other General Stanhope, mounted and sword in hand, headed the marines. His brother, Captain Philip Stanhope, R.N., of H.M.S. *Milford*, marched with him. The line was captured with only forty casualties, but Captain Stanhope had been killed at his brother's side. The following day the French commander capitulated and was granted the honours of war. On October 2nd he and his garrison were embarked on transports and a British garrison of Southwell's regiment and 200 marines occupied Fort St. Philip. The remainder of the force returned to Barcelona.

It is difficult to trace the part played by Wills' regiment in the expeditions to Sardinia and Minorca owing to many pay lists being missing.

On July 26th, Lieutenant-Colonel Walter Palliser embarked at Tarragona on the *Burford* with his company, consisting of Cadet Henry Courtenay and fifty-nine men. The regimental surgeon, Matthew Poumies, accompanied them. The detachment was landed with the other marines for the capture of Cagliari, and after the capitulation all re-embarked and proceeded to Minorca, except Lieut.-Colonel Palliser, who was discharged from the *Burford* at Cagliari. Whether he went home with General Wills, or, being the senior officer in Wills' absence, was shifted into the flagship, we cannot tell.

Besides W. Palliser's company, twenty-two men of Captain Scott's company got to Minorca. There is nothing to show how they got there, but their re-embarkation is recorded. We can also trace ten men of Captain Ord's company, who were disembarked under Sergeant Alexander Denoon with four days' rations. After the capture of Fort St. Philip, Sergeant Denoon re-embarked on the *Nassau*, but with only five of his party. He had, however, managed to pick up Sergeant Crookes and half a dozen men of Captain Davison's company, and Corporal Chester and two men of Captain Bedford's company.

Some men of the regiment were among the 200 marines left in garrison, for, when the fleet left, Surgeon Matthew Poumies and Cadet Henry Courtenay were discharged at Mahon, and when the garrison was re-embarked the following year, we can trace several of Lieut.-Colonel Burston's, Captains Metford's, Saunders' and Davison's companies.

When the marines came to be disbanded in 1713 a Colonel Lee claimed £105 lent by him for marine subsistence at the siege of Port Mahon, and the Commissioners ordered each of the six marine regiments to pay an equal share of that sum, as it was impossible to say from the evidence before them that any one regiment had been more strongly represented at the siege than another.

The Navy was now so badly manned that Government determined as far as possible to put an end to all soldiering on the part of the marines and to confine them to their duty on board ship.

A detachment of the regiment, however, was still in Spain, and saw a little more land service.

DENIA AND ALICANTE

From the time of our capture of Alicante in 1706, a few men of the regiment had always formed part of its garrison, but no company was stationed there until Capt. Hugh Palliser with his company landed in 1707. By the summer of 1708 he had recovered from the wound received at Almanza, and was at Alicante in command of a detachment composed of men from nearly every company of the regiment. His own may have been destroyed at Almanza. At this time the French were overrunning the south of Spain and, when Tortosa fell in July, it was evident that Denia would be the next to be attacked. John Richards, the noble Governor of Alicante, therefore sent whatever he could spare of men and supplies to that place. Among others, Capt. Hugh Palliser, his lieutenant, Thomas Newdigate, and a company, made up from the men of the regiment in Alicante, reinforced the garrison of Denia. Mr. Dalton, and there is no higher authority, says that some of the 30th were present later in Alicante at the heroic death of Colonel Richards, so Palliser did not take the whole detachment.

Denia includes a lower and upper town and a castle. The two latter can be defended after the first has fallen, but the castle is incapable of defence after the upper town is taken. Valero, who had distinguished himself in the defence of Carthagena, was the Governor, and the British troops, who occupied the castle, were commanded by Major Charles Percival. Their strength was 200 marines and fifteen gunners. The officers, besides the two of the 30th, were Lieutenants Mercer of Holt's, Tate of Bor's, Kineton of Seymour's, Capt. Lebors and Lieut. Dilex of the artillery. There were also a dozen officers of the train and acting engineers, among whom was the notorious Capt. Carleton of the *Memoirs*. The strength of the Spanish garrison is not known. The fortifications were out of repair and the place was badly supplied.

On November 1st the French, under d'Asfeldt, invested Denia ; he had 12,000 regular troops with twenty-four siege guns and five mortars. On the 10th the breaching batteries opened fire and two days later the lower town was carried by assault and all in it massacred ; Valero had withdrawn to the upper town with such troops as he could bring off. On the 18th the wall of the upper town was breached, the provisions had come to an end and Valero found himself obliged to surrender, when he and all his garrison became prisoners of war. The defence under the circumstances was considered creditable.

Captain Palliser and Lieut. Newdigate were not released till April 24th, 1712. Forty-nine men of the regiment became prisoners along with Capt. Palliser.

Immediately after the fall of Denia, d'Asfeldt laid siege to Alicante and attacked its strong castle by mining the rock upon which it was built. After three months' hard work his miners had reached a point under the parade ground overlooked by two strong guards. He then summoned General Richards to surrender and allowed him to send two officers to verify the position of his mine. The officers reported that the explosion of the mine would be destructive but not necessarily altogether ruinous to the defence, and Richards determined to hold out. When the hour fixed for the explosion drew near he said to his officers, " Whatever the danger may be we must share it with the soldiers." He then proceeded to walk calmly up and down the parade ground, accompanied by many officers. The issue of smoke from a crack in the rock showing that the slow match to the mine had been lit did not disturb the calm of officers or men. When the explosion came Richards with eleven officers and forty-two men were engulfed. If it had been possible to remove his men from the danger so good an officer would have been sure to do it.

The defenders held out for another six weeks, when Admiral Byng removed them to Minorca. They were mostly Huguenots. Mr. Dalton thinks that the 30th alone of the British army were present ; Mr. Fortescue mentions Hotham's regiment, but not the 30th.

In this year the mayor and justices of Canterbury pointed out to the Admiralty the dangers arising from the marines' pay being so much in arrears.

1709

The following promotions and appointments were made :—

Charles Wills, Esq. to be a Major-General.	1st Jan.
Joseph (D'Acres) Mason to be Captain, *vice* Bedford . .	5th April
John Mohun from Captain-Lieutenant in Barrymore's regiment to be Captain, *vice* Irwyn	24th June
William Cooke to be First Lieutenant to Capt. Mason, *vice* Mason	6th April
Fairfax Burston to be First Lieutenant to Capt. Ord, *vice* Pownall, dismissed	1st Oct.
James Leathat to be Second Lieutenant to Captain W. Palliser, *vice* Fairfax Burston, promoted	,, ,,
The Rev. Alexander Innes to be chaplain, *vice* Merridew .	18th July

Captain Bedford, who retired, was one of the original officers of the regiment, having accompanied Lord Castleton and his son, Captain Saunderson, to Yorkshire to raise men in 1689. He recruited his company about Wakefield and Leeds throughout his service. He had served five campaigns in Flanders (1692–97), including the assault and capture of Namur under Colonel Saunderson ; he commanded the wing of the regiment at the capture of Gibraltar and during the glorious defence under Prince George of Hesse (1704–5). He was present at the capture of Barcelona, and is believed to have shared in the victory of St. Estevan and in the resolute defence of Lerida (1706–7).

Captain Irwyn, who had been recruiting in the north since joining the regiment, applied for sick leave when ordered on service in spring, but General Wills was inexorable and said he must either quit the regiment or embark. The chaplain, Merridew, also left to avoid embarkation. Lieut. William Pownall, son and heir of Colonel Pownall, late 30th, was dismissed for not attending to his duty, but he seems to be the same William Pownall who was captain-lieutenant to Wills when he commanded the Buffs in 1716. He was father of the great Governor Pownall, of Massachusetts Bay.

In the beginning of 1709 the Allied Governments had determined on another effort to capture Cadiz. General Stanhope was to bring a siege train and a few troops from Barcelona, while nine regiments of infantry, one of cavalry and three battalions of marines were to embark at home under General Wills and sail for the Straits, convoyed by Admiral Baker. After a junction was effected in the neighbourhood of Gibraltar, General Stanhope would take command of the troops. The force from England was expected to reach between 6,000 and 7,000 men. The three battalions of marines were each to be 600 strong. The officers were selected by a board of the colonels of the marine regiments, which was now frequently employed by the Admiralty when professional opinion was wanted.

It was decided that the captains who had lately returned from active service in Spain should go on recruiting duty and all others embark. Evidently the colonels were determined no longer to tolerate the scandal of men living quietly at home, year after year, while their brother officers did the fighting abroad. Major Williams was appointed to one battalion, and the Rev. Alexander Innes, the new chaplain, embarked for duty with the expedition. General Wills was ordered to bear in mind that Major Williams was to have the first company which fell vacant, and Lieut.-Colonel Ward the first majority.

The embarkation was interrupted by a distressing incident. The marines had not been settled with since June, 1706, and when ordered to embark, the men asked for the two-thirds of their arrears to which they were entitled. An attempt to compel some of Churchill's regiment to embark without their money ended in a party of about seventy breaking into open mutiny, and starting to march to London to call the attention of their countrymen and the Government to their wrongs. Troops of horse were sent to stop them, but they reached London, where they were made prisoners near St. Clement's Church and lodged in the Savoy prison. They were sent back to Portsmouth, where a court martial sentenced three to be hanged and many to be transported to the Colonies. The sympathy with them was so great that the Governor of Portsmouth, General Gibson, ventured to suspend the execution of the three men till the Queen's pleasure was known, but the Ministers would hear of no mercy, and reprimanded him for annoying the Queen with a petition which he ought to have known she could not grant. The Admiralty, in a letter to the Principal Secretary of State, acknowledge that a refusal to go on board was no new thing if men had not received their two-thirds, and they also addressed a letter to the Lord High Treasurer, pointing out that by the standing regulations of the marines an officer was entitled to be cleared for himself and company on producing his muster rolls, but that in spite of that no officer had received his arrears since June 24th, 1706. It was therefore requested, at the urgent entreaty of the marine officers, that a settlement might be made up to June 24th, 1708, for which nearly all the muster rolls were ready. The letter adds that the officers and men of the marines were under very great uneasiness.

On October 3rd, H.M.S. *Anglesea*, with a party of the 30th on board, under Lieut. Richard Leathat, was cruising off the North Foreland and saw a small French privateer in Sandwich Bay. The privateer ran into shallow water and grounded, and the *Anglesea* was forced to hoist out her boats and send them in to take her. The enemy made a spirited resistance, and both a naval lieutenant and Lieutenant Leathat were wounded in boarding her. As usual about this period, it is impossible to find from the ship's books the number of the 30th engaged, or whether any of the non-com. officers and men were killed or wounded. Lieutenant Leathat's petition for the naval bounty states that he was disabled. It was treated in the usual way—that is, the Admiralty point out that the rules governing the distribution of the naval bounty do not apply to the marines, but Her Majesty is advised to give him the same compensation for his wounds which would be given to a naval officer in like case, that is, a year's pay.

The expedition against Cadiz was delayed by foul weather and, although most of the troops were got on board in April, it was September 9th before the fleet left Cork and Lord Galway, now our Ambassador at Lisbon, warned the commanders that the Spaniards were too well prepared for them. General

Wills, therefore, carried his infantry and dragoons to Catalonia, where he served in command of the infantry under Stanhope.

In December H.M.S. *Pembroke*, 74 (Capt. Rumsey), and the *Falcon*, Pink, were cruising in the Mediterranean off the French coast. Major Williams, Second Lieut. John Walker and forty men of the 30th were on board the *Pembroke*. On the 29th three French ships were met with, *Le Parfait*, 70 ; *Sérieuse*, 60 ; and *Phœnix*, 54. The captain of the *Falcon* nobly refused to save himself, and stood by to see if he could render any assistance to the *Pembroke*. After an hour's action Capt. Rumsey was killed, but the first lieutenant, Berkeley, fought the ship for two hours longer and only hauled down his flag when dis-masted and helpless. The ship's books were lost, and we know no particulars of the fight except that Second Lieut. John Walker was severely wounded, losing his arm. He was awarded a year's full pay of a first lieutenant for his wounds and a half-year's pay in consideration of the expense he was put to in returning home through Germany, not having been able to get a passage by sea. He had lost £300 which he had on board when the *Pembroke* was taken and had to sell his commission to pay his debts. The Lord High Admiral then advised the Queen that as Walker was a North of Ireland man he should be sent over to Ireland with £50 and a recommendation to the Lord-Lieutenant to give him the first vacant commission. For some reason Walker returned to England and in spite of all regulations was given the half-pay of a first lieutenant in the 30th.

In the course of the year the Treasury issued money for a general clearance of marine arrears up to December 25th, 1708, but all men were not cleared, and those who did not get their arrears in this year had to wait till the settle-ment of Christmas, 1710, many even till that of Christmas, 1713.

The question of pay so influenced the life of the regiment for the next few years, that some words about it are necessary. First in fairness it must be said that the sufferings of the marines were not all due to dishonesty and bad administration ; in fact, Lord Godolphin, the Lord High Treasurer at this time, was an exceedingly able financier and man of business. A principal source of the trouble was that after so many years of war the nation had very little money or credit ; the Treasury officials, therefore, after taking care of themselves and their friends, paid nothing which could be avoided or post-poned. All Marlborough's requirements were naturally complied with, but while his army was an object of admiration throughout Europe the marines were constantly in rags and sometimes starving.

The system of payment is hard to understand and must have been harder to work. On shore the marines were entitled to their daily subsistence money, out of which they paid their billet master for board and lodging. If, as often happened, the subsistence money was not issued by the Treasury, the billet master or the officers maintained the men till better times came or the men ran away to swell the number of brigands who infested the roads. On board ship both officers and men were victualled, but received no money payment. The subsistence money was allowed to accumulate till the end of the cruise. The ship's paymaster then sent to the captain of each man's company a charge for whatever debts he had incurred on board. In a twelve months' cruise the marine might have been supplied with sea chest, bedding and cooking utensils at a cost of 30*s*., and underclothing and shoes to an equal amount ; to this would be added 6*s.* stopped for salary of chaplain and surgeon, making a charge of £3 6*s.* due to the ship's paymaster. In addition, the captain of the company was responsible for the stoppage of 12*s.* for Greenwich and Chatham chest, and 10*s.* for company contingent, and a few shillings may have been advanced to the man for tobacco or pocket-money. The man's expenditure during the year would therefore amount to about £4 10*s.*, and as his subsistence money was £9 2*s.* 6*d.* yearly his arrears also would be about £4 10*s.* (These figures are from the account of a private of the 30th.) The captain of the company was obliged to pay all ship charges at once, so that the paymaster's books might be closed without delay, after the ship was paid off. If, as constantly happened, the Treasury did not pay the arrears, the captain of the company could only enter the charge against his man and wait. In the case of men who died he could not hope to recover his money until there was a general settlement; with others he might be more fortunate. The marine, on return from service, started life on shore with no money in his pocket, a heavy debt to his captain, but a larger balance due from the Treasury. At the instance of Prince George, the Treasury had consented to pay, without waiting for a complete muster of the company, two-thirds of their arrears to all men discharged from on board ship. The mutiny in Churchill's regiment, already mentioned, was caused by an attempt to force the men to re-embark before they had received two-thirds of their arrears for their last sea service. This attempt was seldom repeated, but the Treasury got round the regulation by a simple device. Supposing thirty men were wanted for embarkation from Canterbury, a nominal roll of the men detailed was called for with the amount due to each for two-thirds opposite his name. This was presented to the Treasury by the marine paymaster, and if he got the money the men received their due and the embarkation took place. No one got the two-thirds to which he was entitled unless wanted for sea service.

No arrears were paid to officers between Christmas, 1708, and the spring of 1714. This was especially hard on the subalterns, who were constantly on board ship and therefore drew scarcely any subsistence during that period. Many were obliged to sell their arrears to a moneylender for a small sum. In addition to the two-thirds of their arrears of subsistence money the marine on return from sea service was entitled to compensation for savings on victuals. It was customary for the crew to consume two-thirds only of the allowance of victuals and to receive the cash value of the amount saved at the end of a cruise. Prize money was another source of income for the marines. Our superiority over the French was never so great as to drive their ships off the sea and prizes were constantly taken, but we cannot even guess at the amount which came to the men of the 30th. If the Treasury could have contrived to ensure regular payments the marine would have been well off, considering what prices and wages were then.

1710

The following promotions and appointments were made:—

Major C. Williams to be Captain of a company, *vice* Ord, deceased . . . . 30th Nov.

John Roper to be First Lieut. to Ward, *vice* Littleton, deceased . . . . 15th ,,

Johnson Haseltine to be Second Lieut. to C. Davison, *vice* Roper 15th Nov.
Ventris Scott to be Second Lieut. to W. Palliser, *vice* James Leathat, deceased . . . . 16th ,,

Lieut. James Littleton died on board H.M.S. *Tilbury* when it went into Harwich to clean and refit. The ship's log does not say under what circumstances he died, but the Admiralty refused his widow's application for a pension. She in return sent the Admiralty secretary the names of some men of his company whom she proposed to call as witnesses at the investigation, at the approaching Essex Assizes, into the cause of her husband's death. The record of this inquiry has not been preserved, but she was awarded a pension. Nothing is known about James Leathat's death. For two years before he was commissioned his name appears in ships' books immediately below that of Lieut. Richard Leathat.

ANNAPOLIS

The harbour of Port Royal in Acadia (Nova Scotia) had, in French hands, become a refuge for privateers which harassed the trade of New England, and early in 1710 a plan was formed to capture it. Col. Francis Nicholson was appointed to command, with Lieut.-Colonel Vetch as adjt.-general; both had great experience of Colonial and Indian affairs. A battalion of marines was formed of six companies, one from each regiment, under Major Reading, of Churchill's regiment. The company of the 30th, commanded by Capt. Charles Davison, was nominated the grenadier company of the battalion. It was not the grenadier company of the regiment, which was still under Major Saunders. The strength of the company embarked was Capt. Charles Davison, Lieut. William Cooke, Surgeon Matthew Poumies, and seventy-one other ranks.

Of this and other companies, Colonel Nicholson remarked, " Nigh half of these Marines are newly raised men, so 'tis to be feared they will be sickly in their passage, and probably some dye, and they can't be relied on as men that have been in action."

Contrary to usual practice with the marines the battalion was embarked on hired transports, which, along with eighteen merchantmen and some store ships, sailed from Portsmouth on May 8th, under convoy of H.M.Ss. *Dragon* and *Falmouth*. On July 15th the expedition reached Boston, where the Governors of Massachusetts Bay, Connecticut and Rhode Island were assembled. At Boston the Colonial regiments of Sir Charles Hobby, Col. Whiting, Col. Clayton and Col. Tailor, with a party of Indians under Major Levingston and some artillery were embarked and, on September 18th, the expedition, strengthened by three men-of-war, set sail.

On the 23rd it was off Port Royal, situated on the west side of the river Dauphin, which, coming from the north, falls into the sea near the southernmost point of Nova Scotia. The fort of Annapolis was protected on the north-east by the river, and on the south-west by a large morass.

On the 25th, Captain Davison and his company had the honour of covering the landing of the marines on the south-west side of the river near the fort, while some companies of the Colonial troops disembarked on the eastern side.

As soon as the advanced parties had established themselves, Colonel

Nicholson landed his whole force. Sir Charles Hobby's and Colonel Whiting's regiments formed a support to the marines, and the other Colonial regiments encamped on the further side of the river. The landing took place under a harmless fire from the enemy's artillery. The same evening Colonel Nicholson, supported by a bomb vessel in the river, advanced against the fort. The track through the bush was so bad that the men of the advanced party were often obliged to cut their way and march in single file.

On the 26th, the advance was resumed, the marines again leading, with Major Levingston's Indians on the flanks. On getting clear of the bush and in sight of the fort, Colonel Nicholson placed Captain Davison's company and the grenadiers of the Colonial regiments in front, and with his whole force crossed the marsh and advanced on to the upland with drums beating and colours flying.

After a halt for dinner, Captain Davison and Captain Bartlett, supported by another company of marines, were pushed forward to within 400 paces of the fort, and entrenched themselves there. The day's loss had been three marines, who had straggled and were killed or mortally wounded, two marines and a Connecticut man killed in the ranks. The ship's boats had meanwhile been landing guns and stores, and on the 29th half the Colonial troops were brought across the river and employed cutting fascines, and Major Abbot, with a party of 200, relieved Captain Davison. The advanced party under his command had been then on duty for forty-eight hours and had lost three marines and one Connecticut soldier killed and three marines wounded, in addition to the losses on the 26th.

On October 1st the heavy guns were mounted and the fort summoned to surrender. M. Subercase, the Governor, asked for the honours of war, which were granted, and on October 4th the French marched out.

Colonel Nicholson altered the name from Port Royal to Annapolis Royal in honour of the Queen, and sailed for Boston, leaving Colonel Vetch with a garrison of 200 marines, fifty gunners and matrosses, and 250 Colonials. Before sailing, Colonel Nicholson gave Captain Charles Davison a brevet as major. He and his company had led while there was serious work to be done.

The French and Indians closed round the fort as soon as the fleet had gone and the garrison suffered much from confinement and lack of supplies. On one occasion the Indians ambushed and destroyed a party of seventy, of whom a few survivors were released at the peace two years later. The marines were all relieved after the failure of the Quebec expedition in 1711, in which they had no share. They were sent down from Annapolis to New England in trading vessels and from there to England as occasion served. Lieutenant William Cooke, Surgeon Matthew Poumies and twenty men of the regiment came up the river to London on June 20th, 1712, in a merchant vessel of Boston, New England, and they seem to have been the last to rejoin. Their ship had taken three months in coming from the south-west of Ireland.

The occupation of Annapolis is the first occasion on which we hear of men of the regiment on shore drawing rations to be paid for by Government. All supplies had to come from New England, with great risk of capture on the way, and prices were so high that Colonel Nicholson authorized the issue of rations, and in addition directed that the captains of companies should buy for their men to the full extent of the soldiers' subsistence money, "shoes, stockings, brandy, tobacco, and other necessaries." It is needless to say that Government afterwards charged the price of the rations against the captains' pay. In Captain Davison's case the charge was £322 10*s*. The Committee appointed after the peace to examine the debts due to the army, recommended in 1717 that the above sum should be refunded to Captain Davison.

SPAIN

We have seen that in 1709, after the attack on Cadiz was abandoned, General Wills had proceeded to Catalonia. He there joined the army of Count Staremberg on the Segre, 40 miles above Lerida. The army consisted of 25,000 men. General Stanhope commanded the troops in British pay, consisting of 5,600 men, of whom 4,200 were British and the remainder Portuguese. Lieut.-General Carpenter commanded the five regiments of horse, and Wills, now a lieutenant-general in the Spanish service, the nine battalions of foot.

With this force General Wills shared in the successful combat at Almenara, the great victory of Saragossa and subsequent advance to Madrid. For his services at Saragossa Stanhope recommended Wills for the rank of lieut.-general in the British army.

In the retreat from Madrid which followed, General Wills was taken prisoner with the survivors of General Stanhope's force, after the hard-fought action of Brihuega, where the British formed the rearguard, on December 9th, and where the loss of the enemy in killed and wounded was equal to the number of the British troops engaged. In his account of the battle General Stanhope speaks in the highest terms of the conduct of General Wills. " Lieut.-General Wills was during all the action at the post where they attacked with most vigour and which he as resolutely defended." The Rev. Alexander Innes, chaplain of the regiment, was with General Wills throughout the campaign and officiated to the British troops, but it is uncertain if he was made a prisoner.

In this year the regiment was again in money difficulties. In August the Treasury failed to issue subsistence money for officers and men. A crisis was only avoided by the exertions of Colonel Burston and other officers who managed to raise £400 to feed their men. On August 11th the Admiralty secretary wrote to Burston, " The Lords Commissioners desire me to acquaint you that they take it very well of you, your taking up money upon that occasion to subsist the men." The Lords Commissioners also wrote to the Treasury, " If the men can no longer be subsisted, mutiny will follow." The intention of embarking fifty men of the 30th on H.M.S. *Burford* at this time was given up because the men available were so much in credit that the Treasury could not afford to pay two-thirds of their arrears. On August 21st, Burston was forced to send an express to London to urge speed in remitting money.

In this year Lord Godolphin's ministry was dismissed, but Marlborough still remained captain-general. The fall of Godolphin was a misfortune for the marines. Under him things were bad enough, but under the incompetent Harley they became unbearable.

The regiment was supposed to be settled with and cleared up to the end of 1710, but many of the men did not receive their arrears and no officer was paid.

1711

The following promotions and appointments were made:—

Wiltshire Castle to be First Lieut. to C. Davison, *vice* James Davison, who retires . . . . 15th Jan.
John Wiltshire to be Second Lieut. to Mohun, *vice* L. Daws . 1st Feb.
Charles Onebye to be Second Lieut. to Ward, *vice* Walker, retired . . . . 1st May
William Pritchard to be Second Lieut. to Williams, *vice* Clayton, resigned . . . . 23rd Aug.
Ben. Sladden to be Second Lieut. to Scott, *vice* Wills, resigned 24th ,,
Gunsley John Ayerst to be Quartermaster, *vice* Norman, deceased . . . . 30th Dec.

James Davison was hit by a Royal Order forbidding the sale of commissions by officers of under twenty years' service, unless they were disabled. He applied for permission to sell, but was refused. Like his brothers, he had been a fighting soldier. His service included the defence of Gibraltar (1704–5), capture of Barcelona, and the operations which followed in Catalonia in 1705–6. He was taken prisoner at the battle of Almanza in 1707, and so had no share in the defence of Lerida.

It is almost certain that the resignations of Clayton and Wills were due to the same cause, that they were under age. Besides starting a campaign against purchase, Government had made an attempt to lessen, if it could not abolish, the scandal of children holding commissions and drawing pay in the army and marines. A clause of the Royal Order quoted above directs that no one is to have a commission who is under sixteen years of age, except in extraordinary cases. Even Marlborough, though keen for reform, dared not go beyond that. The Admiralty did not venture so far. The following letter was addressed on July 3rd to the colonels of marines:

"All those young gentlemen who have commissions in your regiment and are under thirteen years of age must sell—viz., they that are first lieutenants to sell a second lieutenant's commission and those that are second to dispose of their own, but in such case the first lieutenant must sell to the eldest second."

General Wills was still a prisoner of war in Spain, but Colonel Burston appealed to the Lords of the Admiralty to make an exception in the case of his son. The secretary replied as follows: "I have laid yours relating to M. G. Wills' son being excused from his disposition of his commission before the Lords of the Admiralty and am directed to acquaint you that their lordships can not dispense with it."

In January an expedition was planned for the capture of Quebec and the intention of the Admiralty was that the 30th should take part in it. Lieut.-Colonel David Ward and Lieut. Quarles were detailed for duty and the captain of H.M.S. *Sunderland* was ordered to receive them on board, but his books show no marines, officers or men. The captain of H.M.S. *Edgar* was also ordered to take on board at Deal a party of the 30th and to bring them round to Portsmouth for this expedition, unless the men had already embarked on H.M.S. *Montague* or *Swiftsure*, but no men of the regiment appear on the books of those ships, which received their marines from the 32nd (Bor's) regiment. It looks as if the orders for the 30th were cancelled and the marines required drawn from the western regiments. The fleet touched at the New England ports and taking on board a number of colonial troops sailed for the St. Lawrence. There they encountered such bad weather that many transports were wrecked with great loss of life. The commanders therefore gave up the attempt to reach Quebec and returned home.

This expedition forced the Admiralty to take steps to have the marines clothed; they certainly were in a terrible state. Admiral Sir John Jennings, writing from the Mediterranean, reports that "his marines are all naked and some of them have not been clothed since the taking of Gibraltar (1704), who undoubtedly will perish in the winter for want." Under pretence that the musters were incomplete the Treasury put off paying the clothiers and for more than two years no one had been found to undertake a contract for marine clothing. The Admiralty carried their point, money was issued and orders for clothing given out.

In August Mr. Whitfield, whose conduct as paymaster of the marines had been unsatisfactory, was deprived of his appointment.

On the 22nd of that month Capt. George Ord, late of the 30th, presented a petition to the Admiralty of which this is a summary. His Royal Highness, on December 26th, 1707, allowed his son to succeed him in the company, which he was obliged to purchase from Capt. Andrew Abington, for fear of others passing over his head. His son unhappily died in the West Indies in August last on board H.M.S. *Rupert*. On applying for the pay due to himself and son, is informed that there was £795 respites charged against the pay of the company, of which £198 is an overcharge, as 8*d*. is charged instead of 6*d*. per man.

In consideration of the loss of his son and the fact that his family have ever been loyal, even to the loss of their estates, he prays that the balance, namely £597, may be bestowed upon him. The Admiralty acknowledged that there was an overcharge, but maintained that it was little over £27. The circumstances of Ord's promotion are very obscure. Captain Abington was reported in the books of H.M.S. *Stirling Castle* to have died on February 25th, 1703. Could the Admiralty have allowed the commission to be sold for the benefit of his heirs? Ord is probably right, however, about the respites.

The captain was credited with the subsistence of the full establishment of his company, but this subsistence was respited, that is deducted, for each man absent from muster. This was quite fair, but Mr. Whitfield, paymaster of marines, respited (deducted) the marine's full pay instead of his subsistence. In other words, he charged the captain 8*d*. a day for each man absent, although the subsistence of 6*d*. was all the captain was credited with. Before his death Prince George ordered this unfair exaction to cease and the money improperly taken from the captains to be returned. Lord Pembroke succeeded Prince George as Lord High Admiral, but soon the office was put in commission and Lords Commissioners appointed. The marine officers could get no redress; the paymaster asserted that the money had been paid into the Treasury and could not be recovered. On the return of the marine brigade from Catalonia some of the officers who had been most distinguished in the field presented a petition to the House of Commons praying for an inquiry. The House appointed a committee to investigate. The inquiry lasted over two sessions of Parliament, and in 1711 before it was completed Mr. Whitfield died. His executors acknowledged that the money improperly exacted from the captains of marines had been in his hands all along. The House of Commons ordered that the sum, which was over £6,000, should be restored to the captains. An equal amount due to the chaplains and surgeons of the navy was also found among Mr. Whitfield's effects. The money had been stopped monthly from the marines on board ship for the salaries of those officers, but had got no farther than the paymaster. It was the same with the Chatham chest; disabled marines had been refused relief from the chest because the contribution from the marines had not reached the managers, although the stoppage had been made from the men.

Mr. Whitfield was succeeded as paymaster by Sir Roger Mostyn, who carried on the duty as well as it could be done under most difficult conditions. In looking over his books, Sir Roger was surprised to find that the wife of Capt. John Mohun of the 30th was in receipt of an allowance which was charged to marine funds. A request for information from the Admiralty as to this apparent irregularity elicited the following remarkable letter from the Secretary, October 20th, 1711: "Mrs. Mohun had an allowance by the Queen's particular command, and I had this day a letter concerning her from the Duchess of Ormond. I am sure that what favour you show to the poor gentlewoman will be very acceptable to her Majesty."

There was in this year the usual difficulty in carrying on the service and getting men to embark through want of money to pay the sums due to them. All the marine regiments sent urgent appeals stating that their officers had expended all they had in maintaining their men.

A curious incident occurred on H.M.S. *Shoreham*, on which Sergt. Cooke with eighteen men of the regiment had sailed on July 2nd for North America. Five months after leaving England Private William Moore, of Davison's company, was discovered to be a woman, and was discharged at the first port at which the ship arrived.

On December 31st the Duke of Marlborough was dismissed from all his posts, and the Duke of Ormond appointed Captain-General, a plain intimation that, unless their hearts failed them, the men now in power would try to use the army to upset the Act of Settlement.

At the end of the year the Mayor of Canterbury alleged that the men of the 30th owed their Billet Masters £2760, of which he had guaranteed £500. The case was much the same with all marine regiments. The amount due to them from the Treasury was estimated at £44,400.

1712

The following promotions and appointments were made:—

Thomas Daws to be First-Lieutenant to Williams, *vice* Fairfax Burston, deceased . . . . 21st Feb.

Teddiman Roberts to be Second-Lieutenant to Mason, *vice* Daws . . . . " "

Since October, 1711, informal communications had been passing between the English and French Governments with a view to a peace, and on February 9th the commissioners of the two countries met at Utrecht, but several months elapsed before the conditions of peace took shape. Meanwhile the state of the marines was going from bad to worse.

In February the clothiers declared that unless the payment was put on the same footing as it had been with Marlborough's army they would strike. This action of the clothiers forced the Government to place the marines under the Board of General Officers for inspecting and regulating the clothing of the army, which Marlborough had instituted in 1708, and which had proved of immense benefit to the service.

A good deal of the new clothing for the 30th was already made and Captain Mohun was despatched to the Mediterranean to distribute it among the men serving there. In May Captain Mohun was sent home "express" from Barcelona with despatches from Lord Barrymore. He went by sea to Genoa and rode post from there, doing the journey to the Channel in nine days' riding.

In March Mr. Thomas Blondin, the Mayor of Canterbury, called the attention of all in authority, including the Speaker of the House of Commons, to "the miserable poverty and calamity this town groans under" owing to the treatment of the marines. He pointed out that there were 250 men in the town who had received no subsistence money for three months, that unless money were sent they would either run away or mutiny, and that they would have done one or the other long ago if it had not been for the officers and the credit given them by the victuallers, that the stocks of the latter were exhausted and they had no money to replenish them.

The widows of marine officers also began to make their grievances known. A petition was presented to the Admiralty signed by Judith Burke, Rachel Dobson and Isabella Brown, stating that their pensions had not been paid for more than a year and that some widows were now in prison for debt and others in the lowest extremity. Mrs. Burke and Mrs. Dobson were widows of Lieutenants Burke and Dobson, of the 30th. They understated the case. No widow of an officer of the 30th had received a penny since Christmas, 1708.

Even the muster-masters began to cry out for their pay.

It may seem a small thing for a great nation to pay the poor widows their £20 a year, but to do so would have involved a great deal more. After voting the officers' and men's pay, Parliament expected the army and marines to be in most things self-supporting. Greenwich and Chelsea Hospitals and the Chatham chest were in a great measure supported by deductions from the men's pay. The clothing was paid for in the same way. The money for arms came partly from the pay of fictitious men borne on the muster rolls and partly from a so-called voluntary subscription from captains of companies. In the marines at least, tents to shelter the men on service were supplied to regiments on payment only. The men of the regiment paid the sergeant-major, and the drummers paid the drum-major. The expense of stationery and payment of the sergeant who kept the company accounts was borne by the men of the company. Even the pay of the doctor and chaplain on board ship was stopped out of the men's pay. In like wise the fund for the pensions of officers' widows depended on stoppages from the officers' pay and for men's widows on the pay of fictitious men borne on the pay-list (widows' men). The paymaster-general of marines therefore protested that he could not pay the officers' widows their pensions and arrears or the muster-masters their pay until he had received the arrears of officers' pay, from which the money in both cases was derived. The payment of the officers' arrears was a thing which the Treasury was very loth to do, but the pressure was so strong that at last it consented to pay the officers landing from foreign service two-thirds of their arrears, the same as the men.

This was not carried out any more than the promise to the men, and the widows never really received what was due to them till three years later.

In the beginning of June General Wills, Captain Hugh Palliser and Lieutenant Newdigate arrived in London from being prisoners of war in Spain.

By this time the negotiations for peace had made such progress that the British Government was in immediate expectation of being obliged to occupy Dunkirk, which it had demanded should be handed over by the French as a guarantee of good faith.

On the 6th of the month the Queen was able to announce the terms of peace in Parliament, and on the 27th an order was issued that the strength of the marines on board the fleet in the Downs should be made up to 600 men, the extra number required being furnished by Lord Shannon's regiment and the 30th. One colonel, 1 major, 9 captains, and 18 subalterns were to be selected for duty. An immediate order for embarkation was sent to the two regiments concerned by express at 3 o'clock p.m. on July 1st, but the Admiralty had some doubts about the result, and on the following day the officers commanding at Canterbury and Maidstone were directed that, in order to induce the men to go to Deal and embark, they were to be assured that in case their money did not arrive in time, their debts would be paid to their landlords and the residue given them on their return home. Such an assurance was of course useless. Many of the men had lately landed from service afloat and had large arrears due to them—the adjutant of the 30th estimated that £4,000 was needed to clear the men of that regiment alone. As usual an appeal was made to the regimental officers, who managed to raise £2,500, which satisfied the men's wants for the time and a detachment of 2 captains, 3 lieutenants, 2 second-lieutenants, 11 sergeants, 14 corporals, and 200 centinels of the regiment under Brevet-Major Saunders embarked at Deal and was landed at Dunkirk.

Colonel Burston and Major Williams were thanked by letter by the Lords of the Admiralty for their conduct on this occasion. Fifteen hundred pounds of the money advanced was repaid to the officers in the following year, but the remainder was not recovered for several years.

The detachment remained at Dunkirk till August, when an order was issued to the admiral commanding the Nore fleet to bring the marines from Dunkirk in such light craft as were available and distribute them in the fleet under his command; any men in excess of what he required were to be landed at Deal to march to quarters. The last detachment left Dunkirk at the beginning of September.

Mr. Thomas Blondin, the patriotic Mayor of Canterbury who had on several occasions pledged his own credit to obtain subsistence for the men of the regiment, was now out of office and therefore liable to be arrested for debt, and on October 2nd he addressed the Admiralty as follows:

" The hardships under which I labour at present by serving the Government is grown unsupportable and sufficient to deter others from the like good offices for the future." He goes on to say that he had frequently laid his case before the Lords of the Admiralty and shown them how he had pledged his credit to the amount of £800 to procure subsistence for the men of the regiment in Canterbury ; that now he is daily arrested by the inhabitants for this money and he hopes their lordships will relieve him " that it may not be said of Government that a man was undone for purely serving it without any private view of interest." This must have been rather painful reading to their lordships, but on November 2nd on receipt of a fresh statement from Burston of his want of either money or credit, the Admiralty Secretary wrote to the new mayor to point out to him that his duty was to keep the inhabitants easy, " that Her Majesty's service may not suffer by the men's want of credit in their quarters as the Act of Parliament directs." He does not quote the Act, but the mayor found a sufficient guide for his conduct in his predecessor's experience and resolutely refused to be made a catspaw of the Treasury.

Christmas Eve found a great part of the regiment on the verge of starvation at Canterbury. In a letter, written late in the day to the Admiralty Secretary, Major Williams tells of the sufferings of the regiment and of his vain efforts to get assistance from the new mayor and others and how he and some brother officers had managed to raise a few hundred pounds, which had been spent in subsistence. He then gives the story of the events of the morning (muster day). " The Soldiers to the number of about 300 came this day in a body, in a very mutinous manner to demand their pay. I told them that no Officer had any of their money, nor could we in any way procure (at present) a sum sufficient to support them, so by fair words I obliged them to disperse, desiring them to procure a week's credit from their quarters and I doubted not by that time we should receive money, all which they rejected saying they wanted their pay and their pay they would have, for they were resolved not to starve and added that they would take their arms and go all in a body to London : upon which I made shift to borrow £50 with which we have paid them to Saturday next, if money is not remitted by that time the consequences may be fatal." Christmas time and the New Year passed quietly as far as we know.

RECRUITING OF THE REGIMENT LOCALIZED BY AUTHORITY

We can now speak with fuller knowledge of the recruiting of the 30th. On January 1st, 1708, a new system was introduced, under which the Commissioners for the collection of the land tax paid all expenses connected with recruiting and their accounts in the Marine Register in the Record Office give us the recruits' names and the districts in which they were enlisted. As was to be expected, those for the 30th nearly all came from Lincolnshire. The standard was 5 ft. 5 in. and the bounty £4 for volunteers. There was compulsory service for able-bodied men who did not follow any lawful calling or who had not some other sufficient maintenance. At midsummer, 1709, pressing was stopped, as the ranks were full.

Under the above rules 198 recruits were raised for the 30th in 1708, 1709 and the first quarter of 1710: from Lincolnshire, 161, all volunteers; from

Rutland, 9, 1 pressed; from Nottingham, 4, all volunteers; from Kent, 24, 2 pressed. The accounts fail us after that, but recruiting went on, and in December, 1710, the regiment was for the first time in its history localized by authority. In that month the Admiralty directed the Deputy-Judge Advocate to inform the Board of General Officers of Marines that the interest of the officers of the 30th lay mostly in Lincolnshire and Kent, and therefore those counties should be assigned to them for recruiting; Rutland and Northampton were added afterwards. The Admiralty were at that time working out a scheme of localization for all the marine regiments.

In 1712, when it was evident that peace was at hand, recruiting ceased.

1713

The establishment remained the same as in the previous year, but by not filling up vacant appointments and by the stoppage of recruiting in 1712, the number of the non-com. officers was reduced in the course of the year by about a fourth and that of the men by more than a half

The following promotions and appointments were made:—

John Hobart to be Second-Lieutenant to Captain Charles Davison, *vice* Johnson Haseltine, deceased . . .	19th Jan.
George Burston to be Second-Lieutenant to Captain John Mohun, *vice* John Wiltshire, resigned . . .	23rd March
Lancelot Daws to be Adjutant, *vice* William Dawes, resigned	22nd May

There is nothing to show whether Lancelot and Thomas Daws and William Dawes were related. From 1715 all three names are written with the *e*.

In January new clothing was issued, and Lieutenant Cooke was sent to the Mediterranean with a supply for the men of the regiment on that station: after distribution he was to remain to command the party on H.M.S. *Ormond* in place of Lieutenant Chester Onebye who had been landed sick at Port Mahon and thence invalided home.

On January 6th the marine paymaster wrote to Lord Oxford (Harley) to say that the marines were in greater distress than ever, that those lately landed were at the ports unable to march to Head-quarters for want of subsistence and that those at Head-quarters refused to embark until paid their two-thirds. He foresees a mutiny. Lord Oxford made a note upon this " To be considered when Mr. Caesar is in cash by the loan he is to procure on South Sea stock." Caesar was treasurer of the navy.

Unluckily, events would not wait on Mr. Caesar's leisure. H.M.S. *Ruby* was under orders to cruise against the Sally-men, that is, pirates from Salee, the captors of Robinson Crusoe, who sometimes ventured as far as the Channel. The ship could not sail without marines and an order had been given in December for forty of Wills's men from Canterbury to join her. On January 9th Colonel Burston warned the Admiralty that unless money could be procured it would be impossible to send men to sea. At the same time he sent to the marine paymaster a list of the men warned for duty on board H.M.S. *Ruby* with the amount due to each opposite his name, and asked for a remittance. He followed this up by calling on

Sir Roger Mostyn himself. Birchett, the Admiralty secretary, on receipt of Colonel Burston's warning wrote to Sir Roger Mostyn, and asked him to solicit the Lord Treasurer, Oxford, for money. As was to be expected, the Treasury did nothing, and on January 20th Colonel Burston wrote again to Birchett to warn the Lords of the Admiralty of the dangerous situation. He said "We have stretched our credit to the utmost farthing, the Men are subsisted by us till to-morrow, and the Officers at Quarters write me word that unless their friends will credit them further, they expect the Men to be unruly the first day they shall want their daily subsist, the townsmen have refused these four months past giving them any credit."

Birchett on the same day wrote an urgent appeal to Mostyn to get money and made the confession, "without marines the ships are useless."

Some one at the Treasury was evidently in favour of strong measures, for shortly before this Mostyn had written to the Treasury that the Admiralty thought it too dangerous to use force to put the marines on board, that it had been tried lately in Bor's regiment and provoked a mutiny at once.

By some means another week was got through at Canterbury, and it seemed likely that no attempt would be made to force the men warned for duty on the *Ruby* to embark without at least enough of their arrears to buy sea necessaries. Unluckily the order to embark had not been formally rescinded, and when Major Williams was called upon by the captain of the *Ruby* to hasten the marines detailed for that ship, he felt bound to comply. The men, who had been led to believe that they would be paid before embarkation, naturally thought Major Williams had deceived them. The detachment was paraded on the morning of January 27th, but, as Colonel Burston reports, "when the order to march was given, the Men unanimously cried out one and all not to march unless paid, to which the women of the town cried out 'Knock down the officers and do not march without money.' The officers upon this drew out two or three of the ringleaders and were sending them to gaol, but the rest rushed upon the officers and rescued the prisoners and went and drew up in the green churchyard, but upon the Major's application they have since dispersed and gone quietly to their quarters."

The Admiralty now ordered a detachment of another regiment for duty on the *Ruby*, but found difficulties with it also, and it was a couple of months before the ship was got to sea.

The disturbed state of his regiment brought General Wills on the scene, and one of his first acts was to dismiss the agent, Mr. Potter.

Why General Wills was not called in earlier we do not know, nor how he had been employed since his return from Spain in the previous June.

An order was issued on March 6th for forty men to embark on H.M.S. *Canterbury* on convoy duty to the Baltic, and General Wills thereupon directed the marine paymaster to issue to Colonel Burston the two-thirds for the detachment. The money was paid on April 9th, and the men embarked.

In April the treaty for a general peace was signed at Utrecht. By it we gained Gibraltar, Minorca and Nova Scotia. In the winning of all three the 30th had a considerable share, and in addition to that its good service at Barcelona and in Catalonia, in the capture of the Spanish coast towns in Valencia and on board the fleet, especially at the battle off Malaga, helped to establish our naval superiority in the Mediterranean, the maintenance of which has been a leading principle of British policy ever since.

On June 30th an order was issued for the immediate disbanding of all marines on shore, those serving on board ship to be disbanded as soon as they landed. In answer to a question from Wills and other colonels as to the disposal of the sick and invalids, it was ordered that they should be included in the number discharged. This order for a hurried disbanding could only have been given by some one ignorant of its difficulties. Better considered orders were issued on July 30th and August 12th. By those Sir Stafford Fairborne, Sir William Gifford, and Samuel Hunter, Esq., were named as Commissioners to settle with and disband the marines who were placed under their control by the following order to the colonels of regiments:—

"You are to observe and follow all such instructions as you shall receive from the Commissioners, for paying and disbanding as well as all other matters relating thereto."

Apparently some officers read this as relieving them of the command of their men in all respects, when the Commissioners were present.

Lord Shannon, the senior marine officer, being in bad health, Major-General Wills was appointed to superintend the work of disbanding.

The instructions to the Commissioners were:—

To have all accounts made up from December 25th, 1710, to the day of disbandment.

To pay all legal debts in quarters and to hand the residue to the men.

To see that accounts were adjusted between officers and men, and to make certain that both were satisfied.

The men to receive back their off-reckonings since the last issue of clothing.

Arms issued on indent from the Ordnance Department to be returned to store. (A fusil cost £1 13*s.* 4*d.*)

When the accounts were settled, regiments were to be disbanded by companies or regiments as convenient.

Non-com. officers and soldiers were to retain clothes, knapsack and belt and each private soldier and drummer to receive 3*s.* for his sword or bayonet, which were to be returned to the Ordnanee.

Fourteen days' subsistence was to be issued as a bounty and passes to proceed home, but no more than three were to travel together and no arms were to be carried upon pain of severest punishment.

The above instructions keep up the pretence that the marines were fully settled with in 1710, but there is another order telling the Commissioners that if they found any cases of arrears not paid in 1710 they might deal with them. The Treasury officials must have been well aware that after the pretended settlement of 1710 they still owed Wills's regiment over £3,000. What was not generally known was that there were men who had not received their arrears at the settlement of 1708, and no provision was made for paying them.

The Commissioners determined to begin with the western regiments and to take Wills's at Canterbury and Shannon's at Rochester the last. A difficulty here occurred to them. The Treasury had declared that it would subsist the marines up to June 24th, 1713, but not a day after that till the final settlement. The Commissioners therefore stood a good chance of finding no regiment when they came to Canterbury. As usual the regimental officer was called in to get them out of the trouble, and General Wills through his agent, Mr. Chambers, borrowed £600 of a Mr. Jackson to maintain his regiment for six months.

Jackson was Mr. Whitfield's banker, and still had in his hands the money irregularly acquired by the late Paymaster, so for the concluding six months of its career as Marines the 30th was paid out of money fraudulently detained in previous years from officers and men.

The disbanding of the marines required great care, for the ill-treatment of the last few years had goaded officers and men into a dangerous humour. There was also a general belief that regiments were selected for reduction out of their turn because the men in power believed that they would be an obstacle to the forcible restoration of the male line of the Stuarts. Rightly or wrongly it was supposed that the marines were among those so selected. Mr. Fortescue, in his great *History of the British Army*, says that there was a good deal of heated talk among the officers of the regiments chosen for reduction. It is evident from what happened in the 30th that the non-com. officers and the older soldiers were equally dissatisfied. It may have been from a fear that they would be shut out from the benefits of Greenwich Hospital after subscribing to it for so long. There was one grievance peculiar to the 30th. Many must have remembered and all others must have heard from relations or comrades that, when broke in 1698 under King William, the regiment was marched to Lincoln in the centre of its own recruiting ground and there dismissed, each man receiving a pass and fourteen days' subsistence. Now they were to be discharged with only the same marching money 200 miles from their district, and a great part of the way would have to be marched as if in an enemy's country, for all the rogues in England would be on the alert to intercept some hundreds of men streaming north in unarmed parties of three or less with from five to fifty pounds in their pockets.

The Commissioners reached Canterbury in December and decided to disband the regiment by companies, beginning with the junior, Captain Mohun's, on the 22nd. After settling with it a few days' holiday were to be given for Christmas and work was to commence again on the 29th. It was hoped that the two senior companies, those of Wills and Burston, would be reached on January 4th, 1714, and the payment of all men on shore completed.

When Captain Mohun's company was paraded for settlement on the 22nd, the men disputed the accuracy of the accounts, and only two men signed on that day; the remainder refused their money when tendered to them. On the following day twenty-eight men received their money and signed or made their mark, and before evening all Mohun's men were settled with and they took no part in the violent proceedings of the non-com. officers and men of the other companies. So far as we know there was nothing to foreshadow the outbreak of the following day. No other company could be called up for settlement before the 29th, but the action of the sergeants was so bold and swift, and they had their men so well in hand, that we must believe that it had been long planned. In fact there is every appearance of the outbreak having been hurried on lest another company should be induced to agree to its accounts.

On the 24th the drums beat round the city and the greater part of the non-com. officers and men assembled under arms, the colours were taken from Major Davison's house and the men, leaving a deputation of sergeants to confer with the Commissioners, marched out on the London Road and halted to await the report of the non-com. officers. The deputation to the Commissioners disputed the accuracy of the accounts and demanded the preparation of fresh ones. Their proposals were rejected, as they must have expected, and the sergeants joined their men and the whole marched 16 miles to Sittingbourne, where they lay that night. On Christmas Day they marched to Northfleet, a march of 19 miles. The Medway was crossed at Rochester, where Lord Shannon's regiment, their old comrades in many a fight, was awaiting the disbanding Commissioners. Shannon's was drawn out by Lieut.-Colonel Markham to dispute the passage of the river, but Markham soon found that his men were more likely to join in the march to London than to oppose it; he therefore sent his regiment away before there could be any communication between the two corps and remained with three of his officers to parley with Wills's sergeants. His remonstrances were ineffectual and the march, which had been interrupted for a few minutes, was resumed. On the 26th, Wills's non-com. officers and men marched to Greenwich, where they must have arrived in the forenoon, for the Lieut.-Governor wrote to the Admiralty at 3 p.m.: "Since I wrote to you this morning a body of Marines, about 350 of General Wills's Regiment came hither and requested permission to leave their arms here before they proceeded nearer London. After examining one or two of them I found they had marched in dissatisfaction from Canterbury and therefore, by advice of my officers, I thought it for Her Majesty's service that their arms should be secured, which is done." After handing in their arms the men quartered themselves in Greenwich, Deptford, and the neighbouring places. They had taken no ammunition from Canterbury.

When the non-com. officers and men had marched from Canterbury the Commissioners had sent expresses to London and Windsor, and in consequence General Tatton was ordered to take the Guards, both horse and foot, and any other troops he could lay his hands on, and march out to bar the Canterbury to London road. He was not able to leave the Horse Guards Parade before four o'clock on the afternoon of the 26th, and when he reached Blackheath he found that Wills's men had deposited their arms at Greenwich some hours before, so he returned to London. He had been furnished by the Admiralty with the following proclamation:—

"Whereas we are informed that several non-com. Officers and a great part of the Soldiers belonging to Colonel Wills's Regiment are come from their quarters in a mutinous manner. We hereby do strictly charge and require all the said non-com. Officers and Soldiers immediately to lay down their arms and return in a peaceable manner to their quarters at Canterbury and there obey such orders as they shall receive from the Commissioners appointed by Her Majesty to disband them, as they shall answer the contrary at their peril."

Again the marines are spoken of as being under the command of the Commissioners and not their own officers.

The Admiralty proposed at this time to try Wills and all his officers by a court martial "*or some likelier Judicature to punish them,*" if they should be found not to have done their duty. Another proposal was to pay off the men who had remained at Canterbury and declare the regiment disbanded to save any further charge for the officers and men.

On the 29th, the following unsigned and undated petition was forwarded by the Admiralty to the Secretary of State. It had been left at the Admiralty Office after the non-com. officers and men had delivered up their arms at Greenwich:

"To the Right Honourable, the Lords Commissioners of the Admiralty.

"The Humble Petition of the Marine Soldiers commanded by Major-General Wills,

"Sheweth,

"That your petitioners have considerable sums of money due for their long and faithfull service, and upon their applying for the payment thereof, they find such reductions made by Mr. William Dawes, Sub Agent to the said Regiment, which are altogether so large and unjust that one third part of the pay due is thereby sunk.

"That there is likewise due to your poor petitioners several sums of money on account of short subsistence and clothing; and being driven to great extremity and seeing no likelyhood of being justly relieved at Canterbury, they, in a very orderly manner, marched for London in hope of relief."

No notice was taken openly of this petition, but on the 30th there was a complete change of front on the part of the Government. On the 28th the Commissioners had been instructed by the Admiralty to pay off only the marines who had returned to Canterbury and to declare the regiment disbanded, now instructions were sent for the Commissioners to follow Wills's men to Greenwich and the neighbourhood and pay them there. This was altered on January 1st, 1714, when it was known that the men were returning of their own accord to Canterbury. On the 7th, the Commissioners reported to the Secretary of State that they had "made a review of the Mutineers who are returned almost to a man, being 25 Sergeants, 24 Corporals, 9 Drummers, and 310 Private Men, and begun to pay them." Their pay was computed to the day they marched from Canterbury, and they received Her Majesty's bounty of fourteen days' subsistence. The change in the conduct of the Government dates from the time when the men's petition reached the Secretary of State, and he took charge of the business himself. The total number discharged was 34 sergeants, 34 corporals, 19 drummers, and 482 private men.

It is impossible to say now how far the men were wronged, but if the shortage of subsistence money and clothing, which were such a feature of regimental life for eleven years, was all made good at the last, it was a wonderful feat.

The charges of the billet masters and debts to captains of companies are on a different footing. If debts have been allowed to run on for years there is naturally great resentment when they are forcibly collected. As has been explained before, the debts to the captains were mainly for expenses incurred by the men on board ship which the captain had been forced to pay at once.

The Commissioners pointed out that a mistake had been made in 1712 when a detachment of over 200 men had been sent to Dunkirk. They had received two-thirds of their arrears in full instead of the residue after their debts were paid.

Whatever the rights of the case may be, the conduct of the Treasury had made it certain that the marines would think the settlement an unjust one. In every one of the eleven years for which the marine force had existed there had been a mutiny in one or other of the six regiments, sometimes in all, and these mutinies had been caused by want.

There is no satisfactory evidence to show why the sergeants were so determined to provoke an outbreak. The Commissioners repeat again and again that without them there would have been no trouble and that the ringleaders were Sergeant Stephen Pearce, who had commanded the detachment on board H.M.S. *Roebuck* at Lisbon and Gibraltar in 1704–5, Sergeant Robert Davison, who had been senior sergeant of the detachment on H.M.S. *Salisbury* in the victory over de Fourbin in 1708, and Sergeant J. Martin. The sums due to those non-com. officers were £12 9*s*., £24 1*s*. 6*d*., and £33 16*s*. 8*d*., or an average of £23 9*s*., equal to thrice that amount nowadays. It can have been no light reason which induced men with their long and honourable service to take a lead in such outrageous proceedings and to risk losing their hard-earned arrears.

The Secretary of State ordered that the three sergeants on their return to Canterbury should be arrested and charged before a civil magistrate with riot and misdemeanour.

When this was done the other non-com. officers came in a body to the Commissioners and begged that, as they were all equally guilty, they might all be sent to prison together. We are in the dark as to what took place after that, but in the following October, when George I was on the throne, the three sergeants were paid their arrears and bounty as if nothing had happened.

In addition to the men discharged at Canterbury, a party of eighteen landed at Portsmouth in February, 1714, and marched to Greenwich, where they were paid off, and in July three more men were discharged. The non-com. officers and men were now all discharged except the three sergeants named above.

The settlement with the officers was delayed. On March 16th the staff officers and subalterns of the marine regiments stated in a petition to the Treasury that they had received no arrears since 1708, that some were in prison and all in great distress, that a settlement had been delayed under the pretext of difficulty in adjusting of accounts, but that their accounts were easily adjusted. They had no accounts with their companies, as the captains had, and the only deductions were agency poundage and exchequer fees. The Commissioners backed up the petition heartily. Lord Oxford wrote a minute to ask what the cost would be, but nothing further was done.

Oxford's day of power was now at an end ; his abler and more vigorous rival, Bolingbroke, having displaced him in the royal favour ; but on August 1st Queen Anne died, and owing to the resolution of the Whig leaders and the unprepared state of the Jacobites, George I succeeded without opposition. Oxford was impeached, Bolingbroke and Ormond fled to the Pretender in France. Marlborough again became Captain-General, but General James Stanhope, who was appointed Secretary of State, was the King's chief adviser in military as well as civil affairs.

At the death of Queen Anne the officers of the regiment, although placed on half-pay from the dates on which their companies were disbanded, had not received their arrears, and the captains had not been settled with for their companies. In addition, the Treasury owed to the officers a sum of £2,618 for recruiting, expenses of disbanding, money advanced to detachment ordered to Dunkirk in 1712, and rations supplied at Annapolis in 1710, also their share of £2,935 for purchase and maintenance of transport mules in 1705–7. A sum of £571 was likewise due to widows of deceased officers.

CHAPTER III

REINSTATED IN FORMER SENIORITY IN ARMY. PLACED ON IRISH ESTABLISHMENT. SECOND DEFENCE OF GIBRALTAR. EXPEDITION TO LORIENT. LORD ANSON'S VICTORY OFF FINISTERRE.

1715–1754

THE regiment had been broke in 1713–14 by those who wished to exclude King George from the throne and, now that he had the power, he re-established it in its former place among the forces of the Crown, thus preserving the continuity of regimental life and ensuring to the regiment the precedence it had enjoyed under Queen Anne.

Mr. Dalton has adopted from an old record of the 31st a statement that the King reinstated the three senior marine regiments " on account of their eminent services in the late war." The King may well have said so; it was no more than was due to the three regiments, but the original evidence for this expression of approval cannot now be found.

The expense of raising the 30th was included in the Irish estimates, and while awaiting the sanction of the Irish Parliament for raising the men, the King bestowed a mark of his favour on Major-General Wills by appointing him Governor of Berwick on January 23rd, and at the same time he renewed his commission as major-general. The King also appointed Capt. Joseph Mason Major of the Tower of London in recognition of his wounds and distinguished service at Gibraltar and elsewhere. Appointment dated January 18th.

On March 26th beating orders were issued to General Wills authorizing him to raise volunteers by beat of drum or otherwise in any county of Great Britain for a regiment under his command, for service in Ireland. The regiment was to consist of 10 companies each of 1 captain, 1 lieutenant, 1 second lieutenant, 2 sergeants, 2 corporals, 1 drummer, and 36 private soldiers; 58 officers' servants, and 10 widows' men were included in the 360 privates. The field officers were as usual included in the 10 captains.

The following is a list of the officers who were placed on half-pay at the end of 1713 and beginning of 1714:—

	Commission.	
Colonel Charles Wills (Major-General) . .	13th Oct.	1705
Lieut.-Col. George Burston (Colonel) . .	1st Nov.	1707
Major Charles Williams	12th Aug.	1708
Capt. John Saunders (Major) . . .	10th March	1702

	Commission.	
Capt. David Ward (Lieut.-Colonel) . .	30th May	1702
,, Walter Palliser (Lieut.-Colonel) . .	10th March	1702
,, Hugh Palliser	13th Oct.	1705
,, Charles Davison (Major) . . .	9th Feb.	1707
,, Michael Medford	24th Dec.	1707
,, William Scott (Major)	24th ,,	1707
,, Joseph Dacres Mason	5th April	1708
,, John Mohun	24th June	1709
Capt.-Lieut. William Davison . . .	9th Feb.	1707
Lieut. Rothwell Stowe	29th Sept.	1704
,, Thomas Newdigate	25th Nov.	1705
,, John Thompson	13th May	1706
,, William Dawes	15th Sept.	1706
,, Thomas Burston	10th Dec.	1706
,, Richard Leathat	9th Feb.	1707
,, Henry Long	10th April	1707
,, William Cooke	6th ,,	1709
,, John Roper	15th Nov.	1710
,, Wiltshire Castle	1st May	1711
,, Thomas Dawes	21st Feb.	1712
Second Lieut. Charles Rainsford	24th May	1705
,, ,, Edmund Martin	10th April	1707
,, ,, Henry Aylmer	1st Nov.	1707
,, ,, Edmund Quarles	21st ,,	1707
,, ,, Lancelot Dawes	5th ,,	1708
,, ,, Patrick Aylmer	5th ,,	1708
,, ,, Ventris Scott	16th ,,	1710
,, ,, Chester Onebye	1st May	1711
,, ,, William Pritchard	23rd Aug.	1711
,, ,, Benjamen Sladden	24th ,,	1711
,, ,, Teddiman Roberts	21st Feb.	1712
,, ,, John Hobart	19th Ja	1713
,, ,, George Burston	23rd March	1713
Adjt. Lancelot Dawes	22nd May	1713
Chaplain Dr. Alexander Innes	18th July	1709
Quartermaster Gunsley John Ayerst . .	3rd Dec.	1711
Surgeon Matthew Poumies . . before	Jan.	1705

The dates of Ward's and Walter Palliser's appointments are as given in the commission book and it is impossible to say why Ward is shown as the senior unless his commission was antedated and the alteration not published.

Up to 1715 the titles conferred by brevet for services in the field, although abolished by Royal Order, were constantly used; under King George they disappear.

The rolls given below show the officers by companies on disbandment in 1713–4 and after the regiment was reformed in 1715 :—

1713	1715
Colonel (Major-General) Charles Wills	No Change
Capt.-Lieut. William Davison	,,
Second Lieut. Henry Aylmer	,,
Chaplain Doctor Alexander Innes	,,
Adjutant Lancelot Daws	James Baker
Quartermaster Gunsley John Ayerst	Left Vacant
Surgeon Matthew Poumies	David Hall
Surgeon's Mate	Nicholas Terry
Lieut.-Colonel (Colonel) George Burston	Richard Cobham
Lieutenant Richard Leathat	No Change
Second Lieut. Edmund Quarles	,,
Major Charles Williams	Major David Ward
Lieut. Thomas Daws	John Roper
Second Lieut. William Pritchard	John Hobart
Captain (Major) John Saunders (Grenadiers)	No Change
Lieut. Rothwell Stowe (Grenadiers)	,,
Second Lieut. Charles Rainsford (Grenadiers)	,,
Captain (Lieut.-Colonel) David Ward	Captain Charles Williams
Lieutenant John Roper	John Thompson
Second Lieut. Chester Onebye	Patrick Aylmer
Captain (Lieut.-Colonel) Walter Palliser	No Change
Lieutenant William Dawes	Thomas Dawes
Second Lieut. Ventris Scott	No Change
Captain Hugh Palliser	No Change
Lieutenant Thomas Newdigate	,,
Second Lieut. Edmund Martin	,,
Captain (Major) Charles Davison	No Change
Lieutenant Wiltshire Castle	Thomas Burston
Second Lieut. John Hobart	Teddiman Roberts
Captain Michael Medford	No Change
Lieutenant John Thompson	William Cooke
Second Lieut. Patrick Aylmer	William Pritchard
Captain William Scott	No Change
Lieutenant Henry Long	,,
Second Lieut. Benjamin Sladden	,,

Two junior companies reduced.

Capt. Joseph Dacres Mason	appointed Major of the Tower
Lieut. William Cooke	to Medford's company
Second Lieut. Teddiman Roberts	to Davison's company
Capt. John Mohun	not called up
Lieut. Thomas Burston	to Davison's company
Second Lieut. George Burston	not called up

The clean sweep made of the staff shows that General Wills thought that the affairs of the regiment had not been well conducted during his absence. This does not apply to the surgeon or quartermaster. It is probable that advancing years and the gout, to attacks of which he was liable, made Surgeon Matthew Poumies think that he had done enough soldiering. He had been continuously on service in the years 1705–10, including the defence of Gibraltar, capture of Barcelona, victory of St. Estevan, relief of Barcelona, defence of Catalonia and Lerida, capture of Sardinia and Minorca, and capture of Annapolis. Gunsley John Ayerst was retired because no quartermasters were borne on the Irish establishment.

The case of Colonel Burston differs from that of the other members of the staff, for without doubt his action caused a serious loss of money to General Wills. The story is instructive.

In former years when campaigning in Catalonia General Wills had left with Lord Shannon, commanding the marine regiment at Rochester, a power of attorney to manage the clothing and other financial business of the 30th. Mr. Peters was then the clothier and, as was usual, acted as banker to the regiment ; General Wills left large sums in his hands. In 1710 Lord Shannon was ordered on service and passed on the power of attorney for the 30th to General Whetham. Whetham also was ordered on service, but before going he signed and sealed a blank form of assignment or receipt for regimental clothing, although the clothing had not been received or even contracted for. This blank assignment he handed to Burston, no doubt hoping that he would see to the clothing, fill in the items received and the price, and submit the completed assignment to the Board of General Officers, so that the clothier might be paid and the balance of the off-reckonings handed to Wills or his representatives. If Burston felt that he had no authority to do all this he should have held on to the assignment till he could communicate with Wills. Unluckily he allowed himself to be persuaded by Potter, the agent, to take the worst course of all and handed the assignment, still in blank, though signed and sealed, to Mr. Peters who, as Potter probably well knew, was now insolvent. Peters filled in the assignment (Wills says with fictitious items) above General Whetham's signature and got it discounted, pocketing the whole of the cash. Wills, who also lost £2,000 of private money which was in the hands of Peters, accused Burston of handing the assignment to Peters "perfiduously," but we are not bound to believe that. In any case Burston had to pay Wills £600 for the lost assignment, perhaps the estimated profit on a two years' clothing, and ceased to be lieut.-colonel of the 30th. During the remainder of his life he complained so loudly and unceasingly of vindictive persecution on the part of Wills that many have formed a harsher opinion of the general's character than is fair. All that can be said is that Wills did not bestir himself to forward Burston's interests, which was not unnatural. By the Duke of Marlborough's kindness Burston was appointed lieut.-colonel of Bor's regiment, but was obliged to sell his commission to pay his debts. His son Thomas continued to serve under Wills in the 30th, the younger son, George, was placed on half-pay by the reduction of two companies.

It is impossible not to be sorry for one so brave and witty as Burston, but we may be certain that if it had been possible to serve him the Whig Party would have done so. He had strong claims. Under James II he had been cast out of the army for his religious and political opinions, and he had carried a musket from Honiton to London in one of the Dutch regiments which landed with William of Orange.

Lieut.-Colonel Richard Cobham, who was transferred from Bor's to take Burston's place, was no stranger to the regiment. He had commanded the detachment of Bor's which fought alongside the four companies of the 30th at the storming of Alicante in 1706. He afterwards joined the marine brigade under Wills in Catalonia, and was one of the officers who signed the petition to be allowed to take the field and share in the Almanza campaign.

Charles Williams was not only deprived of his majority, but took the place of David Ward among the captains below John Saunders.

A Royal Order of March 26th, 1715, addressed to Lord Sunderland, the Lord-Lieutenant, directs that the regiment be taken on the Irish strength from March 25th, and gives the following scheme for raising the 30th, 32nd and 33rd:—

" Whereas it is our Royal intention that the expense of raising the said three regiments shall be defrayed by the pay of the non-effective soldiers, which shall be made use of in the place of levy money, we have therefore thought fit that the non-commissioned officers and only one-third of the private soldiers of the said three regiments should be raised by May 24th next, one other third part of the private soldiers by July 24th next and our said three regiments be completed by September 24th next."

The rule at this time was that regiments took rank in the army from the date of their coming on the British establishment, and some doubts must have been expressed about the position of the 30th from its being intended for service in Ireland, for the King defined its rank in an order of April 23rd, 1716: " Whereas we have thought it fit to raise the Regiment of Foot lately commanded by our trusty and well-beloved Lieut.-General Charles Wills, and now by our right trusty and well-beloved George Lord Forrester, which was broke in the months of December 1713 and January following. Our will and pleasure is that the said regiment shall have, hold and enjoy its former rank as if the same had not been broke, notwithstanding any former order, direction or instruction to the contrary."

That rank is recorded in the report of the general officers assembled by Queen Anne to settle the rank and seniority of the several regiments " which now do or lately did serve in the army." The report was presented to Her Majesty on February 19th, 1714, and places the foot regiments from the 29th to the 34th, as follows:—

29th	Lieut.-General Holt's . . .	Broke	
30th	Lord Mark Kerr's . . .	,,	
31st	Lieut.-General Mordaunt's .	,,	
32nd	Major-General Wills's . .	,,	but re-established
33rd	Sir Harry Goring's . . .	,,	,, ,,
34th	Brig.-General Bor's . . .	,,	,, ,,

From General Holt's and General Mordaunt's regiments being broke, and not restored, the regiments of Wills, Goring and Bor became the 30th, 31st and 32nd.

No one has ever been able to suggest a reason why so many junior regiments were placed above Wills's regiment by Queen Anne. Of those regiments the 28th and 29th alone remain. Both were raised five years after the 30th. The three regiments were broke in March, 1698, and re-formed on February 12th, 1702; they were broke again in 1713, and re-formed by King George I. The formation of Saunderson's as marines very likely led to the mistake but, to be consistent, Queen Anne should have placed the present 34th above her marine regiments.

It is not stated where the regiment recruited, but there is no reason to suppose that the officers tried fresh ground. It was a question of not only where they would get the best and most numerous recruits, but where the officers had most friends, and would meet with most hospitality. Of course the popularity or otherwise of the officers with the local gentry re-acted on recruiting.

The regiment was formed at Dublin and the first party of six officers and 150 men was ordered on May 25th to march from London to Holyhead (eighteen marches) for embarkation. The second party of two officers and fifty men was ordered on June 27th to march from London to Bristol (eight marches). Whether it went from there to Dublin by sea or crossed to Wexford and marched up is uncertain. Both parties may very well have come down from Hull by sea as the regimental recruits had done for years past. With the north of England on the point of rebellion it was no time to be marching recruits across country. The reason for the change of the route for the second party is obvious. Government now had cause to fear that if a Jacobite rising in the north had any success, Wales might join the insurgents; the Chester and Holyhead route therefore had become dangerous. On December 24th, after the rebellion was suppressed, an order was issued for a sergeant and twelve recruits to march from London by the original Chester and Holyhead route. This is the last route we have. We know from Burston that when the regiment was broke at Canterbury, General Wills was left with a large number of suits of uniform on his hands which he employed in clothing the regiment when it was re-formed. Perhaps the men were assembled and clothed at Canterbury before going to Ireland.

The lists of officers of the regiment are signed by Lieut.-Colonel Cobham, and it is doubtful if General Wills went to Dublin. He could scarcely be spared from England.

In September the Earl of Mar proclaimed James VIII in Scotland and on October 6th the Jacobites in the north of England rebelled. General Wills was sent to Chester, where he cut off Wales from Lancashire, both strongly Jacobite. He had under him one regiment of horse, three of dragoons and three of foot. The appointment of Wills, who was low down on the list of major-generals, was the work of General James Stanhope, afterwards the first Lord Stanhope, the Secretary of State. Wills had served under him almost continuously from the time of the landing at Barcelona in the autumn of 1705, till they were taken prisoners together at Brihuega in December, 1710. General Carpenter, who had commanded Stanhope's cavalry in Spain, was appointed to command at Newcastle with a force of about the same strength as that of General Wills.

The English rebels who were under Mr. Thomas Foster marched into Scotland after threatening Newcastle, where General Carpenter's troops had not yet assembled, but which was defended by its citizens. In Scotland, Mr. Foster was joined by a few noblemen and gentry and by Brigadier Mackintosh, who had been detached with 1,000 Highlanders from the Jacobite army under Lord Mar and who, after crossing the Firth of Forth in open boats in face of the British fleet, had marched to the English border. The united body now moved south to raise Lancashire, but were met at Preston on the Ribble on December 12th by General Wills, who had advanced from Chester. The rebels had barricaded the southern part of the town, but had neglected to secure the passage of the river. General Wills crossed the Ribble and attacked the barricades without hesitation and a good deal of spirited fighting ensued in which Lord Forrester, who showed great gallantry in leading the Cameronians, was wounded. General Carpenter had followed the rebels from the north with a couple of dragoon regiments, and on his appearance Mr. Thomas Foster and his colleagues surrendered at discretion, to the disgust of a large number of their followers.

On the day on which General Wills attacked the rebels at Preston the Duke of Argyle had defeated the Earl of Mar on Sheriff Muir and 6,000 Dutch troops having landed, the British Government was secure.

After the regiment was formed, the following changes took place:—

The establishment was raised on October 12th, 1715, to fifty private soldiers a company, the other ranks remaining the same.

On November 1st, Major David Ward and Captain Charles Williams left the regiment. Major Ward was succeeded by Major Edward Wolfe, of Dubourgay's regiment, the father of General James Wolfe, the hero of Quebec. Major Wolfe was appointed to the Guards and to the 30th on the same day, but as far as can be judged he served as major in the 30th until promoted lieut.-colonel of it in 1717. Major Charles Williams took Wolfe's place in Dubourgay's regiment. His company in the 30th was not filled up for several months.

In July an incident had occurred which showed that Government was in earnest in trying to put an end to purchase in the army. A Mr. Philip Burean had been appointed second lieutenant in W. Palliser's company in succession to Ventris Scott, but the notification was stopped when it was discovered that it was a case of buying and selling. Burean was appointed two years later.

From this period up to 1724, when purchase was re-established, nearly all vacancies except in the very junior ranks were filled up from the half-pay list, and the officers who left the regiment were not allowed to resign their commissions but were placed on half-pay in order to maintain a reserve of officers. This accounts for the great number of men brought into the regiment from outside.

The following are the promotions and appointments in the regiment from January 1st, 1716, until it embarked for foreign service in the spring of 1718:—

1716

George Lord Forrester from the Cameronians to be Colonel, in place of General Wills, appointed colonel of the Buffs . 5th Jan.

Capt.-Lieut. William Davison to be Captain of the company vacated by Charles Williams in the previous autumn . 17th April

Lieut. Andrew Forrester, from the Cameronians, a younger brother of the colonel's, to be Capt.-Lieutenant . .	17th April
William Harvey to be Second-Lieut. to Capt. W. Davison, *vice* Patrick Aylmer, retired	12th May
Edmund Martin to be First Lieut. in the lieut.-colonel's company, *vice* Richard Leathat, retired	3rd Oct.
John Forrester to be Second Lieut. in Capt. Hugh Palliser's company, *vice* Martin, promoted	,, ,,

In addition to his appointment to the Buffs, General Wills was promoted to the rank of lieut.-general for his victory at Preston. The colonelcy of the 30th was of course a reward to Lord Forrester for his good service in that action.

George Baillie, fourth Lord Forrester, was appointed cornet in the Scots Greys on January 1st, 1707, and fought at Oudenarde and Malplaquet under Marlborough. On November 15th, 1711, he was appointed brevet-colonel of foot. From the 30th he was transferred in 1717 to the Horse Guards. He died in 1727.

On December 15th, the Secretary at War wrote to Lord Forrester to tell him that Lieut.-Colonel Cobham was lately dead, and that an officer from half-pay would be brought into the regiment in succession to him as soon as it was known who would be most acceptable to his lordship. No record has been found of any man being brought in from half-pay, and so far as we know the vacancy was not filled up till the following July.

1717

Brigadier Thomas Stanwix became Colonel of the regiment when Lord Forrester was appointed to a troop of Horse Guards	17th July
Brigadier Andrew Bissett succeeded General Stanwix. .	24th Aug.
Major Edward Wolfe was promoted Lieut.-Colonel of the regiment	10th July
Peter Bettesworth was brought in as Major in succession to Wolfe	,, ,,
Robinson Sowle from the half-pay of Colonel Thomas Chudleigh's regiment became Captain in place of Hugh Palliser, who retired	30th April
Capt.-Lieut. Andrew Forrester succeeded to the company of Captain Michael Medford, who retired	14th Aug.
Francis Pierson came into the regiment as Captain of the company of Walter Palliser, who retired . . .	25th ,,
Lieut. James Baker became Capt.-Lieut. in place of Andrew Forrester	14th ,,
Second Lieut. Edmund Quarles became First-Lieut. in W. Palliser's company, *vice* Baker	,, ,,
Philip Burean, gent., was appointed Second Lieut. to Capt. Walter Palliser, *vice* Ventris Scott, transferred to Holt's regiment	4th May

David Weems (Wemyss), Second Lieut. to Lieut.-Colonel Edward Wolfe, *vice* Quarles, promoted	14th Aug.
Nicholas Smith, Second Lieut. in Captain Michael Medford's company, *vice* Henry Aylmer, who retired . . .	3rd July
Edward Stillingfleet, Second Lieut. to Captain William Davison, *vice* William Harvey, deceased . . .	25th Aug.
Henry Pakenham, Chaplain, *vice* Alexander Innes, who retired	4th May
John Jeffreys was brought into the regiment as Captain in place of Charles Davison, who retired . . .	25th Dec.

Thomas Stanwix was capt.-lieut. of Hastings' Foot in 1691; captain on December 19th, 1692; transferred to Tidcombe's, February 23rd, 1693; transferred to Arran's Horse, February 16th, 1694; to Carbiniers, 1702; lieut.-colonel of Scott's, March, 1704; brevet-colonel, January 1st, 1705; lieut.-governor of Carlisle, April 5th, 1705; served at the battle of Caya, 1709; brigadier, January 1st, 1710; governor of Gibraltar, January 13th, 1711; governor of Chelsea, January 24th, 1715; colonel of a regiment to be raised, July 22nd, 1715; promoted colonel of 30th, July 17th, 1717; to 12th Foot, August 22nd, 1717; died March 14th, 1725.

1718

Major Peter Bettesworth to be Lieut.-Colonel, *vice* Edward Wolfe	7th May
Capt. Robinson Sowle to be Major, *vice* Bettesworth .	,, ,,
Folliott Ponsonby to be Lieut. in the lieut.-colonel's company, *vice* Martin, promoted in Pococke's regt. March 14th, 1717	8th ,,
Charles Rainsford to be Lieut. of the Grenadier company, *vice* Rothwell Stowe, retired	15th ,,
John Purcell to be Second Lieut., *vice* Teddiman Roberts, deceased	21st ,,
Lieut. John Roper to be Captain, *vice* John Saunders, who retired	22nd ,,
Peter Margaret to be First Lieut. to Capt. John Vincent, *vice* Roper, promoted	23rd ,,
John Vincent to be Captain, *vice* Robinson Sowle, promoted .	10th June

The promotions of this year and the appointment of Jefferys as captain in place of Charles Davison in the preceding December are taken from a list which was issued after the regiment embarked for Minorca on May 7th. The books seem to have been hurriedly written up. John Vincent must have been promoted on May 7th, for he embarked as a captain on that date. Martin's vacancy was not filled up for more than a year. There is something wrong about Purcell's appointment as second lieut., for the widow of Teddiman Roberts got a first lieutenant's pension.

On promotion Bettesworth took over the old lieut.-colonel's company commanded by Wolfe, and Sowle took over from Bettesworth the major's company. This was quite a new thing. Formerly all captains kept their old companies on promotion.

We know little or nothing of the life of the regiment during the three years it spent in Ireland, not even how long it stayed in Dublin, nor in what other stations it served, but the duty must at times have been unpleasant. Successive revolutions had disorganized society, there were no police and many of the ordinary processes of law could not take place without the employment of soldiers.

There was a good deal of difficulty in keeping officers with their regiments. Government certainly did not ask too much. The colonel of a regiment could grant leave in Ireland to one-third of his officers, but one field officer had always to be present. For leave out of Ireland, application had to be made to the Lord-Lieutenant.

Colonels were directed to see that their officers had regimental clothes, and that in country quarters no officers mounted any guard but in red or blue clothes. The men also were to be regimentally dressed in their quarters. The officers wore a wig surmounted by a laced hat, a rather long frock-coat worn open so as to show the gorget, cravat of rich lace and waistcoat of the colour of the regimental facings, which, in the case of the 30th, were a pale yellow, knee breeches, white silk stockings and buckled shoes; a scarf or sash was worn over the right shoulder; all the lace and buttons were of silver. Even the officer's horse furniture was embroidered, for the infantry officer rode on the march, and his silk stockings and shoes were then replaced by boots or marching gaiters; he was accompanied by a servant or orderly, who carried his half pike or spontoon. If the regiment passed the general, or any other necessary occasion arose, the officer dismounted, and taking his spontoon in hand fell in with his company. The expense of an officer's kit was excessive: a subaltern's hat, wig, stockings and sash alone cost about £22. A concession was made during the rebellion, when all officers were allowed to mount guard in boots. On the march both officers and men secured freedom of movement by turning back the skirt of the coat and fastening the two points together at the back.

The manner of issuing the pay of officers and men in Ireland and of paying the charges against them was similar to that followed in England, but there were differences in the amounts of both pay and charges. The greatest difference of all arose from the fact that in Ireland soldiers were lodged in barracks and received bedding, fuel, and light. They therefore escaped from paying a billet master, but the privates only received three-fourths of the pay they received in England.

The barracks, if similar to those sanctioned for Scotland in 1716, had rooms 18 feet by 17 feet arranged for five beds, each occupied by two men.

The bedding for two men was: one bed, two pairs of sheets, one blanket, one rug, one bolster. Clean sheets were supplied monthly, but the men paid the barrack master for washing. The custom of two men occupying one bed lasted for another hundred years.

The pay on the Irish establishment was:

	s.	d.
Colonel, as colonel and captain	20	0
Lieut.-Colonel as lieut.-colonel and captain	15	0
Major, as major and captain	12	0
Captain	8	0
Lieutenant	4	0

	s.	d.
Second Lieutenant or Ensign	3	0
Chaplain	6	0
Adjutant and Quartermaster (one officer for both duties) .	4	0
Surgeon	4	0
Mate	2	6
Sergeant	1	6
Corporal	1	0
Drummer	1	0
Private	0	6

The colonel was allowed six servants, lieut.-colonel, major and captains, three each, the lieutenants, ensigns and quartermaster one each ; an allowance of 6*d.* for each servant was given the officers. The servants were clothed as the other men of the regiment, or slop clothing was issued to them at their choice.

In a couple of years after the regiment was formed at Dublin a Royal Order directed that for each year's off-reckonings the soldier should receive a full clothing instead of, as heretofore, a clothing and a half for every two years. A clothing consisted of one large full-bodied coat, one pair of cloth breeches, one waistcoat, hat, shirt, cravat, one pair of stockings and one pair of shoes. Patterns of the material had to be submitted by the contractor to the chief Government Undertaker and to the Board of General Officers, by whom, if approved of, they were sealed : this is the origin of " sealed pattern."

The officers' pay was subject as formerly to a deduction of 1*s.* in the £, nominally for Chelsea, but a first charge against the fund was the expense of the Paymaster-General's office : there was also a stoppage of twelve days' pay a year for widows' fund, and one of 6*d.* in the £ (4*d.* for agency and 2*d.* for chequer fees). There were also charges peculiar to Ireland. Officers arriving in the country had to pay a fee to the Chief Secretary for recording their commissions. A still heavier charge was made for all promotions or fresh commissions. The Chief Secretary also received a fee from all officers going on leave. Even non-commissioned officers going on furlough had to fee the Secretary. All officers had to fee the Secretary and his clerk for passing the quarterly musters for themselves and their companies. On the other hand, if a tradesman wished to sue an officer he had to pay a fee for permission to do so.

The pay of the non-commissioned officers and men was allotted as follows : Sergeants—subsistence, 9*d.* ; off-reckonings, 6*d.* ; clearings, 3*d.* ; total, 1*s.* 6*d.* Corporals—subsistence, 6*d.* ; off-reckonings, 3*d.* ; clearings, 3*d.* ; total, 1*s.* The drummers the same as the corporals, but as their clothes were more expensive, the clearings were only 2*d.* Privates—subsistence, 4*d.* ; off-reckonings, 2*d.* ; total, 6*d.* All ranks had to pay weekly to the company contingent fund ; sergeants, corporals and drummers, 3*d.*, and privates, 1*d.* The drummers had each to pay the drum-major 1½*d.* a week.

This left the sergeant with 5*s.*, the corporal with 3*s.* 3*d.*, the drummer with 3*s.* 1½*d.*, and the private with 2*s.* 3*d.* a week to provide for board, upkeep of kit and pocket-money. As far as we can judge, one pound of bread, half a pound of salt meat with greens and small beer could be supplied daily for 1*s.* 2*d.* a week. If this is right, the private soldier in Ireland would be pretty much on an equality with his comrade in England, who out of his subsistence of 3*s.* 6*d.* paid weekly 2*s.* 4*d.* for lodging and board to his billet master, and 1*d.*

to the company fund. Prices, however, were rising fast in Ireland and in a few years the pay was found to be inadequate. The meat ration for many years was a four-pound piece of salt meat per man issued weekly.

The contingent expenses of a company were part of the unavoidable cost of carrying on a regiment for which no provision was made by Parliament. To find the money required, Government sometimes authorized the captains to draw pay for men who did not exist and cover it by entering fictitious names on the muster rolls (widows' men).

When the 30th came to Ireland one widow's man was borne on the strength of each company, but in 1716 all fictitious names were swept away and the actual strength of the companies reduced by four privates. By this Government got the disposal of the pay of five men per company and the money was allotted as follows: the pay of one man of each company to the widows' fund, one to the colonel, one to the agent, and two to the captain for company contingencies.

It was rarely that the captain could make both ends meet. In the first place, more than half of all the stoppages from men of his company went to hospital and medicine chest, then a third of the two men's pay went to the colonel as off-reckonings. In a bad year the remainder of his fund might be swept up by desertion and its consequences, such as replacing lost kits, advertising deserters, sending after and taking them, or bringing over recruits from England to replace them at a net cost of about £2 per man. Desertion was not of course a constant charge, but the danger was always there. Of the constant charges the heaviest was the payment of fees for musters. Some of the other charges would startle a modern captain. They have been preserved in letters and memorials from captains in different corps. One is for the pay of the sergt.-major. Unfortunately no case has been found in which it is not mixed up with other expenses and the amount given to the sergt.-major is not stated separately. Another constant expense peculiar to that day was shaving the company. For their penny a week the men had a free shave daily. One captain had to pay a sum of £1 0s. 9*d*. out of the company fund for the prosecution of a civilian who had murdered one of his men. Another paid a sovereign for nursing sick and extraordinary expenses in fluxing men. An entry of " 5*s*. for carriage of sick men a horse back, when not able to ride on a car " shows what men suffered who fell sick on detachment and had to be sent to hospital at headquarters, over frightful roads. Apparently His Majesty's mails did not run beyond Waterford, for the company fund is charged with the cost of establishing a post between that city and Duncannon Fort, where there was a detachment. The coinage must have been in a very bad state, for the loss on receiving light gold is a constant charge. Another expense was fetching powder. The Duke of Marlborough, who attached great importance to fire discipline, had sanctioned the issue of four barrels of powder to each regiment for blank firing, but the Ordnance only sent it to the nearest point where the department had an issuer and the captains had to fetch it from there for their companies. There were numerous other expenses, but enough has been done to show how easy it was for a captain to be several pounds out of pocket at the end of the year through no fault of his own.

Swords, bayonets, belts and bandoliers were supplied by the colonel and were expected to last three years.

The first issue of firearms was made free by the Ordnance; they were afterwards kept up by the regiment. It was the same with tents.

The regiment was still recruited in England and the officers employed on that duty were sent over in autumn, returning to Ireland with their recruits in April, so that they might be trained and fit to take their place in the ranks at the general's inspection in autumn, for we have reached a time when inspection was regular and camps of exercise were not unknown. The annual changes of quarters also were ordered to be timed so that the men should reach their new stations early enough to have time for training before inspection. A regiment dispersed in detachments was as far as possible to be assembled for training in the following year.

From the above it seems likely that the 30th left Dublin for dispersed quarters, as the expression was, in the spring of 1716 and was assembled in the spring of 1717 in one of the large southern garrisons preparatory to embarkation in the following year.

In the spring of 1718 the regiment was ordered to embark at Cork for Portsmouth along with the regiments of Charles and James Otway. The transports were small, the largest being only able to accommodate three companies, but at Portsmouth the three regiments were to be transferred to the fleet about to sail under Sir George Byng, with whom the regiment had served so often in the Mediterranean.

Headquarters embarked on May 7th.

MINORCA

In the year 1717, Spain made an effort to regain the position in the Mediterranean which she had lost during the late war. An expedition from Barcelona captured Cagliari and took possession of Sardinia, which had been held by the Emperor since Sir John Leake and General Wills with his marines had captured it for him in 1709. The great Spanish Minister, Alberoni, prepared a yet stronger force for the capture of Sicily in the following year. After the Treaty of Utrecht, it had fallen to Victor Amadeus of Savoy.

The British, French, Dutch and Imperial Governments protested in the strongest manner against the disturbance of the settlement arrived at after the great war, and when it was certain that the attack on Sicily would take place, Admiral Sir George Byng was ordered to take command of a fleet, to be fitted out at Portsmouth, and to proceed to the Mediterranean.

The regiments of Bissett, Charles and James Otway, were brought from Cork in transports, and transferred to the men-of-war of this fleet at Portsmouth, on May 16th, 1718. According to a state furnished at Portsmouth, the strength of each regiment was 434, all ranks included.

Colonel Cosby's regiment, 445 strong, was embarked at the same time.

Sir George Byng thought that the men of the regiments embarked were too young and weak for active service. His opinion on that would be final, and it is a pity that we do not know if his judgment applied to all the regiments, for it would settle the question to what extent the old soldiers had rejoined in 1715. He had little time for inspection.

On July 13th, Sir George Byng reached Port Mahon, in Minorca, and according to his instructions, disembarked the four regiments of Bissett, Charles and James Otway and Cosby, and taking the relieved garrison on board to serve as marines, sailed for Sicily. He there found the Spaniards had captured Palermo and Messina. As the Spanish admiral would give no assurance of a cessation of his enterprise, Sir George Byng attacked on August 11th. The result was the capture or destruction of two-thirds of the Spanish fleet.

In spite of this rough treatment, Spain did not declare war, and Europe was nominally at peace till the end of the year, when England and France issued a formal declaration of war against Spain. This war lasted till the beginning of 1720, but scarcely affected the garrison of Minorca, as the predominance of the British fleet in the Mediterranean was never threatened.

From the day the regiment landed in Minorca, its establishment was raised to 36 company officers (including colonel, lieut.-colonel and major), 5 staff, 25 sergeants, 35 corporals, 12 drummers, 45 grenadiers (one company), 407 privates in eleven companies.—Total, all ranks, 565 in twelve companies.

The two junior companies of Sir Charles O'Hara's regiment (7th Fusiliers) were transferred to Bissett's, the two junior lieutenants (the Fusiliers had no ensigns) being given the option of continuing to serve on ensign's pay until there was a vacancy among the lieutenants or of going home on half-pay. Besides the two companies, a quartermaster, ten drummers, and ten grenadiers were to be transferred. In Ireland one officer had combined the duty of adjutant and quartermaster.

We have no embarkation or disembarkation returns, but from July 13th, the date of landing in Minorca, the company officers and staff were as given below :—

CAPTAINS.	LIEUTENANTS.	SECOND LIEUTENANTS.
Colonel Andrew Bisset, Brig.-General.	James Baker, Capt.-Lieutenant.	William Pritchard.
Lieut.-Colonel Peter Bettesworth.	Folliott Ponsonby.	David Weems (Wemyss).
Major Robinson Sowle.	Peter Margeret.	John Hobart.
William Scott.	Henry Long.	Benjamin Sladden.
William Davison.	John Thompson.	Edward Stillingfleet.
Andrew Forrester.	William Cooke.	Nicholas Smith.
Francis Pierson.	Edmund Quarles.	Philip Burean.
John Jeffreys.	Thomas Burston.	John Purcell.
John Roper (Grenadiers).	Charles Rainsford.	
John Vincent.	Thomas Newdigate.	John Forrester.
Transferred from O'Hara's Regiment (7th Fusiliers).		
James Cockran (Cochrane).	Abraham Meure (Muir).	
Jeffrey Gibbons.	Chichester Hamilton.	

Chaplain Henry Pakenham.
Adjutant James Baker.
Quartermaster John Rogers.
Surgeon David Hall.
Surgeon's Mate Nicholas Terry.

The transfer in Minorca of two companies from the outgoing to the relieving regiment was one of the petty tricks forced on the army by people calling out for economy without taking the trouble to find out where it was feasible and where parsimony only caused suffering and extra expense. By giving the island a garrison on Irish pay one-fourth of the pay of the private soldiers of three or four regiments was saved, but although the men had the old forts to live in the necessaries of life were so expensive that the *4d.* a day scarcely sufficed to feed them. It followed that when a regiment received orders to embark for Minorca there was a great deal of desertion. To meet this, regiments were embarked at short notice, and to make sure of keeping up the strength two companies were detained in the island from every regiment relieved.

The Governor of Minorca, like other governors of that day, was an absentee, and the Lieut.-Governor, Colonel Kane, ruled in his name. He added *2d.* per man daily to the men's subsistence out of the Governor's Contingent Fund, and also gave the regiment paid employment on roadmaking and other public works. His great road across the island is still in use and is called Kane's Road to this day.

The regiment remained seven years in Minorca, and there was no event of the least military importance to relieve the monotony of that time.

In 1720, upon a rumour that Gibraltar was in danger, a detachment of 500 men, taken from all regiments in garrison, was sent to its assistance, but the rumour proved false and the detachment returned.

About this time the orderly officer made his appearance, not a subaltern detailed to visit the quarters of the men of his own company, but the regimental subaltern of the day as we know him. " Barrack damages " too are for the first time the subject of War Office regulations.

In 1722 the officers of the regiment or their representatives were repaid the money which they had advanced in Catalonia in 1705 to equip and maintain the transport of the marine brigade. They had spent a good deal at the same time in the care of their sick men but made no claim on that account.

Between 1720 and 1722 the regiment was rearmed. For those two years two vacancies were allowed in each company to pay for the arms. Swords were not included in the arms. They had fallen out of use, but by Royal Order were now to be worn again by all ranks. To pay for them the annual supply of clothing was put back for some months and the off-reckonings used for that purpose. The free allowance of powder for drill purposes given in England and Ireland was extended to Minorca.

The difficulty of keeping the officers with their regiments increased with the length of their stay abroad, and was the subject of many orders from the King. Mr. Henry Pelham, although as Secretary he signed the King's orders, knew as a " Parliament man," that in ticklish times no minister would refuse a supporter such a trifle as leave for a regimental officer. He says plainly that officers could not be kept abroad as long as they had influential friends at home to importune ministers for leave. He writes, indeed, himself to General Bissett, to say that Ensign Hodge of the general's regiment had been well recommended to him, and he would be obliged by Bissett granting the ensign four months' leave.

At last, whether in jest or earnest, Mr. Pelham declared that as the officers would not go to Minorca, the regiments there had better be relieved.

On June 5th, 1725, he called upon Lieutenant-Governor Kane for a statement of the strength for which transport would be required for the return of the regiments of Bissett, Charles and James Otway, to Ireland, and on August 24th, the regiments of Bissett and Charles Otway actually embarked, being relieved by Handasyde's and Tyrell's.

The following changes took place among the officers of the regiment from the day after it landed in Minorca to the close of the successful defence of Gibraltar:—

1718

Captain John Vincent to be Fort Major of Fort St. Philip. He was not seconded.	17th Sept.

1719

Richard Onslow brought into the regiment as Captain of the company of William Davison, deceased	16th Feb.
LIEUTENANTS—	
Alexander Davison brought into the regiment and posted to Pierson's company, *vice* Edmund Quarles, transferred to Wills's regiment (The Buffs)	9th Jan.
Alexander Davison appointed Adjutant, *vice* Baker resigned.	18th Feb.
William Sherman from the Royal Fusiliers to Roper, *vice* Cooke, who exchanges	25th ,,
Henry Sowle from Stanwix regiment to Vincent, *vice* Newdigate, who retired by sale of his commission	24th April
Joseph Dussaux to Roper, *vice* Sherman, transferred to Cosby's regiment.	20th Nov.
SECOND-LIEUTENANTS—	
Palmer Hodges to Vincent, *vice* John Forrester, retired	16th Jan.
John Horsman to Cochran	29th ,,
Charles Jefferies or Jefferys to Gibbons	,, ,,
James Mossman to Scott, *vice* Benjamin Sladden, placed on half-pay.	25th Feb.
Ventris Scott (returned from Holt's regiment) *vice* Pritchard, transferred to Invalids	6th May

In this year purchase was restored with the very slight restriction that in each case leave must be applied for before buying and selling. The following was the scale of prices fixed for the 30th:—

Colonel	£5,000
Lieut.-Colonel	2,000
Major	1,200
Captain	840
Capt.-Lieutenant	380
Lieutenant	250
Ensign	170
Adjutant	150
Quartermaster	150

Up to this time from the first raising of the regiment the senior subaltern had been captain-lieutenant and the adjutant and quartermaster were chosen for efficiency.

1720

CAPTAINS—

Richard Henley brought into the regiment, *vice* Andrew Forrester, transferred to Horse Guards . . .	1st March
James Baker promoted, *vice* John Jefferies, promoted into Agnew's regiment	21st May

LIEUTENANTS—

Daniel Wemyss promoted to Pierson	15th Feb.
John Horseman promoted	5th April
Transferred to Foot Guards	20th ,,
William Cooke (returned from Royal Fusiliers) to be Lieutenant and Adjutant, *vice* Alex. Davison, resigned .	5th ,,
Daniel Herring to be Capt.-Lieutenant *vice* Baker . .	21st May

SECOND-LIEUTENANTS—

Bryan I'Anson to Bettesworth, *vice* Wemyss . .	15th Feb.
Daniel Herring to Cochran, *vice* Horsman . . .	5th April
Henry Ravenhill to Cochran, *vice* Herring . . .	21st May

CHAPLAIN—

James Auchmuty, Clerk, *vice* Pakenham . . .	30th Sept.

1721

LIEUTENANTS—

James Mossman to	29th March
Robert Throgmorton to Bettesworth, *vice* Martin, promoted in Pocock's regt.	17th May
Charles Jefferies to Vincent, *vice* Henry Sowle, promoted in Tyrell's regt.	1st Sept.
Joseph Jenoure to Bettesworth, *vice* Throgmorton, promoted in Handasyde's regt.	28th Oct.
Charles Janoe de la Bouchetiere to Major Sowle, *vice* Charles Rainsford, promoted Capt.-Lieutenant in Whetham's regt.	5th Dec.

SECOND-LIEUTENANTS—

Joseph Jenoure to Scott, *vice* Mossman, promoted .	17th May
Moses La Porte to Henley, *vice* Hobart, deceased .	5th July
William Orfeur to Bettesworth, *vice* I'Anson . .	12th ,,
Charles J. de la Bouchetiere to Gibbons, *vice* Jefferies, promoted	1st Sept.
George Lovell to Scott, *vice* Jenoure, promoted . .	28th Oct.
Charles Cotterell to Gibbons, *vice* Bouchetiere, promoted.	5th Dec.

1722

CAPTAINS—

William Dawes, *vice* Baker. Captain Dawes is not the former adjutant of the regiment, but a man nearly twenty years younger . . . . 4th May
William Cooke, *vice* Vincent . . . . . 8th Sept.

LIEUTENANTS—

Hon. Henry Vaughan to Bettesworth, *vice* Jenoure, resigned . . . . 4th May
Charles Blunt to Tracy . . . . 3rd Sept.
Charles Jefferies appointed Adjutant, *vice* Cooke, promoted . . . . 7th Sept.
Chichester Hamilton, Quartermaster, *vice* Rogers, deceased 28th April

1723

Richard Barnewall Waller from 3rd Foot Guards, to be Captain, *vice* Richard Henley, who retires . . . 5th March

1724

CAPTAIN—

Daniel Herring, *vice* Tracy, who retires . . . 8th Feb.

LIEUTENANTS—

Henry Ravenhill "to be Captain-Lieutenant of that company whereof he himself is Captain" (meaning the Colonel), *vice* Herring, promoted . . . . . 8th Feb.
William Palmer to Bettesworth, *vice* Vaughan, transferred to Grove's Foot . . . . 16th March

SECOND-LIEUTENANTS—

David Brevet to Bettesworth, *vice* Ravenhill, promoted . 8th Feb.
James Abercromby to Dawes, *vice* Purcell, retired . . 7th May
Thomas Baldwin to Roper, *vice* Smith deceased . . 22nd June

No appointments in 1725.

1726

SECOND-LIEUTENANT—

Alexander Hutchinson to Pierson . . . . . 7th May

Chaplain Henry Smart from Cosby's regiment, *vice* James Auchmuty, who exchanges . . . . . . 21st Nov.
Quartermaster George McLaughlin . . . . . . 23rd March
Surgeon's Mate James Ramsay, date of appointment not known, joined towards the end of December at Gibraltar.

1727

Brigadier Andrew Bissett to be Major-General . . . 3rd March

When the regiment left Minorca, it transferred the two junior companies and thirty-seven men from other companies to Tyrell's regiment along with the following officers, Captains Richard Barnewall Waller and David Herring,

Lieutenant Joseph Dussaux, Second-Lieutenants James Abercromby and Thomas Baldwin and Quartermaster Chichester Hamilton.

The appointments of officers in the years 1715–19 are to Wills's, Forrester's or Bissett's Foot; in 1720–22 to Bissett's Fusiliers; 1723–24 to Bissett's Foot; and in 1726 to Bissett's Fusiliers again.

Among the officers who left the regiment during the period between its landing in Minorca and the siege of Gibraltar were three whose services call for mention.

Captain William Davison, who died in 1719, was the last of five brothers who had served in the regiment, and like all of his family had the knack of being present wherever there was a chance of distinction. He was at the capture of Gibraltar and its heroic defence, at the capture of Barcelona, and as there is no trace of his having gone home, it may be presumed that he was present with his company at the victory of St. Estevan and the relief of Barcelona in 1706. He was certainly serving in Catalonia in the winter 1706–7, and he signed the petition to Lord Galway to be allowed to share in the Almanza campaign, so there is no reason to doubt that he served in the campaign of 1707, including the defence of Lerida. In 1708 he commanded the party of the 30th on board H.M.S. *Salisbury* when it captured the French *Salisbury* in Sir George Byng's defeat of Fourbin in the Firth of Forth; he was therefore not at the capture of Minorca, though some of his (the Colonel's) company were.

Lieutenant Thomas Newdigate, who was allowed to sell his commission, although he had only sixteen years' service, had shared in the capture and defence of Gibraltar and in the capture of Barcelona. In November, 1705, he was appointed to a company employed in guarding prisoners of war in Kent, and it is uncertain whether he went home before St. Estevan in January, 1706, or was present at that action. Towards the end of 1706 his company, now under Hugh Palliser, landed at Alicante, and in 1708 was sent to reinforce Denia, on the fall of which place both he and his captain became prisoners.

Charles Rainsford, promoted into Whetham's regiment in 1721, had been an officer in the 30th for sixteen years, and was now about twenty-three years of age. His later career was worthy of the valiant family to which he belonged. After being many times wounded in action, he finished his life as Major of the Tower, where there is a monument to him in the Chapel of St. Peter ad Vincula.

The Rev. J. Auchmuty alone among Mediterranean chaplains stuck to his post from the day of joining and was created Dean of Minorca. He may have exchanged to remain in the island.

GIBRALTAR

For some time the Spanish Government, which naturally had never really acquiesced in our retention of Gibraltar, had been pressing the British Government to restore it. Their tone gradually became so menacing as to cause alarm for the safety of the place, held by only three regiments each of a nominal strength of 438 of all ranks. The Earl of Portmore, the Governor, and his lieutenant, Colonel Jasper Clayton, were in England.

Orders were therefore sent to Colonel Kane to reinforce Gibraltar and to go there himself and take command.

Our alliance with France, and command of the sea, made Colonel Kane easy in his mind about Minorca, so leaving Handasyde's and Tyrell's to hold the island, he took Cosby's regiment with him to Gibraltar, where he was appointed Deputy-Governor. By his order, Bissett's, which was on its way home, was stopped and disembarked and ordered to complete to Gibraltar strength from Charles Otway's, which continued its voyage to Ireland. The establishment from August 24th, 1725, was to be ten companies of 1 captain, 1 lieutenant, and 1 second lieutenant, 2 sergeants, 2 corporals, 1 drummer, 34 privates, except in the case of the grenadiers who had 2 lieutenants and no second-lieutenant.

When the regiment joined the garrison of Gibraltar, which was on British pay, it was entitled to receive pay at the same rate, and this increase was sanctioned in England from the day after it left Minorca. The Irish Treasury, on hearing this, rightly or wrongly struck the 30th off the Irish establishment, but on March 23rd, 1726, a Royal Order was issued to " replace and continue " it upon the Irish establishment from the date of its leaving Ireland. The plan was that the British Treasury should pay the regiment on the British scale and be repaid by the Irish Treasury the amount it would have received at the Irish rate of pay.

Colonel Kane, with his usual energy, began from the time of his arrival in 1725 to put the fortress in a state of defence and sent home plans for barracks and a hospital, both of which were much needed. Gibraltar had never recovered from the destruction of 1704–5 and the regiment was quartered in the half-ruined houses. On March 28th, 1726, a Royal Order was issued to the Master-General of the Ordnance, to provide bedding for " our regiment of Foot commanded by Andrew Bissett, Brigadier-General, now at Gibraltar, strength, 20 Sergeants, 20 Corporals, 10 Drummers, 340 Privates." This order uses the expression regiment of foot, but on the same page of the entry book Bissett's is termed a regiment of fusiliers.

In December the garrison was strengthened by the arrival of Disney's, Anstruther's and Newton's regiments and 120 men picked out as the strongest of Colonel Fielding's regiment of invalids at Portsmouth. The invalids were distributed among the four regiments already in garrison.

There are no states or musters to show what officers took part in the defence of the place, but a correct list of the officers of the regiment during the siege is given below. All who were not sick or on other duty must have been present. The service was popular at home and numbers of volunteers flocked to Gibraltar to take part in the defence. There is in the Colonial Office a " List of forces in the Garrison of Gibraltar, 27th April, 1727," showing the 30th as having 35 officers, 20 sergeants, 20 corporals, 10 drummers, 340 privates, but that appears to be the establishment, not the number actually present. Thirty-five is the total number of officers in the regiment, including General Bissett, who we know was not present.

FIELD OFFICERS AND CAPTAINS.	LIEUTENANTS.	SECOND LIEUTENANTS.
Col. Andrew Bissett.	Henry Ravenhill. (Capt.-Lieutenant)	Ventris Scott.
Lieut. - Colonel Peter Bettesworth.	William Palmer.	David Brevett.
Major Henry Sowle.	Charles Jefferies, Adjt.	Palmer Hodges.
		George Lovell.

FIELD OFFICERS AND CAPTAINS.	LIEUTENANTS.	SECOND LIEUTENANTS.
Capt. William Scott.	Henry Long.	Alexander Hutchinson.
Francis Pierson.	David Wemyss.	Moses La Porte.
John Roper.	Charles Blunt.	William Orfeur.
James Cochran.	Abraham Muir.	Charles Cotterell.
Jeffrey Gibbon.	Peter Margaret.	Edward Stillingfleet.
William Dawes.	Thomas Burston.	
William Cooke.	James Mossman, Charles Bouchetiere, Grenadiers.	

Chaplain Thomas Smart, Quartermaster George McLaughlin, Surgeon David Hall, Surgeon's Mate James Ramsay.

A great deal of our knowledge of the siege of Gibraltar is derived from "An impartial account of the late famous siege of Gibraltar by an officer who was at the taking and defence of Gibraltar by the Prince of Hesse, of glorious memory, and served in the town during the last siege." London 1728.

As the 30th was the only regiment present in Gibraltar in 1727, and at its taking and defence by the Prince of Hesse in 1704–5, it has been concluded, very reasonably, that an officer of the regiment was the anonymous author of this account.

Two officers of the regiment might possibly have written it but, as far as our present knowledge goes, neither of them exactly fits the description. Captain William Scott was serving in Holt's regiment in 1704–5, but although Holt's served both in the capture and defence of the fortress, de Sidiere's company, in which Scott was a second lieutenant, does not seem to have been there. Scott may, however, have been sent out to the companies at Gibraltar without our knowledge.

Captain Thomas Burston was appointed second lieutenant in Bedford's company at Gibraltar on September 29th, 1704, three months after the capture of the place, and served in the defence. He may when appointed have been serving as a volunteer and so have shared in the capture. The account certainly is an impartial one and shows no more interest in one regiment than another. For the purposes of regimental history a little of Colonel George Burston's cheerful egotism would have improved it. There are several other accounts: among them is one from Colonel Guise of the First Guards, written for the information of General Wills, who was now colonel of that regiment, but none of the narrations varies much from the "Impartial Account."

In the beginning of 1727, the Spanish preparations were complete. An army of 20,000 men was assembled for the siege of Gibraltar, and equipped with a stronger proportion of artillery than had been employed up to that time. The Count de las Torres was placed in command. He had fought against the regiment twenty years before. Towards the end of February, the Spanish army had closed in on Gibraltar, but war had not been declared.

On February 22nd, Colonel Jasper Clayton, the Lieut.-Governor, who had arrived from England with 1,000 men on board Admiral Sir Charles Wager's fleet, wrote to the Count de las Torres to call upon him to desist from forming trenches in front of Gibraltar. Colonel Clayton threatened, in case of non-compliance, to take suitable action.

The answer was considered evasive by Clayton, and at 4 p.m. on that day he caused a shot to be fired over the heads of the Spanish working parties, and this producing no effect, after an hour's grace, he fired into them.

The Spaniards had no guns mounted against Gibraltar, but they opened fire on Sir Charles Wager's fleet in the Bay; at the same time they moved 2,000 men to the east of the Rock, perhaps with a view to an attack over the Middle Hill, as Villadarias had tried in the former siege. The *Dursley* Galley and the *Tiger* fired on this party, and shells and stones were rolled down the back of the Rock, but although the Spaniards suffered severely, they were not dislodged.

On the 24th the first parallel was opened on the North Front and continued to the sea on the east. The artillery of the fortress, which had been strengthened by heavy guns landed by Sir Charles Wager, kept up a constant fire to prevent the advance of the trenches and formation of breaching batteries, but the Spanish commander showed no wish to come close enough to breach until he had subdued the defenders' fire. For this he employed heavy pieces and mortars at long range.

On March 4th he pushed 100 yards nearer the works and formed a battery of twenty-two guns and fourteen mortars, but he was still too far off for breaching. His work had been retarded by heavy rain. One of his difficulties was that the mole running out into the bay in a north-westerly direction had been considerably lengthened since the last siege and mounted with heavy guns which enfiladed any approaches within 1,500 yards of the Rock. Modern guns could rake the mole from the north shore of the bay, but no gun of that date could carry so far.

On March 10th a Spanish battery of twelve pieces was advanced nearer the town to endeavour to dismount the guns on the mole.

Meanwhile, Sir Charles Wager and the British fleet had not been idle. Leaving a few ships to assist in the defence, he swept the whole coast, and cut off the enemy's supplies.

In the following weeks, the Count de las Torres still pursued his policy of trying to damage the defences and dismount the guns of Gibraltar by long range fire and the garrison was much worn with incessant duty and the labour of repairing at night the damage done by day, but on April 21st the Governor, the veteran Earl of Portmore, arrived with reinforcements, including a battalion of the Guards. Lord Portmore was now over 80 years of age. In King William's wars, as Sir David Collier, he had commanded a regiment of the old Scottish army which was often brigaded with Castleton's.

The garrison was now ample, the work no longer pressed heavily on the men and, as extra pay of 6*d*. a day was given, it was cheerfully performed. The policy of playing at long bowls had been fatal to the Count de las Torres, and he had no longer a chance of success.

On April 26th, having completed a number of new batteries, he brought all his artillery into play, and is said for fourteen days to have fired 700 shot and shell per hour, employing 92 guns and 72 mortars. Out of 60 guns mounted in the defence, 23 were dismounted in one week, but they were always replaced, and at the end of his great effort his rapid high angle fire had ruined his artillery. The following description of the fire of the new batteries on the 26th and succeeding days is from the "Impartial Account."

"They broke out upon us all at once with a most terrible fire and continued it with so little intermission that for some time we seemed to live in flames; they dismounted our cannon every day and though by unwearied labour and application we always repaired the damage in the night and remounted our guns, yet their batteries answered so well that we were daily for many hours exposed to the besiegers' fire without being able to return it. . . . But if, as often happened, we could do little else, we always showed them how willing we were to be otherwise engaged, how fain we would have been at 'em. . . .

"They pushed the siege as if the fate, not of Spain only but of all Europe depended upon their success in it. But not all they could offer, not their rage, their threats, their prodigious fire, their projected general assault, though within a few hours of being put in execution, could so much as create an alarm among us or break the established courage of our men, but to the last both officers and soldiers appeared in their posts, they equally exposed themselves in places of the utmost danger and mounted the guard in as sedate and cheerful a manner as they could do at Whitehall or ever did at Gibraltar in times of the most established peace."

On May 28th, King George the First's birthday, the defenders established a marked superiority, the enemy's principal magazine was blown up and the gabions and fascines set on fire.

On June 12th the siege was at an end.

The defence, though not for a moment to be compared to that of 1704–5 under the heroic Prince of Hesse, or to the great defence of Lord Heathfield, 1779–82, was creditable to the defenders, especially to the regiments of the original garrison, which were few in numbers for the work they were called on to do. Nearly 1,200 officers and men were on duty day and night.

The following was the loss of the garrison, which at its greatest strength consisted of twelve regiments and, with artillery and engineers, numbered 5,500 men, and of the 30th which may have had 375 effectives, all told:—

Killed.	*Wounded.*	*Died of Wounds.*	*Deserted.*
60	179	23	14
8	14	4	0

No officers of the regiment were killed, nor, as far as we know, wounded, but from the relative numbers of killed, wounded and died of wounds it would appear that only the very severely wounded are included in the return; those treated in regimental hospitals are omitted.

The Spanish losses are said to have been over 7,000, of whom 5,000 died from sickness and want. Sir Charles Wager's operation had made it difficult to supply the Spanish army, but the terrible loss was mainly due to the apathy or ineptitude of their Government.

Lieut.-Colonel Peter Bettesworth commanded the regiment throughout, and Major Robinson Sowle was second in command.

From the high rate of working pay the garrison had plenty of money, but the exorbitant price of beer seems to have weighed heavily upon them. Small beer was 8*d.* a quart and strong beer of Bristol 16*d.* a bottle of a pint and a half. Bad wine could be got at 5*d.* a pint.

When the siege was raised negotiations were proceeding between the English and Spanish Government and it was hard to say whether the nation were at war or not. In order to have his position clearly defined, the Governor selected Major Robinson Sowle to proceed to Madrid. The war was practically over and the English blockade of the Spanish South American ports had been withdrawn, but the Treaty of Seville was not actually signed till 1729. The death of George I, which took place on his journey to Hanover on June 19th, 1727, may have contributed to the delay. He was succeeded by his son, George II.

When George I died General Wills, who had been appointed a Knight of the Order of the Bath, revived in 1725, was in daily expectation of receiving a peerage and promotion. It is said that his commission of Field-Marshal was made out and only awaited the Royal signature. George II did not carry out his father's intentions.

In consequence of the certainty of Peace the garrison of Gibraltar was reduced, and the 30th ordered to Ireland, where it landed on May 2nd, 1728, and was quartered at Kinsale.

There was at this time a curious expression in use. The companies other than the field officers' and grenadier were "private companies." Thus it might be said that the 30th, consisting of the colonel's, the lieut.-colonel's, the major's, the grenadier, and six private companies had returned to Ireland.

The difficulty in giving a correct succession of officers in the regiment during the reigns of the first two Georges is very great. In 1727 the Secretary of War confessed that the War Office had no lists of the officers of the army with the dates of their commissions. The officers were called on to supply the necessary information within six months, and a promise was made that in future the Secretary of War and the Commissary General of Musters would each compile and keep a correct list. In spite of this, in the year 1735, when there seemed danger of war, the King was told by the Secretary that "it is not possible from the books at the War Office nor from those of the Commissary General to make an exact list of the officers, their commissions not having been regularly entered at either place." He could not even give a trustworthy list of field officers. The information required in order to frame a correct list was called for from commanding officers in all British possessions and the printed army list of 1740 was the result. It is not quite correct but very useful when collated with other sources of information. The MS. army list of 1745 is better as far as the 30th is concerned.

The late Mr. Charles Dalton, who made the subject of our army lists his own, and who at the time of his death was preparing a list of the army of George II, frequently doubted if he had attained accuracy.

The lists of appointments and promotions in the 30th have been compiled with care from many sources, and it is hoped that they are accurate, although most of the information is derived from the books which the Secretary of War condemned in 1735, and their successors, in which there is no very marked improvement up to the date when all appointments were published in the *London Gazette*.

The appointments and promotions from the termination of the siege of Gibraltar to the year 1742, when the regiment was once more warned for active service and its strength increased, were as follows:—

1728

Thomas Gordon to be Major, *vice* Robinson Sowle, who exchanges to the 11th Foot . . . . 12th Feb.
Major Robinson Sowle succeeded to the command of the 11th and led it at Fontenoy. He had then forty years' service, having joined Hans Hamilton's Regt. in 1704.
William Orfeur to be Lieutenant, *vice* William Palmer, deceased 1st ,,
James Ramsay to be Second Lieutenant, *vice* Orfeur. . ,, ,,
Ramsay is probably the Surgeon's Mate, who would lose his appointment when the regt. returned to Ireland.
George Lestanquet to be Second-Lieutenant, *vice* Cotterel . 1st March

1729

Henry Ravenhill to be Captain, *vice* William Scott, deceased 14th June
Thomas Burston to be Capt.-Lieutenant, *vice* Ravenhill . ,, ,,
Edward Stillingfleet to be Lieutenant, *vice* Burston . . ,, ,,
David Hall to be Second Lieutenant, *vice* Stillingfleet . ,, ,,

Captain Scott's services had deserved more advancement than he had received. They included the capture of Barcelona in 1705, victory of St. Estevan and relief of Barcelona in 1706, including special service with a detached company of grenadiers in rear of French army. Promoted captain by Lord Galway, brigade-major of marines in 1707 in Catalonia and during the defence of Lerida, mentioned in despatches and brevet of major. Second lieutenant David Hall must have been nearly related to David Hall, the regimental surgeon.

1730

Ralph Bendish to be Lieutenant . . . . . . 17th March
Peter Burjaud to be Captain, *vice* William Cooke, who exchanges to Second Foot Guards . . . . . 1st Nov.

Captain Cooke had served under Charles Davison at the capture of Annapolis in 1710.

1731

Charles Davenant to be Second Lieutenant, *vice* Lovel . 30th March
Palmer Hodges to be Lieutenant, *vice* Wemyss . . . 19th Aug.
Lord James Maitland to be Second Lieutenant, *vice* Hodges ,, ,,

1732

Richard Harward to be Lieut.-Colonel, *vice* Bettesworth, appointed Lieut.-Governor of Jersey . . . . . 29th May
William Ball to be Second Lieutenant, *vice* Hutchinson, deceased . . . . 1st June
Francis Pierson to be Major, *vice* Gordon, deceased . . 27th Sept.
Thomas Burston to be Captain, *vice* Pierson . . . . ,, ,,
Henry Long to be Captain-Lieutenant, *vice* Burston . . ,, ,,
Ventris Scott to be Lieutenant, *vice* Orfeur, deceased . . ,, ,,
George Jocelyn to be Second Lieutenant, *vice* Scott . . ,, ,,
Moses La Porte to be Lieutenant, *vice* Long . . . 28th Sept.
Thomas Parsons to be Second Lieutenant, *vice* La Porte . ,, ,,
William Sinclair to be Second Lieutenant, *vice* Lestanquet . 20th Dec.

Colonel Bettesworth, who left after commanding the regiment for fourteen years, was a gentleman of good estate and family in Hampshire, who had served his county as Member of Parliament and Justice of the Peace. He was a captain in Evans's Foot as early as 1706.

1733

No appointments or promotions.

1734

Owen Ormsby to be Second Lieutenant, *vice* Parsons . . 1st Sept.
Charles Jefferys to be Captain, *vice* Roper . . . . 1st Nov.
Henry Meggs to be Lieutenant, *vice* Jefferys . . . . ,, ,,

The above is how it stands in the War Office books. Captain Roper had served at Alicante in 1707 and in the expedition to Toulon under Sir Cloudesley Shovel. His retirement was a curious transaction. He was allowed to exchange to half-pay as a lieutenant with Henry Meggs. Jefferys was promoted in his place and Meggs took the place of Jefferys in Burston's company. The State saved the difference between a captain's and a lieutenant's half-pay. Both Jefferys and Meggs probably compensated Roper.

1735

No appointments or promotions.

1736

Peter Margeret to be Capt.-Lieutenant . . . . . 28th Jan.
David Brevett to be Lieutenant, *vice* Margeret . . . ,, ,,
Richard Fitzgerald to be Second Lieutenant . . . . ,, ,,

1737

Peter Margeret to be Captain, *vice* Thomas Burston, deceased 26th Aug.
Abraham Meure (Muir) to be Capt.-Lieutenant, *vice* Margeret ,, ,,
James Ramsay to be Lieutenant, *vice* Meure . . . . ,, ,,
Hayman Rooke to be Second Lieutenant, *vice* Ramsay . ,, ,,
William Stuart to be Second Lieutenant, *vice* David Hall, deceased . . . . 27th Dec.
Robert Hume to be Chyrurgeon, *vice* David Hall, deceased . 17th Aug.

Thomas Burston, the eldest son of Colonel George Burston, was appointed in 1704 to the company of Captain Bedford, an old comrade of his father in King William's wars. Under Bedford he took part in the defence of Gibraltar in 1705 and the capture of Barcelona in 1706. Late in that year he was promoted into the company of Isaac Knight. Knight died at Lisbon in 1707 but the company remained in Catalonia till 1708, and Burston must have seen a good deal of hard service.

1738

Nicholas Romaine from half-pay to be Captain, *vice* William Dawes, who exchanges . . . . 14th Jan.
Abraham Meure (Muir) to be Captain, *vice* Jeffry Gibbons, deceased . . . . 14th Aug.
James Mossman to be Capt.-Lieutenant, *vice* Meure . . ,, ,,
Charles Davenant to be Lieutenant, *vice* Mossman . . ,, ,,
Henry Westenra to be Second Lieutenant, *vice* Davenant . ,, ,,

1739

James Mossman to be Captain, *vice* Cochran, lieut.-colonel in Oglethorpe's regiment	1st March
Charles Bouchetiere to be Capt.-Lieutenant, *vice* Mossman .	,, ,,
Lord James Maitland to be Lieutenant, *vice* Bouchetiere .	,, ,,
Francis Pierson to be Second Lieutenant, *vice* Maitland .	,, ,,
James Gisborne to be Second Lieutenant, *vice* Fitzgerald promoted	1st June

1740

John Wynne to be Major, *vice* Francis Pierson, promoted lieut.-colonel in Otway's regiment	8th Jan.
George Jocelyn to be Lieutenant, *vice* Maitland, promoted in Armstrong's regiment	15th ,,
Samuel Bagshawe to be Second Lieutenant, *vice* Ball. .	,, ,,
William Hammond to be Second Lieutenant, *vice* Jocelyn .	16th ,,
Charles Bouchetiere to be Captain *vice* Meure (Muir), deceased	5th Feb.
Edward Stillingfleet to be Capt.-Lieutenant, *vice* Bouchetiere.	,, ,,
Owen Ormsby to be Lieutenant, *vice* Stillingfleet . .	,, ,,
Richard Harward to be Second Lieutenant, *vice* Ormsby .	,, ,,
Amyas Bush to be Second Lieutenant, *vice* William Sinclair, deceased	10th May

Lord James Maitland succeeded his father as seventh Earl of Lauderdale in 1744.

1741

Henry Ravenhill to be Major, *vice* Wynne	16th Feb.
William Ball to be Captain, *vice* Ravenhill	,, ,,
Edward Stillingfleet to be Captain, *vice* Margeret, deceased .	13th March
Ralph Bendish to be Capt.-Lieutenant, *vice* Stillingfleet .	,, ,,
Hayman Rooke to be Lieutenant, *vice* Bendish	,, ,,
Robert Waller to be Second Lieutenant, *vice* Hayman Rooke	,, ,,
Edward Bermingham to be Second Lieutenant, *vice* Bagshawe, promoted	,, ,,
Peter Chester to be Second Lieutenant, *vice* Westenra, promoted	,, ,,
Richard Harward to be Adjutant	,, ,,
Francis Pierson to be Lieutenant, *vice* Ventris Scott, promoted in Hargreave's regiment	6th June
William Southwell to be Second Lieutenant, *vice* Pierson .	,, ,,
Thomas Margeret to be Second Lieutenant, *vice* William Stuart, promoted in Hawley's regiment	,, ,,
Thomas Stowe to be Second Lieutenant	,, ,,
James Gisborne to be Lieutenant	,, ,,
Captain George Burston from the half-pay of Tyrrel's regiment to be Captain, *vice* Romaine	12th Nov.
William Hammond to be Lieutenant	,, ,,

George Burston, the youngest son of the hero of S. Estevan, returned to the regiment after twenty-six years.

1742

Henry de Grangue to be Colonel, *vice* M. G. Bissett, deceased	2nd Oct.
Louis Marcel to be Captain, *vice* Charles Jefferys, promoted in Battereau's newly raised regiment	2nd April
John Wright to be Lieutenant, *vice* Palmer Hodges, promoted captain in Richbell's regiment	30th ,,
Richard Harward to be First Lieutenant	1st May
Richard Harward to be Captain, *vice* Mossman, promoted	31st Aug.
Thomas Smyth to be Lieutenant, *vice* Harward	,, ,,
Lieutenant James Ramsay to be Adjutant, *vice* Harward	,, ,,

Smyth's promotion does not appear in any commission book, but in a list of officers furnished by Major Stillingfleet to the Irish Government in 1749.

ADDITIONAL OFFICERS ON AUGMENTATION

To be Second Lieutenants—

George (or John) Coghlan, *vice* Bermingham	15th May
Robert Owen, *vice* Hammond	16th ,,
Henry Loftus, *vice* Peter Chester	17th ,,
David Roche, *vice* Stowe	18th ,,
Andrew Paul, *vice* Harward	19th ,,
Charles Chapman	,, ,,

To be Ensigns—

William Davis Bruce	20th ,,
Robert Ross	21st ,,
Henry Monroe	22nd ,,
Henry Rugge	23rd ,,
Apsley Newton	24th ,,
Teavil Appleton	25th ,,
Martin Bennet	26th ,,
Thomas Head	27th ,,

Andrew Bissett joined the Scots army as an ensign in Dumbarton's Foot (Royal Scots) in 1688, but was promoted lieutenant of the grenadier company of the Coldstream Guards in the same year. He was captain and lieut.-colonel on January 1st, 1697; brigadier, January 1st, 1710; colonel of 30th Foot, August 24th, 1717; major-general, March 3rd, 1727; lieut.-general, October 28th, 1735; died August 22nd, 1742, aged 82, and was buried in Westminster Abbey. He had no war service while in the 30th, but had served in Flanders under King William, and was wounded at Landen. He served also in Spain under Lord Galway and commanded the Coldstreams at Almanza. When capt.-lieutenant in the Coldstreams his colonel and captain was Lord Cutts, to whom Steele dedicated *The Christian Hero*, and the ensign of the company was Richard Steele himself.

IRELAND

We know little or nothing of the life of the regiment in Ireland for ten years after its return from Gibraltar, and it would be of no interest if we did.

In all likelihood its discipline suffered the decay which affected the rest of the army under Walpole's long administration. The establishment remained the same, as we can see by a licence to General Bissett to import, from England, duty-free, a full clothing for his regiment of :—

Coats and Breeches	Centinels .	340
	Sergeants .	20
	Corporals .	20
	Drummers .	10
Shirts, Neckcloths, Stockings, Shoes		390
Hats	Centinels .	340
	Sergeants .	18
	Drummers .	10
Grenadier Caps	Sergeants .	2
	Corporals and Centinels .	36

This provides sixteen spare hats for privates. The corporal had a handsomer coat, but wore the same pattern of hat as the private. Recruits received two pairs of stockings, two shirts and a waistcoat. In succeeding years the waistcoat was made from the old coat. The breeches were of kersey. The coat was of cloth, full bodied and well lined. The hat was to be strong and well laced.

There were ten companies as before.

The number of men who were sent to Chelsea Hospital in the first half-dozen years after the regiment returned from foreign service was excessive, but there is nothing to show whether they were admitted to the Hospital or to out-pension, or invalided and discharged without pension.

A great number of recruits must have been required in those years to keep up the strength, but that seems to have presented no difficulty, judging by the standard of 1729. It is laid down that " Men not under 5 ft. 10 in. is a sufficient size for the Horse and Dragoons and that they be chosen with good countenances, good limbs and broad shoulders and that the size of men for the Foot Guards be 5 ft. 9 in. and for the Marching Regiments 5 ft. 8 in. with shoes."

In 1734 we have the first signs that the peace was drawing to a close. The Government did not meddle with the establishment in Britain, but the 30th and nine other regiments in Ireland were ordered to increase their strength by 1 sergeant, 1 corporal, 1 drummer and 26 men a company, and to be ready for transfer to England. The danger blew over and the extra men were reduced, and in 1737 General Bissett was paid £185 as compensation for the consequent loss of off-reckonings.

In 1739 it became evident that, unless Spain allowed freedom of trade to British subjects in the Southern Seas, war was inevitable. In June, the 30th was ordered to raise 1 sergeant, 1 corporal, 1 drummer and 36 privates additional per company. On October 19th war was declared in London against Spain.

In 1740, the fictitious names which had again crept back on to the muster rolls were forbidden. It is really extraordinary how they disappeared and returned again every half-dozen years or so for a century. The expense they were intended to provide for never disappeared. The pay of two men a company was now allowed for the widows and, as before, two men for the captain, one for the colonel and one for the agent.

On Christmas Day, 1741, General Sir Charles Wills died, aged 76. He was buried in Westminster Abbey. He was colonel of the first regiment of Guards, Lieut.-General of the Ordnance, Governor of Berwick, Member of Parliament for Totnes, and a Privy Councillor.

In this year Frederic the Second, King of Prussia, had attacked the Austrian Empire, whose possessions had been guaranteed by all the European Powers and a general war ensued. The British and Dutch Governments determined to fulfil the obligations of the treaties binding them to support the Empress of Austria. In 1742, Walpole was turned out of office and a force was sent to the Continent to act against the Prussians, and practically against the French, although we were still nominally at peace with them. This again led to an augmentation of the 30th.

In June the following addition was ordered :—

1 Lieutenant of grenadiers.
1 Second Lieutenant to every other company.
1 Sergeant
1 Corporal
1 Drummer
30 Private men
} To each company.

At the same time the regiment was directed to give to a regiment which Colonel Baterau was raising 1 corporal and 5 privates from each company, so that the newly-raised corps might not want men qualified to be non-commissioned officers.

The number of men in the regiment before augmentation commenced in 1739 could not have been more than 20 sergeants, 10 corporals, 10 drummers, 240 privates, for there were 50 widows' men included in the establishment, and 10 corporals and 50 privates were given to Baterau's regiment, 20 sergeants, 30 corporals, 20 drummers, and 710 privates, or 780 recruits in all had therefore to be found. The hardest part for the commanding officer was to find the 20 sergeants and 30 corporals out of the 240 privates who were left to him, and who for the most part were uneducated. It is unlikely that Colonel Harward was so pedantic as to insist upon all the corporals being able to read and write, but even in those days it was considered a useful accomplishment and one necessary for sergeants.

Unless the more energetic officers and non-commissioned officers were scattered far and wide to look for them the recruits could not be got, and without these officers, training would make small progress. There were also nineteen young gentlemen, who had received their first commissions in 1741–42, to be broken in. Mr. Henry St. John, when Secretary of War, more than thirty years before, had pointed out the injury done to the service by officers being urgently required for two different lines of duty at the same time, but neither he nor any one of the able men who filled that office ever took the trouble to devise a remedy.

The regiments in England could be made fit for the field in a much shorter time, for they started with twice as large an establishment.

In both countries a commanding officer was lucky if he did not lose a number of his most experienced sergeants as soon as there was a prospect of service. To save expense men were kept with the colours long after they should have been discharged to pension, and the first result of an order to increase the strength might be the disappearance of a number of pay-sergeants and of the regimental staff.

The new accoutrements were paid for by the State. The muskets were almost certainly paid for by vacancies. The Irish Government had been buying muskets with iron ramrods since 1726, so the new arms must have been of that pattern.

The following promotions and appointments were made from the augmentation of the regiment in 1742 till it was brought on the English establishment in 1755 :—

1743

Brigadier Charles Frampton to be Colonel, *vice* de Grangue . 1st April

de Grangue was a cavalry officer and was shortly after this commissioned to raise a regiment of cavalry.

1744

John Young from Lord Stair's regiment to be Captain, *vice* Charles Bouchetiere, deceased	26th June
Thomas Maurice (or Morris) to be Ensign.	24th ,,
John Gustavus Hancock to be Ensign	25th ,,
John Whiting to be Quartermaster (in consequence of the transfer of the regiment to England)	26th ,,

1745

Henry Ravenhill to be Lieut.-Colonel, *vice* Richard Harward, who retires	20th June
Louis Marcel to be Major, *vice* Ravenhill	22nd ,,
Ralph Bendish to be Captain, *vice* Peter Burjaud, deceased .	21st ,,
David Brevet to be Captain, *vice* Marcel	22nd ,,
James Christie from Murray's regiment to be Captain, *vice* John Young, appointed major of Bolton's regiment .	29th Nov.
Moses de la Porte to be Captain-Lieutenant, *vice* Bendish .	21st June
Amyas Bush to be Lieutenant, *vice* de la Porte . . .	,, ,,
Teavil Appleton to be Lieutenant, *vice* Brevet . . .	22nd ,,
Charles Chapman to be Second Lieutenant, *vice* Bush . .	21st ,,
Henry Bromley to be Ensign	,, ,,
Thomas Stowe to be Ensign	,, ,,
Robert Compton to be Quartermaster, *vice* Whiting, resigned.	9th Dec.
Henry Rand to be Chaplain	21st March

1746

Edward Stillingfleet to be Major, *vice* Marcel, who keeps his company	20th March
It is doubtful if Marcel ever joined the 30th. He was engineer of the train of artillery in Ireland in 1728. As a major he was doing duty as an engineer.	
James Gisborne to be Captain, *vice* Ball, invalided . .	26th Feb.

Henry Rugge to be Lieutenant, *vice* Gisborne . . . . 26th Feb.
John Hutchinson from Invalids to be Lieutenant, *vice* Davenant . . . . 17th Nov.
Robert Ross to be Second Lieutenant . . . . . 20th March
Thomas Price to be Second Lieutenant . . . . . 9th April
William Snell to be Ensign, *vice* Rugge . . . . . 26th Feb.
John Craven to be Ensign, *vice* Ross . . . . . 20th March
Henry Fayling Giles to be Ensign, *vice* Bromley . . 9th April
John Hare to be Ensign, *vice* Hancock . . . . . 16th ,,
John Kelly to be Ensign . . . .
William Hughes to be Ensign, *vice* Monroe, retired . . 30th ,,
John Mackenzie to be Surgeon's Mate . . . . . 25th June
Hugh Bayley to be Quartermaster, *vice* Compton, resigned . 31st Jan.

The date of Kelly's appointment is uncertain, but his place in the order of seniority was between Hare and Hughes.

1747

Major-General Charles Frampton to be Lieut.-General . . 12th Aug.
Timothy Bridge to be Captain, *vice* John Christie, transferred to 3rd Foot Guards . . . . 22nd May
James Ramsay to be Captain, *vice* George Burston, invalided. 13th July
George Lewis from Dalyell's to be Lieutenant, *vice* Hutchinson, deceased . . . . 9th Jan.
Andrew Paull to be Lieutenant, *vice* Ramsay . . . . 13th July
Thomas Head to be Second Lieutenant, *vice* Paull . . ,, ,,
Thomas Armstrong to be Ensign, *vice* Stowe, deceased . 19th March
David Ross to be Ensign, *vice* Robert Ross . . . . 9th May
John Minnitt to be Ensign, *vice* H. F. Giles. . . . 29th ,,
William Challoner to be Ensign, *vice* Thomas Head . . 13th July
Charles Hewett to be Surgeon . . . . 22nd June
Andrew Paull to be Adjutant, *vice* Ramsay. . . . . 13th July

1748

William Hammond to be Captain, *vice* David Brevet, who retires 11th May
Moses de la Porte to be Captain, *vice* Hammond, deceased . 11th Oct.
Hayman Rooke to be Captain Lieutenant, *vice* la Porte . ,, ,,

This name is written la Porte, Laporte or de la Porte.

George Coghlan to be Lieutenant, *vice* Hammond . . . 11th May
William Snell to be Second Lieutenant, *vice* Coghlan. . ,, ,,
John Lay to be Ensign, *vice* Bennett, retired . . . 7th ,,
Richard Ward to be Ensign, *vice* Snell . . . . . 11th ,,

1749

Brig.-General John Earl of Loudoun to be Colonel, *vice* Lieut.-General Charles Frampton, deceased . . . . . 1st Nov.
Captain James Robertson from half-pay to be Captain, *vice* Ralph Bendish, who exchanged . . . . . . 15th April
Lieutenant William Stiel from half-pay to be Lieutenant . ,, ,,
Second Lieutenant Robert Edward Fell promoted Lieutenant in the 49th company of marines . . . . . ,, ,,

1750

Sir William Boothby Bart. from the 43rd Foot to be Lieut.-Colonel, *vice* Ravenhill, retired . . . . . . 19th March

1751

Captain Richard Harward to be Major, *vice* Stillingfleet, retired . . . . 2nd May
Captain James Gisborne to be Major, *vice* Harward, deceased 19th Aug.
Lieutenant Francis Pierson to be Captain, *vice* Harward, from half-pay . . . . 2nd May
Captain James Rich from half-pay to be Captain, *vice* Gisborne 20th Aug.
Lieutenant Thomas Smyth to be Captain, *vice* Louis Marcel, resigned . . . . ,, ,,

To be Lieutenants—

Second Lieutenant William Skipton, *vice* Bush, resigned 9th Feb.
,, ,, Robert Owen, *vice* Pierson, promoted 2nd May
,, ,, James Moutray, *vice* Andrew Paull, resigned . . . . . . 20th Aug.
,, ,, Henry Loftus, promoted . . 29th ,,
,, ,, Thomas Price, promoted . . 20th ,,

To be Second Lieutenants—

John Augustine Jevers, *vice* Owen, promoted . . 2nd May
John Woodward, *vice* Moutray, promoted . . . 20th Aug.
Peter Dumas, *vice* Loftus, promoted . . . . 29th ,,
Simon Digby . . . . 20th ,,
John Wright to be Adjutant, *vice* Paull, resigned . . ,, ,,

1752

No appointments made in this year.

1753

Lieutenant Henry Rugge to be Captain, *vice* Timothy Bridges, retired . . . . 20th June
Second Lieutenant David Ross to be Lieutenant, *vice* Rugge, promoted . . . . ,, ,,
William Lushington, gent., to be Second Lieutenant, *vice* Ross, promoted . . . . ,, ,,

1754

Lieutenant George Lewis to be Captain, *vice* Francis Pierson, resigned . . . . 12th March
Second Lieutenant William Snell, to be Lieutenant, *vice* Lewis, promoted . . . . ,, ,,
Second Lieutenant Godfrey Knuttal from the half-pay of Boscawen's regiment to be Lieutenant, *vice* Ross, promoted captain lieutenant in the 5th Foot . . . . . 4th Sept.
Henry Norton Jevers, gent., to be Second Lieutenant, *vice* Snell, promoted . . . . 12th March
William MacKerral to be Surgeon, *vice* Hewett, deceased . 4th Sept.

In 1743 numbers were assigned for the first time to regiments, and clear instructions given as to the colours to be carried. Each regiment was to have two colours only, the King's and the regimental, both of silk. The King's colour was to be The Union throughout, formed of the crosses of St. George and St. Andrew. The regimental was to be of the colour of the facings, which in the case of the 30th were pale yellow or, as they were sometimes described, primrose yellow, and to have a small union in the upper corner nearest the staff. In the centre of both colours the regimental number was to be embroidered or painted in Roman characters surrounded by a wreath of roses and thistles on one stalk. The colours still were furnished by the colonel, and were his property, so we may assume that Brigadier-General Frampton, who was appointed colonel on April 1st, gave to the regiment in this year, the first colours to bear the XXX. When appointed he was commanding a brigade under Lord Stair in Germany, but returned home shortly afterwards.

At the beginning of 1744 orders were issued for the 30th to cross to England, but still to remain on the Irish establishment, the British Treasury as usual paying the difference between British and Irish pay.

On March 10th the regiment was mustered and embarked on several vessels. It was seldom at that time that any one of the transports employed could accommodate three companies, and then only when the companies were very weak. The muster showed that the strength on parade was 141 short of the full complement of 1,000 privates.

From the detached duties upon which it was employed and other causes the 30th can seldom have moved in battalion after leaving Ireland till the day, more than two years later, when it came under fire of the enemy. One of the difficulties in training a regiment was that few towns could accommodate 500 men and that in the greater part of England a company's billets might be spread over two or three parishes and those of a regiment of 1,000 men be scattered over a dozen miles of country. Until summer camps of exercise were formed, ten or twelve years later, the divisional, brigade and even battalion training of the army only began after it had taken the field.

ENGLAND

We do not know when or where the several companies landed, but from April to June, 1744, the bulk of the regiment seems to have been about Reading and Newbury, with one strong detachment at Chester and others at Northampton, Bromsgrove, Uppingham and Oxford. A pay warrant dated March 22nd, 1746 (two years later), tells us that Lieut.-Colonel Ravenhill was in command at the muster of March 10th, 1744 (he was in reality major till 1745), and that " the regiment landing in different parts of England, occasioned its not being mustered for the muster ending April 24th, 1744." The warrant says that the regiment was really nearly complete when it crossed, that a number of men appeared immediately after muster and that officers and non-commissioned officers with a great number of recruits joined it on arrival in England and that the balance was made up by men in hospital, on duty or on furlough. This is the first mention of men of the regiment receiving furlough in appreciable numbers.

On March 29th, 1744, the British Government published the long expected declaration of war against France.

The move to England may have been hastened in anticipation of the new scheme of recruiting and localization which Lord Stair, the commander-in-chief, ordered to come into force on April 11th. By it the counties of Berks, Oxford and Northampton were assigned to the 30th for recruiting, but this could only be a temporary arrangement for, although on the English rate of pay, the regiment was still on the Irish establishment, and when it returned to that country would draw its recruits from the English and Welsh counties washed by the Irish Channel, according to Lord Stair's scheme.

Lord Stair's order was accompanied by a revival of an old Act of Queen Anne's compelling all men to serve who had no visible means of subsistence. The standard was 5 ft. 5 in. for volunteers and 5 ft. 4 in. for pressed men. All pressed men were to be divided into three classes according to height. The tallest were to be given to the marching regiments, those next in height to regiments in Minorca, Gibraltar or America, and the smallest to the marines. To gather in its pressed men the 30th sent, in May, a guard of an officer and eighteen non-commissioned officers and men to each of its three counties, one to Northampton, one to Islip (Oxford) and one to Newbury (Berks). A field officer accompanied the first and last parties. The recruits for Ireland from Cumberland, Lancashire, Cheshire and North Wales were ordered to assemble at Chester for embarkation and the 30th and Richbell's regiment had each a guard of 100 officers and men there to look after them. The pressed men deserted as fast as they could and on June 11th pressing was stopped by Lord Stair's orders. Volunteers who offered themselves at the drumhead were still to be received. The ranks were now nearly full.

An integral part of Lord Stair's scheme of reform in recruiting was the establishment of two depôt companies in each regiment, but no attempt to form them in the 30th was made till 1757.

On June 1st camp equipage was issued to the regiment, and by the end of the month all detachments were called in to the Reading and Newbury district.

On August 31st an order was issued for two companies to march to Southampton and embark for Jersey. Little is known of those companies, but they must have got well on their way by September 11th, for on that day a letter was written to the Governor of the island telling him to place a proper guard on his French prisoners as soon as the two companies arrived.

Later in the day on August 31st an order was sent to the regiment directing it to march forthwith to the east coast where the smugglers and owlers had overpowered and beaten the civil magistrates, constables, and customs and excise officers, had created riots and appeared openly in arms. Similar outbreaks had taken place in Sussex and Hampshire, and a regiment had been despatched to each of those counties.

One person usually combined the trades of smuggler and owler. On his outward voyage in his capacity of owler he took a cargo of wool (owl), the export of which was forbidden by law. He returned as a smuggler with his vessel laden with French wines, brandy and lace and sometimes a cargo of arms for the Jacobites. The fact that they were both opposed to Government formed a strong tie between the Jacobites and the smugglers and the disturbances all round the coast in this year were a sign of the activity of the former, whose hopes were very high at this time. In 1743 there had been an abortive attempt at invasion by the Young Pretender and Marshal Saxe.

The distribution of the regiment in the east of England was, Norwich 2 companies, Swaffham and Dereham 1, Yarmouth 1, Lynn 1, Thetford 1, Bungay and Beccles 1, Holt 1.

The officer commanding the regiment was given power to move his men in his district as seemed best for the public service, but was enjoined not to repel force by force except in a case of necessity or when he had the sanction of the civil magistrate. As a matter of fact the companies could not be kept together. Captains had to split them up into small parties scattered in hamlets along the coast according to the requirements of the customs officers, a most demoralizing state of affairs.

1745

On April 2nd an order was issued for the regiment to find from the 8 companies in Norfolk and Suffolk a detachment of 2 captains, 8 subalterns and 400 non-commissioned officers and men for marine service. Of these, 100 were to march to Gravesend for duty on H.M.S. *Royal Sovereign*, 200 to Chatham for H.M.Ss. *Royal George* and *Prince George*, and 100 to Deal for H.M.S. *Duke*.

The detail of the detachment on board the *Prince George* only is given here, but the service and discharge of the others were similar except in the case of the *Royal Sovereign*, guardship at the Nore.

Prince George—Lieutenants Thomas Smith and John Gustavus Handcock embarked on April 25th, 1745, with 119 non-commissioned officers and men, sixty-one were discharged during the following eight months at different Channel ports, the two officers and the remaining fifty-eight men were landed at Portsmouth on December 8th under circumstances to be presently narrated. They were marked as " discharged to regiment."

The regiment had come from Ireland under Lieut.-Colonel Harward, but in the summer of 1745 he retired on 12*s.* 4*d.* a day and was succeeded by Major Henry Ravenhill.

The movements of the regiment during 1745 and the spring of 1746 were governed by the attempt of Prince Charles Edward Stuart to regain the throne of his ancestors and his invasion of England at the head of a small army of Highlanders. The following is a summary of events :

1745

August 19th. The Prince raised the standard in Glenfinnan.

August 31st. Seven British and six Dutch regiments landed in England. Fourteen new regiments ordered to be raised.

September 17th. Prince Charles captured Edinburgh.

September 20th. He routed the Royal army at Preston Pans.

October 18th. The Duke of Cumberland assumed command of the Royal army. He brought from Flanders one battalion of the Guards and eleven foot regiments.

November 17th. Prince Charles captured Carlisle.

November 27th. Reached Preston on the Ribble.

December 3rd. Reached Derby.

The Prince was now nearer London than the Duke of Cumberland's army, which had formed at Lichfield.

December 6th. The Highland leaders having a superior Anglo-Dutch army under Marshal Wade and Prince Maurice of Nassau in their rear in Yorkshire and the Duke of Cumberland, also in superior force, on their flank at Lichfield, commenced their retreat and were pursued by the Duke of Cumberland.

December 18th. Skirmish at Clifton, where Cumberland's advanced guard was checked.

December 20th. Highlanders re-enter Scotland.

1746

January 17th. The notorious General Hawley, who had succeeded Cumberland in the active command, was worsted in an action at Falkirk and made a disgraceful retreat upon Edinburgh. The Duke of Cumberland ordered to return to Scotland and take the command from Hawley.

January 31st. Highlanders retired to their own country; 5,000 Hessians under Prince Frederick reinforced Cumberland.

April 16th. Final defeat of Highlanders at Culloden.

September 20th. Prince Charles, after five months' wandering in the Highlands with a reward of £30,000 on his head, escaped to France.

On September 19th, 1745, the eight companies of the regiment in Norfolk and Suffolk were ordered to concentrate in Essex, three companies at St. Edmondsbury, three at Ipswich, two at Sudbury. From the terms of the order it is plain that the men serving on board ship had not embarked by companies but had been drawn from all.

On September 30th the regiment was ordered to march to Dover for duty there, crossing the Thames at Grays, below London. On arrival at Canterbury, however, it was halted for nearly two months; the news of the rout of Cope's army at Preston Pans had reached London and the times were too critical for regiments to be stationed where they would not be immediately available.

By the middle of November the Duke of Cumberland had assembled 8,000 men at Lichfield, and now that London was covered by the army which had fought so gloriously if unsuccessfully at Fontenoy, the King's Government seems to have thought that troops could be spared for the fleet. The sailors were deserting in numbers, and while at Canterbury Frampton's men had been urged to exert themselves in apprehending all seamen straggling and a reward had been promised for every one captured.

On November 24th the eight companies were ordered to march from Canterbury to Portsmouth, there to await orders from the Admiralty. We have not the order for their embarkation, but they, or a portion of them, certainly went on board ship.

Meanwhile the Pretender had marched south; when the regiment left Canterbury he was already in Lancashire and on December 3rd he was at Derby. Every available man was called up to defend the Capital and a hurried order was sent to the regiment to disembark forthwith and march for London with all speed, carriages to carry the men being obtained by pressing or hiring.

It disembarked on the 8th, and marched to London, the eight companies being strengthened by the detachments embarked in April on the *Prince George, Royal George* and *Duke*. The detachment of the *Royal Sovereign* did not disembark.

The regiment occupied Hammersmith, Chiswick and Turnham Green on the extreme left of the line of defence from Finchley to the Thames.

On December 18th, the Highland army having retreated and being now 300 miles from London, the 30th moved a little way to Southwark, Dartford and Gravesend, and on the 22nd still further to Sevenoaks and Croydon.

By the end of January, 1746, it was evident that the rebellion was practically over, though the Jacobite clans were still in arms and Culloden was yet to come, and from Croydon three companies marched to Lewes to assist in putting down disorders caused by owlers and smugglers who had taken advantage of the Civil War to conduct operations on a great scale, frequently appearing in arms. The companies took tents and camp equipment as if for a campaign. The tents were made of sail cloth with three coats of oil paint and were small, being adapted for five men only.

On March 4th those companies were ordered to march to Portsmouth and embark as directed by Admiral Stewart, taking their camp equipment with them. At the same time the remaining seven companies were ordered to march to Gravesend to embark on transports there.

The two companies in Jersey had evidently been brought over for the defence of London, but the orders have not been found. The detachment on the *Royal Sovereign* was relieved and joined for embarkation. The regiment was now under orders for North America.

In 1745 the British Colonial troops had captured Cape Breton, and as soon as the King's Government was relieved of its fears from the Highlands, it resolved to follow up this success by the despatch of a force against Quebec. Eight thousand men were to be raised by the American Colonies and six regiments to be embarked at Portsmouth on board transports and sail for the St. Lawrence under convoy of a strong fleet. Whether from the inefficiency of the Government or its dread of parting with such a large number of troops, the expedition was not ready to start till too late for an American enterprise and the object was changed to Lorient, the headquarters of the French West India Company in France.

A part of the force, however, had sailed in the spring of 1746 for America, and had landed in April at Louisburg, then held by two regiments formed from the New England troops, which had captured it so gallantly in the previous year. The regiments landed were the 29th and 45th and the three companies of the 30th which had moved from Lewes to Portsmouth at the beginning of March. Those companies had with them the regimental clothing for 1746. They were the colonel's, the grenadiers, and that of Major Gisborne, who was in command. The detachment remained in garrison at Louisburg for two years, and suffered severely from a want of warm clothing and of sufficient food. In 1748, when it became certain that the 30th would not arrive in America, orders were issued for the effective men of the three companies to be drafted into the 29th, and for the officers and non-commissioned officers to be sent home. In the spring of that year sixty-one suits of centinels' clothing were landed in Ireland for the use of three companies of Frampton's regiment, so we may assume that by that time some officers and non-commissioned officers had returned to Ireland to receive recruits and re-form the companies and that others proceeded to England to enlist the men. Though employed in England and on the fleet, the regiment was still on the Irish establishment.

To return to the Lorient expedition. The Headquarters of the regiment and seven companies had been embarked in April, 1746, at Gravesend and the transports had assembled at Spithead, but on June 26th General Frampton was informed that His Majesty had thought fit to employ his regiment "in the nature of marines" on board such men-of-war as the Admiralty should appoint as soon as the men had disembarked from the transports. This evidently marks the time when the idea of an American expedition was given up. An order of July 3rd cancels the above and directs that all the companies which had landed at Portsmouth should go on board the transports again. This marks the resolution to attack Lorient.

On July 12th, General St. Clair, who was appointed to command, went on board H.M.S. *Superb*. On August 23rd, the transports sailed under convoy of a fleet commanded by Admiral Lestock, but anchored again in Plymouth Sound. The expedition sailed finally from there for Lorient on September 14th. Long before the fleet sailed, Admiral Anson wrote to a friend that every one was sick of the expedition except the general and would rejoice if it was given up, and that from the long confinement on board ship the 30th was suffering from scurvy. This was no wonder, for the seven companies had been five months on board.

When it appeared that the fleet was at last going to sail, the field officer of the Guards in Staff Waiting was ordered to send a proper escort to the Savoy prison, to take from it fifty deserters of various corps incarcerated there, march them to Plymouth and hand them over to the 30th regiment on board the transports. This was successfully performed, the men being put on board on the morning of sailing, when there was no chance of their slipping away, but apparently some were so troublesome that it was felt they would be better under naval discipline, for the books of H.M.S. *Devonshire* record that on September 14th she took off from the *Winchester* transport twenty-four deserters of Frampton's regiment and kept them till October 13th, when the expedition returned to Portsmouth. The *London Post* had a story from its Plymouth correspondent that the night before the fleet sailed for Lorient some thirty of Frampton's regiment endeavoured by making a hole in the back wall to escape from the main guard, and doubtless those are the men put on board the *Devonshire* on the following day.

It was an anxious duty conducting a large number of deserters through England. The sympathies of the people were with the prisoners and attempts at rescue frequent. Desertion was not so common among men who had enlisted at the drumhead as among the pressed men, who besides the idlers without visible means of subsistence contained a sprinkling of poachers, smugglers, and men guilty of slight offences. Naturally too there were then, as always, a number of criminals of the worst class to be found among deserters. The risk of an outbreak among the prisoners assisted by sympathizers among the populace forced the military authorities to send escorts with deserters far in excess of what is now necessary, and the march of fifty prisoners from London to Plymouth must have been like the move of a couple of companies.

LORIENT

Lorient, which was unfortified, is situated about 3 miles from the open sea on the northern shore of the estuary of the river Blavet. The mouth of the estuary, three-quarters of a mile broad, is protected by the fortress of Port Louis on the southern shore. The object of the British commanders was to avoid Port Louis, land further up the coast to the north and make a dash for Lorient, which it was thought was incapable of defence.

On September 19th, the fleet and transports anchored in Quimperlé Bay, about 9 miles N.W. of Port Louis, and the alarm was given by the French all along the coast by flags and bonfires. On the following day the signal was given for the troops to land and the men were passed into the boats. Three 40-gun ships and the lighter vessels of the fleet were to cover the landing. The French had been able to assemble only some militia and armed peasants with a few guns and dragoons from Port Louis. After a little manœuvring General St. Clair was able by suddenly pulling down wind to reach an unoccupied spot and land 600 men before the French could come up. The militia then fled and the landing was completed without opposition.

General St. Clair, who headed the landing, was anxious to reassure the peasantry and induce them to lay down their arms and bring in supplies, but before he could reach the nearest village the sailors forming the boats' crews, broke away and plundered every house within reach. The British force, 4,500 strong, advanced a few miles before nightfall to a village called Guidel. On the 21st the advance was continued in two columns to Plœumur from which a road runs to Lorient, 4 miles distant. General St. Clair himself commanded the right column, which was protected by the sea and our fleet on its right and on its left by a column composed of Harrison's, Frampton's and Richbell's regiments, under General O'ffarel. It might have been expected that the left column on the exposed flank would have been carefully covered by scouts and flanking parties, but nothing of the sort seems to have been attempted. The road led through a country covered with thickets and hedges, and the column was surprised by a fire at close quarters from a hidden enemy. The men fell into great disorder and opened an irregular fire which killed some of our own men; a few threw down their arms and fled. Luckily Harrison's did not share in the panic to the same extent as the other regiments and the grenadier company of the 30th stood fast. Order was quickly restored and the enemy, who consisted of militia and armed peasants, infuriated by the conduct of our seamen, took to flight in their turn. Great care should have been taken to begin well with such inexperienced soldiers as formed a large part of our force.

On the 22nd the army advanced to within a couple of miles of Lorient and summoned it. The Governor was willing to capitulate, but General St. Clair, being assured by his engineers that the place could not hold out for twenty-four hours, insisted on an unconditional surrender.

As the navy had not been able to pass light craft up the estuary, and so cut off Port Louis from Lorient, the position now was that General St. Clair was three times as far from his base in Quimperlé Bay as Lorient was from Port Louis, and that while the Governor of Lorient had water carriage from Port Louis, all supplies for St. Clair's army had to be carried or dragged by soldiers; there were no draught animals. On October 24th a battery for two 12-pounder guns and a 10-inch mortar was completed and on the morning of the 25th fire was opened, but to the discomfiture of our artillery and engineers the French replied with seven guns and a mortar of greater calibre. Two more 12-pounders were brought up by the British on that day, but the roads were deep, and the men who had been continuously under arms for five days,were exhausted and went sick in great numbers; the ammunition therefore came up but slowly and the attackers' fire was intermittent and much inferior to that of the defenders who received guns, ammunition and gunners from Port Louis. During this farcical siege the detachment at Guidel near the landing place was attacked, and after slight loss made a shameful retreat to the shore. The officer in command of it was tried by court martial and cashiered. This incident is mentioned because it was for a time erroneously believed that the detachment belonged to the 30th.

Admiral Lestock now found out that his fleet was anchored in too exposed a position and that a south-west gale would drive it ashore. The Admiral was an infirm, worn-out man, and he was said to be ruled by his mistress, who was on board and did not like the south-westerly swell.

On receipt of this statement by the Admiral a Council of War was held which decided that the best course was to re-embark. On the evening of September 26th the four guns and the mortar were spiked and the force retreated to the seashore. The embarkation was completed by the 28th without interference from the French. The total British loss was a dozen killed and forty missing. General St. Clair, as he had been the first to land, was the last to re-embark.

On October 1st, the fleet left Quimperlé Bay without a rendezvous being appointed. On that day a gale dispersed the ships, and many transports steered for Falmouth. The whole fleet would have returned there, having no pilots for the French coast, but later in the day a tender joined the flagship with pilots from the Channel Islands. Quiberon was then given as a rendezvous to all ships within reach, and on the following day a great part of the fleet anchored at the entrance to Quiberon Bay. The troops present, whose number amounted to only 2,500 men, were landed on the isthmus of Quiberon, where works were thrown up and armed with French guns found on the peninsula. After a week, nothing having been heard of the missing transports, the guns were spiked and the troops re-embarked.

A month after it had sailed from Plymouth, the 30th disembarked at Portsmouth, where it remained for duty as marines in the fleet.

The expedition returned covered with disgrace, and it cannot be denied that the greater part of the troops, including the 30th, behaved very badly. The only thing that can be said in palliation is that the Government seems to have laboured to create a bad army. Colonel James Wolfe, the hero of Louisbourg and Quebec, wrote to his father General Wolfe in September, 1755: "I have but a very mean opinion of the Infantry in Courage. I know their descipline to be bad and their valour precarious. . . . Their shameful behaviour in Scotland, at Port Lorient, at Melle, and upon many less important occasions, clearly denoted the extreme ignorance of the Officers and the disobedient and dastardly spirit of the men." Wolfe's abhorrence of any misconduct under arms led him to speak strongly, but perhaps not too strongly;

the state of the army was deplorable, but we shall see how quickly it rallied under new leaders and a vigorous Government; four years after Wolfe used those terrible words, he fell in the moment of victory at Quebec, and the mutual confidence and affection between him and the army he led was a main element in their success.

Seven companies of the regiment served on the Channel Fleet during the winter of 1746–7.

About this time a good deal of thought was given to the subject of hair and wigs. An order issued in January, 1747, directs that no soldier is to wear a wig. In June, men whose hair was long enough were ordered to tuck it up under their hats. In August, men who from age or infirmity could not wear their own hair were directed to provide themselves with wigs made to turn up like the hair and wear them on mounting days. Perhaps this only applied to the Guards in London. All officers were to mount guard with queue wigs or with their own hair in queue.

FINISTERRE

In the spring of 1747 our Government learned that a great fleet of treasure ships, store ships and merchantmen was on the point of sailing from the French ports for their East Indian possessions under convoy of about a dozen men-of-war, and Admiral George Anson was ordered to endeavour to capture it. He sailed from Plymouth with a powerful fleet on April 3rd and on May 3rd he sighted the French fleet of thirty-eight sail off the Spanish coast, near Cape Finisterre. Thirteen ships of Admiral Anson's fleet carried detachments of the 30th, of which the distribution will be given later. Of the French fleet only nine were line of battle ships and they were so hopelessly inferior in strength to the British, that the most they could hope to do was to fight a delaying action to give the valuable ships under their convoy time to escape. The nine French line of battle ships therefore shortened sail, while the others, escorted by six frigates, crowded on all sail to escape. The British fleet pursuing in line and hampered by some slow sailing ships had not managed to come into action by 2 p.m. and the French line of battle ships began to make more sail and draw off. Admiral Anson then made the signal for a general chase, and so gave freedom of action to the captains of his faster ships. The *Centurion*, Captain Denis, was the first to come into action, about 4 p.m. She was supported by the *Namur*, *Defiance*, *Windsor Castle* and *Pembroke* and a warm engagement ensued with the five sternmost French ships. The *Centurion* lost her topmast and dropped astern, the four other British ships, seeing that the French ships opposed to them were crippled, pushed on to head the enemy's fleet. The remainder of our ships were now coming into action. Rear-Admiral Warren in the *Devonshire* came up with the French Commodore de la Jonquière in the *Sérieux* and made him strike; he then attacked Commodore St. George in the *Invincible*; the *Bristol* at that moment came into action and with one broadside dismasted the *Invincible* and drove her men from their guns. The remainder of the French men-of-war now tried to escape, having done all that was possible, but most of them and the ships under their convoy were captured. Among the last to be overtaken were the *Vigilante* and *Modeste*, which had on board three chests of gold and thirty of silver. They became prizes of H.M.Ss. *Nottingham* and *Monmouth*.

From the above slight sketch it appears that the British ships which were in close action were the *Namur, Defiance, Windsor Castle, Bristol, Pembroke, Centurion, Devonshire*, and to these may be added the *Princess Louisa*, which we learn from a letter in the *London Post* had two killed and thirteen wounded. The four last-named ships had detachments of the 30th on board. There were also detachments of the regiment on board eight other ships, two of which, the *Monmouth* and *Nottingham*, added enormously to the success of the day by capturing after a smart chase and action the two richest treasure ships.

The distribution of the parties of the regiment is a little hard to follow, because after the fleet put to sea there were frequent transfers of soldiers and marines from ship to ship. The following is a list of the ships and the detachments serving on them :—

H.M.S. *Devonshire*—Lieut.-Colonel Ravenhill, Surgeon William Graves and 50 other ranks.
,, *Centurion*—Lieutenant Andrew Paull and 30 other ranks.
,, *Pembroke*—Lieutenant Thomas Smyth, Ensign Wm. Hughes and 49 other ranks.
,, *Princess Louisa*—Captain David Brevet, Ensign Thomas Head and 53 other ranks.
,, *Monmouth*—Lieutenant H. Loftus, Ensign John Craven and 48 other ranks.
,, *Nottingham*—1 sergeant and 18 privates.
,, *Falkland*—Lieutenant Francis Pierson, Surgeon's Mate John Mackenzie and 30 other ranks.
,, *Eagle*—Lieutenant Teavil Appleton and 24 other ranks.
,, *Inverness*—Lieutenant Thomas Price and 27 other ranks.
,, *Kent*—Lieutenants Wm. Hammond and G. Coghlan and 60 other ranks.
,, *Prince Frederick*—Ensign John Hare and 21 other ranks.

After the action the fleet returned to Portsmouth with its prizes, and by July the seven companies had disembarked, some at Portsmouth for garrison duty, some at Southampton as guard on the French prisoners of war. The booty had been immense : a private's share was £7 5*s*. 4*d*., an ensign's, £33 4*s*., and a captain's £293.

Besides what went to the general prize fund, some boxes were broken open by the sailors and found to contain hundreds of dozens of ruffled shirts, and for days after the return of the fleet Portsmouth was full of sailors dancing, whooping and waving their cudgels and wearing ruffled shirts over their clothes.

The loss of the British in the action was 520 killed and wounded ; the regimental loss was not stated. Admiral Anson was made a peer for his victory. Towards the end of the year he reported that, as might have been expected from its long term of sea service, the 30th was suffering severely from scurvy. It was therefore ordered into country quarters. On January 1st, 1748, the officer commanding the seven companies was told to send four companies to Salisbury and three to Winchester, but to leave a detachment of the strength of two companies at Southampton to guard the French prisoners.

The Winchester companies had scarcely reached that place when fresh orders directed them to march to Salisbury. By July the Southampton detachment had been drawn in and the seven companies marched to Exeter where they were joined by the three companies which had embarked for Cape Breton in 1746, and which were by this time re-formed.

The French Government now found it convenient to make peace for a time, although it had by no means given up its intention of ousting us from India and North America. The treaty of peace was signed at Aix-la-Chapelle on October 1st, and was followed by unusually reckless reductions of the British army. The 30th was ordered to be placed on a peace footing and to return to Ireland. It was to be reduced by 10 subalterns, 1 quartermaster, 20 sergeants, 20 corporals, 20 drummers, and 710 privates, leaving 10 companies, each of 1 captain, 1 lieutenant, 1 second lieutenant, 2 sergeants, 2 corporals, 1 drummer, and 29 privates. The reduced officers were to be placed on half-pay. In the case of the other ranks, after the commanding officer had chosen the 340 he wished to keep, drafts were to be formed of the best men in sufficient numbers to complete the regiments of General Howard and Colonel Bocland and all in excess of that were to be disbanded, each man receiving 3*s*. for his sword and fourteen days' subsistence to carry him home. As usual the men were allowed to take their belt, clothes and knapsack.

How many were drafted to Howard's and Bocland's regiments and how many weakly men were turned adrift is not known, but a grant of land and a free passage to Nova Scotia were offered to such of the married men as were willing to emigrate. Ten ensigns were placed on half-pay of 1*s*. 10*d*. a day, namely Thomas Morris, John Craven, John Hare, John Kelly, William Hughes, Thomas Armstrong, James Minett, John Ley, Richard Ward and John White. The quartermaster, Hugh Bayley, received half-pay of 2*s*. a day.

The military emigrants, about 4,000 in number, founded the city of Halifax, Nova Scotia.

On November 17th the officer commanding the 30th was ordered to call in all his detachments and to march five companies to Barnstaple, three to Bideford, and two to Torrington and to await the arrival of transports to convey the regiment to Ireland. Government hoped that the regiment would arrive in Ireland before December 25th and be taken on the Irish strength on that date, but through stress of weather the landing in Ireland was delayed till the middle of February, 1749.

IRELAND

On February 13th, 1749, the regiment landed at Waterford. We have no knowledge of where it was quartered for five years after that, for, although it was inspected by General de Grangue in July, 1749, the place is not given.

On November 1st, Brigadier-General John Earl of Loudoun was appointed colonel, *vice* Lieut.-General Charles Frampton, deceased.

General Frampton began his military career in the Horse Guards, in which he was serving as lieutenant in February, 1711. He was appointed captain and lieut.-colonel in the 1st Guards on April 8th, 1712, lieut.-colonel of that regiment February 16th, 1739, brig.-general February 18th, 1742. He commanded the Brigade of Guards in Germany till appointed colonel of the 30th on April 1st, 1743, when he handed the brigade over to the Duke of Marlborough. He was promoted major-general in 1743 and lieut.-general in 1747. His lieut.-colonel, Henry Ravenhill, complained that he could get no response to his repeated reports to the general of the bad state of the regimental equipment on the return of the 30th from service on Lord Anson's fleet, which looks as if General Frampton had fallen into bad health long before his decease. The *Gentleman's Magazine* has an obituary notice of him dated September 23rd, several weeks before Lord Loudoun's appointment : " Death—Lieut.-General Frampton, at Butley Abbey, Suffolk : remarkable for his integrity and honour as well as great humanity to all mankind."

From his want of success when holding the chief command in North America, Lord Loudoun has been assigned a place by historians very much below what he deserves. Although not a great general, he was a good and energetic officer and was far too great a soldier to neglect his regiment. There is evidence that he visited it and took a personal interest in it. During the twenty-one years of his command the equipment was always excellent, and when on service all effective officers were present with the colours, and the 30th won the good opinion of all the generals under whom it served both in quarters and in the field. Lord Loudoun was fortunate in his field officers, perhaps it should be said in his choice of field officers. Lieut.-Colonel Sir William Boothby was a man of great vigour both of body and of mind, and a born leader. His successor, John Jennings, and the two majors, James Gisborne and James Ramsay, were above the average of their day.

On July 1st, 1751, the orders of 1743 regarding the colours and dress of the army were repeated and a little amplified, but there was no change which affected the colours of the 30th. The cords and tassels were ordered to be of crimson and gold mixed, the length of the pike and the size of the colours were to be the same as those of the Foot Guards.

With regard to dress, the drummers were ordered to be clothed in yellow faced with red, as they had been for nearly fifty years, and the front of the drums and the bells of arms were also to be painted yellow. The front of the grenadiers' caps was to be of the same colour as the facings of the regiment with the King's cypher embroidered and the crown over it, the little flap to be red with the white horse and motto over it, " Nec Aspera Terrent," the back part of the flap to be red, the turn-up to be the colour of the front with the number of the regiment in the middle part behind. The Royal regiments and the six old corps had different orders.

Eight regiments had blue facings, nine had green, eight had buff, four had white, one had red and one orange. Eighteen had yellow, and of these the 20th, 26th and 30th had pale yellow. There were only forty-nine regiments of the line in all.

In 1752 new colours were presented to the regiment, which were carried for ten years. When the colonel put his own device on the colours he naturally presented a new stand on appointment, but from now, as the colours were fixed by regulation, a newly-appointed colonel took them over from his predecessor exactly as he did the rest of the equipment.

In the spring of 1754 the regiment was in Cork, and here we have an example of the duties which fell to the military before the constabulary was raised. Lieutenant Teavil Appleton gives a full report of the matter to Major James Gisborne, who was in command at the time.

"A report of the detachment sent under my command to Beerhaven, May 2nd, 1754.

"On Friday the 3rd we sailed out of Cork harbour about one o'clock in the morning but wind coming against us we beat to windward all day and at night it blew so hard that we were obliged to come to an anchor in Glandow Bay. Care however was taken to observe your orders in keeping the soldiers from being seen, for I made them lie down in the boat and covered them with a sail while it was daylight and did the same when any boats passed us at sea.

"On Saturday the 4th the wind coming fair pretty early in the morning, we stood so far out to sea that we could not be observed from the land and when it grew duskish stood in for Beerhaven and landed at the late Mr. Puxley's house a little before midnight. After staying an hour to refresh the men we marched over the mountains to Murtogh Oge Sullivan's house, near the river Kilmore, though it rained all the time excessively hard. But as this was an inconvenience to us, so I thought it might make his scouts, who were said to be constantly on the watch, less alert in their duty.

"About 3 on Sunday morning the 5th, we surrounded the house in which were Murtogh Oge Sullivan, Daniel Connel, John Sullivan, commonly called little John Sullivan, and as well as I can learn five or six of their accomplices, before they had any suspicion of our design. The Civil Magistrate's demand to surrender themselves was answered by the discharge of several blunderbusses and musquets from the house, which continued very brisk till one of the soldiers was shot through the shoulder, another had his thumb shattered, another slightly touched, though all possible care was taken to keep them under shelter. The Civil Magistrate then finding it absolutely necessary to repel force by force, ordered the house to be set on fire, which was done as expeditiously as the excessive rain would permit. At last the flames obliged Sullivan and his accomplices to quit the house. As soon as they came out they presented at the soldiers who stood near the door, but luckily their flints were so worn with the continual fire from the house that few of them went off and those from their confusion did no mischief. This fire was by order of the Civil Magistrate instantly returned by the few pieces that would go off (most of our powder being quite wet) by which Murtogh Oge Sullivan and one of his accomplices were killed on the spot and Daniel Connell and John Sullivan were taken prisoners. By the darkness of the night and the care taken to secure these two men the rest escaped, though we heard afterwards that two of them died of the wounds they received.

"The soldiers from what they endured at sea and the terrible wetness of the night were so fatigued that I was obliged to return to Puxley's house and on Monday the 6th went in search of Sullivan's sloop, which we found with holes bored in her bottom which made it impossible to bring her off, so that no way remained of securing but by burning her which was accordingly done.

"I have brought back my whole party and the two prisoners Daniel Connel and John Sullivan who are lodged in the County gaol—and Murtogh Oge Sullivan's body in order to prove the identity of the person before the Coroner's inquest.

"(Signed) Teavil Appleton,
"Lieut. in the Earl of Loudoun's Regiment.

"*9th May*, 1754.
"CORK."

About the middle of May four companies, 1, 2, 5 and 9, marched to Duncannon Fort; two, numbers 3 and 7, to Enniscorthy; and four, numbers 4, 6, 8 and 10, to Wexford. The companies are numbered according to the seniority of their captains.

The distribution of the officers by companies at this time is given in "Millan's Irish Establishment":—

	Captain.	Lieutenant.	Second Lieutenant.
1.	Col. The Earl of Loudoun	Hayman Rooke (Capt.-Lieut.)	Peter Dumas
2.	Lt.-Col. Sir Wm. Boothby, Bart	Teavil Appleton	John Augustine Jevers
3.	Major James Gisborne	Henry Loftus	Charles Chapman
4.	Capt. James Ramsay	John Moutray	Simon Digby
5.	,, Moses de la Porte	William Snell	Henry Norton Jevers
6.	,, James Robertson	John Wright	
7.	,, James Rich	William Skipton	Thomas Margaret
8.	,, Thomas Smyth	Thomas Price	
9.	,, Henry Rugge	Godfrey Knuttall	John Woodward
10.	,, George Lewis	William Stiel	William Lushington

Chaplain: Henry Rand.
Adjutant: John Wright.
Surgeon: Charles Hewett (died September 4th).
,, William McKerrall.
Surgeon's Mate: John Mackenzie.

CHAPTER IV

TRANSFERRED TO THE BRITISH ESTABLISHMENT. EXPEDITIONS TO ROCHEFORT, ST. MALO, CHERBOURG, ST. MALO A SECOND TIME, AND ACTION AT ST. CAST. CAPTURE OF BELLEISLE. GIBRALTAR, BRITAIN, IRELAND.

1755–1781

BOTH in North America and in India the French, in spite of the peace, were pursuing their policy of hemming us in by a girdle of fortified posts and by alliances with native chiefs, and the strength of our army had to be raised in consequence. The new establishment of the 30th was: 1 colonel, 1 lieut.-colonel, 1 major, 7 captains, 11 lieutenants, 9 ensigns, 1 chaplain, 1 adjutant, 1 quartermaster, 1 surgeon, 1 surgeon's mate, 30 sergeants, 30 corporals, 20 drummers, 700 privates.

On March 25th, 1755, the regiment landed at Liverpool and marched to Leicester. On August 7th a route was issued for it to march forthwith to Colchester. At that station Major-General Stuart made the annual review and inspection of the regiment, the strength was 29 officers, 20 sergeants, 30 corporals, 20 drummers, 665 privates, of whom 21 were sick. Colonel Lord Loudoun was on duty in North America.

General Stuart's inspection report concludes: "A good regiment, well clothed and disciplined, well appointed, fit for service." The state of the regiment reflected great credit upon the Lieut.-Colonel, Sir William Boothby, as he had 440 men in the ranks who had less than 12 months' service.

There are other points of interest in General Stuart's report. He notes that the drummers have white bearskin caps, and that the grenadier company has two fifers. The grenadiers' caps, too, had been altered as well as the drummers' and they were now of black bearskin. The sergeants had sashes. The officers of the grenadiers carried a light fusil, and the other officers the spontoon. The muskets had iron ramrods.

The regiment paraded in marching gaiters; these were of brown linen with black buttons and much shorter than the white spatterdash. When the men were ordered to parade in marching gaiters, the officers wore boots.

On October 28th a route was issued for five companies to proceed at once to Gravesend, Northfleet, and Dartford.

On November 14th an escort under a captain was ordered to march from Dartford to Woolwich and receive twelve field-pieces and bring them back to Dartford. The regiment was to keep two pieces for its own use and hand the remainder over to Major-General Stuart's regiment, which would keep two and pass on the remainder to another regiment, which would keep two and pass on the remainder; all the regiments near Woolwich received similar orders, and in this way every regiment in the kingdom was equipped with two short 6-pounders manned by a party of Artillery.

The regiment was now increased to twelve companies of 4 sergeants, 4 corporals, 2 drummers, and 100 privates, the officers to each company remaining as before. The two new companies were to form a depot and relieve the service companies of the burden of recruiting, a reform Lord Stair had advocated for many years.

On November 27th the regiment was ordered to march five companies to Croydon, and one each to Mitcham, Streatham, Tooting with Wimbledon, Carshalton, Sydenham with Beckenham South, Bromley and Lewisham. From the extended quarters, concentration of the company must have been difficult and battalion exercise almost impossible.

In January, 1756, the Seven Years' War commenced. England and Prussia in alliance were pitted against France, Austria, Russia, Sweden, and Saxony.

The defeat of General Braddock and other failures in North America led to the appointment of Lord Loudoun to command H.M. Forces there, in the spring of 1756, with the local rank of general, and at the same time he was given the Royal American Regiment, to date from the previous Christmas. He still remained colonel of the 30th.

Early in the summer a route was issued for the 30th to march on June 16th, and to encamp within the new lines thrown up for the defence of Chatham Dockyard. The other regiments in camp were: Lord George Bentinck's, The Earl of Home's, Lord Robert Manners', and Major-General Stuart's.

The people of England now realized to what a condition their long-continued neglect of their army and navy had brought them and a wild panic was the result. Partly as a means of defence and partly to quiet the people, the detestable expedient of bringing in foreign troops was resorted to, as it had been in 1715 and 1745, and considerable forces of Hanoverians and Hessians were landed to garrison the country. Chatham and other places were fortified and the troops assembled in camps which served not only for instruction but as points of concentration to resist the invasion which was feared.

About the middle of October the camp broke up, and the 30th marched to Colchester and the adjacent district.

Many new regiments were now being raised, and the Secretary for War could not resist the temptation offered by the newly-formed depot companies. They were all swept into the fresh corps. Those of the 30th went to form the 52nd, afterwards known as the 50th, to which 1 captain, 1 lieutenant, 2 second lieutenants, 10 sergeants, 10 corporals, 4 drummers, and 147 privates were transferred. The Guards also furnished 10 men, and the Royal Fusiliers 110.

On November 8th an order was issued for the regiment to march to Canterbury and district. Two companies were dropped at Gravesend and the grenadier company was given charge of the French prisoners of war at Cranbrook. In December the number of companies at Cranbrook was increased to five.

In March, 1757, four companies were sent for a time to Sandwich for some reason not mentioned. It may have been fear of invasion, which kept the coast towns constantly crying for protection. In this month Admiral John

Byng was shot by sentence of court martial for not having done his utmost to save Minorca which the French had taken from us in the previous year.

Some months before this Mr. Pitt had come into power and under him the spirit of the people had begun to rise again. He sent away the foreign troops whose presence degraded the nation and began to form a militia to release the regular army for service abroad. Though driven from office in the spring of 1757, he returned in June and the orders to the 30th show with what effect.

The summer camps were formed as usual and the 30th joined the one commanded by the third Duke of Marlborough on Barham Downs on June 21st, but in July the training was cut short by an order to prepare for active service. Britain under Mr. Pitt's guidance was once more about to assume the offensive. The point selected for attack was Rochefort, the great French naval and commercial port on the Charente, and by Mr. Pitt's order the ten regiments best fitted for service were to be employed.

Sir John Mordaunt, Wolfe's friend, was to command the army, General Conway was to be second in command and James Wolfe, the lieut.-colonel of Kingsley's, was appointed quartermaster-general. His battalion, on which he had bestowed so much care, was to form part of the army, and although the others were all picked regiments, it is probable that in his alone firing with ball cartridge had been practised. In 1755 he had recommended it to his friends as teaching the men "to level incomparably, making the recruits steady and removing the foolish apprehension which seizes young soldiers when first they load their arms with bullets."

The orders of July 18th, 1757, direct the march to the Isle of Wight of the 5th, 15th and 30th from Barham Downs, the 25th, Colonel Hodgson's and Brudenell's from Chatham, the 1st Battalion Buffs, 1st Battalion 8th (King's), and the 1st Battalion 20th from Dorchester Camp, the 1st Battalion 24th from Amersham Camp. The troops were to take their camp equipment and encamp every night on the march. The vacancies in the camps were to be filled up by newly-raised battalions of militia. It all sounds very methodical and shows real progress, but although on August 10th the little army was concentrated in the Isle of Wight, the transports were not ready and it was September 7th before the expedition started.

The fleet with the troops on board returned to England without having made any real attempt to carry out the purpose for which it was sent. For nearly a week (September 23rd–27th) it had lain off the Charente. Councils of war were unceasing; once the troops were got into the boats for landing, but after three hours were recalled to their ships, and we can well believe that officers and men were furious at what appeared to be a want of decision of their leaders. The commanders of both land and sea forces were men of ability, and it may be that from the information in their possession they were right in not risking their troops. The only cheerful thing in the whole business was the evident eagerness of the soldiers and seamen to meet the enemy.

If the troops had landed, the 3rd, 8th, 20th, 25th, and 50th regiments would have formed the first brigade and the 5th, 15th, 24th, 30th, and 51st the second. The disembarkation was to be covered by the grenadier companies of the army which had been formed into one corps under command of Sir William Boothby, lieut.-colonel of the 30th.

On landing at Portsmouth, the regiment was ordered (October 7th, 1757) to march four companies to Reading, three to Newbury and one each to Elsley and Wallingford. The grenadiers landed later and rejoined Head-quarters at Reading.

A new establishment was sanctioned from Christmas, 1757. The regiment was to consist of 9 companies each of 1 captain, 1 first lieutenant, 1 second lieutenant, 1 ensign, 4 sergeants, 4 corporals, 2 drummers, 100 privates; the grenadiers to have 3 lieutenants. Apparently this order did not take effect in the 30th till April 13th, 1758.

In November, 1757, the distress in Gloucester among the weavers, who were at the point of starvation, had reached a head and serious riots occurred. The 30th sent two companies to Gloucester, two to Cirencester, and one to Lechlade. Luckily peace was restored without the troops having to use their arms, and in March, 1758, the regiment re-occupied its former quarters in and near Reading.

In the spring of 1758 the regiment was warned to be ready for active service.

Mr. Pitt, undeterred by the failure at Rochefort, was determined to try again with greater strength to destroy the enemy's naval arsenals and shipping and so to make a powerful diversion in favour of our ally, Frederick the Great, King of Prussia.

About 13,000 troops, including a few light cavalry, with sixty guns, were placed under the orders of the Duke of Marlborough for an expedition to the French coast.

Lieut.-Colonel Sir William Boothby received orders to march the 30th on May 6th in three divisions to Southampton for the Isle of Wight, all to be in camp in the Isle of Wight on May 10th.

Captain James Ramsay at this time succeeded Major John Jennings, who had been appointed lieut.-colonel to raise and train the 62nd, one of Mr. Pitt's new regiments.

The troops were all embarked from the Isle of Wight before the end of May and sailed from St. Helen's on June 1st. Admiral Howe commanded the convoy and a fleet under Lord Anson covered it to the west.

The infantry was divided into three brigades:—

1st, The Guards, formed of battalions from each of the three regiments under General Drury.

2nd Brigade, Bentinck's, Manners', Talbot's and Home's, under General Mostyn.

3rd Brigade, Loudoun's (30th), Wolfe's (8th), Kingsley's (20th), under Lord Waldegrave.

Major-General Edward Wolfe, who had been major and lieut.-colonel of the 30th in 1715–18, was colonel of the 8th (King's). His son, James, the lieut.-colonel of Kingsley's in the previous year, was now engaged in the reduction of Cape Breton.

The grenadier companies of the line regiments were formed into two battalions, under Lieut.-Colonel Sir William Boothby of the 30th. Major Ramsay commanded the 30th.

On June 4th the fleet anchored off Cancale, 4½ miles from St. Malo. On the 5th the Guards landed without opposition, and the remainder of the troops were put on shore on the following morning.

On the 7th the third brigade commenced to entrench to cover the landing of artillery ; the other two brigades, under Lord Ancrum, advanced towards St. Malo. The Guards Brigade had been strengthened by the two battalions of grenadiers under Sir William Boothby. The advance of Lord Ancrum's column was through a closed country and sometimes of necessity in single file. It was late therefore when the two brigades encamped near St. Servan.

At nightfall Brigadier Elliot took with him 200 horsemen, each with a grenadier mounted on the croup of his horse, and rode down to the harbour of St. Servan where, with their hand grenades, they fired the crowded shipping. They thus destroyed 1 50-gun ship on the stocks, 1 ship of 36 guns, 24 privateers, 70 merchantmen, 40 small craft.

The staff now decided that the siege of St. Malo was beyond the means at our disposal and the army was re-embarked. The weather was very bad and the fleet remained at anchor till the 16th, when it moved along the coast.

Granville was reconnoitred, but judged too strong. Stress of weather forced the fleet to anchor again in Cancale Bay. On the 21st it ventured to sea again and a gale blew it over to the Isle of Wight. With better weather it stood over again to the French coast, and on the 29th anchored 2 miles from Cherbourg. All dispositions were made for landing and some men were actually in the boats when the wind sprang up again. The landing was countermanded and the fleet had difficulty in getting to sea against the wind blowing directly on shore.

The expedition returned to England and landed the soldiers for refreshment on the Isle of Wight on July 2nd. After three weeks they were embarked ; General Bligh was now in command. His first object was to be the destruction of the military port of Cherbourg, and if successful there, he was to " carry a warm alarm " along the north coast of France. Commodore Howe was again to command the fleet.

The army was accompanied by " H.R.H. Prince Edward Duke of York, with many volunteers of noble extraction and gentlemen of merit and fortune," as the newspapers of the day put it.

The expedition sailed on August 1st and anchored off Cherbourg.

At 9 a.m. the transports entered the Bay of Marais and landed their men in flat-bottomed boats, the Guards and Sir William Boothby's grenadiers leading. They were under General Drury. The French, who were about 3,000 in number, did not behave very well. General Drury marched up to them, sustaining three fires without replying, and firing at close quarters, routed them. The Guards and grenadiers had only a score of casualties, and there was no more fighting. The arsenal and docks were destroyed, some hundreds of cannon thrown into the sea, and twenty-five brass pieces carried off as trophies. The Light Horse laid the country under contribution and brought in £18,000.

Great wine magazines were discovered and such disgraceful drunkenness followed that the army might have been destroyed. The troops re-embarked on August 16th and reached England two days later. After the success at Cherbourg another landing in France was determined upon, and as the enemy were supposed to be in strength about Brest, General Bligh decided to land again near St. Malo.

On the evening of September 3rd, the fleet anchored in the Bay of St.

Lunaire and on the 4th the army landed without opposition, but the sea was so high that several men were drowned. Next day St. Malo was reconnoitred and Lieut.-Colonel Sir William Boothby with the grenadiers burned thirteen ships in the harbour.

Commodore Howe now refused to join in an attack on St. Malo, or to remain at anchor in such a dangerous spot. As the sea ran too high to risk a re-embarkation General Bligh agreed to march across to the Bay of St. Cast, where Commodore Howe undertook to meet him with the transports.

On September 8th camp was struck and the army marched for St. Briac and encamped half a mile from the village. Throughout the night the men were kept under arms by parties of French hovering round the camp. The tidal river Arguenon had to be crossed the following day. One low tide was lost and, when the army did cross, the water was up to the men's middles and numbers would have been drowned but for parties with tent poles who pulled the struggling men to shore. Many firelocks were lost. About thirty officers and men were killed and wounded. As the fire came mostly from armed peasants General Bligh ordered all villages to be burned. The crossing was not completed till nightfall on the 10th, but the army encamped without molestation.

On the 11th the army moved, covered by an advanced guard of Sir William Boothby's grenadiers. The French regular troops were now assembling fast, and though there was no opposition in front, the flanking parties lost thirty men, and the French killed all prisoners, in retaliation for the burning of their villages. The army camped at Matignon, the French army, estimated at 12,000 men, being six miles off, half militia, half regulars.

General Bligh was now within reach of his transports at St. Cast and had the option of fighting or of re-embarking at the risk of being caught in the act. He chose to avoid an action.

At 1 a.m. on September 12th the French were heard to beat to arms, but "the general" was not beat by the British till near daybreak, and it would have been better if silence had been observed, as Bligh's only wish was to reach the boats without interruption.

He formed a rearguard of 1,000 of Boothby's grenadiers and four companies of the Foot Guards, the whole under General Drury.

By nine o'clock the leading men had begun to enter the boats; about an hour later the French appeared and commenced an action with the rear-guard, but the main body of the army got on board with little loss. The difficulty was to draw off the rear-guard. General Drury formed his men into two bodies and ordered them to attack and throw back the advancing French. The attack was only partially successful and there was nothing left for it but to make for the boats. The French did not follow to the shore but lined the heights with musketry and got some field pieces to a plateau half-way down. The sea had got up, and but for the gallantry of the sailors not a man of the rear-guard would have escaped. Commodore Howe was conspicuous in his barge encouraging his men. Many boats were destroyed and at the end numbers of men were left on the beach or in the water, at whom the enemy continued to fire volleys long after the last boat had gone. General Drury and Sir William Boothby having done their duty to the end reached the shore after the boats had left. Drury was wounded and supported by a grenadier.

They attempted to swim off, but Drury was too weak and was captured. Sir William Boothby was picked up by a boat after a prolonged struggle. The accounts of the day say he was two hours in the water.

A tall column on the heights above the Bay of St. Cast, in view of the modern golf links of St. Briac, preserves the memory of the defeat of the English, and the peasants love to narrate the events of that great day, which few of the golfers have ever heard of.

After this mishap our troops returned to England, and the 30th landed in the Isle of Wight. On October 1st it marched to Cowes and crossed to Portsmouth, and thence marched to Dartford. It left nine wounded who were admitted to hospital in the Isle of Wight on October 2nd and appears to have picked up some recovered men. The losses of the grenadier company are not known, but Lieutenant Williamson was killed and Lieutenant Price, 1 sergeant, 2 corporals, and 25 privates were taken prisoners. Many must have perished; the prisoners of whom we have knowledge are those only who lived to return in April, 1759.

Prince Edward, Duke of York, who was a spectator of the resistance offered by the rear-guard, appointed Sir William Boothby his Master of the Horse.

On its arrival at Dartford the regiment was met by an order to march to Canterbury where it spent the remainder of the year with, for a part of the time, two companies detached at Sandwich, Wingham and Ash. A return signed by Major Ramsay shows that 119 men were wanted to complete to the establishment. Ten officers' parties were recruiting at Chester, Norwich, London, Hatfield, Chelmsford, Nottingham, Basingstoke, Walsingham, Sudbury and Warrington. Lord Stair's localization of fourteen years before had quite disappeared.

In March, 1759, all detachments were ordered to rejoin and the regiment to move to Brampton, Rochester, Stroud and Finsbury. They arrived in their new quarters at the end of the month. In June the regiment moved into camp on Dartford Heath with two companies detached at Dartford.

In Dartford Heath camp, the regiment only wanted thirty-seven men to complete to its establishment of 900 rank and file, but it sent a draft of twenty-one men to our army in Germany. Most of the recruiting parties were now withdrawn, but three were left, at Chelmsford, Romney and Sudbury. Seventy sick were at Rochester in charge of the surgeon's mate. Sir William Boothby was in command.

On September 18th the regiment was reviewed within Chatham lines by Major-General The Earl of Effingham. He reports that the corps is in general a very good regiment and very fit for service.

The custom as to mounting guard at inspection seems to have varied. On this occasion the ordinary guards were mounted: Quarter guard, 1 subaltern, 1 sergeant, 1 corporal, 1 drummer, 46 privates. Rear guard, 1 sergeant, 1 corporal, 12 privates. Hospital guard, 1 corporal, 5 privates.

Other general officers put the guards in the ranks for purposes of inspection, as far as was possible.

It appears from the recruiting account that £3 extra was disbursed by the recruiting officers for three grenadiers. No doubt they were of exceptional height. The craze for very tall soldiers started by the father of Frederick the Great seems to have reached the 30th.

On the camp being closed for the winter the 30th marched to Hilsea Barracks, which it reached on November 19th. The officers were all present except Captain Rich and Captain Smyth, sick in London, Lieut. Moutray, sick in Bristol, and Captain Wright, A.D.C. to General Howard. Four parties were out recruiting at Huntingdon, Reading, Walsingham and Winchester. Fifty-three men were wanting to complete.

At the beginning of 1760 the regiment had 96 men on the sick list. In February another draft of eighteen men was sent to Germany under Ensign Clarke and a third of the same strength was sent under Lieutenant Loftus in May.

By that time Lieut.-Colonel Sir William Boothby had left to take up a Staff appointment in Germany, and Major Ramsay was in command. Sir William Boothby never rejoined. In October he was promoted colonel of the 63rd Foot. He died in 1787, colonel of the 6th Foot and a major-general.

In July the regiment joined the camp at Sandheath under Major Ramsay. The sick list had fallen by nearly a half. Lieutenant Digby and one man had died.

In October King George II died and was succeeded by his grandson, George, Prince of Wales.

On October 24th Colonel John Jennings returned from the 62nd to take command, in succession to Sir William Boothby. As lieut.-colonel of the 62nd he had in the spring of the year an annoying experience. While training the new levies at Carrickfergus, near Belfast, he and his unarmed recruits had been captured when the notorious Thurot made his descent on the North of Ireland.

On October 31st, 1760, Lieut. Vincent Cunningham, of the 30th Foot, at Guildford, was ordered to cause the convalescent men of his regiment in the hospital at Guildford to march from time to time to Chichester to join their regiment. The 30th was probably moved to Chichester on the closing of the camp at Sandheath. From Chichester the regiment must have moved in the autumn to Salisbury, for on January 5th, 1761, Lieut.-Colonel John Jennings was ordered to march thence to Basingstoke, Hartley Row, Hartford Bridge, Mappledoor, and Wellhatch. It is doubtful if this order was carried out. The recruiting parties were now at Norwich, Newbury, Winchester, Petworth, Walsingham and Sudbury. Ninety-four men were wanting to complete.

In February, 1761, the regiment proceeded to Taunton and Bridgwater. On St. Patrick's Day it was ordered to return to Chichester and Petersfield, but a final order caught it at Shaftesbury on the second day's march from Taunton. It was, in spite of all previous orders, to march to Hilsea Barracks and await orders from General Hodgson. It had been selected to serve under him in the Belleisle expedition.

BELLEISLE

Belleisle, about 9 miles in length from N.W. to S.E. and 5½ wide at its greatest breadth from N.E. to S.W., lies off the French coast abreast of Quiberon Bay.

The British Government determined early in 1761 to occupy it, and the

16th Light Dragoons, the 9th, 19th, 21st, 30th, 67th, 69th, 76th, 85th, 90th, and 98th regiments were selected for this duty; the latter five were newly-raised corps.

The garrison of the island was about 2,500 men, without counting the local militia, the coast was precipitous, the landing place fortified, and the works round Le Palais, the capital of the island, had been recently strengthened. The town and citadel were within one general line of defence, but the citadel was capable of defence after the capture of the town. Major-General Studholme Hodgson was appointed to command the British troops and Commodore Keppel the navy.

Mr. Pitt had in his selections gone pretty far down the army list to find men to command, as he had done in 1758–59 in appointing to the commands in America. Under Major-General Hodgson, Major-General Crauford, Brigadiers Howe and Carleton commanded brigades, and Brigadier-General Desagulier commanded the artillery.

The transports assembled at Spithead early in March and were joined by eight line-of-battle ships. On the 29th the expedition sailed.

A week before sailing, General Hodgson received an excellent order from the King, that the 5*d.* hospital stoppage was not to be exacted from his men while on board a hospital ship, but that a sum equal to the amount of such stoppages was to be expended in the supply of such refreshments as are necessary, but not usually supplied in hospital.

It is difficult to say what rations the men received on the transports, but the following is the contract to supply a company on a transport between Leith and London in the same year :—" To be issued to each mess of six men on every day, 4 lbs. of bread and 2 pints of brandy. *In addition,* on Sunday, Tuesday, Thursday, Saturday, 8 lbs. beef, 2 pints of peas. Monday, Wednesday and Friday, 4 pints oatmeal, 1 lb. cheese, 1 lb. butter."

The ship provided bowls, spoons, candles, coppers and furnaces and fuel to dress the provisions. An allowance of vinegar was supplied also. Probably the men for Belleisle received the same or its equivalent. The beef of course was salt, and the bread, biscuit.

On March 29th, the day the regiment embarked, General Hodgson had written to the Secretary at War to say that he had appointed Hayman Rooke, captain of grenadiers in the 30th, his aide-de-camp, that he knew him to be a good officer who had had hard luck in the service.

On April 7th the expedition arrived off Belleisle, and the flat-bottomed boats were got out while the general and admiral reconnoitred the shore.

On the following morning the signal was made for part of the troops to land and three line-of-battle ships with two bomb vessels opened fire on some works near Locmaria, the southern point of the island. The guns of the men-of-war soon silenced the batteries and the boats pushed for the shore. The attack failed; about sixty of the grenadiers of the 69th (Erskine) forced their way to the top of the precipitous heights, but were destroyed or thrown down. Our loss was 400, among whom were Major Purcell of the 98th, 1 captain, 2 lieutenants killed. Over 300 prisoners remained in the enemy's hands. The 30th were not engaged, although the grenadier company was in the boats, but for some reason General Hodgson was anxious to do the regiment a good turn. He not only appointed Captain Hayman Rooke to the majority in the 98th *vice* Major Purcell, killed in action, but on his representations to the Secretary at War the promotion was allowed to run right through the 30th, so that Capt.-Lieutenant Loftus, Lieutenant Thomas Price and Ensign Clarke got a step without purchase and Surgeon's Mate Lewis Lewis received a commission as ensign, all through an officer of another regiment being killed in action. After the repulse of the 8th inst. a demonstration was made of landing at Sauzon on the north of the island, but heavy weather suspended operations till the 22nd, when at five o'clock in the evening the men-of-war opened fire on all the enemy's batteries and entrenchments and feints at landing were made at different points. The real attack, however, was in the south and entrusted to Captain Sir Thomas Stanhope of H.M.S. *Swiftsure* and Brigadier Lambert (lieut.-colonel of the 19th regiment) who had with him his own regiment, the grenadier company of the 30th and parties of marines and of Stewart's and Crauford's regiments.

At two o'clock in the morning of the 23rd Sir Thomas Stanhope made the signal for the boats of the men-of-war and transports of his command to assemble at H.M.S. *Swiftsure*, Commodore Barton at the same time assembling his division. About an hour later they moved off, Commodore Barton leading. In this order they approached the shore near Locmaria and moved along it as if looking for a landing place; presently they reached a rocky point which Commodore Barton rounded, making a feint at landing beyond it. As previously arranged, Sir Thomas Stanhope, instead of following Barton, turned and made direct for the shore on the near side of the point. When close in the leading boats with the grenadiers of the 19th on board lay on their oars to allow the rest of the division to close up; they then ran their boats ashore and landed; one man, probably Captain Patterson, climbed to the summit of the precipitous heights. On his signal his men followed and the whole division hastened in support. The French had not been entirely deceived and the grenadiers of the 19th were attacked before they reached level ground. The French fire was very heavy, but our men escaped a good part of it by squatting among the rocks to reload. Reinforced by some marines, the grenadiers of the 19th drove back the French opposed to them and occupied a wall behind which the enemy had sheltered. The Regiment de Bigorre, supported by the fire of six guns, now advanced to retake the wall, but General Lambert, bringing up his own regiment, the grenadier company of the 30th and the remainder of the marines, ordered the troops to move briskly to the attack, which they did with great spirit, and the top of the hill was cleared. It was now daylight (5 a.m.) and the whole British force was being landed. General Hodgson and Admiral Keppel in the admiral's barge moved about the bay giving orders and urging on their men. The French supports too now became exposed to the fire of our ships. A British naval officer wrote that "the broadsides of the *Hampton Court*, *Swiftsure* and *Essex* mowed down rank and file quicker than a fowling piece does starlings from the tops of the Lincolnshire reeds." Making allowance for the exaggeration caused by the writer's laudable pride in naval gunnery, he is right in thinking that the fire of the ships had a great deal to do with the hurried retreat of the French to le Palais which now took place. Their rear-guard fired one final round from their field pieces as our men approached and then spiked and abandoned them. The honours of the day were with the grenadiers and the 19th.

After landing, the infantry regiments of the army were brigaded as follows :—

1st Brigade	9th, 67th, 1/76th, 85th.
2nd „	19th, 30th, 90th, 85th.
3rd „	21st, 2/76th, 69th, 85th.

The 85th was broken up and a party given to each brigade to bring it up to a strength of 2,085. The regiments not brigaded were posted at de Andro and other points along the coast. Lieut.-Colonel Jennings was given command of a brigade and Major Ramsay took command of the 30th.

General Hodgson now invested le Palais, within which the French commander, General St. Croix, had assembled the local militia as well as the whole of his regular troops. The total strength was about 4,000 men.

Stormy weather prevented the landing of the artillery for some days, but at the end of April the British got some guns ashore and began their approaches, and on May 2nd fire was opened from a few mortars. On the following night the enemy made a vigorous sally with a body of 400 men headed by one of their best officers. They penetrated into our most advanced work and there made prisoners of Major-General Crauford and his aide-de-camp with many men. They then pushed on and attacked the trenches at a part defended by the 30th. Their leader was heard to call to his followers: " Allons mes enfans, mes soldats, allons ! Voila, c'est là voye à l'honneur, à la gloire, à la victoire ! " He then leaped into the works but was instantly pierced through the heart by the thrust of a bayonet from a grenadier of the 30th. The attack was repulsed, but the French carried off with them in their retreat the garrison of our advanced work. The damage was quickly repaired and the siege progressed without interruption till further approach was stopped by a line of redoubts with which the French engineers had covered the main body of the place. It was determined to carry those redoubts by storm on the morning of the 13th, and this duty was entrusted to a body of marines, supported by a wing of the 30th. The attack was preceded by a heavy fire from all our guns, after which the marines assaulted the nearest redoubt. The French fought gallantly and it was only when the marines were joined by the 30th, and after " an obstinate dispute with the bayonet," that the redoubt was carried. The two corps combined then rushed with comparative ease the remaining redoubts in succession, supported by Colvill's regiment which General Hodgson had pushed forward as soon as he saw that the 30th had joined the marines. Some of the assaulting party even reached the town. The enemy had about 150 killed and wounded and 75 prisoners (5 officers); our loss was 15 killed, 29 wounded. The citadel now lay open to attack. An incessant fire was kept up night and day, which was vigorously returned by the defenders till May 25th, when their fire slackened and it was evident that the end was near.

In the last days of May and beginning of June the Buffs, 36th, 75th, and remaining companies of the 85th arrived, and four brigades were formed each with a battalion of marines attached.

1st Brigade,	The Buffs, 30th, 75th, marines.
2nd „	9th, 69th, 76th, marines.
3rd „	19th, 36th, 90th, marines.
4th „	21st, 67th, 85th, marines.

At noon on June 4th, in honour of the King's Birthday, the fleet fired three rounds and every gun and mortar on shore did the same against the citadel. On the 7th a breach was reported practicable and the place summoned. The Governor was in a hopeless position, and saved his men from being put to the sword by capitulating.

In recognition of its gallant defence the garrison was allowed to march out through the breach with the honours of war, drums beating, colours flying, etc. They were landed on the French coast a few days later.

The total loss of the army in the capture of Belleisle was :—

	OFFICERS.	SERGTS.	DRUMMERS.	RANK & FILE.	TOTAL.
Killed .	11	9	2	261	283
Wounded	24	16	6	410	456
Total .	35	25	8	671	739

That of the 30th was :—

Killed .	—	—	—	22	22
Wounded	1	1	—	40	42
Total .	1	1	—	62	64

From the large proportion of men killed, more than one-third of the total casualties, it is probable that only the wounded who were sent to a general hospital were included in the returns. The loss was principally in the marines, who landed 1,800 men, and in the older regiments, the 9th, 19th, 21st, 30th and 67th.

Lieutenant Henry Norton Jevers, who was the officer of the 30th reported wounded, lost an arm. There is a portrait of him with an empty sleeve at Castle Jevers, in Ireland.

Major Hayman Rooke and Captain Barton, R.N., were given the honour of taking home General Hodgson's and Commodore Keppel's despatches, and they arrived in London on June 13th, where the news of the fall of Belleisle was received with great satisfaction. On May 29th, Mr. Pitt had written to General Hodgson concerning the capture of the redoubts by the marines and 30th regiment. " His Majesty has learned with particular pleasure that the redoubts have been attacked and captured with such distinguished spirit and happy success. I have only to assure you of His Majesty's entire approbation of your conduct and of the King's perfect satisfaction in the signal bravery and resolution shewed by the officers and soldiers under your command."

The capture of Belleisle certainly was a good piece of work, and a few years earlier would have driven the country wild with delight, but coming after the great victories of Clive, Wolfe, Amherst, and Sir Edward Hawke, it scarcely received the attention it merited. General Hodgson's difficulties were not known to the general public. In a letter to the Secretary of War he says : " The number of absent officers is intolerable, fifteen in one regiment, among whom are the Major and five Captains." In another letter somewhat later he says : " I can reckon very little on seven battalions that are with me, they are quite undisciplined and ignorant of their duty."

We have a state of the 30th dated Belleisle, August 31st, 1761. Three officers only are absent, namely, Lieut.-General Lord Loudoun, Lieut. Vincent Cunningham, with the sick in England; Lieutenant Henry Norton Jevers, wounded.

The non-commissioned officers, rank and file fit for duty are 651. There are 70 sick with the regiment and 55 in general hospital, 37 sick in England, of whom 26 had been sent home from Belleisle. These numbers include wounded.

Captain Rich, who for years had been sick in London, had retired on full pay, and on July 15th the Secretary of War wrote to General Hodgson that no time had been lost in complying with the general's desire that the succession to Captain Rich should run in the regiment. Capt.-Lieutenant Price had been promoted captain, the eldest lieutenant, Moutray, captain-lieutenant, the eldest ensign, Johnson, lieutenant. General Hodgson could appoint the ensign from among the volunteers serving with the army. He selected Thomas Wilkie, the second surgeon's mate, of the 30th. Mr. Robert Primrose was appointed mate in his place. The appointment of two mates to commissions was so unusual that for more than a year Lewis and Wilkie are shown in the musters both as surgeon's mates and ensigns. This was the second time that General Hodgson had interested himself in the regimental promotion, and no doubt when the commissions were "wetted," according to custom, many bumpers were emptied to his health.

In January, 1762, General Hodgson embarked with five regiments, of which the 30th was one, to return to Portsmouth.

After leaving Belleisle very heavy weather was met with and the fleet of transports and its convoy were scattered, but on January 10th, 1762, General Hodgson was able to report that all the transports had arrived except the *Britannia* and *Hopewell*. The *Hopewell* had on board Lieut.-Colonel Jennings, Captain Loftus, 1 lieutenant, 3 ensigns, 8 sergeants, 8 corporals, 4 drummers, 153 privates of the 30th. Nearly an equal number of Manners' regiment were on board the *Britannia*.

A fortnight later it became known that the two transports, separated from the others and somewhat battered in the gale, had been picked up by three Dunkirk privateers of 16, 12, and 4 guns. The transports were not armed and resistance seems to have been impossible. The *Britannia* was taken into Dunkirk, but the *Hopewell* on account of the severe weather had to run into Dieppe. The system for exchange of prisoners was working well and orders were issued at once for two cartel ships to sail from Portsmouth with a suitable number of French prisoners to Dieppe and Dunkirk. On May 30th the exchange had been effected and there landed at Dover, Lieut.-Colonel Jennings, 1 captain, 1 lieutenant, 3 ensigns, 8 sergeants, 4 drummers, 151 rank and file. Five men had died at Dieppe. Four are reported to be "seduced by the French."

Lieut.-Colonel Jennings and his party rejoined the regiment by march at Hilsea Barracks.

The seven companies of the regiment which had returned with General Hodgson landed at Portsmouth under Major Ramsay and marched to Salisbury, arriving on January 14th and 15th. One company was detached to Wilton, Lieutenant Henry Norton Jevers was still absent in consequence of wounds, and the regiment seems to have been 165 privates under strength.

In the following month, Lord Loudoun wrote to the Secretary at War, Privy Garden, London :—

" As Lieutenant Jones is promoted to a regiment in Ireland I beg leave to recommend Henry Rooke the eldest ensign to succeed him, and also Thomas Jennings, nephew to Lieut.-Colonel Jennings, to succeed as ensign."

These appointments were made.

On February 20th, Lord Loudoun, who had not been employed since relieved in 1758 of the command-in-chief in America by Major-General Geoffrey Amherst, was informed that he was to command the British force which it was proposed to send to the Peninsula to defend our ally the King of Portugal against the Spaniards, with whom war had been declared on January 2nd, 1762. The regiment did not accompany its colonel to the Peninsula, but he took with him Lieutenant Muller. A Mr. Muller did good service at the capture of le Palais in Belleisle as acting engineer and it is unlikely that there were two of that name in the expedition. Among the preparations for Lord Loudoun's expedition there was an important suggestion made by Doctor Pringle, P.M. officer of the army, that during the fever months every man in the force should take daily a prescription containing quinine. Mr. Townshend directs Lord Loudoun to submit the proposal to the " Physical gentlemen with the Army."

The regiment moved to Hilsea Barracks in February, 1762, where it was joined by Lieut.-Colonel Jennings and his party from Dieppe, in May.

On July 15th new colours were presented to the regiment of the same pattern as those issued ten years before.

In the autumn the regiment moved to Canterbury. In December the negotiations for a general peace were so far advanced that preparations for bringing home and reducing our armies were begun.

A new establishment was issued for the 30th dated December 25th, 1762, by which it was to be reduced to nine companies of 3 sergeants, 3 corporals, 2 drummers, and 47 privates ; the grenadiers were as usual to have 2 fifers. The junior captain was to remain on full pay and do duty with the colonel's company until absorbed, but the extra lieutenants were to be swept away. Under this order eight lieutenants were placed on half-pay. There was very little economy in this reduction. One of those officers, David Maxwell, drew half-pay till 1831.

No orders were at this time given for paying off the men in excess of the establishment.

In February, 1763, peace was signed at Paris with France and Spain. We gained Canada, Cape Breton and Florida and had become the ruling power in India ; we restored to France Martinique, Guadeloupe, Goree, Pondicherry and Belleisle.

The land and sea forces received the thanks of Parliament for their services in this victorious war.

The regiment marched from Canterbury to Winchester about this time, probably early in March.

There was still in England a party opposed to the maintenance of a standing army, and among those belonging to it were some who gave vent to their feelings by insulting the officers and ill-treating the soldiers when they thought they could do so with safety. Even magistrates too often disgraced themselves by assisting in this cowardly persecution. The Secretary of War between his duty to the army and his desire not to offend people who were politically influential was pulled in different directions. The following correspondence gives a good example of the disputes which were common all over England.

In March the Secretary wrote to Lieut.-Colonel Jennings at Canterbury regarding " A dispute and some odd behaviour which passed between the Mayor and Lieutenant Nunn of your regiment and that the Lieutenant had or intended to bring an action against the Mayor for false imprisonment." The Secretary goes on to say that he is solicited by some friends of the Mayor who are " men of consequence " to effect an amicable arrangement " which indeed I wish to be done." In other words, to oblige some " men of consequence " the Secretary tries to prevent an officer who had been wronged from obtaining legal redress.

In the absence of Lieut.-Colonel Jennings, Major Ramsay answered the Secretary in a letter of admirable temper.

" I have communicated to Lieutenant Nunn your request that the affair between him and the Mayor of Canterbury should be accommodated in an amiable manner, which was all along our desire had not the obstinacy of ye latter prevented it and Lieutenant Nunn has by this post wrote again to his attorney to drop the prosecution if the Mayor will pay the costs and make such an acknowledgement as one Gentleman ought to another, which I imagine you will think are easier terms than Lieutenant Nunn could have expected had he been the aggressor."

The dispute seems to have been settled on the lines laid down by Major Ramsay, as the official correspondence ceases.

On March 31st, Lieutenant-Colonel Jennings wrote from Winchester to the Secretary of War that he had that moment received his order to march to Exeter and that the first division would set out in a few days and the second two days later. He continues : " As it would be a great hurt to those men you have signified your desire to have discharged at the reduction, to go to the length of Exeter, I should be glad to know your pleasure concerning them and if I may discharge them on Monday next before the last division moves out." As few of the men of the regiment had been enlisted from further west than Basingstoke this was a kind and thoughtful proposal on the part of Colonel Jennings. It has been said on very high authority that the British officer of that day neither cared nor knew anything about his men. The evidence as to whether this applies to the 30th is very scanty, but what little there is seems rather to point the other way. We will put beside this letter of Colonel Jennings one from Colonel Gisborne (afterwards Major-General Gisborne) who had joined the regiment in 1739 and whom Jennings had succeeded as Major in 1755. Colonel Gisborne in his letter is fighting for a Chelsea pension for an old 30th man.

" STAVELY, DERBYSHIRE,"
Nov. 4th, 1764.

" One Richard Brennan served as a private Grenadier and Corporal under my command in the 30th Foot during the war of 1741 and towards the conclusion of it being disabled by a broken leg, was admitted to a Chelsea pension. In 1756 he found me out in Ireland and told me he was recovered and as a new war was broke out, he wished to spend the remnant of his strength in His

Majesty's service. I took him into the 10th regiment of foot and made him a sergeant, where he served till the other day when I find by a letter from London he was discharged in Ireland as unfit for further service and upon his application to be re-admitted to Chelsea pension, he is refused and told he must take up with the Irish pension of Kilmainham."

There certainly seems nothing wrong in the relations between officer and soldier here. Those letters are in no way picked, but come from two officers who held the rank of regimental major in succession.

Soon after writing the letter quoted above, Lieut.-Colonel Jennings went on leave for the recovery of his health and apparently never rejoined.

Major (Brevet Lieut.-Colonel) Ramsay in a letter of April 13th, announces to the Secretary of War the arrival of the regiment at Exeter, and the receipt of orders to carry out the reduction.

As a consequence of the peace, Lord Loudoun's army had been ordered home from Lisbon and was now landing. Some months before leaving Lisbon, Lord Loudoun had written home that he lost men every week, that the Portuguese, whom they had come to defend, murdered every man of his whom they could catch alone. Some of his officers now brought the news that Lieutenant John Muller of the 30th, who had accompanied his colonel to Portugal from Belleisle, had been assassinated in the streets of Lisbon by a party of Portuguese. Lieut.-Colonel Ramsay, who seems to have been quite unfettered in command, recommended that the vacant lieutenancy should be given to the senior ensign, Lewis Lewis, who had served six years in the regiment, two of them as ensign, and had always behaved as a good officer. This recommendation was followed.

The reduction of the 30th may have been delayed till the War Office was quite sure how far the House of Commons would go in disbanding the army, and ultimately the 30th and three other old regiments which were in no danger of being broken up were chosen to form the garrison of Gibraltar and their strength fixed at 529 of all ranks. The strength of the regiment in November, 1762, had been 695 rank and file, of whom 161 were sick, and it is to be feared that a good many weakly men were turned adrift. We can trace the discharge of only twenty-eight men on reduction at Exeter, but an enormous number of sick had been left at Winchester under Surgeon Wright and none of them rejoined. The order for the 30th to hold itself in readiness to embark for Gibraltar was issued on May 21st, and soon afterwards it moved to Plymouth Barracks to await the arrival of the transports. Contrary winds delayed the transports till July 31st, when Colonel Ramsay reported that the 19th had embarked successfully, but that the tenders with the 30th on board could not reach the transports, and in any case until the wind shifted they could not get out of Plymouth Sound.

The transports were delayed at Cork and the regiment did not disembark at Gibraltar till September 3rd.

The strength on embarkation was—Nine companies of 25 sergeants, 22 corporals, 19 drummers, and 335 privates.

The officers were Brevet Lieut.-Colonel James Ramsay, Captains Peter Dumas, Christopher Maxwell and John Augustus Jevers; Lieutenants Vincent Cunningham, Loftus Nunn, John Winter, William Prosper Popple (Adjutant), John MacLellan and Thomas Jennings; Ensigns Lewis Lewis, Thomas Wilkie and Jonathan Warner Gibbes; Chaplain Edward Thomas; Quartermaster Robert Marshall; Surgeon's Mate Robert Primrose.

GIBRALTAR

The third tour of duty of the regiment in Gibraltar was quite uneventful, Lieut.-Colonel Paston Gould, who had succeeded Jennings in March, 1764, joined some time after midsummer in that year and took over the regiment from Major Ramsay.

In 1768 new clothing regulations were issued. The change of most importance was in the spatterdash coming above the knee. It was still retained in its old form and was of linen as before, but the colour was changed to black instead of white, so that the pipe-claying and consequent dampness were got rid of.

The officers' buttons and lace were still of silver, but the latter had a sky-blue stripe. The coats of the other ranks were looped with worsted lace and their hats bound with white tape. Bearskin caps were to be issued to the grenadiers and drummers when required.

On February 26th, 1770, Colonel Boyd, Lieut.-Governor, who was in command at Gibraltar, was ordered to reduce each regiment to nine companies of 3 sergeants, 3 corporals, 2 drummers, and 42 privates apiece, with 2 fifers in addition for the grenadiers.

The attitude of the Spanish Government caused an increase of the army in December, when the colonels of 66 regiments were ordered to enlist 20 men extra for each company and to bring the number of companies up to ten by the formation of a company of light infantry. The increase in the strength of the companies was countermanded in the following year, when there was an amicable settlement with Spain, but the light companies remained. That of the 30th may be said to date from January 25th, 1771, when the officers were appointed, though no further steps were taken in its formation till the arrival of the regiment in England. The officers were Captain John Augustus Jevers, Lieutenants Lewis Lewis and Sir Robert Stuart, Bart., of Tillicoultry.

On March 15th, 1771, the 30th embarked at Gibraltar for home, along with the 19th. Lieut.-Colonel Gould had gone on leave in the previous year, and Major Ramsay brought the regiment home. The orders to be observed on board ship were :—

Every morning men to be brought up on deck, berths cleaned and bedding brought up to air.

No smoking between decks.

Men to be on deck as much as possible.

No man to land at any port without a pass.

3*d*. a day stoppage from every officer and man for provisions.

No one to be suffered to vend or distribute drams or spirituous liquor; no gaming.

BRITAIN

The regiment landed in April, but we have no knowledge of its movements afterwards. On the 6th of the month the War Office was uncertain whether it would land at Plymouth or Portsmouth. For some reason the muster for June 25th was not held until July 10th, when the regiment was at St.

Albans. Major Ramsay was in command. Most of the officers were on leave or recruiting, as were the sergeants and drummers; the officers' servants were with their masters, 29 men were on furlough, and the effectives on muster parade were :—9 sergeants, 13 corporals, 5 drummers and 111 privates.

The regiment had brought over 250 privates from Gibraltar, but 111 men had been discharged on June 28th, of whom about a fifth had been recommended for pension. A number of recruits had been collected to await the arrival of the regiment, but when the order for augmentation was cancelled, they had been discharged, and only nine joined.

Immediately after being mustered, the regiment moved to Lincoln, where it received new buff leather accoutrements for all the companies, including the light infantry. The latter were also supplied with a hatchet and case to be carried, usually on the knapsack, but sometimes in one of the two frogs on the waist belt, the bayonet being carried in the other. In place of the long spatter-dashes, the light company were given gaiters coming up to the calf of the leg. The officers and sergeants of the company carried fusils instead of the spontoon and halberds.

A new establishment was now sanctioned consisting of 1 colonel, 1 lieut.-colonel, 1 major, 8 captains, 11 lieutenants, 8 ensigns, 1 chaplain, 1 adjutant, 1 quartermaster, 1 surgeon, 1 surgeon's mate, 20 sergeants, 30 corporals, 12 drummers, 360 privates. The capt.-lieutenant was now commissioned as a captain. The sergeants include the drum-major, who was no longer paid by the drummers. He was clothed as a sergeant with drummer's appointments. The twelve drummers include the two fifers for the grenadiers. The privates include nine pioneers with axes and saws and leathern aprons, and also eighteen warrant or widows' men.

In the spring of 1772, the 30th marched under Lieut.-Colonel Paston Gould to Ferrybridge, where it was reviewed by Major-General Alexander Mackay, on May 18th. One hundred and ten men were wanting to complete the establishment and four captains and four lieutenants were on recruiting duty. The recruiting stations were Lincoln, Gainsborough, Boston, Sheffield, Nottingham, Leicester, and Edinburgh. The first five of these stations were in the district in which the regiment was raised, and from which it drew its recruits till 1714.

An entirely new feature of the drill at inspection was the skirmishing of the light company out in front to cover the formations.

Major-General Mackay, who was considered hard to please, made a very favourable report.

The regiment moved from Ferrybridge to Berwick and Tynemouth, and was again inspected by Major-General Mackay; it was now complete, consequently more than two-thirds of the men on parade had joined in the preceding eighteen months. Major Peter Dumas was in command, Lieut.-Colonel Paston Gould being on two years' leave.

In the inspection state Lieutenant John MacLellan is shown as absent on important business. The fact was that he had just won his claim in the House of Lords to the title of Kirkcudbright. His family had taken an active part in the civil wars from 1642 to 1660 and had fallen into such poverty that the title had been unclaimed for some generations. He returned from leave as Lord Kirkcudbright.

After inspection, the regiment moved to Edinburgh and then to Perth to be employed in making roads in the Highlands. A stone was erected by the men on the road to preserve the memory of their work and a photograph of it, taken 130 years later, was kindly presented to the officers' mess by General Sir Archibald Hunter, then commanding in Scotland.

At Perth on June 11th, the regiment was inspected by the Duke of Argyll, who expressed his approval of everything except the physique of the men, especially the recruits. He ordered the regiment to be diligent in recruiting, and to get rid of their weakly men as fast as others could be got. The Duke's orders do not go to the root of the matter, which was that wages and the standard of living in civil life had improved without any corresponding increase in the soldiers' pay. The recruiting sergeant had therefore to take what he could get, and every regiment had a number of men who from age or some physical defect were undesirable. When the regiment moved to Ireland, where wages were lower and living cheaper, recruits came in readily, but as they were mostly Catholics, whose enlistment had been forbidden for nearly a hundred years, the young men did not realize the seriousness of the step they were taking and joined one day and walked off the next if the whim took them.

In the autumn of 1774 the regiment moved to Glasgow, and it would seem to have had a detachment at Paisley, for Lieutenants James Lee and William Rochfort joined the old Masonic Lodge there. In the spring of 1775 it embarked for Ireland.

IRELAND

The 30th landed in Ireland on April 19th, 1775, and was quartered in Galway, where it was inspected on June 9th by Major-General Gisborne, who had served sixteen years in it and naturally expressed a high opinion of it.

The strength was 378 non-com. officers and men, of whom 266 were English, 98 Scots, 12 Irish and 2 foreign. The recruiting parties were at Boston, Nottingham, Leicester and Coventry.

General Gisborne notes that the officers' uniform consisted of scarlet coats, silver laced buttonholes, faced with pale yellow, white lining, round sleeve and cut pocket, buttons numbered and waistcoat and breeches white; the whole made exactly conformable to the men's; a silver epaulette and a silver laced hat.

The remainder of the year was spent at Galway and Athenry.

An order dated August 7th, 1775, raised the strength of each company by 1 sergeant, 1 drummer and 17 privates.

In 1776, the 30th occupied Belfast, Killough, Carrickfergus, Armagh, Drogheda and Charlemont. Augmentation always did knock a regiment to pieces, for a time, because the officers and non-com. officers were absent recruiting when they were most wanted with the battalion. The plight of regiments broken up in detachments was, of course, much worse than that of those united at Headquarters.

We are fortunate in having a copy of a recruiting poster issued by Captain Jevers at Boston.

"ALL GENTLEMEN VOLUNTEERS

"Who have spirit, inclination and ability to serve His Majesty King George the Third, in the King's Own Yellow Regiment of Foot, the Thirtieth, whereof Lieut.-General John Parslow is Colonel, and in the Company of Light Infantry commanded by John Augustus Jevers Esq., Captain in the said Regiment; are hereby invited to repair to Captain Jevers at his station at Boston, in the County of Lincoln, or to his recruiting party in this town, where they will receive His Majesty's bounty, and enter into present pay and good quarters, and when they join the Regiment now fixed in the plentiful and flourishing Kingdom of Ireland, they will receive new clothes, arms, and accoutrements, and every appointment necessary to complete a Gentleman Volunteer.

"Ireland is the only part of His Majesty's Dominions where the price of provisions has not been raised. Beef and Mutton are sold at twopence a pound throughout the year and every other article proportionably cheap. Vegetables of all sorts are in such profusion that for one penny as much can be bought as will serve six men. His Majesty has been pleased to allow convenient and comfortable barracks for all his troops in every part of the Kingdom and to supply them with beds, bedding, kitchen utensils, coals and candles without any charge.

"All young men who are inclined to embrace this opportunity of entering into so desirable a service, shall meet with every encouragement and indulgence that their officers can grant. They shall have the free exercise of their respective trades and occupations in a country where there is the greatest demand for, and the highest prices given to manufacturers, artificers and workmen; and when discharged from the regiment they will be entitled to the privilege of setting up in their different professions in any town in Great Britain or Ireland without expense, hindrance or molestation.

"No apprentice, seafaring man or militiaman will be admitted."

The drummers who beat round the town daily were dressed in yellow, the colour of the regimental facings, and the drums they beat were of the same colour, which explains the flourish about "The King's Own Yellow Regiment."

In 1776 there was some rather amusing correspondence over Captain Christopher Maxwell's promotion. Major Peter Dumas informed the regiment early in the year that he was willing to go, but that he expected 3,000 guineas, the price he had paid for the majority. Captain Maxwell was first for purchase, but could, with the assistance of his friends, only find 1,000. Lieutenant Wilkinson, the seventh lieutenant on the list, then offered 2,000 guineas for Maxwell's company, and the lieutenants senior to him agreed to allow him to pass over their heads, but the Irish Government upset the arrangement by refusing to give a company to so junior a lieutenant as Wilkinson, who was not twenty years of age. In this difficulty Captain Maxwell invoked the assistance of the Secretary of War. He reminded him of the services of his family to the Crown and his own long service, and implored the Secretary's good offices with His Majesty. Maxwell forgot one thing, which was, that although the War Office knew all about the over regulation prices given for commissions, the officials pretended on all occasions to be quite innocent of any such knowledge. The Secretary of War was therefore placed in rather an awkward position but he got out of it by endorsing Maxwell's letter—"Acknowledge," etc. :—

"I have nothing to do with the promotion in Ireland but I am surprised any officer in the King's service should presume to receive or pay more for a commission than the King's regulated price and that he should own it in a letter to the Secretary of War."

After this there was nothing left for Maxwell but to make another arrangement, and we are in doubt whether Major Dumas got his price. Wilkinson had to wait till he was twenty-three before he got his company. The regulated price of commissions was now : Lieut.-colonel, £3,500 ; major, £2,600 ; captain, £1,500 ; capt.-lieutenant, £800 ; lieutenant, £550 ; ensign, £400. Major Dumas therefore asked for £550 over regulation, not an excessive demand.

In 1777 the regiment was assembled at Dublin and the previous year's increase and decrease are very interesting. The number enlisted was 415, but 17 died, 43 were discharged, and 155 deserted, leaving a net gain of 200.

Lord Cavan inspected the regiment on September 11th, and expressed his opinion with great frankness. The men, their accoutrements and clothing were dirty, they were not well set up, "wanting that carriage of body so as to give them any degree of consequence ; there was no sprightly appearance among them." He adds, "as the regiment consists of a great number of young soldiers, some allowance must be made." This is to put the case very mildly ; the regiment was really in a state of flux. Not every one of the 155 deserters can have been clothed, but enough uniforms must have been lost to run every company heavily into debt.

Lord Cavan ordered new colours to be applied for in place of the old ones, which were quite worn out after fifteen years' service. He also ordered the buff cross belts of the light company to be replaced with tanned leather.

The increase of the regiment was entirely in English and Irish, the number of Scots was almost stationary. Two hundred and fifty men were still wanting to complete, for the establishment had been raised to 4 sergeants, 4 corporals, 2 drummers, and 77 privates in each company.

In 1778 the 30th had the misfortune of being broken up again in detachments, Youghal, Dungarvan, Waterford, Lismore, Arklow, Clogheen and Wexford being the stations. It was brought together for the drill season in the camp at Clonmel, where it was inspected on October 15th by General Gabbett. The number enlisted since last inspection was 284, but 8 had died or been discharged, and 76 had deserted, leaving a net gain of 200.

The officers absent from inspection were Captain Henry Roper, A.D.C. to the Lord-Lieutenant ; Captain Robert Hobart, on service in North America ; Lieutenant Daniel Patterson, D.A.Q.M.-General in the northern district, England, and three officers recruiting.

General Gabbett's remarks show that in spite of the constant influx of recruits, the regiment was getting into shape again. He says that the officers made a good appearance, the non-com. officers were clean, well dressed and alert, the men were not well dressed or properly set up but "they levelled and fired well and by attention would soon be a fine corps ; they are fit for service."

The light company was at this time in camp at Kinsale. A distinct advance had been made in training troops for war by the formation there of a "Camp of Instruction" under Lieut.-Colonel Alexander Stewart of the Buffs with Captain Marjoribanks of the 19th as second in command with the rank of major of light infantry. The light companies of the Buffs, 11th, 19th, 30th, 32nd, 36th, 66th, 67th, and 68th were assembled at this camp, where in addition to the usual movements the troops were practised in skirmishing and constantly worked in half companies or smaller bodies in order to place the junior officers and the non-com. officers in positions of responsibility. The skirmishing was usually at four paces interval.

In 1779 the regiment was scattered in detachments at Waterford, Dungarvan, Wexford, and Arklow; the light company was still at Kinsale.

In the summer the grenadiers and the battalion companies were assembled for drill in the camp at Carrigaline, where they were inspected by General Mocher on September 30th. Captain Erskine had gone and had been succeeded in the command of the grenadiers by the adjutant, Thomas Campbell Lieutenant Tenison Edwards had been appointed adjutant. Since last inspection 194 had enlisted, but 23 had died or been discharged, 10 had been transferred to other corps and 85 had deserted, leaving a net gain of 76.

On March 17th, 1780, Captain Lee, commanding the light company, exchanged with Captain George W. Ramsay of the 94th. The command of the light and grenadier companies was of special importance now in view of the hard service they were soon to experience.

At the end of the drill season the regiment was assembled at Kinsale with a detachment of three companies under Major Maxwell at Charles Fort. The new colours to take the place of the worn-out ones of 1762 were probably presented at Kinsale in October.

As the showing of fictitious names in the muster rolls to enable pay to be drawn for contingent expenses was now about to disappear, it is worth while to give a few of them. In the colonel's company they appear as Thomas Hail, John Rain and Richard Frost; in the lieut.-colonel's as John Rain, William Hail and Thomas Snow; another has Thomas Winter, John Summer and William Evergreen; another has John Slack, Arthur Letgo, and Thomas Fall.

In a letter dated October 1st, 1780, the Secretary of War informed the Irish Government that the King had selected the 3rd, 19th and 30th regiments of foot for foreign service, but that they were not to be increased by two recruiting companies. Three regiments, of which the 49th was one, had been brought home after very severe service in America and their recruiting companies were to be employed in filling up the 3rd, 19th and 30th before enlisting for their own corps. In return for this good office the 30th as soon as it was ready for service was to transfer nine lieutenants and ten sergeants to the 49th.

In the month of March, 1781, the 3rd, 19th and 30th were reported complete, and the following transfers were made from the 30th to the 49th: Lieutenants Daniel Patterson, Andrew Rock, James Dumas, Alexander Hervey, Jonathan Rodgers, William Charles Lynam, Xtopher Irwine, John Henry Gamble, George Byng, and ten sergeants.

CHAPTER V

WAR OF AMERICAN INDEPENDENCE. BATTLE OF EUTAW SPRINGS. CREATED THE CAMBRIDGESHIRE REGIMENT. SUPPRESSION OF TWO RISINGS OF NEGROES IN DOMINICA

1781–1791

ON March 17th and 18th, 1781, the Buffs, 19th and 30th regiments embarked at Monkstown in transports under command of Colonel Paston Gould of the 30th. The following is the strength embarked by the 30th: the officers are given by companies:—

General Parslow's company: Capt.-Lieutenant Brereton, Ensign Thomas Mitchell Brown, Adjutant Tenison Edwards, Quartermaster John Marshall, Surgeon James Bowen, Mate John Grove.

Lieut.-Colonel (colonel) Paston Gould, Ensign George Gladstone, Major Christopher Maxwell, Ensign George Hall (sick).

Captain (bt.-major) J. A. Jevers, Lieutenant Jeremiah Palmer, Ensign Wilfred Wilkinson.

Captain John Winter, Lieutenant Wm. Green, Ensign Charles Ruxton.

Captain Thomas Campbell (grenadiers), Lieutenants Joseph Shearman, Elias Best.

Captain G. W. Ramsay (light), Lieutenants Thomas Anketel, Isaac Montgomery.

Captain Charles Cracroft, Lieutenant Thomas Wilkie, Ensign Piers Rowe.

Captain Wm. Rochfort, Lieutenant Thomas Gowan, Ensign Richard Fitzgerald.

Captain Wm. Wilkinson, Lieutenant John Hodnett, Ensign Edward Gibson.

Total effectives: 31 officers, 30 sergeants, 40 corporals, 22 drummers, 623 privates; about 25 sick were also embarked.

A volunteer of the name of Nicholas Metcalfe embarked with the light company. There is little doubt that this was Lieutenant Nicholas Metcalfe who had retired from Jevers' company in the previous November, and who rejoined when he found that his old regiment was going on service.

SOUTH CAROLINA

Some of the leading events in our war against the revolted American colonies, France, Spain, Holland, the Mahrattas and Hyder Ali in the twelve months previous to the arrival of the 30th in South Carolina are here given.

On May 13th, 1780, Charleston, the capital of South Carolina, surrendered to General Sir Henry Clinton, the British Commander-in-Chief in North America. Sir Henry, leaving Lord Cornwallis with 4,000 men in South

Carolina, returned to our Headquarters in New York, where the British army was confronted by the American under General George Washington. On Clinton's departure, Lord Cornwallis pushed columns into the interior and established posts on the upper waters of the rivers. Fort 96 was the strongest and the most distant of those posts. In June, General Washington sent two regular regiments from his army to South Carolina under General Gates. Guerilla bands sprang up to aid them throughout the State. In New England a French army of 6,000 men with a siege train landed in July and fortified a harbour in Rhode Island. In August, Lord Cornwallis routed General Gates. The autumn and winter were unlucky for us; in India, Hyder Ali destroyed a force under Colonel Baillie at Conjeveram; in South Carolina, Ferguson's rangers, 1,000 strong, were destroyed by American mounted riflemen; in December, Holland joined the coalition against Britain and the neutral fleets of Russia, Sweden and Denmark were fitted out as a threat to us; in January, 1781, our irregulars, under Tarleton, were defeated and the loyalist militia broken up throughout South Carolina. In spring things took a turn for the better; in India, Colonel Camac defeated the Mahrattas, and on March 17th Lord Cornwallis defeated General Greene, the new American commander in South Carolina. In April, Lord Cornwallis left Lord Rawdon with very scanty forces to defend South Carolina and commenced his march to Virginia.

With 800 men Lord Rawdon again defeated General Greene, who had 1,200 with him. In spite of this victory Rawdon decided to contract his lines, give up all the country which he could not effectively hold and concentrate near the coast. The difficulty was to withdraw from Fort 96, our furthest and strongest post, and to provide for the safety of Augusta, the capital of the neighbouring State of Georgia. To achieve this, Lord Rawdon sent repeated orders to Colonel Cruger, commanding at Fort 96, to fall back upon Augusta; he would thus avoid General Greene's force and secure the safety of Augusta. Rawdon himself retired on Monk's Corner, 30 miles from Charleston. Unluckily, all his orders were intercepted by American partisans; Colonel Cruger stood fast at Fort 96 and was besieged by General Greene, who was able to spare sufficient force to besiege Augusta also.

The relief of Fort 96 seemed hopeless, but on June 3rd the situation was entirely changed by the unexpected arrival of the Buffs, 19th and 30th, who were supposed to have been ordered to land at New York. The regiments were surprisingly healthy, in spite of the long voyage of seventy-eight days. In the 30th the officers who had embarked at Monkstown were all effective, the strength of other ranks disembarked was 742. Two volunteers now joined, Mr. Thomas Legge, who was placed in the colonel's company, and Mr. William Jevers, who served in the company of his father, Bt.-Major Jevers.

Augusta fell two days after the 30th landed, but Lord Rawdon at once took steps to relieve Fort 96. He formed the flank companies of the newly-arrived regiments into a battalion under Major Marjoribanks of the 19th regiment, under whom so many of the officers and men had served in the camp of instruction at Kinsale, and added it to his field force; he then advanced, leaving the 19th at Monk's Corner and the Buffs at Orangeburgh, and pushed on to Fort 96. Colonel Paston Gould, who, after Lord Rawdon, was the senior officer in South Carolina, remained in command at Charleston and kept the battalion companies of the 30th with him.

Fort 96 is 150 miles north-west of Charleston, Monk's Corner 30 miles north and west, and Orangeburgh midway between Charleston and Fort 96; Lord Rawdon's great danger was therefore that he might arrive just too late and find that Fort 96 had fallen and that he was in presence of a superior force under General Greene with no support nearer than the Buffs, 75 miles away at Orangeburgh.

Colonel Cruger had under him at Fort 96 over 500 excellent soldiers of the New York and New Jersey loyalists and was himself a capital officer, so that the siege was carried on slowly and with considerable loss to the besiegers. Things were not nearly ripe for storming, but hearing of Rawdon's approach, Greene determined to risk it The assault was made on June 18th and failed badly. General Greene thereupon sent off his heavy baggage, and on the 20th fell back across the Saluda, and continued to retreat northwards till he had passed the Enoree. Lord Rawdon arrived at Fort 96 on the 21st and followed across the Saluda, but General Greene had too good a start, the heat was terrific and after crossing several parallel streams Lord Rawdon gave up the pursuit. Colonel Cruger was ordered to escort with his own troops and half of Lord Rawdon's all the loyalists in Fort 96 and its neighbourhood, who might wish to withdraw to Charleston. Meanwhile, Lord Rawdon covered his retreat with 800 infantry and fifty sabres. The operation would have been safe enough if the Buffs had been where Lord Rawdon had posted them at Orangeburgh, but by some mistake they had been withdrawn, and as usual the letter from their commanding officer containing a warning of this to Rawdon had fallen into American hands. General Greene, who had turned and followed Lord Rawdon the moment the latter retreated, now heard that the small British force was isolated, and pushed on, while Lee's Legion occupied the roads in Rawdon's rear, breaking down the bridges over the creeks and holding the fords.

From the boldness with which the Americans came on, driving in his cavalry and capturing his foragers, Rawdon judged that Greene's whole army was at hand; he therefore retired at once on Orangeburgh, brushing aside the Legion which tried to stop him. He was joined there the following day by Lieut.-Colonel Stuart and the Buffs, who had been ordered back, and he also received word that Colonel Cruger had reached Charleston with the loyalists and their families and was preparing to join him in the field.

That evening, General Greene made a show of reconnoitring Rawdon's position, but decamped in the night, passed Rawdon and marched against the 19th regiment at Monk's Corner. He at first had some success, but Lieut.-Colonel Cotes occupied some houses and sent to Charleston for support. On the approach of Lieut.-Colonel Paston Gould with 700 men Greene fell back to the north, and marched to the high hills of Santee, 40 miles from Orangeburgh. The heat during these operations had been deadly; Lord Rawdon's force alone had lost fifty men from sunstroke. He himself was incapacitated and was forced to take advantage of the leave to Europe which had been granted him long before, but which he had declined to use till he had put things in South Carolina on a proper footing. He was a leader of a very high class.

The departure of Lord Rawdon left Colonel Paston Gould, lieut.-colonel of the 30th, the senior officer in Carolina; he was a colonel of 1777 and was shortly to be given local rank as major-general. Being in hourly anticipation of the arrival of Lieut.-General Alexander Leslie, he awaited his coming in Charleston, leaving the command of the field force at Orangeburgh with Lieut.-Colonel Stuart of the Buffs.

After being joined by Colonel Cruger, Stuart had a slight superiority over General Greene and followed him to the hills in August, thus reversing Lord Rawdon's policy. Both the latter and Lord Cornwallis had beaten the enemy handsomely in the field, but had found it impossible to maintain themselves far from the coast.

About the time of Stuart's advance, Greene received an important reinforcement of three regiments of regular (continental) troops and began to move, but as he had no means of crossing the Santee he marched up to Camden, apparently in retreat, and crossed there, then turned south towards Stuart on the Congaree. Stuart, to meet a convoy which he expected, fell back upon Eutaw Springs, about 70 miles from Charleston and 20 north-west of Orangeburgh. During this time the American irregulars were so active that Stuart's knowledge of Greene's movements was very slight. Like most British commanders he was hampered by want of cavalry, having only fifty mounted men with him.

On September 8th General Greene arrived at Burdell's plantation 7 or 8 miles from Colonel Stuart's position at Eutaw Springs. A fortnight before he had been only 15 miles distant from the farm, but had made a circuit of 70 miles to reach it. He resolved to attack next day; he had with him over 2,000 men, of whom 1,200 were regular soldiers.

Colonel Stuart's force was probably 300 less, he was very much weaker in cavalry and his guns were of smaller calibre. He had already begun to find out the difficulty of maintaining such an advanced position, being out of breadstuffs and dependent for vegetable food on rooting parties who were sent out daily and whom he had not sufficient cavalry to protect. He must have been well aware by now that even a victorious action, unless General Greene's army was actually broken up, must be followed by retreat, but he appears to have been unconscious of the enemy's proximity.

Starting from Burdell's plantation at four o'clock on the morning of the 9th, Greene's cavalry came upon a party of unarmed men belonging to the British force gathering sweet potatoes, and 2 subalterns, 1 sergeant, 4 drummers and 140 men were captured, of whom 1 subaltern and 62 rank and file belonged to the flank battalion. The few mounted men who were with the rooting party galloped back to Colonel Stuart's position 3 miles in rear and gave the alarm.

The road on which the Americans were advancing led through high forests with here and there clearings and farm houses. Stuart's position was in one of the largest of these clearings where he had room to deploy his line, now less than 1,600 strong. The ground was not, however, entirely cleared, a number of thickets being left.

On the British right was the battalion of the flank companies of the Buffs, 19th and 30th; its right was about 200 yards from the Eutaw creek, which ran at right angles to the line. The actual strength of the battalion in action was 2 majors, 5 captains, 10 lieutenants, 1 adjutant, 1 quartermaster, 1 surgeon, 11 sergeants, 12 drummers, 234 rank and file. This is after deducting the men captured in the morning. On the left of the flank battalion were two guns and in rear of them the New York and New Jersey loyalists who had defended Fort 96 so well under Colonel Cruger. The line was continued by the remains of the 63rd and 64th who had served with great honour throughout the war. The Buffs were on the left. A small stream or ditch ran along the British front.

As soon as General Greene had room to deploy, he formed his army in three lines.

The first line was entirely composed of militia under Marion, Pickens and Malmady with two guns in the centre.

The second line consisted of regulars, and stood from right to left, North Carolina regiment under Sumner, Virginian under Campbell and two Maryland regiments under J. S. Howard and Hardman. Lee's Legion covered the right flank of the militia and Henderson with some Carolina light troops the left. The second line also had two guns with it. The cavalry under Colonel Washington and a Delaware regiment were in the third line.

The American guns were 6, and the British were 3-pounders.

The militia up to this time had not behaved with any great steadiness in a regular action, though they had shown great gallantry and enterprise in detached fighting. The action commenced with an artillery duel in which each side had a gun dismounted. The American militia then advanced to musketry range and showed unexpected firmness, some corps burning fourteen or fifteen cartridges without giving way. The North Carolina men alone, who were opposed to the Buffs, broke after half a dozen discharges. Their regular battalion advanced at once to support them, but also fell back; the fugitives of the militia probably embarrassed them. The Buffs on this unluckily charged, but Colonel Campbell wheeling up the Virginian regiment fired a volley and advanced on them with the bayonet while they were in disorder. They were forced to give way and carried with them the 63rd and 64th. General Greene now prepared for a general advance and to clear his left sent Washington with the cavalry to dislodge the flank battalion. Washington advanced very gallantly between the British right and the creek and forming to his right tried to roll up the battalion from right to left, but Major Marjoribanks, who showed remarkable skill and alertness throughout the day, received the American cavalry with such a fire that Washington was wounded and taken prisoner, and every officer with him was killed or wounded. An attack by the reserve regiment of infantry was equally unsuccessful. The New York and New Jersey loyalists had at the first sign of confusion been ordered to occupy the brick farm buildings in rear of their position and Marjoribanks now fell back in line with them, occupying some fences and thickets on their right. The rest of the British force rallied on its centre and right, and the position was now stronger than ever. To add to General Greene's difficulties the British camp had been uncovered by the retreat of the army to a new position and some liquor found, and some of his regiments which had behaved so well in the morning dispersed in whole or in part to plunder. One desperate effort was made by the Americans to win the day, two 6-pounder guns with two smaller ones captured from the British were brought up to batter the brick buildings at close quarters, but the fire from the houses and gardens was so hot that the gun detachments were shot down; the loss also among the American officers in trying to re-form and bring up their straggling men had been excessive. General Greene now saw that it was useless to persevere and prepared for retreat. The remains of the American cavalry were ordered to the front, but were nearly annihilated by the fire from the house and garden. Major Marjoribanks saw the time had come, and bringing up his right shoulder, opened such a fire on the flank of the Americans that Greene fell back, leaving in front of the farm house the four guns and a mass of killed and wounded.

There was no pursuit beyond the clearing. The American army retreated the 7 miles they had advanced in the morning to Burdell's plantation. General Greene claimed that he had gained the victory, but had to retire for want of water, but there was abundance of water at Eutaw and Greene had only to drive Stuart from his position to get it ; if he failed in that, he failed in everything. He had carried off one British 3-pounder gun and left two of his own 6-pounders on the field.

Colonel Stuart in his despatch alleges that he was the victor and that the American army was completely routed, and that he only wanted cavalry to complete its destruction. The truth seems to be that Colonel Stuart had distinctly the best of the day's fighting, but that he did not gain such a victory as would enable him to maintain his advanced position.

Eutaw was the hardest fought action of the war and the losses were heavy on both sides.

On that of the British 3 officers and 82 men were killed, 16 officers and 335 men wounded, 10 officers and 247 men taken prisoners. General Greene admitted the loss of 60 officers and 494 men killed, wounded, and missing. No doubt his regular troops returned correct statements of their loss in action, but the militia returns are incomplete and misleading. Colonel Stuart alleges that 200 dead were left by the Americans on the field, and he can scarcely be mistaken about that. Probably Greene's loss in men was more in proportion to his loss in officers than he allows.

The loss of the flank battalion was : Killed, Lieutenant Hickman, light company, 19th regiment, 2 drummers, 12 rank and file ; wounded, Lord Edward Fitzgerald,[1] light company, 19th regiment, Lieutenant Anketel, light company, 30th regiment, 5 sergeants, 81 rank and file.

Among the rank and file wounded was volunteer Nicholas Metcalfe of the light company of the 30th ; neither he nor Lieutenant Anketel was ever able to serve again. Metcalfe was probably maimed, for he received compensation for wounds on the same scale as an ensign. General Parslow, colonel of the regiment, forwarded a memorial from Lieutenant Anketel to the Secretary at War, and said he would be infinitely obliged if the Secretary could procure some relief for that officer, whose case was uncommonly distressing, but we do not know the result of this application.

Two drummers were missing ; with that exception the flank battalion lost no prisoners in action. Among the rank and file two out of every five men were killed or wounded.

It is impossible not to feel that the co-operation of Stuart and Marjoribanks and the lessons learned in the Camp of Instruction at Kinsale had a good deal to say to the favourable course of the action. One-half of the flank battalion was for two years in that camp under these officers.

[1] This is the Lord Edward who came to so sad an end in 1798.

On the day following the action neither side showed any disposition to renew the fight, but in the evening Colonel Stuart, leaving seventy men behind, who were too severely wounded to be moved, fell back towards Charleston. General Greene followed, but made no attempt to harass the British rear. In fact his only object seems to have been to write his despatch from some place in advance of where the action was fought. In the despatch, written at Martin's Tavern, he announces his intention of retiring again to the high hills of Santee as soon as he had rested and refreshed his men. Meanwhile, Major-General Paston Gould (he had received local rank) had marched to meet Colonel Stuart's force, taking with him the battalion companies of the 30th. The junction was effected on the 12th at Monk's Corner, where General Gould took command. On the 14th he heard that General Greene had reached Martin's Tavern, 12 miles distant. Two days were spent in sending the sick to Charleston and resting Stuart's men who had been overworked and underfed for a week. On the 16th, General Gould advanced to bring Greene to action, but the latter fell back hastily to the Santee, crossed it in three divisions and destroying or hiding the boats, returned to the hills from which he had started a month before.

On September 30th General Gould wrote to the Commander-in-Chief from Maham's Plantation on the Santee that he had no intention of following General Greene further, but " will fall down the Santee River, living on the country to spare the stores in Charleston." He adds that he does not care to be far from Charleston as Colonel Nesbitt Balfour sends him word from there that there is a French fleet on the coast. Apparently this is the first news received of the great French fleet under de Grasse which finished the war and gave their independence to the United States.

Lord Cornwallis had united with the British troops in Virginia in June, and after some indecisive fighting with a force under the Marquis de la Fayette on the James River had occupied Yorktown and Gloucester on August 2nd. The first of these places is on the south and the second on the north bank of the York River close to where it falls into Chesapeake Bay. On the 30th of the month the Comte de Grasse arrived in the Bay with twenty-eight French sail of the line and on September 5th Admiral Graves attacked him with a fleet from New York of twenty-four British line-of-battle ships. We claimed that the action was indecisive, but it was one of the decisive battles of the world. Graves had to retire to New York to refit, eight more French ships joined de Grasse, bringing 4,000 troops and a siege train under the Comte de Rochambeau from Rhode Island. The French troops were landed and joined by de la Fayette's army from the James River. Admiral de Grasse managed also to send vessels to the head of Chesapeake Bay which brought 5,000 of Washington's army. General Washington accompanied his men and joined de Rochambeau on September 14th. Lord Cornwallis had little but field artillery to oppose to the French siege train and on October 19th, having only one or two guns which could be fired, he surrendered with 6,630 men, of whom 2,000 were in hospital and nearly 1,000 detached across the water at Gloucester and unable to join in the defence. The capitulation was signed by de Grasse for the French navy, by de Rochambeau for the French army and by General George Washington for the American army. Five days too late our fleet returned from New York with 5,000 troops on board under the Commander-in-Chief, Sir Henry Clinton ; he found the place deserted and returned to New York.

This practically ended the war. It is true we had only lost a detachment, though an important one, and still held the winning cards, but the party in the House of Commons opposed to the war received such an accession of strength that on February 27th, 1782, a motion was carried to stop all offensive operations against the American colonists. The majority against Government was nineteen in a House of 449.

In South Carolina there had been no operations of importance in October. Lieut.-General Leslie had arrived and taken command at Charleston ; General Greene was still on the high hills of Santee, trying, in spite of the exhaustion of the colony, to equip a force to march on Savannah. His means did not justify any attempt on Charleston, although the American partisans now and then showed themselves close to it, and there were frequent small skirmishes.

There was a good deal of sickness after the summer campaign. At Christmas time the eight battalion companies of the 30th had thirteen officers and 439 other ranks fit for duty ; 103 were sick. The returns of the flank battalion do not give the number of the regiment except when there is a casualty among the officers, so we do not know the losses of the grenadier and light companies of the 30th in the field, but we know that in each company eighteen out of seventy-seven non-commissioned officers and men were killed or died of disease in six months, and that at the close of the campaign each company had two non-commissioned officers effective out of the seven with which it started.

After the surrender of Cornwallis at Yorktown the French turned their strength against the British possessions in the West Indies, where they hoped to be able to repay themselves for their exertions on behalf of our revolted colonies. The 30th felt the effect of this. On March 14th, 1782, the Commander-in-Chief ordered the Buffs, 19th, 30th, 82nd, 84th and a detachment of Jægers to be ready to sail from Charleston to the Leeward Islands. This move was to be in concert with the despatch from home of a fleet under Sir George Rodney to the West Indies. The influence of General Greene's threatened movement against Savannah was now seen. A council of war assembled on April 27th by General Leslie declared that unless we evacuated Savannah, no troops from the south could be spared for the West Indies, and that the 30th had already embarked its baggage and was on the point of sailing for Savannah. The following day, however, Major-General O'Hara arrived from the Commander-in-Chief with orders to take the Buffs, 19th and 30th to the relief of Jamaica. Lieut.-General Leslie gave O'Hara only two regiments, the 19th and 30th. The flank companies were ordered to embark with their regiments.

The strength of the 30th on embarkation was 1 major, 6 captains, 7 lieutenants, 5 ensigns, 1 adjutant, 1 quartermaster, 1 surgeon's mate, 27 sergeants, 18 drummers, 487 rank and file, fit for duty ; 1 sergeant, 1 drummer, 33 rank and file sick ; 98 women and 64 children. The state shows 2 drummers, 41 privates, "remained behind." They were 2 drummers and 34 privates prisoners of war, 5 officers' servants, 1 cadet, son to Major Jevers, and 1 in prison.

Major-General Paston Gould did not embark with his regiment ; he had been sick in the previous summer and autumn, but had recovered sufficiently to do duty in the winter months. With the approach of the hot weather and the rains, his health failed rapidly and the surgeons ordered him home. Captain Campbell, whose own health had suffered from the climate, embarked with him for England on July 13th. General Gould died at Axminster on November 7th in his fifty-second year. He had commanded the 30th for eighteen years. Bt.-Major Jevers was also sent home sick and his son was given a furlough to accompany him ; the latter did not rejoin. Surgeon Bowen had been invalided six months earlier ; he died at home. Three officers died in South Carolina during the war, Lieutenant William Green on July 27th, 1781, Ensign Charles Ruxton on September 14th, and Lieutenant Thomas Wilkie on October 7th. Wilkie was a very old subaltern, having been given a commission from surgeon's mate for good conduct at Belleisle in 1761.

When General O'Hara, with the two regiments, left Charleston on May 6th, no news had arrived of the great victory gained by Sir George Rodney on April 12th over the French fleet under the Comte de Grasse near Martinique. After the regiments had sailed the position at Charleston remained much as before. In the constant skirmishing which took place, Ensign William Lockhart, of the 33rd, afterwards lieut.-colonel of the 30th, was severely wounded in November, 1782.

WEST INDIES

The transports with the 19th and 30th on board touched at Barbados on June 18th, 1782, and on the 25th arrived at Antigua, the Headquarters of General Matthew commanding in the Leeward Islands. He wrote on July 2nd to the Commander-in-Chief at New York that Rodney's victory had changed everything in the West Indies. Jamaica was no longer in danger and that he had despatched Sir Charles O'Hara with his two regiments to St. Lucia that day.

After leaving Antigua the fleet of transports with the 19th and 30th on board, met with head winds which were too much for some of them to face. A victualler and two transports with Major Maxwell and three companies of the 30th and a like number of the 19th on board, had to run for it. They were supposed to have foundered, but after running 800 miles they reached Jamaica. The others made a better fight of it and at least saw St. Lucia. They sighted the Sugar Loaf hill on that island on July 5th 40 miles to the S.E., on the following day it was 6 leagues to the N.E., on the 7th, the island was 5 or 6 leagues to the E.S.E. ; on the 8th, 11 leagues to the S. ; on the 9th, 18 leagues to the S.E. ; on the 10th, S. by E. 7 leagues ; on the 11th, 8 leagues E. by S. ; on the 12th, the fleet was beaten back and was off St. Vincent ; on the 13th it was close to Martinique, and on the 14th to Guadeloupe ; on the 15th it anchored again in English harbour, Antigua, which it had left a fortnight before. Such were the good old days before steam.

General Matthew was much disappointed ; he was anxious for the safety of St. Lucia, and transferring five companies of the 30th to H.M.S. *Amsterdam* and two to a stout privateer, he sent them off again. They sailed on the 21st and reached St. Lucia on the 26th. The 19th followed a little later. Sir Charles O'Hara took command of the island. The seven companies of the 30th were commanded by Brereton, the captain-lieutenant ; the strength was a little over 300 officers and men.

Preparations were now made to transfer our army from North America to the West Indies, leaving behind a sufficient garrison to hold the ports, and to sweep the French out of the islands. On further consideration, this was given up and the idea of a general peace was entertained by our Government. On November 30th, a treaty of peace with the United States was signed, but its provisions were not to take effect till preliminaries had been exchanged with their allies, Holland, France and Spain.

On January 2nd, 1783, the seven companies of the 30th and the 19th regiment were embarked by General O'Hara at St. Lucia for Jamaica. On landing, they were joined by the prisoners of war and other casuals whom they had left six months before in what had now become the United States of America. Our troops had evacuated Charleston in December. The preliminary articles with France and Spain were signed on January 20th, and the war might be considered at an end. The general peace was finally settled in September, 1783. The net result of the fighting from 1775 to 1783 was that the American colonies which had revolted from Britain secured their independence, France got the Island of Tobago, Spain got Minorca and Florida, the Dutch got nothing. Britain had preserved Canada, Gibraltar, her Empire in India, and her West Indian islands except Tobago. All had been strenuously attacked by one or more of her enemies.

The 30th were united at Stoney Hill Barracks under Lieut.-Colonel Maxwell, who had succeeded Colonel Paston Gould, deceased, from November 7th, 1782, but the official notice of his appointment as lieut.-colonel and that of Bt.-Major J. A. Jevers as regimental major did not reach Jamaica till well on in 1783. As usual after a war there was a reduction and rearrangement of the army and a rush of officers to return to civil life. Apparently there was a false report that Major Jevers, who was at home on sick leave, had retired or died, for a Major John MacGill was appointed to the regiment by the general at Jamaica. He acted as major and captain for some months, but went home on leave in June and was no more heard of regimentally. The general commanding in the West Indies was likewise convinced that Jevers had retired as a captain and gave his company to the Capt.-Lieutenant, Robert Brereton, and brought into the regiment as capt.-lieutenant a Lieutenant John Thomas Layard from the 54th. When the official news arrived that Maxwell, Jevers and Brereton had succeeded to the vacancies caused by the death of Paston Gould, all the appointments made locally were cancelled.

Lieut.-Colonel Maxwell went home in June on six months' leave, which was extended from time to time till nearly seven years later.

Soon after the 30th had reached the West Indies it had become a county regiment. On August 31st, 1782, a list of county titles was published for fifty-eight out of the existing seventy regiments. The 30th was now the Cambridgeshire regiment, but, as was the case with most regiments, the title was seldom used except in official correspondence.

The reduction of the 30th was ordered to take effect from August 24th, and the establishment fixed had a new feature, for, in addition to eight service companies, the officers of two reserve companies were to be maintained so that in case of need the strength of the regiment might be quickly increased. The officers of the reserve companies, who were termed "officers en second," were available for duty with the service companies. All companies were to have the usual three officers, and the staff was unchanged. The service companies were to have 2 sergeants, 2 corporals, 2 drummers, 48 privates, and 2 fifers for the grenadiers, a total of 434 non-commissioned officers and men.

As Captain Campbell was on leave at home, the regiment was commanded by Captain Brereton for eighteen months after Lieut.-Colonel Maxwell left it.

The first step in reduction was the discharge, on August 24th, of all men whose term of service had expired, both the old soldiers who had enlisted before the war and the men who had enlisted for the war since Christmas, 1775. Nine sergeants, 17 corporals, 6 drummers, 277 privates were discharged, which left the regiment with 18 sergeants, 13 corporals, 11 drummers, and 137 privates, but on the succeeding day recruiting commenced. The standard was 5 ft. 7 ins. and men could be taken up to thirty years of age. A bounty of a guinea and a half was offered. It was not much, as the regiment was only two years out from home and had evidently a long spell of West Indian service before it. The result, however, was satisfactory: about half of the discharged men "took on" and the regiment was completed to within sixty of its complement. Almost all the non-commissioned officers were either not discharged or they re-enlisted, an immense benefit to the regiment, for they were very good. Later on seven of them received commissions and two left as staff sergeants.

DOMINICA

Under the new distribution of the army the 30th was sent back to General Matthew for duty in the Leeward Islands. It sailed for Antigua on November 10th under Captain Brereton, and after a short stay in that island went on to Dominica, where it landed on March 17th, 1784.

Dominica, where the regiment stayed for seven years, is an island of 29 miles in length from north to south and from 12 to 16 in breadth; the coast is precipitous and the only good landing places, both on the west coast, are Prince Rupert's Bay in the north and Roseau, the capital, in the south. There is also a harbour for small craft at Grand Bay on the south-east. The interior consists of lofty mountains cleft by wild gorges and ravines which are almost impassable owing to the tropical vegetation; there are numerous waterfalls. On the coast and on the flanks of the hills were the coffee plantations, mostly owned by Frenchmen and cultivated by about 14,000 negro slaves. Roseau had less than 500 houses and the total free population was about 1,200 whites and 500 blacks. The climate was not good, the duties were heavy and it would have been an undesirable quarter but for the extremely friendly relations between the 30th and the inhabitants.

A recruiting company must have been given to the regiment about this time and soon abolished. All we know of it is that Captain Maddison, Lieutenant Piers Rowe and Ensign Will. Geikie are shown on the half-pay of it in the Army List for 1785.

On landing in Dominica a party of twenty-five men was detached to Prince Rupert's Bay, and it is said that nearly every man, woman or child of the 30th and Royal Artillery on that detachment died in the succeeding twelve months.

In June, 1784, the strength of the regiment in Dominica was Captain

Brereton (in command), Captain Marshall (acting both as capt.-lieutenant and quartermaster), 6 lieutenants, 16 sergeants, 24 corporals, 18 drummers, and 291 privates. Six months afterwards orders came to reduce the drummers to 1 and the privates to 42 a company, but the grenadiers were still to keep their two fifers.

The Governor of the island inspected the regiment at Roseau on June 17th, 1785. Major Campbell, who had succeeded Major Jevers by purchase, had returned from England and was in command. He had under him Captains Brereton, Ramsay and Marshall, 5 lieutenants, 4 ensigns, 1 surgeon's mate and 301 other ranks. The adjutant, Thomas Gowan, and the surgeon were on sick leave. Gowan did not rejoin; he never recovered from the campaign in South Carolina. His duties were performed by the sergeant-major, John Russell, who was appointed adjutant on August 3rd.

According to the returns, fifty-seven men had enlisted in the previous twelve months. They were probably from the 49th. The MS. history of the 30th in the R.U.S. Institution says that the regiment received a company from the 49th. This does not appear to be quite correct. The 49th had two companies reduced and the 30th received a captain, a subaltern, and a number of men, but they did not come as a company but a draft. The captain, Joshua Roche, was appointed en second and exchanged to half-pay before joining. The subaltern, Ensign the Hon. William Gore, served with the 30th for a few years. The dates of their appointments given in the army list differ and appear to be wrong. To add to the confusion the name of the captain transferred is often given as Andrew Rock instead of Joshua Roche. Andrew Rock was one of the lieutenants given by the 30th to the 49th in 1781.

Before Governor Ord's inspection, Ensign William Murray joined from being a prisoner of war. He had been appointed to the regiment in February, 1782, and Sir Guy Carleton, Commander-in-Chief in North America, had given him leave of absence from the September of that year till he should cease to be a prisoner of war. We know nothing of how or where he was captured.

Ensign Thomas Legge was at this time shown as absent without leave. He had served as a volunteer throughout the campaign of 1781 and had been granted a commission. When the regiment left Jamaica he was allowed to remain on employ in the island but fell into bad health and died there.

On November 15th, 1786, General William Roy succeeded General Parslow as colonel of the 30th. General Roy had served almost entirely in the quartermaster-general's branch of the staff and had been present at Rochefort in 1757, and at Minden in 1759, and was deputy-quartermaster-general to Lord Granby in the campaign which followed. In 1761–2 he served on the staff at home, but returned to active service in Germany in 1763, when he was promoted lieut.-colonel. In 1783 he was appointed colonel of Engineers. He is best known now by his triangulation to connect the French maps with our own and to determine the actual positions of the observatories of Paris and Greenwich and by his valuable work on military antiquities. He was the first to suggest the use of the barometer in ascertaining the heights of mountains. Many of his surveys are in the British Museum. He took a great interest in his regiment and seems to have been universally liked and respected.

On August 29th, 1787, a great hurricane swept the Leeward Islands; the barracks occupied by the 30th at Bruce's Hill were blown down, Captain

Masterton, Ensigns Cuyler and Gore and thirty-seven men were badly injured and the health of the regiment suffered from exposure.

Ensign Patrick Garret had died at the beginning of the month, but we do not know under what circumstances.

New barracks were completed and occupied by the regiment early in 1788 and the men's health improved, but the detachment at Rupert's Bay continued to be very sickly. In the course of the year there was fear of a rising of the black slaves, and one-half of the 30th was constantly on duty in the unhealthy interior of the island. Although this trouble was dignified by the authorities with the name of the Negro War, our men appear to have been able to keep the peace without using their arms, but many of them died and more lost their health on this service.

There were now ten service companies under an order issued at home in 1787, and the officers en second had been posted to the 9th and 10th. A number of other changes had taken place which came to nothing, such as raising the establishment of privates to 560, then lowering it again to 370. The actual strength of the 30th was under 350 non-commissioned officers and men at the beginning of 1788 and about 400 at the close.

Captain G. W. Ramsay exchanged to half-pay in March, 1789, and the light company was now commanded by Captain W. Urquhart. Captain Ramsay distinguished himself on the Staff in the West Indies in the great war and died a lieutenant-general.

In 1789, General Matthew was ordered to make a tour of inspection in the Leeward Islands in the following year, and Lieut.-Colonel Maxwell rejoined at the end of 1789 or beginning of 1790. On visiting the hospital he was shocked at the appearance of the sick. The surgeons all agreed that the regiment was exhausted by long service in a trying climate, by the hardships consequent on the hurricane of 1787 and the fatigue of the Negro War. The garrison surgeon warned Colonel Maxwell that unless an immediate and plentiful supply of wine was given he would lose some of his men. Colonel Maxwell in a letter to General Roy says: "I told the doctor to buy wine and I would pay for it as I could not let a regiment suffer for want of twelve or fourteen guineas I had been happy in thirty-three years."

Three months later General Matthew arrived and sanctioned an unlimited supply of wine for the 30th hospital.

On May 25th, 1790, General Matthew held a review of the regiment at Roseau, where Colonel Maxwell had under him on parade, 4 captains, 4 lieutenants, 6 ensigns, all the staff except the chaplain and 349 non-commissioned officers and men. General Matthew reported that in appearance and discipline the regiment was inferior to none. Governor Ord, however, wrote to the Secretary at War that while conveying his complete satisfaction with the state of the regiment at present, General Matthew warned him that looking to the age and long service of the men, it might be on the verge of a decline. This raises the question what steps had been taken to introduce young blood. As the effort to establish a recruiting company had been given up in 1788, one-half of the officers were scattered over England, recruiting in the old way, but they sent their recruits to a depot battalion in Chatham Barracks to learn their drill and to be forwarded to their regiments in drafts. This battalion was formed to serve the thirty regiments in Gibraltar, Jamaica and the Leeward

Islands. Two captains and six subalterns were detailed by those regiments for "stationary duty," as it was called at Chatham.

Recruits had already begun to join in Dominica before General Matthew made his inspection. The services of the men of the regiment were 244 men of over eight years' service, 36 men with from two to seven, and 126 with less than two years' service.

This is the first inspection we know of where trousers were worn instead of breeches and gaiters or the spatterdash. The last was irksome in a hot climate, but like its diminutive the spat it kept the shoe on. Marching in trousers only, one-half of the men lost their shoes.

General Matthew says nothing about the head-dress. Throughout the army the three-cornered cocked hat had been given up, and the brim before and behind was now looped up, forming a two-pointed cocked hat worn "athwart ship." If in consideration of the climate, the regiment was allowed to wear the hat with the brim not cocked or looped up, General Matthew would most likely have mentioned it. We know that the grenadier company took their bearskins to Dominica, for they reported them a total loss in the hurricane of 1787. Not one was saved. It sounds a little like a fairy tale. There is a hint elsewhere that they had been damaged by moths.

Since the last inspection by a general officer of which we have a record many changes had been introduced which General Matthew does not mention, but of which we would have been glad to hear, although some are trivial enough. Had the officers laid aside the spontoon which they had carried for a hundred years and had they equipped themselves with a regulation cut and thrust sword with a silvered hilt and a sword knot of crimson and gold? Was the hair of all ranks worn in the queue or clubbed in the latest fashion? it was powdered of course. Was the order enforced for the officers, when on duty, to wear the sash over the left shoulder and the sword belt over the right, outside the coat, but when off duty to wear no sash and the sword belt under the coat and over the waistcoat? The regiment no doubt fell in in three ranks and marched in ordinary time at seventy-five steps with, when required, a quick time of 150 steps a minute, but it may have had a drill peculiar to itself, for Colonel David Dundas had only commenced, in Dublin, the experimental drills which were to give to the army one uniform system.

We have reason to believe that one great reform had been introduced in the 30th and that the men were practised in firing with ball as well as blank cartridge, and that great attention was paid to the handling of arms and the firing exercise with a view to increased rapidity of fire. It was to enable the men to keep up this rapid fire that the number of rounds carried was increased so much that they could no longer be carried on the waist belt. This led to the use of the crossed shoulder belts and a pouch and magazine to carry fifty-six rounds.

After the inspection Lieut.-Colonel Maxwell returned to England and Major Campbell resumed command of the regiment.

In July, General William Roy died and was succeeded as colonel of the 30th by Sir Henry Calder.

The service of the regiment in the West Indies was now drawing to a close. On November 13th the adjutant-general informed the War Office that the 30th would be relieved at Dominica by the 15th regiment and would be brought home.

On January 20th, 1791, the 15th had disembarked at Roseau and the 30th had taken its place on board ship.

Before embarking, Major Campbell had been presented with an address signed by eighty of the leading people of the island, including the Lieut.-Governor and the Speaker of the House of Assembly. The address dwelt upon the benefits which the community had derived from the superior good order and discipline of the regiment. It went on to lament the fact that owing to the House of Representatives being prorogued the representatives were a second time deprived of the power to show in their legislative capacity such public marks of esteem as the conduct of the 30th demanded from a colony between whom and the regiment a mutual harmony and esteem had existed for many years. The address desired to express in the strongest terms the sense of the whole community of the service rendered by the 30th when the lives of the inhabitants were in danger and desired Major Campbell and the officers to communicate the good opinion and thanks of the community to the non-commissioned officers and men who had deserved the title of good soldiers and good citizens.

The House of Assembly had not been sitting at the time when the disturbances among the negro slaves terminated in 1788 and was not sitting now, in fact the struggle for supremacy between the elected Assembly and the nominated Council gave the Governor so much trouble that he dispensed with the assistance of the elected body as often as he could. In his difficulties the Governor was accustomed to consult Captain Brereton of the 30th and described him to the Secretary for the Colonies as a confidential officer from whom nothing was hid and who could, on arrival at home, give by word of mouth a fuller account of the difficulties of the government of the island than was possible in a despatch. Brereton was adjutant-general of the militia and brigade-major of the regular troops, and as assistant-engineer was carrying out a scheme of fortification sanctioned by the home Government. As a youth he had studied at a military school and acquired a knowledge of French and fortification. Major Campbell was one of the committee supervising the fortification. He also had a military education as a cadet in the Royal Artillery.

The discontent among the slaves in the West Indies which had been repressed in 1788 was now showing itself again. News of the agitation in Britain for the suppression of the slave trade, and ultimately of slavery, had naturally excited the blacks. In Dominica the slaves were possessed with a belief that Governor Ord had instructions, if not to set them free, at least to limit the amount of service due to their masters to three days a week, and that he had suppressed the instructions from fear of the planters. Towards the end of 1790, the blacks had begun to steal away from the plantations and to form bands in the mountains and forests of the interior. On January 13th, 1791, the day on which the address was presented to Major Campbell, a gentleman wrote to the Governor that he heard from faithful slaves that the embarkation of the 30th was to be the signal for a general insurrection of the blacks. Sir John Ord did not stop the embarkation of the regiment, but he delayed the sailing of the ship and told Major Campbell to be ready to land again at a moment's notice and to have detachments prepared to march at once on landing and to operate in the interior of the island. Sir John informed the Colonial Office that in case of disturbance he preferred to employ the 30th rather than newly-arrived troops on account of their thorough acquaintance with the island and the experience they had gained in suppressing the troubles of 1788.

At ten o'clock on the night of the 20th news reached Roseau that the black slaves, under the leadership of free mulattoes, were in full revolt and that the white people were fleeing from the plantations. The 30th was landed and the light company under Captain Urquhart marched north for Rosalie, while Captain Marshall was sent by sea with a stronger detachment to Grand Bay on the south-east to advance from there. We had a small fortified post at Grand Bay occupied by a detachment of Royal Artillery under an officer. Urquhart's subalterns were Lieutenants Thomlinson and Cuyler; Marshall had with him Lieutenants Copley and Gray and Ensign Brook.

The plan was that Marshall on landing should strike north into the French district near Point Mullatre, where it was supposed that the leaders of the blacks were trying to assemble their men on the plantation of a Mr. Sorhaindo. Urquhart's march would prevent the disaffection from spreading to the west, where the planters still trusted their slaves.

Urquhart had a severe march and reached Rosalie at 11 p.m. on the 21st, most of the officers and men having lost their shoes in scrambling in the dark along the rough bed of a watercourse. He had been accompanied as guide and adviser by a local magistrate. After nightfall, as they drew near to Rosalie, word was brought by a plantation overseer that the owner's house was occupied by insurgents. This gave new spirit to the men and although very much fatigued they marched on briskly. Urquhart approached the house under cover of some patches of sugar cane and before taking further steps asked the magistrate if he was quite certain that the inmates of the house were insurgents. On being assured that there was no doubt of it he surrounded the place in silence. As soon as the soldiers were perceived the negroes gave a great shout and endeavoured to escape, but recoiled from the bayonets. As they did not appear to have firearms, Urquhart ordered a party to enter the house and secure them. At this moment the magistrate, running up to the front door, called out: "They are all runaways, kill them, my lads, let not one escape." The party in front of the house fired a volley, dashed into it and with their bayonets killed or wounded the whole. Urquhart and his subalterns, although convinced they were insurgents, in vain tried to save them. He said he could have captured the whole without wounding one if the magistrate had not interfered, but he thought the conduct of the latter, though rash, was all in good faith. It turned out that although not actually insurgents the men killed and wounded were only waiting for one of the rebel bands to arrive in order to join them; meanwhile they were drinking the planter's liquor. Captain Urquhart was shot through "the sidelock of his hair" in striking up the muzzles of his men's pieces. A glance at a portrait of the period will show the importance of the sidelock and the care bestowed upon it.

Marshall after landing at Grand Bay advanced to a plantation owned by a Mr. Bertram; there he left Ensign Brook with Corporal Bird and six privates to keep his communication open. He himself pushed on to one of Mr. Sorhaindo's plantations, where he arrived about 7 p.m. on the 21st. Soon after his captain had quitted him, Brook was told by Mr. Bertram that the negroes were in great force in front of Marshall, who would want every man he could get. Brook on this started to follow Marshall, but was overtaken by darkness when close to him, and having no guide fell into a ravine occupied by rebels. Marshall heard the firing and took a part of his force to the edge of the ravine, but in the darkness could do no more. Brook with two men managed to get through to Marshall, but two men were taken prisoners and Privates Farroll, Slater and Patterson were killed. They were all old soldiers of fifteen years' service or more. The two prisoners, Corporal Bird and Private Mackenzie, escaped and by hiding by day in a stream reached the coast where they were picked up greatly exhausted.

Captain Marshall sent to Headquarters a request that a party might be sent to keep the line clear in rear of him, and at the same time he arranged for a combined attack by himself and Urquhart on the runaway slaves, who it was now certain were in his vicinity. A company of the 15th had already been sent to Grand Bay under Capt. Combe, who was senior to Marshall. He behaved in a most unselfish manner and sent up his subaltern, Lieutenant Robinson, with fifteen men to strengthen Marshall, but left the latter in command.

At noon on the 23rd, Marshall attacked the negroes, who occupied a steep hill on which they had collected piles of large stones to roll down on the troops. They had only about thirty muskets. The moment Marshall appeared in their front, Urquhart who had marched at 2 a.m. fell upon their rear, and the whole thing was over in five minutes. The leader of the rising, a man called Palinaire, escaped, though Urquhart tells us that Private Phillips, the best marksman of the light company, had two shots at him but was too much out of breath from running fast up hill to hit. Unless there had been a good deal of ball practice, Urquhart would not have been so pat with the name of his crack shot. Marshall found the bodies of the three soldiers killed and buried them. Lord Carhampton stated in the House of Lords that the negroes had captured three soldiers and cut them up into pound pieces, but he seems to have been mistaken. None of the horrors which usually accompany a revolt of slaves had taken place, perhaps because the action of the troops had been so swift that the rising had been suppressed almost as soon as started.

When the outbreak took place, the Governor had called up the House of Assembly in order that he might get its sanction for the expense incurred in restoring peace. On January 25th, 1792, he informed the House that news had come that Marshall had forced the post held by the negroes and that the rebellion was practically at an end. He added: "Too much credit cannot be given to the zeal and activity of the troops who have rendered essential service to the colony. You, Gentlemen, without any recommendation of mine will best know how to reward them."

As soon as the House of Assembly had been called up the strife between it and the Council had recommenced. In answer to the Governor's message, the House proposed that the thanks of Government should be given to all the troops in the island for their services in the late troubles. In addition, a Bill was introduced and passed by the House to present the officers' mess of the 30th regiment with a piece of plate of the value of £500 as a memento of the seven years' friendship between the regiment and the inhabitants of the island, of the undeviating good conduct of officers and men and of the invaluable services rendered in the suppression of two risings of the blacks.

The Council amended the Bill and voted £500 for the presentation of plate to all the troops in the island, the amount given to each corps to be in proportion to its strength. The discussion between the two bodies became so heated that the Governor was obliged to prorogue the House of Assembly and the Bill was lost. Had the Council said outright that it thought the presentation of plate was extravagant and uncalled for, it would have been on strong ground. The position it took up, that the troops who had scarcely landed had earned the gratitude of the inhabitants as much as those who had served the colony for seven years, seems rather weak.

In addition to the three men actually killed the regiment lost Captain Marshall and several men who died on the way home from their exertions in an unhealthy locality during the late operations.

Peace having been restored in Dominica by Captain Marshall's operations, aided by the laws passed for the protection of the blacks from the cruelty of some of their masters, the 30th was at liberty to return home.

On February 4th, 1792, it sailed for England on H.M.S. *Sheerness*. Major Campbell was in command and had under him Captains Brereton and Urquhart and Capt.-Lieutenant and Quartermaster John Marshall (sick), 4 lieutenants, 5 ensigns, the adjutant, John Russell, and the surgeon, 16 sergeants, 10 drummers, 343 rank and file, including 41 sick. The establishment had been again raised to 560 privates. The officers absent with leave were the colonel, Lieut.-Colonel Maxwell, Captain McCulloch, Lieutenant Murray, Ensigns Wallace and Lucas, and, as usual, the chaplain. The officers recruiting were Captains Wilkinson, Torriano, Satterthwaite and Wilson, Lieutenants Brown, Lockhart, Salisbury, Pilcher, Balfour and Bultiel, and Ensign Hamilton. Colonel (Major-General) Sir Henry Calder had died on February 2nd, and had been succeeded by Major-General Thomas Clarke.

CHAPTER VI

FRENCH REVOLUTIONARY WAR. DEFENCE OF TOULON. CAPTURE OF S. FIORENZO. BASTIA AND CALVI. HOTHAM'S NAVAL VICTORY OFF HYÈRES. FOUR COMPANIES DRAFTED AND SENT HOME TO RECRUIT.

1791–1798

ON March 13th, 1791, the 30th landed at Portsmouth and occupied Hilsea Barracks, with detachments at Fort Cumberland and Forton. The strength was the same as on leaving Dominica, except that Capt.-Lieutenant (Quartermaster) John Marshall and seven men had died, the result of the rough campaign in the mountains in the previous month. Lieut. Hall succeeded Marshall as capt.-lieutenant and Quartermaster-Sergeant Ninian Craig became quartermaster. On May 7th Major-General Hyde reviewed the regiment. It was beginning to gain strength, but the old soldiers were coming forward for discharge. Very nearly one-half of the men were over thirty years of age and an equal proportion were 5 ft. 7 ins. in height.

On September 5th the 30th was ordered to march to Liverpool and to furnish a detachment of three companies at Whitehaven.

The establishment was now 10 companies, 1 colonel, 1 lieut.-colonel, 1 major, 7 captains, 1 capt.-lieutenant ranking as captain but on lieutenant's pay, 11 lieutenants, 8 ensigns, 1 chaplain, 1 adjutant, 1 quartermaster, 1 surgeon, 1 surgeon's mate, 1 sergt.-major, 1 quartermaster-sergeant, 20 sergeants, 30 corporals, 12 drummers including 2 fifers, 400 privates.

In 1792 an effort was made to stop the desertion which in late years had assumed enormous proportions, sometimes amounting in Ireland to one-sixth of the strength. Since the American War a small allowance had been given the soldier weekly to meet the constantly rising price of bread. This allowance was now advanced to 10½*d.* a week with good results. Neither Government nor the House of Commons were yet educated to the point of giving the soldier an adequate ration of bread and meat at a fixed sum and so making him independent of the fluctuations in their price.

In Ireland the troops were of course in barracks. In Great Britain nearly all were in billets, but Mr. Pitt's great scheme to cover Britain with barracks took shape this year.

Towards the end of 1792 there were signs that war was at hand. Britain had watched with hope and sympathy the revolution which was in progress in France and had held aloof from the Royalist coalition against that country, but when French troops overran Flanders in November and December, we were touched in the most sensitive spot and Mr. Pitt's Government felt constrained to protest, and as a precaution, to embody one-half of the militia. This was followed by orders for the raising of 100 independent companies and, the danger of war still increasing, the establishment of the 30th was raised to 32 sergeants, 30 corporals, 22 drummers, and 570 privates, the commissioned ranks remaining as before.

As the French Revolution affected so powerfully the life of the regiment, a few dates are given to recall to memory the course of events.

July 11th, 1789. Fall of the Bastille.
July 14th, 1790. France declared a Limited Monarchy.
June 21st, 1791. The Royal Family attempt to escape.
June, 1792. The Empire and Prussia make war on France.
Sept. 20th, 1792. France declared a Republic.
Dec., 1792. French conquer the Austrian Netherlands. British militia embodied.
Jan. 21st, 1793. Execution of King Louis XVI.
Feb. 1st, 1793. France declares war on England.
May 31st, 1793. Proscription of Girondins. Reign of Terror established, followed by revolt of Southern France.
July 28th, 1794. Robespierre guillotined. End of Terror.
Nov. 1st, 1795. The Directory.
Nov. 10th, 1799. The Consulate. Buonaparte First Consul.

The French Republic had at the end of 1792 not only occupied the Austrian Netherlands in spite of our protests but had threatened an attack on all monarchical governments. The British Ambassador was therefore withdrawn from Paris and France declared war against Britain on February 1st, 1793. The 30th was concentrated at Liverpool and the non-commissioned officers and men of the three independent companies of Captains Archibald Campbell, Matthews and Thomlinson were ordered to be incorporated with it. The officers were to be attached for duty only. The companies should have been over a hundred strong, but they only brought from seventy to eighty men apiece, of whom a fourth deserted at once. A dozen other independent companies sent each a few men to make up the numbers.

Britain was quite unprepared for war. Our Government had at its disposal a body of capable naval officers and a number of ships, most of which were laid up in reserve, but the seamen and marines were very few. It had also in the army good officers and, if we may judge of the whole by our own regiment, excellent non-commissioned officers. There were however only 2 sergeants, 3 corporals and about 30 trained men in each company.

Obviously the fleet had to be got to sea, whatever else was left undone, so the ships which had been laid up were fitted out with all speed, what seamen could be collected were put on board and the deficiency in numbers made good by enlisting and pressing landsmen. In order that the newly-commissioned ships might have a disciplined nucleus on board the army had to be sacrificed. Nearly every regiment left in Britain, after the Guards and a few line regiments had been sent to Germany, was broken up and distributed by companies through the fleet to serve as marines. The 30th was one of the corps so broken up.

On March 14th, 1793, the regiment was ordered from Liverpool to Portsmouth in three divisions: it was relieved at Liverpool by the Warwick militia.

Orders were received on the line of march for the three divisions to regulate their marches so as to arrive at Portsmouth on April 10th and the days following and to embark as directed by the Lieut.-Governor.

By the 13th six companies had embarked :—

On H.M.S. *Terrible* (grenadiers), Captain Robert Brereton, Lieutenant Wm. Maxwell, Ensign Wm. Murray and 76 other ranks.

On H.M.S. *Audacious*, Captain Wm. Wilkinson, Lieutenant Alexander Hamilton, Ensign Humphrey Forster and 78 other ranks. This company was transferred to the *Robust*, Captain Keith Elphinstone, R.N., on May 8th.

On H.M.S. *Alcide* (light), Captain Charles Torriano, Lieutenants Leonard Brown and Hay Livingstone and 76 other ranks.

On H.M.S. *Suffolk*, Captain Archibald Campbell (independent), Lieutenant Ralph Smyth and 77 other ranks.

On H.M.S. *St. Albans*, Captain John Delves Broughton, Ensigns Charles Barrett and Wm. Wood and 57 other ranks.

On H.M.S. *Princess Royal*, Captain Wm. Thomlinson (independent), Lieutenant Henry Cuyler, Ensign Eyre Peter Shewbridge and 100 other ranks.

Thomlinson's case shows how unfair the system was of allowing officers to enlist men for rank. He was a junior lieutenant in the regiment when he got permission to raise a company. On February 9th he reported his company complete and became a captain. At his request the company was incorporated with his old regiment and he joined for duty as a captain over the heads of eight subalterns to whom he had been junior a very few months before. In the state of March 1st he is shown as absent ; nothing is said about his having leave, but a note " Gone to Gretna Green but to return immediately " shows that Colonel Maxwell knew all about it.

Of the officers not detailed for service on the fleet at this time, Colonel Christopher Maxwell exchanged with Lieut.-Colonel Charles Green of a regiment which General Edmiston had been ordered to raise. Colonel Maxwell had joined the 30th in 1755, had commanded it as a major on service in South Carolina in 1781 and been lieut.-colonel of it for eleven years.

Major Thomas Campbell, who had joined the Royal Artillery as a cadet in 1760 and the 30th in 1763, was transferred to an independent company ; the transaction was rather curious. He was allowed to sell his majority, still retaining the rank of captain. Captain Brereton paid him the difference between the value of a major's and a captain's commission and became regimental major. Captain Catlin Craufurd from an independent company took over Campbell's company in the 30th without purchase and Campbell was given an independent company. He died in 1809 and in consideration of his meritorious service his widow received a compassionate allowance.

Captain Urquhart's company was broken up to strengthen those ordered on board ship and he was employed recruiting till April 4th, 1794, when he exchanged with Wm. Linnaeus Gardner, captain of an independent company.

Captain James Clarke Satterthwaite, M.P. for Haslemere, was attending Parliament when the regiment was broken up. He sold his company on April 1st to Lieutenant Wm. Lockhart.

Captain McCulloch sold his company to Lieutenant Gilbert Congleton of the 28th Foot. Congleton did not join till July 4th ; till then the company was commanded by Captains Arch. Campbell and James Smith Bailey in succession, both from independent companies. Congleton left in a few months to act as A.D.C. to his uncle, Sir Gilbert Elliot, Governor of Corsica, and in December Lieutenant Ralph Smyth succeeded to the company.

Lieutenant Lyster, who had been allowed to raise an independent company, was made captain of it. He was succeeded by Ensign Roger Montgomery, who joined Smyth's company at Cork in April, 1794.

Lieutenant Philip Vaumorel from the 57th became capt.-lieutenant by purchase from Hall, who retired. Hall had been in bad health for some time.

The above changes resulted in the regiment having a younger and more alert body of officers to face the active service ahead of it, for it must be confessed that the long commands of Lieut.-Colonel Gould and Lieut.-Colonel Maxwell, however good in themselves, had been detrimental from their very length. The two officers had between them ruled the regiment for twenty-nine years.

A similar change had taken place among the men. Since the regiment landed in 1791, most of the men enlisted before the American War had been discharged to pension.

The distribution of the regiment by the end of April was, as nearly as can be judged, for the returns do not quite agree :—

		sergeants	corporals	drummers	privates
Effectives	Embarked .	14 sergeants	19 corporals	7 drummers	414 privates
	at Hilsea .	18 ,,	12 ,,	7 ,,	20 ,,
Recruiting . . .					19 ,,
Command . . .					19 ,,
Sick in Hospital . .					23 ,,
		32 ,,	31 ,,	14 ,,	495 ,,

The number wanting to complete the establishment was 6 drummers and 75 privates.

These numbers include the men of the independent companies incorporated with the regiment.

We have no knowledge of how far the regiment had been trained for war. Ball firing however had been carried out to some extent, and in this particular the British troops were alone in Europe.

From the correspondence about Captain Wilkinson's company we know what one of the companies looked like, and there is no reason to doubt that the others were similar.

In his company there were thirty-two trained men of the regiment correctly dressed in cocked hat worn athwart, red coat, waistcoat, breeches and gaiters, four regimental recruits, picked up on the march from Liverpool, clothed in garments borrowed from an independent company, namely, red jacket, linen trousers with flannel waistcoat and drawers, nineteen of Captain Archibald Campbell's independent company, headed by a sergeant who had only a pair of regimental trousers to show his connection with the army, and fourteen men of Captain Thomlinson's independent company. The greater part of Campbell's and Thomlinson's men had received a red jacket, and linen trousers, but of so bad a quality as to be already nearly worn out. All wore the round hat or " chimney-pot " which in civil life was superseding the cocked hat.

Such was the trained body which was not only to enforce discipline on board ship but teach it by example. Things, however, looked worse than they were. The officers were excellent, Wilkinson died a lieut.-general and a G.C.M.G., his lieutenant, Alexander Hamilton, was *the* Hamilton of Toulon, Egypt, the Peninsula, and Quatre Bras, and Forster the ensign was mentioned in despatches in his first engagement. The non-commissioned officers, as has been said, were good, and with thirty disciplined men in the company, a little time only was wanted to put everything straight.

The ship's company resembled the soldiers; there were excellent officers, with Captain Keith Elphinstone at their head, and doubtless many good petty officers and able seamen, but eighty were rated as landsmen, having never been to sea before, and most of the ordinary seamen were little better. Man for man the soldiers were physically the stronger throughout the fleet.

After the embarkation the command of Headquarters at Hilsea had devolved on John Russell, who signed the returns as ensign, adjutant, and commanding officer 30th regiment. He had with him some newly-joined subalterns, the surgeon and quartermaster, some officers' servants and the sick, but by the middle of September he was able to scrape together forty-seven rank and file to send on board ship; Lieutenant John Copley, withdrawn from recruiting, took command with Ensign Woolridge, lately joined, to help him. The party embarked on H.M.S. *Boyne* on September 16th, but was transferred to the *Bellona* on October 13th.

The interest in the history of the regiment at this time lies chiefly in the doings of the companies engaged in the defence of Toulon and in the operations in Corsica, but the two companies embarked on H.M.Ss. *Suffolk* and *St. Albans*, and the detachment on H.M.S. *Bellona* were employed on an even more essential service, in the Channel Fleet and on convoy, though they had not an equal chance of distinction.

The company on H.M.S. *Suffolk* was transferred to H.M.S. *Swiftsure* on November 27th. The ship cruised in the Channel and off the south coast of Ireland till May, 1795, when it sailed with a large convoy to the West Indies. It remained there for a year and on May 1st, 1796, the company was transferred to H.M.S. *Africa*, which landed it at Sheerness on September 23rd; strength, Captain K. Smyth and fifty-seven other ranks. On landing, the company marched to Chatham.

Captain Broughton's company on H.M.S. *St. Albans* sailed on convoy duty to the Mediterranean and returned to Portsmouth on December 2nd. Here Captain Broughton left it to raise men for rank, in which he was so successful that on September 18th, 1794, he became lieut.-colonel of the 106th Foot. The company was transferred to H.M.S. *Ruby* and sailed on it on convoy duty to Cadiz, returning to the Downs on July 27th, 1794. The company was transferred on February 27th, 1795, to H.M.S. *Sampson*, which landed it at Portsmouth on April 15th; strength, Ensign Wm. Wood and thirty-eight other ranks. From Portsmouth the company marched to Newmarket to join the depot of the 30th which was being formed in Cambridgeshire.

The detachment on H.M.S. *Bellona* saw some fighting. The ship was in Lord Howe's fleet but was detached when the battle of "The Glorious First of June" was fought in 1794. The small squadron of which she formed part encountered the defeated French fleet on the day after the battle, but only a slight engagement followed. On October 13th the *Bellona* sailed for the West Indies and on November 15th was at Guadeloupe. For some time we had been losing ground in the West Indies, where thousands of our men had died of yellow fever. Half of Guadeloupe was in the enemy's possession, but General Procter still held out in Fort Matilda with a garrison reduced to less than 600 men. On November 19th Lieutenant Woolridge, one sergeant and as many soldiers and marines as could be spared were sent ashore from the *Bellona* to assist in the defence of the fort and on December 2nd a party of thirty men followed. The fire of the enemy's batteries increased daily and on the 10th General Procter found that the place was no longer tenable. His sick and wounded had been sent off from time to time and now he determined to save the remains of the garrison. Shortly after midnight the boats of the fleet pulled in-shore and the men were taken off almost without loss, although it was broad daylight before all the boats were out of range and the enemy opened a smart fire from his batteries and from a gun brought down to the beach. The defence had been very creditable to General Procter and his garrison. The strength embarked on the 11th was 481 effectives, of whom eight belonged to the 30th. They were under Lieutenant Woolridge. The other corps represented were the 15th, 21st, 35th, 39th, 56th, 60th, marines, black corps, legion, artillery, engineers, and a body of seamen.

The *Bellona* returned home on convoy and sailed again for the West Indies, where she served through the winter 1795–96. In June the detachment was transferred to H.M.S. *Minotaur*, which landed Lieutenant Woolridge and thirty-two other ranks at Portsmouth. There they were picked up by Captain Hamilton on his return from the Mediterranean with the skeletons of the four companies which had served at Toulon, in Corsica, and in Admiral Hotham's victory off Hyères. Hamilton conducted the whole party to the regiment at Bandon.

TOULON AND THE MEDITERRANEAN

The companies which sailed in Lord Hood's fleet and saw so much service in the Mediterranean in 1793–96 were those of

Brereton (grenadiers) on H.M.S. *Terrible*, Captain Campbell, R.N.
Wilkinson on H.M.S. *Robust*, Captain Keith Elphinstone, R.N.
Torriano (light) on H.M.S. *Alcide*, Captain Woodley, R.N.
Thomlinson on H.M.S. *Princess Royal*, Captain Purvis, R.N.

The strength of all ranks was 341, the detail by ships has been given before. None of the regimental staff embarked.

Lord Hood sailed from Spithead on May 24th, and reached Gibraltar on June 19th. He found that the south of France, headed by the three great cities of Lyons, Marseilles and Toulon, had risen against the Government in Paris, and that Toulon in particular had declared for King Louis XVII and the constitution of 1790. After some negotiation Lord Hood agreed to take over the arsenal and fleet in Toulon and to hold them in trust till the end of the war and the restoration of constitutional government in France.

On August 28th Lord Hood occupied Toulon and landed the companies of the 11th, 25th, 30th and 69th regiments with what marines he had, making in all about 1,500 men.

About 4.30 a.m. on that day, the companies of the 30th assembled on board the *Robust* and at 9 a.m. she stood in for Toulon Roads, towing her barge, launch and pinnace and accompanied by the *Egmont, Colossus, Courageux, Meleager* and *Tartar*. At 11 a.m. the ships hove to and the boats pushed off, landing their men at St. Marguerite on the extreme east or right of our line of defence for the protection of the town and harbour of Toulon. Captain Elphinstone, R.N., was in command of the troops, but he had no engineers and no Staff, and all that could be done was to place companies at the points which were obviously of most importance and leave their officers to make the best dispositions they could.

The principal forts protecting the outer harbour had already been occupied by the French Constitutionalists and the position was made more secure by the arrival of 1,000 of our Spanish allies under Admiral Langara with a promise of 3,000 more to follow. Langara's second in command, Admiral Gravina, was appointed Governor of Toulon and Captain Keith Elphinstone Governor of the great fort La Malgue to the S.E. of Toulon. Captain Wilkinson of the 30th was Town Major.

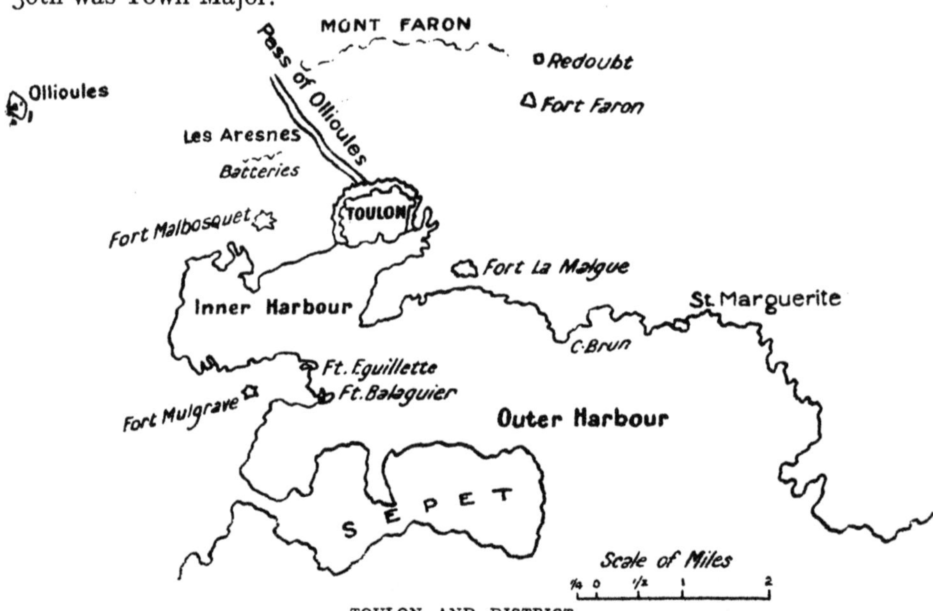

TOULON AND DISTRICT.

Toulon is surrounded to the north and west by rocky hills. Among those which most affect the defence of the place and the command of the harbour is Mont Faron, immediately to the north and running east and west; its highest point is 1,700 feet. We occupied it from the first. The defences of Toulon being planned to resist an attack from the sea rather than from the land side, Faron had a fort on the southern slope, but none on the northern or on the summit. As soon as labour was available we placed a redoubt on the summit and established a picquet of sixty men in advance to watch the "Pas de la Marque," a zigzag path by which the enemy might ascend the steep northern slope two abreast.

At the western end of Mont Faron is the pass of Ollioules, through which the road runs north into the interior of France.

To the west of the pass are the heights of Aresne, which we never had strength to occupy. A ceaseless artillery duel was carried on throughout the siege between the French on those heights and the ships detailed by Lord Hood to answer them.

To the south-west of Toulon and bounding the western side of the harbour is the peninsula of La Grasse, rising to a height of 250 feet and having on the water's edge Fort Eguillette at its northern point and Fort Balaguier at the southern. At first we occupied these forts only but were soon forced to seize and fortify the heights above them.

It is difficult to get a clear view of the share of the regiment in the defence of Toulon, for the reason that it was the work of detached companies or parts of companies. The companies again had no fixed position in the line of defence, which was 11 miles from east to west. No man was sure whether on the following day he would be at Cap Brun on the right or on the heights of La Grasse on the extreme left. There were no tents and probably the most comfortable position was that of the men on guard at La Malgue, but they were not secure of a quiet night even there; a movement of the French might cause a counter move on our part, the leader of which would insist upon having at least fifty British bayonets with him and the guard would be relieved by Spaniards. Then might follow a march or a night in an open boat in the drenching rain, the landing at dawn, a slight skirmish with one or two casualties and an immediate return.

The cause of this incessant duty was the absolute necessity of British troops being present in every undertaking of attack or defence. Our Allies, with the exception of the Sardinians and Piedmontese who were as good as our own men, were quite untrustworthy. In this there is no intention of saying anything against the courage of two gallant nations, but soldiers know how easy it is for a government to destroy the fighting power of its army; instances might be cited from the history of our own army and regiment, but there was a striking example in this very war where the French army from political causes fought very badly in 1793, and for some time afterwards.

When Lord Hood occupied Toulon the enemy in front were weak and disorganized, the main Republican army of the south was engaged in the siege of Lyons, but Marseilles had already surrendered, and 10,000 troops from there were on the march to form the siege of Toulon. On August 31st 750 men of this body advanced through the Pass of Ollioules, but were attacked by Captain Keith Elphinstone with 500 men from La Malgue, and totally routed. Apparently no companies of the regiment were engaged, but Captain Elphinstone in his report says of Ensign Forster, 30th regiment, his A.D.C:—" His conduct was such as to give me the fullest satisfaction and I hope will make him an object of attention."

The following is a diary of the principal events of the siege:—

Sept. 6th, 1793.—Lord Mulgrave arrived and took command of the British force on shore. He found that he had to defend a line of 8 miles long (the heights of La Grasse not yet being occupied) with about 1,200 men of the four British regiments and a few sailors and marines, assisted by 3,000 Spaniards, the refuse of their army mixed with galley slaves.

Sept 16th.—Napoleon Buonaparte arrived to command the enemy's artillery.

Sept. 18th.—The French reinforcements enabled them to extend their line to the right and open fire from some heights to the north of La Grasse, thus forcing the defenders to occupy the latter peninsula.

Sept. 21st.—By diminishing the garrison of La Malgue and leaving one relief only for the duties of Toulon, Lord Mulgrave was able to muster 150 British under Major Brereton, 30th regiment, and 350 Spaniards under Colonel Rafael Echavura. Embarking at midnight the detachment landed between Eguillette and Balaguier at two in the morning and advanced at once.

La Grasse consists of a ridge with three wooded knolls at the top, separated from each other by shallow valleys. Lord Mulgrave found that from the western end of the ridge he had a complete view of the whole of the enemy's position to the west of the Pass of Ollioules and that, the ground falling rapidly towards the enemy, the position was very capable of defence. Being in small force he contented himself for the time with occupying the easternmost and lowest knoll, 500 yards from the landing place.

The enemy in the evening occupied the high ground to the west and attacked in a half-hearted way. The skirmish ceased in an hour and the enemy withdrew; the British loss was 1 rank and file killed, 1 captain, 3 rank and file wounded. The 30th had one man killed and one wounded.

Lord Mulgrave reported that at nightfall if it had not been for the steadiness of Major Brereton the Spanish officer in command would have shamefully deserted his post with his men.

As a consequence of this reconnaissance the western end of the heights of La Grasse was fortified as strongly as the limited time and labour allowed and heavy guns from the ships were mounted. The work was called Fort Mulgrave and was supported by another called La Grasse, between it and the landing place at Balaguier. As reinforcements reached the Allies the garrison of the Peninsula was strengthened till at the end of the siege it reached 3,000 men. Our men called the position "the little Gibraltar."

Lord Mulgrave now divided the British troops into two corps, one under Major Brereton of the 30th, the other under Captain Moncrief of the 11th, both of whom he recommended for some mark of His Majesty's favour.

In the despatch in which he mentions those officers he represents the cheerfulness and alacrity of the soldiers in suffering incessant fatigues in posts in which they might be considered constantly on duty; Lord Mulgrave further expresses his fear that such overwork may lead to sickness.

Sept. 27th.—Eight hundred excellent Piedmontese grenadiers and chasseurs, and 2,000 indifferent Neapolitan troops arrived with a promise of 4,000 more to follow.

Oct. 1.—At nightfall, September 30th, General Lapoype, the French Commander-in-Chief, assembled a column of 2,000 men with supports to the north of Mont Faron. Under Victor, afterwards Marshal and Duke of Belluno, they ascended the rugged hill by the Pas de la Marque. The British picquet of 60 men at the top discovered them at daybreak and fell back on the redoubt of Faron, which should have been held by a Spanish garrison. The Spaniards, however, had abandoned it and the picquet retreated to the Fort of Faron some way down the southern slope. The enemy occupied the redoubt of Faron and the whole summit and began to press the garrison of Fort Faron. A reinforcement of 92 men of the 30th under Captain Torriano was at once sent to the garrison by Captain Elphinstone, R.N., from La Malgue, and a council of war was assembled in the town. It was evident to all that Toulon was lost unless Mount Faron was recovered and steps were immediately taken with that object.

Lord Hood landed a strong force of sailors to enable every soldier to take the field, and by relieving the British troops on guard Lord Mulgrave was able to collect 300 British rank and file, of whom he sent 50 under Captain Beresford, 69th regiment, to join Captain Elphinstone, R.N., at Fort Faron.

With the remainder and 300 Piedmontese, Lord Mulgrave prepared to ascend Mount Faron from the west.

Admiral Gravina led a column of two companies of Sardinian Chasseurs, 183 Spanish rank and file, 400 Neapolitan grenadiers and 100 French loyalists of the Regiment de Bourgogne to ascend from the south-west. On the advance of the Allies 200 of the enemy on the western summit of Mount Faron fired hastily at long range and retired, and Lord Mulgrave and Gravina reached the top having suffered no obstruction except from the heat and the precipitous nature of the ground.

Lord Mulgrave having taken the left of the line with the British and Admiral Gravina the right, the Allies prepared to advance along the summit of the hill towards the east. The ground was bare of vegetation and broken by transverse ridges of rock, very sharp and difficult to march over.

The enemy had formed his line with his right on the crest of the northern slope, but Lord Mulgrave noted that the Pas de la Marque, the path by which Victor had ascended in the morning, was undefended and between the opposing forces and that the enemy's left, which was thrown back at right angles, did not reach the edge of the southern slope of the mountain. By Lord Mulgrave's orders Captain Moncrief, 11th regiment, advanced rapidly to his left front and seized the Pas de la Marque, thus attracting the enemy's attention to his right. Admiral Gravina, an excellent officer, meanwhile passed the greater part of his force over the southern edge of the mountain with orders to file along the slope out of sight as well as they could till opposite the enemy's left flank and then to form to the left and charge.

Lord Mulgrave advanced with as much show as he could make and opened fire at long range and the enemy weakened his left more and more. At the decisive moment Gravina brought up his right shoulder and on the extreme right Captain Elphinstone, R.N., sent Captain Torriano, 30th, and Captain Beresford, 69th, up the rocks with their handful of men, while Lord Mulgrave rapidly closed with the French right. The French, outflanked and surprised as they were, made a gallant stand, but presently the whole line gave way and fled down the hill to the north and east. In his despatch, Lord Hood says: "The action was short but hot, the enemy had upon the heights from 1,800 to 2,000 men, the flower of the eastern army, not a fourth of which we are well informed ever returned to Headquarters, for what did not fall by the bullet or bayonet broke their necks in tumbling over the precipices in their flight."

Captain Elphinstone reported to Lord Mulgrave: "Captains Torriano of the 30th and Beresford of the 69th had infinite merit for the intrepid manner they led their men up an almost inaccessible mountain under a heavy fire," and Lord Mulgrave thanked the party for this steady, active and gallant conduct.

Attacking as he did almost at right angles to the enemy's flank, Captain Torriano's loss was slight, amounting to 1 killed and 6 wounded of his 90 men.

Admiral Gravina was unfortunately wounded and left no one among his countrymen fit to take his place.

Oct. 7th.—Notwithstanding Lord Mulgrave's exertions the attack on Toulon was gradually getting the better of the defence. Reinforcements were pouring into the French camp and the artillery attack developed day by day. By October 7th we had lost eighty men in the floating batteries and gunboats opposed to the French batteries on the heights of Aresne.

The fire of the batteries attacking Fort Mulgrave had become so intolerable that Major Brereton, 30th regiment, was ordered to plan and lead a sortie from that work.

From the way the sortie was carried out it is evident that Major Brereton called to mind his old experiences in America and the surprise visits of Marion's irregulars to Monk's Corner and to the outskirts of Charleston. The following is his account of the affair to Lord Mulgrave:—

"Fort Mulgrave. Hauteur de Grasse, October 9th, 1793.

"Yesterday morning at half-past 12 at night we moved from this post and having formed a junction in the bottom with the marines and Piedmontese from Cepet, we marched off from our right in one Column in order to ensure the greater regularity in a night attack. Our march to the top of the height where the new battery had been constructed by the enemy was conducted with all possible order and expedition, the troops observing the greatest silence by which with the aid of the French deserter, who answered the Sentinels of the enemy as we passed them, our advanced party arrived at the entrance into their first battery perfectly undiscovered. The first Sentry having been put to death the advanced party composed of the Grenadiers and Light Infantry of the line of British under command of Captain Stewart of the 25th very gallantly rushed in and put every man to the bayonet who opposed them. The remains of the enemy retired to their second battery and made a sort of stand, but the greater part of the detachments having by this time taken positions for attack, the enemy were soon routed. The enemy's guns were spiked and destroyed. From the precipices and narrow broken paths we had to descend it was impossible to carry them off."

Brereton retired from his left. By advancing from his right and retiring from his left he kept the British flank companies between the enemy and the remainder of his force. The loss was entirely in those flank companies. He reported the casualties of the night as: Killed—1 corporal and 1 private; wounded—1 corporal and 6 privates. A private of the 30th must have died of wounds after the report was written, for the loss in Brereton's (grenadier) and Torriano's (light) companies was: Killed—1 corporal and 2 privates; wounded—1 corporal and 3 privates; that is seven out of the nine casualties in the force. Evidently Brereton used his own men, as was natural.

On October 9th Lyons fell and a great part of the Republican army was released to join in the siege of Toulon. The numbers of the besiegers now amounted to 40,000 men. The defenders received a reinforcement of 3,000 Neapolitans and 200 British soldiers were landed from the fleet.

On October 15th it was considered necessary to establish a post at Cap Brun, at the eastern end of Mount Faron, and Captain Elphinstone sent 100 men of the French Regiment Royal Louis there with orders to entrench themselves. They were attacked on the following morning and Elphinstone moved to their relief with 100 of the 30th, 100 Neapolitans, 50 Spaniards, and the remainder of the Royal Louis. The 30th, who were on the left, outflanked the enemy who retreated, and Captain Elphinstone occupied the post with Captain Thomlinson's company; the remainder of his force was in support, but some if not all were employed in strengthening the position and in bringing up guns when the Republicans returned with reinforcements and made a sudden and successful attack. The 30th lost Captain Torriano killed; Lieutenant Alexander Hamilton and nine rank and file wounded; Lieutenant Shewbridge and four rank and file missing.

In his despatch Captain Elphinstone says: "I am sorry to inform you that in both affairs many brave officers and men fell. In the person of Captain Torriano, 30th regiment, His Majesty has lost one of the most respectable officers in Europe. Were I to enter into the conduct of officers and men I should fail to do justice to their merits; the 30th did all that men could do and the Royal Louis, though only a few days formed, fought with determined valour." The gallant conduct of the non-com. officers after the detachment had lost its officers, led to two of the sergeants being recommended for commissions. One, whose name we do not know, was found unsuitable (probably from age) and was awarded a handsome gratuity to be paid when he claimed his discharge. The other was Sergeant Samuel Bircham, now twenty-four years of age, who was born in the regiment at Gibraltar, and enlisted in 1782 as a drummer in Colonel Paston Gould's company of which his father, Sergeant James Bircham, had long been senior sergeant. James Bircham was a corporal of the grenadier company in 1761, and must have been among those who headed the attack when General Hodgson's force scaled the heights at Locmaria and forced a landing in Belleisle.

Sergeant Samuel Bircham was gazetted ensign on November 20th, and rose to be a major in the regiment.

On October 27th, 750 of the Royals, Royal Irish, and artillery arrived from Gibraltar under Lieut.-General Sir Charles O'Hara with Major-General David Dundas, the future Commander-in-Chief, as second in command.

On this day also Captain Elphinstone was directed to convey to the troops serving under him the satisfaction His Majesty had derived from their spirited and successful exertions in his service.

Lord Mulgrave, who had done splendid service and who had kept on excellent terms with Admiral Lord Hood, returned home on the arrival of O'Hara. Before leaving, he had communicated to Lord Hood his misgivings about the safety of Toulon.

Generals O'Hara and Dundas, as soon as acquainted with the situation, repeated Lord Mulgrave's warning and advised Lord Hood to prepare for the evacuation of Toulon, and the destruction of the French arsenal and fleet. It must be borne in mind that although the general officers commanded the troops while on shore, the greater part of the British force, namely the detachments of the 11th, 25th, 30th, and 69th, formed part of the crews of Lord Hood's ships and could be withdrawn by him.

On October 30th Lord Hood received orders from Mr. Henry Dundas, the Secretary of State for War, to send the 300 of the 30th regiment to Gibraltar for service elsewhere. It is probable that Mr. Dundas intended to unite the detachment from Toulon with those on the *Bellona, Ruby* and *Suffolk* for service in the West Indies as a battalion. Whatever his intention may have been, it was frustrated by Lord Hood's positive refusal to part with a man. His ships were undermanned and the soldiers were too important a part of the crew to be spared.

November 15th.—General Dugommier, a good officer if somewhat aged, took command of the Republican army and the superiority of the attack became pronounced. A determined attack was made on Fort Mulgrave, but repulsed by the Royals with heavy loss to the besiegers.

November 29th.—The increasing fire from the north on both sides of the Pass of Ollioules between the western end of Mount Faron and the heights of Aresne induced General O'Hara to order a sortie, which was led by Major-General Dundas who forced his way through the pass and nearly reached the artillery park beyond. Meanwhile 400 British on the left carried the heights of Aresne and spiked the guns in battery. A counter-attack led by Dugommier drove the Allies back through the pass and their mixed force fell into great disorder.

O'Hara, a man of extreme gallantry, dashed into the crowd to rally them, but was severely wounded. He was taken prisoner, and the British lost 200 men and 12 of their officers, who were already all too few for the work. Few of the 30th were engaged and the regimental loss was only three men wounded. Napoleon Buonaparte was actively engaged against us and was near O'Hara when he fell. After this blow the British could not take the same share in the defence as before. Fatigue and exposure were telling heavily on the over-worked soldiers of the Allies, who had 4,000 sick and wounded in hospital. The French now numbered 40,000 and Buonaparte had his siege train ready.

Dec. 16th.—For several days a heavy fire had been kept up along the line by the besiegers, but especially on Faron to the north and La Grasse and Fort Mulgrave to the south-west. At 2 a.m. on the 16th five batteries opened fire on Fort Mulgrave and on the 17th it fell.

The following account of its fall is given by Captain Vaumorel, 30th regiment, who with his detachment was the last to hold out:—

"FALL OF FORT MULGRAVE.

"The strength of the detachment did not exceed 500 men, the attack was led by Victor, afterwards Duke of Belluno. The brunt of the attack was met by sailors, artillery, 2nd Royal Scots, Royal Marines, 30th Foot, and Piedmontese, who all took an active part in giving their utmost support. The attack was preceded by five hours' unceasing fire from five batteries under Napoleon Buonaparte, by which 2 sergeants and 20 rank and file were killed, 3 sergeants and 42 rank and file were wounded. Nine engineers at rest from fatigue were killed by a single shell.

"The critical point was on the south side where the marines, 30th, and Piedmontese were.

"The first attack was between eleven and midnight. A second attack was repulsed with a loss to the enemy of 1,200 men. Six British mortars and five 36-pounder guns inflicted great loss. This attack was about 2 a.m. 17th December.

" Orders in writing had been received on the morning of the 16th by the officer commanding the fort to do his utmost in defending the same for the protection of that part of the fleet anchored in the inner harbour, and should the ammunition be expended to receive the enemy on the bayonet. Captain Vaumorel's predecessor, at variance with this order, abandoned the post during the first attack without being followed by a single British soldier, or even by his own servant, in consequence of which Captain Vaumorel decided to defend the place to the last extremity.

" After daylight on the 17th, the enemy made a third attack and succeeded in getting possession of a Spanish battery of five 24-pounder guns divided by an epaulement from the British side. After a feeble defence the Spaniards, 700 men under a lieut.-colonel, gave way and retired to the shore at Balaguier; the enemy entered through the embrasures and the British were obliged to surrender half-an-hour after daybreak."

In confirmation of the foregoing account there is a letter written many years later to Colonel Vaumorel, commanding 30th regiment, by General Sir Thomas Hislop, G.C.B., when Commander-in-Chief in Madras; he had as a captain been deputy-adjutant general of the British force defending Toulon in 1793. In this letter General Hislop certifies to Captain Vaumorel's merits and services at the defence of Fort Mulgrave and says: " On that occasion you and the brave soldiers with you who nobly did your duty fell into the hands of the enemy while the officer who was intrusted with the command of the post abandoned it early in the contest and soon after quitted the service."

As soon as Fort Mulgrave was captured the attack on Faron was commenced. The Neapolitans and Spaniards fled and the British detachments were overpowered.

Early on the 18th a council of war decided that the lost positions could not be regained. The troops were ordered to concentrate at La Malgue preparatory to embarkation and the inhabitants of Toulon, who had reason to dread the vengeance of the Republicans, were offered a refuge on the fleet. About 14,000 took advantage of this offer.

On the night of the 18th the fleet moved to the outer harbour and the destruction of the French arsenal and fleet commenced, but the Neapolitans and Spaniards having without warning to their Allies abandoned their positions and embarked, the enemy was able to seize several forts, especially Malbosquet, and open fire upon our seamen. The Spaniards, too, whose fleet should have co-operated in setting fire to the French ships, were so slack as to incur the suspicion of treachery.

On the following morning the British troops and the French Royalists embarked under the direction of Captain Wilkinson, 30th regiment, the town major.

As far as we can trace, the companies of the regiment embarked as follows: Major Brereton with Ensign William Murray and 45 other ranks of his company on their own ship, the *Terrible*. Lieutenant William Maxwell had been transferred to another regiment.

Captain Wilkinson with Lieutenant Hamilton (wounded), Ensign Forster, and 46 other ranks of the company on their own ship, the *Robust*.

Lieutenant Livingston with 47 other ranks of the light company joined the *Robust* at Hyères. Their own ship the *Alcide* was detached.

A few of Thomlinson's company re-embarked on the *Princess Royal*. Thomlinson had been transferred to another regiment and Vaumorel had taken command of the company. Lieutenant Cuyler, all the non-com. officers and nearly all the men who survived were prisoners of war along with him. The company had left England 100 strong and must have been reinforced from the other companies for the defence of Fort Mulgrave.

The British troops embarked with only the worn-out clothes they had on, the remainder of their kits were lost with the baggage which had to be abandoned. Compensation was given by Government at the rate of £2 to sergeants and £1 to drummers, rank and file.

The weather was fine, which was fortunate, for some of the men-of-war designed for a complement of 700 had 2,000 souls on board.

December 20th.—Lord Hood anchored a few miles off in Hyères Bay and lost no time in sending off the foreign troops to their own country and the refugees to a place of safety. The British troops were distributed to their proper ships and he was ready for a fresh enterprise, for although he may not have managed quite well at Toulon, he was really a great man and of too firm a mind to be discouraged.

From Hyères Sir David Dundas was able to write his despatches, dated December 20th. In the first he states that the defence of Fort Mulgrave under the senior British officer was gallant but unsuccessful, but that owing to the hurried embarkation on different ships no returns or details had reached him.

In a second of the same date he forwards the casualty returns, signed by the Deputy Assistant Adjutant-General, Captain Thomas Hislop. The returns for Fort Mulgrave give :—

MISSING.

30th Regiment	2 Officers	5 Sergeants	3 Drummers,	140	rank and file
Marines .	2 ,,	—	—	60	other ranks
Navy . .	1 ,,	—	—	26	,,
Other Corps .	— ,,	—	—	50	,,
	5 Officers.			286	of other ranks.

The 30th officers were Captain Vaumorel and Lieutenant Henry Cuyler.

In his despatch General Dundas says of Captain Vaumorel and his companions : "The fate of the above officers and men is not nor can not be known, it is much to be apprehended that they fell before daybreak gallantly defending the post they were entrusted with when abandoned by other troops."

General Dundas had in his mind a decree of the French Convention declaring Mr. Pitt, the British Prime Minister, an enemy of the human race and ordering that no quarter should be given to British soldiers. We know now that he was mistaken, and that in spite of the orders of their Government the French gave quarter. On September 27th, 1794, Captain Vaumorel wrote from Romans, France, that he and the other prisoners were well treated. How many of the detachment perished we do not know, but the two officers of the

30th, after being struck off the strength as dead and their places filled up, appeared towards the end of the following year and were reinstated.

Included in the missing are the sixty-seven sergeants, rank and file whom Captain Vaumorel gives as killed and wounded by shell fire before the actual attack.

The attack was planned and conducted by Napoleon Buonaparte; the next time the 30th met him was at Waterloo.

Lord Hood and General Dundas soon had an opportunity of using their force.

The Corsican National Party had risen against the French and their sympathizers and on January 4th, 1794, Paoli, the leader of the Nationalists offered to Lord Hood to place the island under British protection. An agreement was come to with the Corsican National leaders, and Lord Hood proceeded to the Gulf of San Fiorenzo to secure a safe harbour for his store ships and transports. The force had been strengthened at Hyères by the 50th and 51st regiments, under Lieut.-Colonel John Moore.

On February 8th, at the entrance to the Gulf, Lord Hood's progress was delayed by a solid masonry tower, 15 feet thick and mounting two 18-pounder guns, placed on a point jutting out from the western shore called Mortella or Myrtle, which we have changed into Martello. This fort completely got the better of a line-of-battle ship and a frigate which were ordered to silence it. Sir David Dundas therefore landed with 1,400 troops, among whom were the 30th and some heavy guns, which after two days' fire forced the garrison of the tower, thirty-eight in number, to surrender.

This opened the way for an advance to the head of the Gulf, where the fortifications, especially a redoubt called the Convention, were found to be too strong to be carried by assault. The troops occupied the mountains in rear of the Convention redoubt while the sailors with incredible labour dragged some ships' guns to a position above it. Fire was opened on the 16th, and the redoubt was stormed on the 17th. The detachment of the 30th which was in second line had one sergeant killed, two rank and file wounded. Lieutenant Hamilton was in command of it.

The other fortifications were abandoned by the French, who fled to Bastia, 8 miles off over the mountains.

Sir David Dundas now prepared to advance on Bastia, to which there was a fair road.

On the 23rd he crossed the mountains, taking with him Colonel John Moore and some Corsican troops and followed by the 51st, 52nd and 69th; he noted that although Bastia was more strongly fortified than San Fiorenzo, it had the same weakness in being commanded by the neighbouring heights. Having selected positions for his batteries, General Dundas ordered the Corsican troops with him to fortify them, moving up the British to their support. Before the British troops could get up, the French drove off the Corsicans and began to entrench. As a result of the day's operations General Dundas announced to Lord Hood that having failed to secure the positions to the west, which commanded Bastia, it was now necessary to abandon the idea of a siege and to trust to blockade for the capture of the place. Lord Hood replied that in his opinion a siege was quite a possible operation and ordered on board ship the detachments of the 11th, 25th, 30th and 69th, which he had received originally as marines. General Dundas, who had already received leave from the Secretary at War on account of ill-health, now left for England. Lord Hood after a vain effort to induce General Daubant, who succeeded Dundas, to attack Bastia from the west, sailed round the north of the island and anchored to the north-east of the town to besiege it himself.

On April 4th the detachments of the four regiments serving as marines were landed under Lieut.-Colonel Villettes, with Major Robert Brereton as second in command. By adding seamen the force was made up to 1,400 men and Captain Horatio Nelson, R.N., was placed in command. Meanwhile Bastia was closely blockaded to the west and south by the Corsicans, supported by the troops under General Daubant. After a week's labour, during which he had very bad weather, Nelson opened fire with thirteen guns placed in battery on a ridge 1,600 yards from the citadel. At the same time Lord Hood sent in the *Proselyte* gun vessel to support him. The *Proselyte* was quickly set on fire and destroyed by red-hot shot. She had served at Toulon, and both then and on this occasion she had on board a party of the 30th who no doubt had been trained to work the guns. The crew were rescued by the boats of other ships. It has been found impossible to learn more of the services of this party, all records being lost.

Colonel John Moore recorded in his diary on April 5th his opinion that the fire of Nelson's guns would silence the weak redoubt of Campanelli but that there our success would end: that Bastia could only be taken by blockade.

The siege and blockade dragged on till May 19th, when the garrison, 4,000 in number, having consumed its provisions, capitulated as Sir David Dundas and Moore had foretold. If Lord Hood and Nelson were mistaken about the siege, they were excellent judges of behaviour before an enemy and their praises of the troops under them are gratifying. Lord Hood says in his despatch, "I am unable to give due praise to the unremitting zeal, exertion and judicious conduct of Lieut.-Colonel Villettes, who had the honour of commanding His Majesty's troops, never was either more conspicuous. Major Brereton of the 30th regiment, and every officer and soldier serving under the lieut.-colonel's orders are justly entitled to my warmest acknowledgments. Their persevering ardour and desire to distinguish themselves cannot be too highly spoken of."

Nelson says of his men, "We are few but of the right sort," and again of the soldiers, "Their zeal is unexampled." It would have been strange indeed if there had been any slackness under Nelson.

The detachments of the four regiments were now very weak, and their casualties in the siege were under twenty. Those of the 30th were four killed and four wounded. As far as we know only one officer was wounded, a captain in the 69th.

At the conclusion of the siege the detachment of the 30th re-embarked.

Lord Hood now heard that the French fleet in Toulon Harbour which we had failed to destroy had been fitted out and was preparing to put to sea. He therefore sailed with all available ships to bring the enemy to an engagement. The French managed however to keep their ships under the protection of the batteries on the islands off the coast near Antibes and Lord Hood gave up for the time all hope of an action. He detached Nelson in the *Agamemnon* with a squadron of ships to assist General Sir Charles Stuart, our new Commander-in-Chief in the siege of Calvi.

Sir Charles was at this time assembling his troops on transports in Martello Bay, and there the Headquarters Staff of the 30th reported its arrival, in his command, from Hilsea where it had been left in 1793. The history of the part of the regiment serving in England had been as follows. :—

ENGLAND

Soon after the service companies had embarked Captain William Lockhart had rejoined and taken command at Hilsea. On October 1st, 1793, he had under him 5 lieutenants, 3 ensigns, 1 adjutant, 1 quartermaster, 1 surgeon, 22 sergeants, 10 corporals, 10 drummers, and 43 privates.

For the first time in regimental history there now appears an attempt to establish a connection with Cambridgeshire and a recruiting party was sent to Newmarket. Its strength was Captain Leonard Browne, Lieutenant Thomas Brudenell, Ensign David Maxwell, 4 sergeants, 4 drummers and 22 rank and file. There was a dearth of recruits owing to the great increase in the army and navy, but 77 recruits were raised in six months. Some may not have been enlisted in Cambridgeshire. David Maxwell for instance enlisted some in Dumfries through the interest of his father, David Maxwell of Cardoness, who was always willing to help his old regiment.

In the autumn of 1793 the Secretary at War had tried to withdraw the four companies of the 30th serving at Toulon from Lord Hood's command. This was found to be impossible at the time but he seems to have formed a plan later, to form the four companies into a battalion for service in Corsica.

In December Captain Lockhart was ordered to be ready to embark for that island with the regimental staff and every effective man not employed on recruiting.

In March 1794 Lieut.-Colonel Charles Green took command at Hilsea but left again very soon.

In May Captain Lockhart embarked with the Headquarters. We have no embarkation return but he seems to have had about 70 privates, 3 drummers, 12 corporals and 12 sergeants. The officers under him were Captain Catlin Craufurd, Lieutenants Ninian Craig and John Russel (Adjutant), three ensigns, of whom David Maxwell was one, and Quartermaster Henry Craig. The party arrived in Martello Bay on June 15th, and was sent on to Bastia.

We do not hear of Lieut.-Colonel Green in Corsica until 1795, when he was appointed Inspector of Forces in the island.

SIEGE OF CALVI

The troops under Sir Charles Stuart consisted of the 50th and 51st Regiments and the reserve which was composed of the remains of the Royals and the flank companies of all the battalions in Sicily. Major Brereton still commanded the grenadiers, and Lieutenant Livingstone the light company of the 30th.

The reserve was commanded by Lieut.-Colonel John Moore, with Major Brereton as brigade major, but after the siege was formed Colonel Moore divided the reserve into two corps, one of light infantry under Major Brereton, and the other of grenadiers under himself, but he still retained command of the whole.

On July 7th our batteries opened fire upon Mozello, a pentagon fort of stone about 800 yards from the town. On the 13th a breach began in Mozello, but Captain Nelson of the *Agamemnon* was wounded, and lost the sight of an eye. On the 19th Mozello was stormed by Moore and Brereton. Sir Charles Stuart wrote of this attack, "The troops under Lieut.-Colonel Moore of the 51st and Major Brereton of the 30th proceeded with a cool steady confidence and unloaded arms towards the enemy, found their way through a smart fire of musketry, and regardless of live shells falling into the breach, or the additional defence of pikes, stormed Mozello."

Batteries were now formed against the town and opened fire on July 31st. Lieutenant Hay Livingstone, commanding the light company of the 30th, was wounded by the return fire. On August 10th Calvi surrendered.

The crown of Corsica was now offered to King George III by the inhabitants and the island was incorporated with the British Empire.

After the fall of Calvi, Lord Hood called upon Sir Charles Stuart for the men who had been landed for duty and to whose services as marines he was entitled. Sir Charles supplied the fleet with the proper number of men but took advantage of the arrival of a draft from England to send on board younger and fresher men. Those who had served on the island suffered greatly from fever.

The muster of December 24th, 1794, was signed at Bastia by Major Robert Brereton, commanding the regiment, Captain William Lockhart, paymaster, Lieutenant John Russel, adjutant. At this time the second in command always acted as paymaster.

The other officers reported present were Lieutenant Ninian Craig, Quartermaster Henry Craig and Surgeon Tagart. The other ranks were 15 sergeants, 12 corporals, 10 drummers, and 96 privates. Lieutenant-Colonel Green is not mentioned; shortly afterwards he is shown as Inspector of Corsican troops. Major Wilkinson (Brevet-Major September 9th) had gone home on leave, Captain Congleton was A.D.C. to his uncle the Viceroy of Corsica. Lieutenant Hay Livingstone had gone home wounded, Lieutenant Robert Copley, who had gone home sick after the fall of Toulon, left the service soon after this muster, as did John Copley who had been sent home sick from the West Indies. The remainder of the regiment are shown as "on command," that is on the fleet, except the recruiting party in Cambridgeshire, and a small one at Chatham. The total strength was 33 officers and 550 other ranks. One hundred and forty men were said to have been discharged without pension; no doubt the men of Vaumorel's party are meant.

Lord Hood had returned home in the autumn and had been succeeded by Admiral Hotham. By the beginning of 1795 the French Mediterranean fleet was decidedly stronger in ships and men than ours, and although their officers well knew where the real superiority lay they were ordered to sea by their Government.

On March 12th Admiral Hotham heard at Leghorn that the Toulon fleet was out and sailed at once to meet it. On the 14th the fleets came in sight of one another and the French tried to avoid an action, but the *Ça Ira* carried away her fore and main top masts. She was taken in tow by another ship and the French sailed so well that Admiral Hotham could not bring them to an action. Nelson alone was able to close with the *Ça Ira*, but he had to be recalled as it was impossible to support him.

On the following morning the *Ça Ira* was seen about 3½ miles from the British fleet in tow of the *Censeur*, and trying to regain the protection of the French fleet, from which they were a mile and a half distant. Both fleets made sail, the British to cut off the *Ça Ira* and *Censeur*, and the French to defend them, and an action ensued. The main body of the French fleet made but a half-hearted effort to cover the *Ça Ira* and *Censeur*, which after fighting gallantly and losing 500 men apiece were forced to surrender. Admiral Hotham unfortunately refused to listen to Nelson's advice to pursue the French fleet and force it to a decisive action. This unhappy decision was one of the causes of our losing Corsica and having to quit the Mediterranean.

The 30th had present in this action :—On H.M.S. *Princess Royal*, Captain J. Catlin Craufurd, Lieutenant J. Gordon, Ensign David Maxwell, and 84 other ranks ; the ship's casualties were 3 killed, 8 wounded. On H.M.S. *Terrible*, Captain Linnaeus Gardner, Lieutenants Alexander Hamilton and Samuel Bircham, 85 other ranks ; the ship's casualties were 1 killed, 5 wounded. On H.M.S. *Egmont*, 1 sergeant, 15 rank and file. In justice to Admiral Hotham it should be stated that his ships were weakly manned. He with 14 ships carrying 7,700 men beat the enemy with 17 ships and 17,000 men and took two prizes, but he wasn't a Nelson.

In the spring of 1795 the Regimental Headquarters were moved to San Fiorenzo and remained there or in the neighbouring village of Oletta till ordered home.

On April 15th Major Brereton addressed a memorial to Sir Gilbert Elliot, the Viceroy, in which he prays that he may be given a commission as Lieut.-Colonel to raise one of the proposed Corsican regiments. Sir Gilbert, who had been a witness of Brereton's services at Toulon and in Corsica, sent the memorial home with a statement from himself that Major Brereton was highly deserving of His Majesty's favour and countenance. As a consequence of this recommendation and of the repeated mention of his name in despatches, Major Brereton was promoted on May 27th, 1795, to be Lieut.-Colonel, without purchase, of H.M. 63rd Regiment of Foot, and Captain Wilkinson became major of the 30th. Lieut.-Colonel Brereton's chance came to him at a time when most men are thinking of retiring, and he seized it with the energy of youth. When he landed as a captain at Toulon he had thirty-two years' excellent service as a company officer ; he died in 1816 as a Lieut.-General after holding several commands on service and at home ; the last of his employments was the command of the Dublin district.

Lieutenant Hay Livingstone was soon afterwards promoted without purchase into another regiment in recognition of his conduct at Calvi, so both the officers commanding flank companies in the 30th received rewards for their bearing in the capture of that place.

The Viceroy's efforts on behalf of the regiment were not always so successful.

On December 5th, 1794, Mr. Henry Dundas sent orders to Sir Gilbert to draft the detachments of the 11th, 30th and 69th regiments, and to complete the Royals, Royal Irish, 50th and 51st, the officers and non-com. officers of the drafted regiments to return home to recruit. To this Sir Gilbert replied on January 17th, 1795, that he would prefer to keep the 30th and 69th regiments, which were physically the stronger, and had not been so long abroad, and to draft the Royals and Royal Irish into them. He forwarded in support of this the opinion of Lieut.-General Trigge, and mentioned that he had promoted that officer and made him Commander-in-Chief. When this letter reached home there was a storm. On February 28th Mr. Dundas wrote, pointing out that the Viceroy had usurped the King's prerogative and that he must at once revoke all his commissions, especially that in favour of Major-General Trigge. With regard to the drafting of regiments Sir Gilbert was informed "that His Majesty's choice was determined by considerations connected with a general plan of the services to be provided for and not by the situation of the regiments in Corsica alone."

Sir Gilbert was not dismayed. He threw himself at His Majesty's feet (by letter), and apologised for his presumption, but he still maintained that great inconveniences would arise, especially with regard to the sea service, if the 30th and 69th were drafted. It was impossible, however, for Mr. Dundas to draw back after his statement about the plan, so Sir Gilbert was ordered to draft the three regiments at once, giving a bounty to the men if he found it necessary.

On May 1st, 1795, the 30th had at Fiorenzo 111 rank and file, of whom 68 were drafted into the Royal Irish. The officers and non-com. officers, and the remaining rank and file, were ordered to return to England with the fleet of merchantmen, which was to sail as soon as the portion from the Levant arrived. The detachments on H.M. ships were not drafted and continued to serve on the *Princess Royal*, *Terrible*, *Egmont*, and *Berwick*.

At the end of July, Captain Lockhart embarked with the Headquarters Staff on board the *Fidelity* transport and sailed for Gibraltar where the great fleet was to assemble: the staff of the 11th and 69th regiments were in company on the *Spencer* and *Harmony*.

The party on the *Fidelity* consisted of :—Captain Lockhart, Lieutenant Ninian Craig, Adjutant John Russel, Quartermaster Henry Craig, Surgeon Edward Tagart, 12 sergeants, 14 corporals, 11 drummers, 35 privates.

There was a long delay at Gibraltar, but on September 24th 300 merchantmen and transports sailed for Britain under convoy of His Majesty's ships *Censeur* of 74 guns, *Fortitude* 74, *Bedford* 74, *Argo* 44, and *Juno* 32. The two latter were frigates.

On board the *Censeur* was Lieut.-Colonel Brereton going home to take command of his new regiment.

Nearly one-half of the merchantmen parted company immediately after sailing and it was fortunate for them that they did so. Our irresolution in not utterly destroying the French fleet in Toulon harbour and in not following up Hotham's victory of March 14th was now bearing fruit. The French Mediterranean fleet had been reinforced from Brest and was more than holding its own.

On October 2nd the garrison of Gibraltar had the chagrin of seeing a French fleet of six line-of-battle ships and three frigates pass the straits, evidently in pursuit of the convoy which had sailed for home a week before. Off Cape St. Vincent the French squadron, commanded by Admiral de Richery, came up with the convoy and the signal was at once made for the merchantmen and transports to disperse and save themselves while the three line-of-battle ships formed line to meet the enemy.

De Richery's squadron consisted of the *Victoria* 80 guns, *Barras* 74, *Jupiter* 74, *Berwick* 74, *Resolution* 74, *Duquesne* 74, and three frigates.

The action commenced at one o'clock in the afternoon. The odds were more than two to one, and to make matters worse the *Censeur,* which we had captured on March 14th, rolled her foretopmast overboard at the commencement of the action, and could not keep her station. The enemy concentrated their fire upon her, and although it was returned with great spirit and her two consorts did all they could, by half-past two she was totally dismasted and forced to surrender. The other British men-of-war made off, and the French ships pursued the scattered merchantmen and were able to take forty of them before nightfall. The transport *Fidelity* was captured, but the *Spencer* and *Harmony* escaped. Lockhart, who had just got his majority, and his party were taken as prisoners to Bordeaux, Lieut.-Colonel Brereton to Cadiz.

The " Terror " was over in France and a cartel had been arranged for the exchange of prisoners of war under which the whole party was landed at Portsmouth on January 14th, 1796, from the *Fidelity,* which was now acting as a cartel ship. On landing, the Staff of the regiment marched under Major Lockhart to Colchester, where there was a strong party of the 30th of which Wilkinson, now second lieut.-colonel of the regiment, had lately taken command.

The movements which led to the establishment of the Headquarters of the 30th at Colchester were as follows :—

When the regiment was broken up to serve as marines in 1793 a strong recruiting party was left in Cambridgeshire as well as others at Chatham and elsewhere. In 1794, when Headquarters were transferred to Corsica, an order was issued for all detachments in England to concentrate in Cambridgeshire. The detachment at Newmarket was therefore joined not only by the scattered recruiting parties but by the remains of Captain Broughton's company which landed under Ensign Wood from H.M.S. *Sampson* at Portsmouth on April 15th, 1795, on return from the West Indies. There was no barrack at Newmarket, not even a hospital or guard room, and many of the men who had returned from the Indies were sickly. Lieutenant Brudenell represented that if left uncontrolled in billets their health would never improve and as a consequence the detachment was moved to Thetford in May, and to Colchester Barracks in July. Captain Robert Douglas, lately appointed to the regiment, joined there in July and applied for contingent allowance as the strength under him was already greater than that of a company.

In September the regiment was given a second lieut.-colonel and a second major and the position of the field officers requires a word of explanation. Major Wilkinson became second lieut.-colonel on September 22nd. T. V. Reynolds from the Scots Brigade and Brevet-Major Clinton from the 1st Foot Guards were brought in as majors. A fortnight later Major Clinton was promoted into another regiment and Captain Lockhart got the step, without purchase. Thus Lieut.-Colonel Charles Green, Governor of St. Vincent, and Major Reynolds, a staff officer, were respectively lieut.-colonel and captain and major and captain, while Wilkinson and Lockhart were borne on the strength of the Colonel's (General Clarke's) company, as "Field Officers without companies," although one was really commanding-officer, and the other second in command.

In the autumn the prisoners lost in the defence of Fort Mulgrave began to return.

The first of the non-com. officers and men to return was Sergeant John Stalliard, who joined at Colchester on September 18th. He was followed in December by several non-com. officers. It is nearly impossible to trace the privates or to say how many of them rejoined. Those who came back are mixed up with the recruits who joined.

Whether the bulk of those added to the strength were old soldiers returning from France or recruits, the losses from discharges and other causes nearly equalled the gains. From the increase of the army and navy and from our dreadful losses from sickness in the West Indies, the supply of recruits in Britain and Ireland did not meet the demand.

In March, 1796, there was a change in the staff of the regiment, when Quartermaster Henry Craig resigned the appointment and was succeeded by Quartermaster-Sergeant Arthur Poyntz. Sergeant Poyntz had served the campaign in South Carolina in 1781.

In June Capt.-Lieutenant Hay Livingstone, who had been wounded at the siege of Calvi, was given a company without purchase in the 13th Foot.

In May the regiment was ordered to proceed to Ireland by way of Liverpool. The march was a leisurely one. On June 1st Lieut.-Colonel Wilkinson signed the monthly returns at Dedham, and on July 1st at Daventry. On July 16th he embarked at Liverpool for Ireland.

Strength: 1 lieut.-colonel, 1 major, 3 captains, 4 lieutenants, 1 ensign, 1 quartermaster, 1 surgeon, 1 mate, 35 sergeants, 18 drummers, 111 rank and file.

While on the march the regiment received an order to abstain from wearing powder in the hair for the future. At this time an extra company was added to the regiment for recruiting. Had there been barrack accommodation in Cambridgeshire it would in all probability have been established there, but it accompanied the regiment to Ireland and was stationed at New Geneva near Waterford. Two companies were ordered but only one appears in returns.

On landing, the Headquarters marched to Bandon, where shortly afterwards they were joined by Captain Hamilton with the remains of the detachments which had been left behind in the Mediterranean, and that which had served under Lieutenant Thomas Woolridge on H.M.S. *Bellona* on the West Indian Station. The following had been the history of the detachments left behind on board ship in the Mediterranean:—

When Sir Gilbert Elliot was ordered to draft the 30th in Corsica, Mr. Dundas had intended that the detachments on board ship should be drafted as well as those actually in the island, but, probably to oblige the Navy, the order was interpreted as only covering the men at Headquarters.

The detachments on board continued to serve as before with little change, except that Lieutenant Eyre Peter Shewbridge, who had been taken prisoner at Cap Brun on October 13th, 1793, was released in September, 1795, at Leghorn and was kept for service on the fleet. In July, Captain Craufurd had gone home on promotion to another regiment, and Captain Gardner to retire.

Captain Gardner found scope for his adventurous spirit in the service of native rulers in India. It is impossible here to give even a sketch of the career of this remarkable man, but it may be said that although daring to the verge of rashness, he was a man of modest and gentlemanly manners and that if Mr. Thackeray's " Major Geogehan " is intended for him it is a libel.

None of the ships on which the 30th served was in serious action after March 14th, 1795, although the *Princess Royal* and *Terrible* were present at the destruction of *L'Alcide* on July 13th. The Navy was at that time not only undermanned, but badly manned, and it was of the utmost importance to keep up the strength of the marines. Mutinies were frequent, and Captain Hamilton's party had been instrumental in suppressing a mutiny on the *Terrible*, and had been thanked by Captain Campbell, commanding the ship.

When, in the spring of 1796, Mr. Dundas returned to his original idea of drafting all the men of the 30th in the Mediterranean he decided that this time the Navy should have the drafted men. One hundred and thirty-two privates were therefore transferred to the marines between April 1st and June 8th, and the officers and non-com. officers returned home, landing at Portsmouth from the *Princess Royal* on September 2nd under Captain Hamilton.

The men drafted had been transferred to ships in the fleet under Sir John Jervis. It was a disgrace to the administration that drafting was necessary, but at any rate in this case the men went where they could best serve their country, and we may hope that they fought stoutly under Jervis at St. Vincent and stood by the great admiral loyally in the still sterner conflict he waged to restore discipline in his fleet.

The evacuation of Corsica and our retreat from the Mediterranean which took place in October owing to the great successes of Buonaparte in Italy, to the increasing discontent in Corsica and to the falling away of Spain from the Coalition, affected Lieut.-Colonel Green, Inspector of Corsican Forces, and Captain Gilbert Congalton, A.D.C. to the Viceroy. Lieut.-Colonel Green was appointed Governor of the island of St. Vincent in the West Indies and Captain Congalton rejoined the regiment.

Captain Hamilton re-embarked at Portsmouth for Ireland on September 2nd, having under him, including Lieutenant Woolridge's party, Lieutenant Sam Bircham, Lieutenant Thomas Woolridge, Lieutenant David Maxwell, 9 sergeants, 3 drummers, 97 rank and file. On landing in Ireland the detachment marched to Bandon.

The only company now afloat was Captain Ralph Smyth's. It landed at Sheerness on September 23rd, and from there proceeded to join Headquarters at Bandon ; its strength was Captain Ralph Smyth, 1 sergeant, 1 drummer, 54 rank and file.

On Christmas Day the strength of the regiment under Lieut.-Colonel Wilkinson at Bandon was, 50 sergeants, 20 corporals, 22 drummers, 291 privates.

The regiment had no sooner assembled at Bandon than it was in danger of being overwhelmed by a French force under the celebrated General Lazare Hoche.

The expedition to Ireland had been long preparing, but had been put off so often that the British Admiralty seem to have thought it useless to guard against it, especially as negotiations for peace were going on in Paris. The

French fleet sailed, however, on December 15th, from Brest with 18,000 troops on board, and three days later, when the French Government was sure that it had got away, Lord Malmesbury, the English envoy, was dismissed from Paris.

On December 20th the expedition was at the mouth of Bantry Bay, but the ship which carried Lazare Hoche was missing and with him was the naval Commander-in-Chief. Baffling winds, and probably a want of resolution owing to the absence of its leaders, kept the fleet near the mouth of the bay, and on Christmas Day the greater part of it was blown out to sea, the remainder followed and the whole returned to Brest, carrying with it the naval and military commanders whom it met on the way.

The utmost exertions of Major-General Dalrymple, who commanded at Cork, had only enabled him to collect 2,000 men at Bandon on Christmas Day, and Cork could not have been saved had the French been able to land at the head of Bantry Bay.

It was probably constant fear of another attempt on the part of the French which caused the regiment to stand fast at Bandon for two years and saved it from a share in the troubles of 1798. From Bandon not only could Bantry Bay be watched, but also the smaller bay of Clonakilty, still nearer Cork. A small detachment was usually kept at Clonakilty. For a few months the regiment was in camp, for the remainder of the time in barracks.

Numerous recruiting parties were out and a considerable number of recruits were raised at Headquarters. Nine officers were employed in recruiting in addition to Captain Freeman Lane's recruiting company at New Geneva, near Waterford.

While at Bandon changes of importance were introduced, the chief of which was an increase of pay of the subalterns and lower ranks and the grant to the latter of a ration of one pound of bread and three-quarters of a pound of meat for a fixed stoppage of pay irrespective of the fluctuations in the market price.

All pay was to be issued when due without delay as arrears and the stoppage of 1*s.* in the £, and the deductions for hospital and agency were abolished. Poundage had been abolished for privates, and the stoppage for surgeon and chaplain for all ranks in 1771.

The new rate of pay was: Colonel, £1 2*s.* 6*d.*; lieut.-colonel, 15*s.* 11*d.*; major, 14*s.* 1*d.*; captain, 9*s.* 5*d.*; lieutenant, 5*s.* 8*d.*; ensign, 4*s.* 8*d.*; adjutant, 8*s.*; quartermaster, 5*s.* 8*d.*; surgeon, 9*s.* 5*d.*; assist.-surgeon, 7*s.* 6*d.*; sergeant-major and quartermaster-sergeant, 2*s.* 0¾*d.*; sergeant, 1*s.* 6¾*d.*; corporals, 1*s.* 2¼*d.*; drummers, 1*s.* 1¾*d.*; privates, 1*s.* a day. Five pints of small beer were still issued daily in barracks. All other allowances were abolished.

The authorised weekly stoppage from the soldier was 4*s.* for messing, and 1*s.* 6*d.* for upkeep of kit. The remaining 1*s.* 6*d.* was to be paid into his hand. He paid for cleaning materials and washing.

The net increase to the soldier's pay was 2*d.* a day and he had at last a fixed sum due to him and a fixed ration irrespective of prices.

The daily increase of pay to the subalterns was 1*s.* a day.

The surgeon's mate now became a commissioned officer as assistant surgeon and was no longer the nominee of the colonel.

The latter's other nominee, the regimental chaplain, was abolished. Few of them had ever seen their regiments. In their place army chaplains under the War Office were appointed.

The staff-sergeants now appear in their proper place in the pay lists. Previously they had been entered and paid as sergeants. The extra pay of the sergt.-major and drum-major had been paid by the privates and drummers; who paid the quartermaster-sergeant does not appear.

A paymaster without other duties was appointed to each regiment. Ensign Michael Jones was the first paymaster of the 30th. His commission was dated August 1st, 1797.

The system of drill drawn up by Sir David Dundas was still adhered to, but the formation in two ranks instead of three was now allowed and soon came into general use.

The standard for recruits was still 5 ft. 5 ins., with one inch less for growing lads and still another inch less for men and lads from the county to which the regiment belonged.

By 1796 the modern hat shaped like a chimney-pot with a brim had superseded the cocked hat in civil life, and the army followed the fashion. The hat seems at first to have been of leather, afterwards of felt. That of the officers of the grenadier company had a white plume, the light company a green one and the other companies a red and green one. When on duty officers wore a crimson and gold cord round the hat with a rosette brought to the edge of the brim.

The gorget was ordered to be gilt instead of silver and to be tied with ribbons of the colour of the regimental facings.

In 1798, Colonel Wilkinson and Lieutenant Samuel Bircham were on duty with the second light battalion and Major Lockhart commanded the regiment. The light battalions were formed of the light companies of the Irish Militia and General John Moore had selected Colonel Wilkinson to discipline the Second.

CHAPTER VII

OCCUPATION OF MESSINA. SIEGE AND CAPTURE OF VALETTA. EXPEDITION TO EGYPT.

1799–1802

THE strength of the regiment gradually rose till the latter part of 1798, when it was considered fit for service, and was ordered to embark at Cork along with the 89th for Minorca, to serve under Lieut.-General Sir Charles Stuart.

The embarkation took place on December 30th, 1798, under Major Lockhart. Lieut.-Colonel Wilkinson was still on duty commanding the second light battalion.

The muster of December 31st, 1799, shows on board ship :—

Major William Lockhart, Captains David Maxwell, Philip Vaumorel (grenadiers), Ralph Smyth, Alex. Hamilton (light), John Collins.

Lieutenants Arch. Campbell, Christopher Williamson, Francis Sharkey, Philip Bulkeley, John Russel, Christopher Maxwell, Dan Mansergh, Edmond R. Boggis, Will Ed. Fitzthomas, Elias Malet, John Amory, Robert Murray.

Ensigns : Peter Dumas and James Sullivan.

Staff Ensign and Adjutant, Alex. Young ; Q.M., Arthur Poyntz ; Surgeon, Edward Tagart ; Assist.-Surgeon, John Price ; Paymaster, Michael Jones.

Thirty sergeants, 30 corporals, 18 drummers, 456 privates. These numbers include a draft of 95 non-com. officers and men from the recruiting company at New Geneva.

Captains B. Roche and Roger Montgomery ; Capt.-Lieutenant Baynton ; Lieutenant Lahiff, Ensigns J. B. Lynch, J. Goodwin, and E. P. Shewbridge were on recruiting service. Captain Freeman Lane was commandant at New Geneva. Lieutenant T. Woolridge was a brigade-major in Ireland, Lieutenant Sam. Bircham was with Colonel Wilkinson attached to the second light battalion, and Lieutenant Will Fawcett was an A.D.C. in Ireland.

Lieut.-Colonel C. Green was Governor of Grenada and Major T. V. Reynolds was on the staff in England.

Twenty sergeants and 5 drummers were recruiting.

Although the two regiments embarked on December 30th, 1798, they were kept in the Cove of Cork till the end of January, 1799 : foul weather was the alleged cause.

Meanwhile, they were urgently wanted in the Mediterranean. We had reversed the policy of 1796 and re-entered that sea in force. On August 1st, 1798, Sir Horatio Nelson had destroyed the French fleet after a fierce action in Aboukir Bay (Battle of the Nile) ; and in November, General Sir Charles Stuart had taken Minorca from the Spaniards. The French, however, had invaded

Naples and the King and Queen had fled, under Nelson's protection, to Palermo, December 22nd, 1798. Nelson, who was well aware of the weakness of Stuart's garrison in Minorca, expressed his regret to him in a letter of February 16th, 1799, that we could not have 1,000 British soldiers at Messina to make Sicily secure.

Almost as soon as this letter reached Sir Charles, the arrival of the transports with the 30th and 89th enabled him to comply with Nelson's wish, though the two regiments did not appear to have been all he could desire. On March 1st he wrote to Mr. Dundas from Port Mahon :—

"The regiments from Ireland are arrived with nearly 200 sick, and out of their reduced numbers 250 from the jails guilty of offences among which rebellion appears the slightest taint ; with these men I shall endeavour to assist Lord Nelson." (Nelson had been created a viscount for his victory at the Nile.) Sir Charles gave orders for the regiments not to disembark and sailed with them for Palermo on the following day.

On March 10th, the transports anchored for a few hours in Palermo Bay and then sailed for Messina, the general proceeding there by land to gain some knowledge of the country and people. On arrival at Messina he was received by a crowd of 50,000 people and was so impressed by their loyalty and the hardy, active appearance of the inhabitants of the mountainous country he had ridden through, that he determined to get rid of the Neapolitan regiments in Messina and to complete the 30th and 89th, whom he placed in the Citadel, to 1,000 men apiece with Sicilian recruits. Apparently on further acquaintance with those regiments he had satisfied himself as to their power of absorbing and training queer material.

Leaving his adjutant-general, Colonel Stewart, in command he returned to Minorca, paying a visit to Malta on his way.

In trying to estimate how far the regiment was prepared for the active service before it, we have only the returns to guide us.

To take physical health first, there is no appearance of any sickness, beyond the normal, up to embarkation, and it may be safely said that the sickness on arriving in the Mediterranean was entirely due to long confinement (78 days) on unhealthy transports. As has so often happened in the regiment under similar circumstances the disease was violent, but when the men were landed and had fresh air, light, exercise and decent food, those who were not too far gone recovered quickly. The deaths in each month were : From embarkation on December 30th, 1798, to February 24th, 1799, *none* ; to March 24th, 5 ; to April 24th, 16 ; to May 24th, 7 ; to June 24th, 3. There still were 44 sick in July, but Messina is not a very healthy place in summer. Eighty-four sick had been landed in March.

In composition the regiment, as far as we can judge, was pretty equally divided between men of long service and men of two years' service and under. Many of the sergeants were veterans of the American War. The recruits enlisted by the regiment seem to have been satisfactory, desertion was entirely confined to those who had not joined or had only served a few weeks. No man of over four months' service had deserted in the two years preceding embarkation.

There remains to be considered the draft of ninety-five men from New Geneva which arrived just before embarkation, and it is there that we must look for our proportion of the 250 scoundrels spoken of by Sir Charles Stuart. It is certain that the greater part of them were prisoners taken in Wexford and the neighbouring counties, who had been given their choice between enlisting and being hanged for their share in the rebellion. The regiment has often suffered from the practice of the War Office in collecting riff-raff and throwing it on board at the last moment to fill up numbers in the returns, but on this occasion it has been impossible to trace those men, except in one instance. The probability is that they were lads who had gone astray in the troubles which tore Irish society to pieces in 1798 and that they recovered themselves under discipline, or if there were any among them irreclaimably vicious that they did not find in the regiment a congenial field for the display of their qualities.

Sir Charles Stuart was a man of the very highest capacity in civil and military affairs, and from his noble character, was the idol of the Army and Navy, but he wrote his letter from Port Mahon in a state of high indignation with (in his opinion) an incompetent Minister, and when he was nearly worn out by the sufferings which caused his too early death. It is permissible, therefore, to believe that he accepted an exaggerated report as to his reinforcements.

On May 2nd Colonel Thomas Graham, with whom the regiment had served at Toulon, assumed command at Messina.

In his instructions to Colonel Graham, Sir Charles Stuart varied his original scheme by limiting the number of Sicilians to be admitted into the 30th and 89th to a proportion of two to three ; unfortunately, however, he had to return to England at the end of May in ill-health and the scheme fell to the ground. It would have been a most interesting experiment.

Nothing of interest occurred during our occupation of Messina. Lieut.-Colonel Wilkinson rejoined and took command. Captain Hamilton was fort major, and Lieutenant Archibald Campbell, fort quartermaster.

On November 7th, General Robert Manners succeeded General Clarke as colonel of the regiment.

MALTA

In June, 1798, Buonaparte, on his way to Egypt, had occupied Malta and left a garrison there, but after the Battle of the Nile, Captain Alexander Ball had, by Nelson's order, blockaded the island with a fleet of British and Portuguese ships. In August, 1799, the Maltese rose against the French and drove them into Valetta, and in September Colonel Graham proposed to the General Commanding in the Mediterranean that, leaving Major Lockhart at Messina with a depot, he should take the 30th and 89th to Malta to aid in the siege of Valetta. The general left the decision to Lord Nelson.

On November 28th, Captain Sir Thomas Troubridge, R.N., arrived with his lordship's orders to embark, and on December 6th the two regiments marched from the citadel and went on board H.M.Ss. *Culloden* and *Northumberland* at Maritimo.

On December 10th the regiments landed in St. Paul's Bay, Malta.

Major Lockhart had been left at Messina to protect the heavy baggage and to " keep up the name of the British Government." He had under him Lieutenants James Sullivan and John Amory with 54 rank and file.

The strength of the regiment embarked at Messina and landed in Malta was :—

Lieut.-Colonel Wilkinson, Captains Vaumorel, Smyth, Hamilton, B. Roche and David Maxwell, Capt.-Lieutenant George Grey, Lieutenants Chris. Maxwell, Campbell, Bircham, Bulkeley, Murray, Fitzthomas, Malet, Mansergh, Hare, Boggis, Love, R. B. Lynch, Lahiff, and James ; Ensigns, Goodwin, J. B. Lynch, Peter Dumas, Gore, Rogers and Garnon. Adjutant Young, Quartermaster Poyntz, Surgeon Tagart, Assist.-Surgeon Price ; and Paymaster Jones. Sergt.-Major Peter Wallace, Quartermaster-Sergt. McPherson, Paymaster-Sergt. Pat. Matthews, 31 sergeants, 19 drummers, and 381 rank and file, 23 soldiers' wives with 15 children.

In the foregoing return Capt.-Lieutenant George Grey appears for the first time. He had been appointed *vice* Baynton, March 12th, 1799, and had joined at Messina. Captain Grey's name is frequently spelled with an " a " in official documents, but he invariably signs " Grey." Captain Hamilton had been appointed brigade major before sailing.

Brigadier-General Thomas Graham and Surgeon-General Alexander Jameson sailed in the *Culloden.*

The 30th, on landing, were quartered at Bicascara and the 89th at Nasciar and Cazel Lia. Soon after landing two brothers who had been implicated in the Irish rebellion in Wexford deserted from an outpost of the 30th, and General Graham took a very serious view of the matter, but this was the only case of misconduct we hear of which could be traced to disaffection.

It was a time of great anxiety for General Graham. The Maltese, although brave and willing, had no officers, and the men who took the picquets at night had usually spent the day in cultivating their fields. It was therefore to be feared that a bold sortie on the part of the French would be successful.

On December 30th, General Graham was able to write to the Duke of York that, some more marines having landed from the fleet, he was able to concentrate the 30th and 89th on Zeitun and the ridge opposite the Cotonera and Ricasoli : that was the vital point ; if Zeitun were lost Graham was prepared to embark at Marsa Sirocco. One advantage of the concentration on the ridge was that it removed the troops from the marshy ground in which the Maltese alone could work with impunity.

Both the besieged and besiegers anxiously looked for intelligence of a French relieving force from Toulon, and in February, 1800, a seventy-four, three frigates and a corvette had all but reached Malta, when they unexpectedly fell in with a squadron under Lord Nelson, who was on his way to visit the island. The seventy-four and a frigate were captured, the others escaped.

One of General Graham's first endeavours had been to raise a Maltese regiment officered from the British army, and by April several companies had been formed. The officers of the 30th employed on this duty were Lieutenants Richard Hare, W. E. Fitzthomas, Philip Bulkeley and Peter Dumas. Sergt.-Major Peter Wallace, Sergeant Robert Thompson, and a few corporals and privates were also lent as instructors.

Lieutenant Archibald Campbell acted as assistant-engineer until his death on August 19th.

In June Sir Ralph Abercromby, who now commanded in the Mediterranean, was able to send from Minorca a reinforcement of 1,500 men under

Major-General Pigot. Brigadier-General Graham, on being superseded, left the island, but General Pigot kept Captain Hamilton as brigade-major.

The French held out till September 4th, 1800, when General Vaubois offered to treat, and on the 5th, Ricasoli, Tigne and Floriana were occupied by our troops, Colonel Wilkinson being deputed by General Pigot to receive the keys of Valetta.

Some changes occurred in the regiment during the siege. Ensigns James Blake Lynch and Peter Dumas had been transferred to the 20th Foot before the regiment left Messina, but they had landed in Malta and served for some time there before receiving notification of the transfer; at the same time information was received that Captain Freeman Lane had been transferred to the 20th in January 1800.

On May 24th, Lieutenant Edward Boggis was transferred to the 56th regiment.

On June 25th, Sergeant James Glen, of Douglas's company, was appointed quartermaster-sergeant in place of Charles McPherson, discharged. The latter had served as corporal in the light company at the action of Eutaw Springs in 1781, and had been senior sergeant of Torriano's company when the latter was killed in action on October 13th, 1793, at Cap Brun.

The Messina detachment joined the regiment in Valetta after the occupation.

EGYPT

During the year 1800 the number of troops in the Mediterranean under Sir Ralph Abercromby had been raised to over 20,000 men by successive reinforcements and he received orders to drive the French from Egypt: he was to employ 16,000 men on this service, the remainder of his army being despatched to Portugal to defend it from the Spaniards.

About one-fourth of the troops under Sir Ralph were enlisted for service in Europe only, and he gave 5,000 of those limited service men to Lieut.-General Sir James Murray Pulteney to take to Portugal, and ordered the two battalions of the 40th which were likewise enlisted for limited service to relieve the 30th and 89th regiments in Malta, and those two regiments, which were enlisted for general service, he ordered to form part of the expeditionary force under his own command. The 89th is said to have embarked and joined Sir Ralph at Gibraltar; the 30th did not.

The troopships and transports were in urgent need of repair, which could not be effected at Gibraltar, and in the second week of October, 1800, the first division of Sir Ralph's force sailed for Minorca and in the following week the second division left Gibraltar for Malta. The troops were landed in both islands and the repairs of the vessels taken in hand.

On November 26th, Sir Ralph reached Malta, and on that day issued an order for the 30th to prepare for immediate service. Æneas Anderson, the historian of the Egyptian expedition, tells us that on the following day Sir Ralph inspected on the Floriana, amidst a great concourse of spectators, the 30th, the 48th, and the second battalion 45th regiment. "After the review Sir Ralph was pleased to express himself in terms of great satisfaction as to the appearance of the regiments, distinguishing at the same time the 30th with peculiar expressions of approbation." We hear now, for the first time, of the use of the bugle by the drummers of the light company. The queue had been abolished in the light company by a general order in the previous year, but the men still wore their hair long. It was brushed up under the hat behind and fixed with grease or soap.

By the end of the month the repairs to the ships were completed and the fleet assembled. On December 14th the troops embarked.

The 30th had received new clothing, including great coats, before embarkation, but the great coats were left at Malta with the heavy baggage. Probably Sir Ralph thought, like Wellington after him, that without them the load which the men had to carry was as much as they could bear. His orders lay down that on landing, each man was to carry 60 rounds of ball cartridge and 2 spare flints, 3 days' bread, 3 days' pork, cooked, and a canteen full of water ; 2 shirts, 1 pair shoes, 2 pair socks ; a blanket ; entrenching tools and camp kettles ; 3 days' rum was to be in charge of the quartermaster.

It is uncertain if the new clothing included the chaco, a bad and uncomfortable headdress of felt with a heavy brass plate in front, which had lately been introduced in the Army. Probably the regiment embarked in the old round hat.

After embarkation the wind was contrary for nearly a week, but on December 20th the fleet sailed for the Levant.

Lieutenant Samuel Bircham remained in the island as acting captain of the Maltese regiment, Sergeant-Major Peter Wallace and Sergeant Robert Thompson as acting ensigns in the same corps, and Quartermaster-Sergeant James Glen in charge of the heavy baggage of the regiment.

Among those who arrived in time for the expedition was Captain John Douglas, who since his appointment in 1797 had been shown every month as " absent without leave, but supposed to be serving in the East Indies with his old regiment the 78th." In those days a man might well serve for a year in India in ignorance of his promotion and transfer, and he might perhaps be another year in joining his new corps.

The fleet anchored in Marmorice Bay, on the coast of Asia Minor, on December 30th, 1800. The sick were landed at once and then the effectives of the army.

The strength was, officers 805, other ranks 15,526.

The 30th landed 30 officers, 480 other ranks, effective, 6 sick were present, and 6 had been left at Malta. The health of the regiment was thoroughly re-established and contrasted favourably with that of the regiments which had been long on board ship.

The officers present at the beginning of the year or who joined in time for service in Egypt were :—

In the Commandant's company : Lieut.-Colonel William Wilkinson and Major William Lockhart (field officers without companies), Capt.-Lieutenant George Grey, Lieutenant David Crowe, Ensign Henry Craig, Adjutant Alexander Young, Quartermaster Arthur Poyntz, Surgeon Ebenezer Brown, Assist.-Surgeon Henry Dewar, Paymaster Michael Jones ; *Officers en second :* Lieutenants Dan. Mansergh, Richard Hare, R. B. Lynch, Thomas Lahiff, Christopher Williamson (sick), James Sullivan, Thomas James, Ensign C. W. Barry.

The Lieut.-Colonel's company : Lieutenant Philip Bulkeley.

The Major's company : Lieutenant William Edward Fitzthomas.

Grenadier company: Captain Philip Vaumorel, Lieutenant John Russel, Ensign Arbouin.

Smyth's company: Captain Ralph Smyth, Lieutenant Ninian Craig, Ensign James Adam Dolling.

Light company: Captain Alexander Hamilton, Lieutenants Robert Murray and Elias Malet.

Roche's company: Ensign Thomas Rogers (Captain Benjamin Roche who was on leave in England sold the company on April 17th to Lieutenant Frederick Holbrooke of the 13th Foot, who got out to Egypt in time for the latter part of the campaign).

Douglas' company: Captain John Douglas, Lieutenant John Amory.

Maxwell's company: Captain David Maxwell.

Montgomery's company: Lieutenant Edward Jennings.

The officers recruiting were Captains Christopher Maxwell and Roger Montgomery, Lieutenants William Fawcett and Archibald Bertram, Ensign Noblet J. Lindsay.

The lieut.-colonel, Charles Green, was Governor of Grenada; the major, T. V. Reynolds, on the staff in England; Lieutenant Thomas T. Woolridge, brigade major in Ireland.

The following were on leave: the colonel (General Manners), captain, en second, Hen. Sellick (sold to Lieutenant Thomas Lahiff, March 12th), Lieutenant Eyre Peter Shewbridge (sold to Ensign J. R. Dolling, March 27th. Shewbridge had been long in bad health).

On the return from Egypt Sergeant David Glass was acting sergeant-major in the absence of Wallace, and was confirmed in the appointment, so it is reasonable to conclude that he acted in Egypt.

It will be seen that the colonel's company had on its strength seven lieutenants and one ensign en second, that is, at the commanding officer's disposal to supply the place of officers detached on the staff or recruiting.

On landing the army was brigaded as follows:—

Guards: Major-General Ludlow, strength in rank and file, 1,338.

1st Brigade: Major-General Coote, 2/1st, 54th (2 battns.), 92nd, strength 2,586.

2nd ,, ,, Craddock, 8th, 13th, 18th, 90th, strength 2,497.

3rd ,, ,, Lord Cavan, 50th, 79th, strength 1,970.

4th ,, Brig.-General Doyle, 2nd, 30th, 44th, 89th, strength 1,681.

5th (Foreign) Brig.-General J. Stuart, Minorca Regt., De Roll's, Dillon's, strength 2,067.

Reserve: Major-General Moore, Brig.-General Oakes, 23rd, 28th, 42nd, 58th, 4 companies 40th, Corsican Rangers, strength 3,283.

Cavalry: Brig.-General Finch, 1 troop 11th, the 12th, 26th, and Hompesch's Light Dragoons.

Field Artillery: 24 light 6-pounders, 4 light 12-pounders, 12 medium 12-pounders, six 5½-inch howitzers.

A siege train of 24 guns, 12 howitzers, 40 mortars.

500 Maltese pioneers accompanied the force.

As there were only boats for about 6,000 men, the reserve, the Guards and the 1st brigade, to whom Sir Ralph had given the honour of being the first to land in Egypt, were practised incessantly in disembarking and rapidly forming on the beach.

The army was for eight weeks at Marmorice. This long stay was necessary firstly to receive information about Egypt and the enemy's forces; secondly to remount the cavalry and artillery, and procure horse boats, and thirdly to arrange joint operations with the Turks.

In the only fragment of regimental history which we possess, and which was probably written by Colonel Philip Vaumorel, nearly twenty years later, the great storm of February 8th, 1801, is the most conspicuous feature. It seems to have impressed the narrator more than the landing and fighting in Egypt. He says: "The hail, or rather the ice stones, as big as large walnuts lay in the camp two feet deep, and pouring from the mountains swept everything before them. The horses broke loose, the men were in the dilemma of remaining still to be frozen or facing the elements, while the fleet in the bay suffered equally from driving, etc., and one ship (the *Swiftsure*) was struck by lightning."

On February 20th the army re-embarked and on the 23rd sailed for Egypt. During their stay at Marmorice, the conduct of the troops had been so exemplary that not a single complaint had been preferred by the inhabitants.

The good conduct of the troops was facilitated by the action of the Turkish Government in sending to Marmorice a Pasha of high rank accompanied by a gentleman of the name of Campbell, who forty years previously had been obliged to quit his native Argyllshire for having killed his man in a duel and had since risen to the rank of general in the Turkish service. On the first conviction of a native for overcharging a British soldier, General Campbell caused the erring tradesman to be bastinadoed, and announced that he would cut off the head of the next offender. A common cause of friction between the troops and the natives was thus removed, or at least lessened.

On the voyage Sir Sydney Smith joined the expedition and gave it the benefit of his experience of the Egyptian and Syrian coast. He informed Sir Ralph Abercromby that the French force in Egypt could not be reckoned at less than 30,000 men, that is twice the number estimated by the British Government and nearly twice the number Sir Ralph had under his command. Sir Sydney was nearer the mark than the British Government for the enemy's strength was somewhat over 25,000.

The French were commanded by General Menou, who had succeeded to the command on the death of Kléber, whom Buonaparte had left as his successor when he returned to Europe.

The main body of the French was about Cairo, with weak detachments on the coast and others on the Nile in support of them. On March 2nd the fleet anchored in the Bay of Aboukir, 7 miles from the beach, and Sir Ralph and Major-General Moore, with Sir Sydney Smith, rowed close in to reconnoitre the shore and look for a landing place. On his return the general issued orders for a landing on the following morning, but a gale sprang up and the operation had to be postponed. On the 7th the wind and sea fell, and orders were issued for the attempt to be made a little after midnight.

The troops detailed to force a landing were, from right to left, the reserve, the Guards and the 1st brigade (Coote's), except the 92nd, for which there was no room in the boats.

Major-General Moore commanded the whole.

A rocket from the flagship gave a preparatory signal at 2 a.m., and within a couple of hours the boats were filled with men and had started for their long row to the rendezvous, where mark boats had been placed a mile from the shore, on which to form line. The naval arrangements were under Captain Cochrane —he and General Moore arranged the boats as they arrived, in three lines.

In the first were 58 large flat-bottomed boats at 50 paces interval, in the second 84 cutters covering the intervals of the first line, and in the third 7 launches. Fourteen launches, each with a field gun, followed under Sir Sydney Smith to be landed if the attacking force made good their footing on shore.

As soon as the boats had landed their men, they were, without waiting for the issue, to pull back the 6 or 7 miles to the fleet for the remaining troops.

About 9 a.m. on March 8th the order was given to the lines of boats to advance, and they almost immediately came under fire.

As the 30th was not in Moore's force and was not engaged on that day, it is sufficient to say that Sir Ralph Abercromby's careful preparation ensured the exact execution of his plan, that the leading of General Moore and others was superb and that the regiments showed not only courage but surprising discipline and quickness in forming and in re-forming when, as necessarily happened, there was some slight confusion from sinking of boats and other causes.

The loss of the nine regiments engaged was 31 officers and 621 men killed, wounded and missing. The Navy lost 100 officers and men killed and wounded.

The French, who were commanded by General Friant, brought only 2,000 men and 15 guns into action; they lost 8 guns and over 300 men.

By nightfall our remaining troops and the field guns had been brought ashore and the army bivouacked 2 miles in advance of the scene of conflict; this advance had isolated the French detachment in the Castle of Aboukir, and "The Queen's" and another regiment were detailed to blockade it.

The right of the army rested on the sea, and the left on Lake Maadieh, which was entered by the British gunboats.

Water was luckily found in sufficient quantities, for there was such a heavy sea on the 9th and 10th that it could scarcely have been landed from the ships.

On the 11th the horses and supplies were landed and by using Lake Maadieh the distance the stores had to be hauled was much shortened, a matter of the first importance, for the few horses the army possessed had to be assigned to the cavalry and artillery, and the soldiers and sailors had to do the work of the transport train, and the sand was deep.

On the 12th Sir Ralph advanced with his right close to the sea and his left on Lake Maadieh. A march of 5 miles would bring the army to the western end of Lake Maadieh and there the left would strike the canal from Ramanieh on the Nile to Alexandria. The embankment along this canal was the main line of communication between the French in Alexandria and the rest of Egypt. On the south of the canal was Lake Mareotis, partially dry at this season. General Friant, who had been reinforced by Lanusse with his division and had about 5,000 men and 21 guns, determined to fight for his communication with his commander-in-chief, who still loitered near Cairo with the main body of his army. Although weak in numbers, Friant had a great superiority in guns, and they were well horsed; moreover, his 600 cavalry were excellent.

The position occupied by the French was a line of low hills from the sea to within 500 yards of the canal, the bare ground in front of the hills formed a glacis for the play of their artillery. From some remains of a Roman palace near the sea the position received the name of the Roman Camp.

On seeing that Friant meant to risk an action Sir Ralph halted and bivouacked about a mile and a half from the French position, giving orders for the attack on the following morning. The space between the sea and the canal is sufficient for three brigades deployed, and Abercromby's force was drawn up in three lines. Doyle's brigade was on the left of the Guards in the third line. Moore with the reserve formed the right of the army.

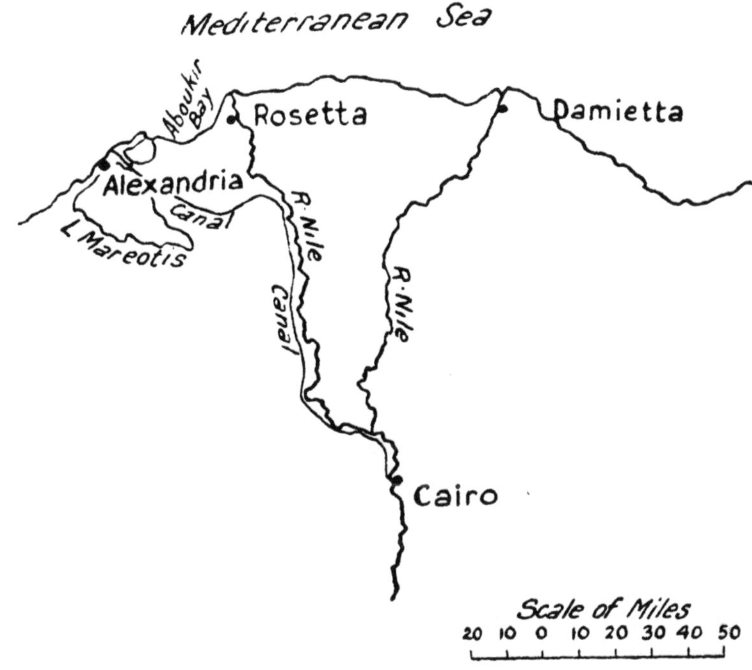

LOWER EGYPT.

By half-past six on the morning of the 13th, the army advanced from the left of brigades forming three parallel columns at deploying distance, the few mounted men were between the reserve and Craddock's brigade, and the fifteen field pieces were drawn by sailors. The 90th and 92nd formed the advanced guard.

The advance was through deep sand and halts were frequent to close up the column and allow the artillery to keep its place; eventually the advanced guard got too far ahead and appeared in front of the French line without any apparent support.

Leaving a small force to hold Moore in check, Friant at once left his ground to attack the 90th and 92nd, but though hard pressed the two regiments were not broken, and the British had time to deploy. The French were slowly driven back, our men suffering severely from the fire of their artillery, which was well handled and took up successive positions in retreat to cover their infantry.

Friant had so weakened his left that Moore without much difficulty gained the hill near the sea, called the Roman Camp, when the whole French line retired about 2 miles to the fortified heights of Nicopolis. Sir Ralph followed them into the plain in three columns as in the morning, but halted half a mile west of the hills he had captured and directed Major-General Hutchinson, his second in command, to take the 3rd (Cavan's), 4th (Doyle's) and 5th (Stuart's) brigades to attack the French right, while Moore assailed their left.

General Hutchinson resolved to extend his line to the left across the canal, and by advancing part of his force along the bank of Lake Mareotis, where the water had receded, to outflank the French right. The execution of this turning movement was entrusted to Doyle's brigade and the point of crossing selected was a bridge about a mile in advance which appeared to be commanded by a conspicuous green hill a little to the right of it and somewhat nearer to where our line had halted.

It has to be kept in mind that our commanders had no maps and no trustworthy information except such as could be given by Mr. Baldwin, who had formerly been Vice-Consul in Alexandria, or which had been gleaned by our few engineer officers, who made some daring reconnaissances.

Doyle's brigade had been, up till now, in third line and, like the rest of the army, in column of companies left in front. The canal here makes rather a wide sweep to the south and General Doyle, reversing the order of march, left the canal and advanced into the plain, right in front, with the 30th leading, towards the right of the green hill. In the advance some swampy ground was passed over. About 1,000 yards from the enemy's main position he halted the brigade and sent the 89th up the hill. As soon as the hill was occupied the 44th passed in rear of it and moved to the left to attack the bridge, which was carried with great gallantry, the French being in force and having a howitzer with them.

It has been said that the bridge was cleared by the light companies of the 30th and 55th, but this probably only means that it was carried in the good old way, the light companies skirmishing in front to keep down the enemy's fire and the 44th passing through them and giving the assault in column of subdivisions with fixed bayonets.

From the mention of the light companies of the 30th and 44th as acting together, it seems probable that the 30th followed the 44th in support towards the bridge, where it could cover the gap between the hill, still held by the 89th, and the canal which were now found to be 450 yards apart, more than twice the effective range of the musket. This opinion is strengthened by the fact that in 1882 a party of the 38th (South Staffordshire) regiment, while digging trenches near the canal, turned up two skulls and an old 30th breastplate, such as was worn at the beginning of last century.

General Doyle's operations had been a pattern of an old-fashioned field day. In those days a brigade, or regiment, right in front had to deploy to the left and one left in front to the right; nothing else was morally possible. General Doyle, after following the line of the canal all the forenoon, left in front, was no sooner ordered to attack than he changed to right in front to take advantage of the ground, and keep out of musketry range from the canal and any French skirmishers who might be on the further bank of it. When he came within striking distance he deployed to his left, his centre regiment occupying the green hill and his left the bridge. He was now ready to cross the bridge and advance again left in front as laid down in Army Orders.

A footing having been gained beyond the canal, an advance of a few hundred yards would place Doyle's brigade on the extreme right of the French line and the success of an attack seemed certain.

These operations, however, had not escaped the enemy's attention, and from the moment the 89th had been seen on the green hill, every available French gun had been brought to bear on the brigade. General Hutchinson was now able to see that the enemy was closely supported by the fortifications of Alexandria and he doubted if the lines could be held when captured. He therefore halted his force and sent to Sir Ralph for fresh instructions. Doyle's regiments took what cover they could and Cavan's and Stuart's brigades, which had followed in support, were placed in close columns on the reverse slope of the green hill, which was little more than 700 yards from the enemy's works.

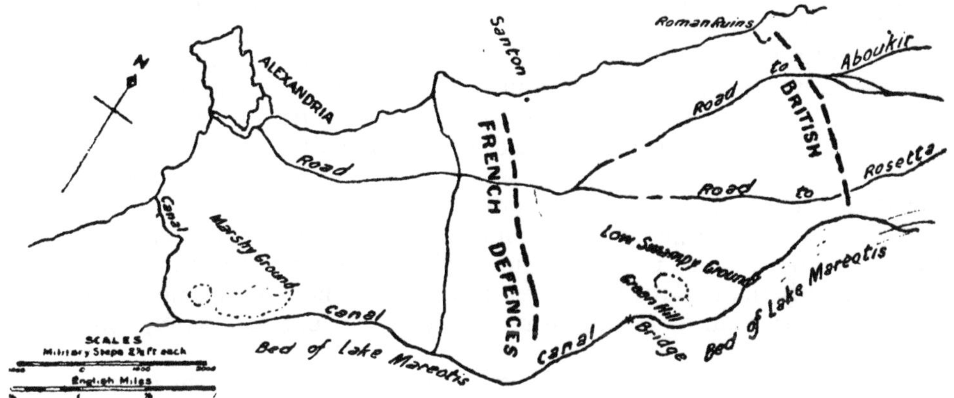

ACTIONS OF 13TH AND 21ST MARCH (BATTLE OF ALEXANDRIA) AND BATTALION ACTION AT THE GREEN HILL, 17TH OF AUGUST, 1801.

The Commander-in-Chief sent Colonel Hope, his adjutant-general, to report upon the position, and upon his advice the enterprise was given up. In fact there was little alternative, for it was now evening, two hours having been consumed in consultation and reference. We now know that the heavy forts in rear of the French army were incomplete, and that General Hutchinson must have succeeded had he gone on, but he could only decide from what he saw, or thought he saw. He was a fine, resolute soldier, and it was not from want of courage, either physical or moral, that he halted. Like Sir Ralph he was very short-sighted. The troops retired to the position taken from the French in the morning, Craddock's brigade covering the movement.

In Doyle's brigade the 44th had suffered most severely, no doubt in the attack on the bridge. It had 3 officers, including the commanding officer, wounded ; 2 sergeants, 2 rank and file killed, 20 rank and file wounded, in all 27 casualties.

The 30th had :—

Killed : Ensign T. Rodgers, Sergeant Ross, Private Thomas Craddock, all of Roche's company.

Wounded: Captain John Douglas, Lieutenant Duncan (21st regiment, doing duty with the 30th) and 6 rank and file. Eleven casualties in all.

The 89th had 1 sergeant, 6 rank and file wounded.

The total loss of the army was 78 officers, 1,068 other ranks killed and wounded. The French lost 500 men and 5 guns. The hill occupied by the 89th is of interest to the 30th, for it was *the* Green Hill which was the scene of the gallant action under Lieut.-Colonel Lockhart later in the campaign. After the action of March 13th, the French fortified the hill and later on flooded the swampy ground which Doyle's brigade had crossed on that day.

Sir Ralph now decided to establish himself firmly on the ground he had won. He was at last aware that, although he was greatly superior in force to the troops immediately in his front, he was far outnumbered by the French army in Egypt. His present position had the advantage that transport by water was available to within a mile of its centre.

The position was fortified as well as means allowed and thirty-six guns were mounted. For a week the army was employed in throwing up entrenchments and bringing up ammunition and stores. Although the distance was short the work was very severe; the firewood too had to be dragged from a distance, and being entirely date palm the smoke was pungent and hurtful to the eyes. The water, however, was good and abundant. The tents too had been brought up and they were very welcome, for if the days were intensely hot, the nights after the heat had gone out of the sand were very chilly.

During the week succeeding the action of the 13th inst. the sick list in the army ran up very fast, but the 30th kept its former excellent health.

Meanwhile, General Menou, the Commander-in-Chief of the French army in Egypt, had determined to do what he should have done at first and confront the British with the bulk of his forces, though even now he did not seem to realize that every man he could muster would not be one too many, and left 7,000 men at Cairo.

Sir Ralph being in possession of the route along the canal from Ramanieh on the Nile to Alexandria, Menou had to seek a way across Lake Mareotis. The water was so low that in many places the bottom was exposed and he was able to cross, bringing his guns and cavalry with him.

On the 19th he reached Alexandria and his troops raised the French force to 10,000 men. This was about equal to the number Sir Ralph Abercromby could bring against him, but included in the French numbers were 1,400 excellent cavalry and forty-six guns well horsed. Trusting to these and fearing the early arrival of a Turkish army, and of a force under Sir David Baird, despatched from India, Menou determined to attack the British in their entrenched position.

Sir Ralph Abercromby had occupied the ground with the reserve under Moore on the right, on the hill close to the sea known as the Roman Camp. To the south of this hill there is a depression through which the road or track runs from Alexandria to Aboukir. This depression is about 500 yards in breadth, and to the south of it a hill runs from north to south for three-quarters of a mile and sinks into the plain 500 yards short of the Alexandria Canal. The Guards held the central hill on Moore's left and Coote's brigade with its left thrown back prolonged the line to the canal. Craddock was on the left of Coote with his left thrown back almost at right angles till it touched the western end of Lake Maadieh. A couple of batteries were placed to cover the low ground between the hill and the canal.

In second line, from right to left were Stuart's foreign brigade, Doyle's, Finch's cavalry, and Lord Cavan's with its left thrown back parallel to Craddock's in the front line.

Doyle's brigade stood from right to left—the Queen's, 44th, 89th and 30th.

As the Roman Camp was slightly in advance of the central ridge the army stood with its right advanced and left refused. On the right the British gunboats could approach near enough to throw an effective cross fire along the front of Moore's position.

The plan of attack adopted by General Menou was to force the British right and centre in succession, and it was hoped that once driven from the heights their discomfiture would be completed by his fine cavalry, of whom 900 were to be kept in reserve till the infantry had done its work. To render harmless the fire of the gunboats and the British guns, the attack was to commence before daybreak, and to alarm General Abercromby for his left the Dromedary Corps and some cavalry were to cross the canal by the bridge which Doyle's brigade had captured and abandoned on the 13th, and to push on as far as Lake Maadieh.

As yet, Sir Ralph was in ignorance of the arrival of Menou with reinforcements, but in General Orders of the 20th he warned his men to sleep fully accoutred and to parade half an hour before daybreak on the following morning; the brigades were, therefore, assembled and ready to move into their positions when the attack came.

The false attack on our left was well conducted and the enemy driving in the pickets captured a fortified post on the canal and turned the 24-pounder in it on Craddock's brigade. They were, however, checked and forced to retire by the fire of a redoubt in rear. No change was made in the disposition of our troops, for almost simultaneously with the false attack a heavy fire broke out on the right and all knew that it was there that a real battle was to be fought.

The first French attack by the division of Lanusse, nearly 3,000 strong, was made with great gallantry and at first had some success, an advanced work and a gun being captured; but the summit of the hill was too well fortified and held and the division had to fall back with the loss of its leader mortally wounded. The attack on our centre by Rampon's division of 2,000 men should have followed the attack of Lanusse on the right, but in the darkness Rampon had inclined too much to his left and one of his brigades overlapped the right of the division of Lanusse and had to fall back along with it.

The British pressed on in pursuit, and to extricate his infantry General Menou had to order the leading brigade of his cavalry reserve to charge. The charge was made with vigour and many men who had lost their order in the pursuit were sabred, but our regiments recovered themselves with wonderful quickness and the French cavalry getting into broken ground were almost destroyed.

The French troops were too good to accept defeat easily, and their two infantry divisions were rallied and led again to the attack, but this time Rampon marched straight upon his proper objective, our central hill held by the Guards. He was met with a steady fire, but being supported by his colleague Reynier, who with 3,000 men had been ordered to observe and contain the British left, Rampon succeeded in throwing back the left of the Guards. He was checked, however, by the Royals, the right battalion of Coote's brigade, which moved to its right, giving relief to the left companies of the Guards. At the same time Sir Ralph moved up Stuart's foreign brigade to the hollow between the Roman Camp and the central hill on the right of the Guards. It arrived just in time, for Menou, seeing his infantry attack at a standstill, ordered General Roize to charge with the three regiments which remained of his cavalry. The charge could not have been executed more boldly, though the men (500) were too few for their work. The three regiments galloped through the hollow between the Roman Camp and the central hill, overthrowing everything in their way. In the confusion Sir Ralph Abercromby was wounded and for a few moments a prisoner. Some of the French horsemen even penetrated to the rear of the redoubt on the Roman Camp, but they fell fast before the musketry from the ruined buildings, and their destruction was finally completed by Stuart's foreign brigade, but not before the Minorca regiment had been ridden over and 13 officers and 200 men killed and wounded.

This was the last attack of the French, but they still held their ground and kept up a fire of skirmishers on our position. On our right there was a lull, both Moore and his opponents having run out of ammunition. On this part of the field the combatants actually threw stones at each other. When fresh ammunition had been brought up and the British guns opened all along the line, Menou retired to the heights of Nicopolis.

Half an hour before the close of the action Sir Ralph Abercromby had ceased to command.

When the last charge was made by the French cavalry and he was involved in the mêlée, he received a severe contusion on the breast, but he made no mention of a bullet which had lodged in his thigh. After the repulse of the cavalry he posted himself in a battery on the central hill from which he could command the whole plain, but towards evening he fainted, and it was then for the first time discovered how seriously he was wounded. He was carried on board ship and lingered till the 28th, when he died.

Sixty years after his death a story was still told in the regiment of how, when he lay on the field of battle, he felt his head raised, and asked: "What is that you are placing under my head?" The answer was: "Only a soldier's blanket." The old hero replied: "Only a soldier's blanket! It is of great importance to him, see that he gets it." The affection Sir Ralph felt for his men was fully returned by them, and under him every man was at his best. Those who wished to find fault blamed him for exposing himself too freely to danger when in action. He was succeeded in the command by General Hutchinson.

The loss of the army in the battle of Alexandria was 74 officers and 1,400 men killed and wounded. John Moore, as usual, was among the badly wounded, and so was Colonel James Hope, the adjutant-general.

On this day as the left regiment of Doyle's brigade, the 30th regiment stood immediately in rear of the left of the Guards, and when the left companies of the latter were pressed back, the regiment covered the gap between them and the right of Coote's brigade. This accounts for the casualties in the 30th being greater than those in the other regiments of Doyle's brigade.

The losses of the brigade were :—30th, killed : 4 privates, John Melcome, Pat. Nally, Ed. Connor, Hugh Fleet ; wounded : Captain Ralph Smyth, Lieutenant Thomas James, Paymaster-Sergeant Pat. Matthews, Sergeant Wm. Robertson, Sergeant David Crutchley ; 21 rank and file. Thirty casualties.

The 44th had 16 casualties and the 89th had 11. The Queen's was detached to the right.

The French losses had been very heavy ; probably amounting to 3,000 men, and their magnificent cavalry had ceased to exist.

General Menou retired to Alexandria with about 5,000 men, and some 2,000 retreated across Maréotis to the Nile, where they were joined near Ramanieh by troops from Cairo.

This was the third time in fourteen days that the French had been pitted against the British with inadequate forces through the fault of their commander, and they never again fought with the same spirit. Their conduct had been worthy of the " Army of Italy."

General Hutchinson, who succeeded to the command on the death of Abercromby, resolved to hold the French in Alexandria with 6,000 men entrenched on the peninsula of Aboukir, while he, himself, dealt with the other bodies of French in Egypt. For some days he was busy entrenching the chosen position, but being joined on the 25th by the Capitan Pasha and 4,000 Turks, preparations were made for an advance. On April 6th Colonel Brent Spencer was sent with a few British and Turkish regiments to seize the mouth of the Nile at Rosetta, and as our fortifications in front of Alexandria grew in strength, regiment after regiment was despatched to join the force on the Nile.

On the 13th the dyke was cut and the sea admitted to Lake Mareotis, thus completely isolating the French in Alexandria.

The 30th and 89th left the main army on the 17th and reached Rosetta on April 19th.

On the 26th, Hutchinson, leaving Coote with 6,000 men in the fortified position in front of Alexandria, took command of the army on the Nile and prepared to advance on Cairo. The Nile gave him water carriage, but it was falling and nothing was taken with the army which was not absolutely necessary ; in particular the heavy rum casks were left behind and water was the only drink for officers and men.

" Never, indeed, had an army been before so abstemious and, consequently, so well conducted," as is said in the fragment of regimental history in the 30th Orderly Room books.

On May 5th the advance southwards from Rosetta commenced. The main body of the army marched on the western side of the Canal and Doyle's brigade was on the right of all except a small body of British cavalry which covered the right flank. On the 9th the army was in front of Ramanieh where the French appeared to be inclined to risk an action. A skirmish ensued, but Doyle's brigade was not engaged. On the eastern bank of the river, where our force was mainly composed of the Capitan Pasha's troops, there might have been serious trouble if General Hutchinson had not sent the 89th across. Their steadiness put heart into our Allies. The French decamped at night and marching with great celerity were no more seen till our army reached Cairo. The force in the Delta to the eastward of the Rosetta branch of the Nile remained under command of Lieut.-Colonel Stewart of the 89th during the advance. On the 17th, at Algam, word was brought from the Arabs that they were trying to hold up a French convoy from Alexandria which was crossing the desert on our right to reach Cairo. General Doyle ordering his brigade to follow, galloped off with our few mounted men to intercept the convoy, and found it successfully defending itself against the Arabs. His brigade was far in rear, toiling through the deep sand, but he called on the convoy to surrender, and it did so on an assurance being given that the men would be sent home to France. Nothing could show more clearly how homesick the French were, and how disgusted with the war and their leaders.

Captain Hamilton with the light company of the 30th was ordered to escort the 400 men of whom the convoy consisted to the coast for embarkation, and the 600 camels they had in charge were a welcome addition to our transport.

On the same day news was received that the Grand Vizier, who had advanced across the desert from Palestine, was now coming within supporting distance to the eastward. To strengthen the communication with him the 30th was sent across the river on June 6th to join the 89th and the force in the Delta under Colonel Stewart. This force advanced parallel to the army till the occupation of Cairo.

On June 15th a summons was sent to General Belliard commanding the French in Cairo, which was disregarded, and on the 21st the Turks were in front of Cairo and the British were at Ghizeh on the opposite bank of the Nile. The Allies were connected by a bridge of boats, which was in charge of the 30th and 89th.

On the 27th, General Belliard capitulated on condition that he and his army with their arms and artillery should be sent to France, and on the following day the 30th occupied Fort Zulkowsky, half a mile from Cairo. The convention covered the whole of Egypt outside Alexandria and included 13,000 French soldiers, of whom 8,000 were effective. General Hutchinson had under him in front of Cairo only 4,000 men, and the army of our Turkish Allies, although perhaps twice as numerous, had little fighting value. On July 9th the 30th occupied the Citadel of Cairo.

Although the march to Cairo had been nearly devoid of incident, it had been very trying. The marches were of necessity short, for it was difficult to struggle through more than a few miles of deep sand in the great heat, and there were difficulties too about transport. General Hutchinson did what he could for his men, and was particularly careful in insisting that they should never be kept uselessly under arms ; but he could seldom procure fresh provisions or vegetables, and even the sick had little to eat but ration pork and biscuits. The regiment ceased to have the small sick list which had distinguished it at the beginning of the campaign, but it contrasted favourably with others. Since landing, two officers and fourteen men had died from disease.

Of the two officers who died, the first was Matthew Arbouin, ensign of the grenadier company, who joined from home in May and died on June 13th. The second was Ensign Charles Wynne Barry, attached to Montgomery's company. He died two days later than Arbouin.

Generals Hutchinson and Craddock went sick after the capitulation of the French, and General Moore, who had recovered from his wound, took command of the army at Cairo. General Doyle, also, was sick for a short time, and Colonel Brent Spencer commanded the brigade. After reaching Cairo, General Hutchinson had been able to sanction the issue of a daily ration of spirits to the troops, and he presented a puncheon of Sicilian wine to the officers' mess of each battalion.

On July 15th General Moore marched north, having under him 8,000 French soldiers convoyed by 3,500 British, about an equal number of Turks, and a few Mamelukes. Doyle's brigade made part of the British force.

Lieutenant Hare was left sick in Cairo, all the other sick seem to have been sent down to Rosetta.

On July 30th the French and the escorting force reached Rosetta and the embarkation commenced. In ten days the last man had been got on board, and General Hutchinson was able to move in force against the 5,000 of the enemy who yet remained under Menou in Alexandria.

Before advancing, the army was re-distributed in brigades, and Doyle's, which was now the 5th brigade, consisted of the 30th, 50th, 89th, and 92nd. It marched into camp before Alexandria from Rosetta on August 9th.

The trying march from the coast to Cairo and back had forced Lieut.-Colonel Wilkinson and many others on to the sick list. Lieut.-Colonel Lockhart commanded the regiment. Colonel Wilkinson suffered from both dysentery and ophthalmia, but as long as he was able to get about he visited the hospitals daily, especially the plague hospital, where a few of the sufferers belonged to the 30th.

Reinforcements from Minorca had raised the British army to 16,000 men, and by the 15th it was concentrated near the scene of its victories in the preceding spring and facing the eastern defences of Alexandria. Unremitting labour on the part of the French had made those defences so strong that General Hutchinson determined that his main attack should be from the western side, which Menou had partially neglected.

On the evening of August 16th three brigades under Major-General Coote were embarked on Lake Mareotis with orders to land to the west of the city and attack from that side.

To occupy the enemy General Hutchinson ordered that at daybreak on the morning of the 17th General Moore on the right and General Doyle on the left should drive the French from the outworks on the eastern side, but not risk serious loss by attacking the main defences.

It will be remembered that on March 13th General Doyle's brigade, after passing over some swampy ground, had occupied and then retired from a green hill in front of the right of the French lines; this is the ground which the brigade was ordered to attack on the morning of August 17th. The French had erected some redoubts on the hill, and when General Hutchinson had admitted the sea into Mareotis their engineers had taken advantage of the increased height of the water to flood the swampy ground. General Coote, however, had erected a dam to restrain the inundation, and it was along this dam that Doyle's brigade was ordered to advance against the hill. According to the maps the dam was only partially successful and there was water on both sides of it.

The following is General Doyle's order for the attack:—

"Camp before Alexandria, August 16th, 1801. The regiments will parade to-morrow morning at two o'clock in order to be at the dam of the inundation at the hour appointed, they will cross the dam by files from the right of regiments in the following order, 30th, 50th and 92nd. The 30th will oblique to the right for the attack of the French redoubts in that direction, the 50th will oblique more considerably to the left against the enemy's works near the lake, allowing the 30th regiment, which has a greater distance to march, ten minutes law, that their movements may commence at the same time, the 92nd will form a reserve as near the centre as possible. As the attack will be made before daylight, firing will only create delay and perhaps confusion. The redoubts must, therefore, be carried with the bayonet. The General knows too well the troops he has the honour to command to feel any anxiety on the score of their conduct before the enemy, but he must impress on them the necessity of the most perfect silence and that no man shall upon any account attempt to load without positive orders from the officer commanding the detachment."

Lieut.-Colonel Lockhart was commanding the 30th in the absence of Lieut.-Colonel Wilkinson, who was still on the sick list.

The attack was perfectly successful; an officer of the regiment, writing to his wife, says that at 3 a.m. "we ran into them at the point of the bayonet without loading, like a pack of foxhounds, and soon made the enemy give way."

When the hill was gained three companies of the 30th were sent to the front to occupy a small eminence. As soon as the French saw that we were in possession of their outworks they opened fire from the forts and our men were allowed to seek protection in the ditches and behind ridges, under cover of the three companies in front. While they were thus scattered, at seven o'clock in the morning, General Menou launched against them a column of 600 men who advanced without firing a shot. Lieut.-Colonel Lockhart, on seeing them leave the French works, which were about 700 yards distant, asked permission from Colonel Brent Spencer, commanding that portion of our line, to advance to meet them instead of awaiting their attack. Colonel Spencer gave a ready assent, upon which Lieut.-Colonel Lockhart caused the assembly to be beat, and as soon as he had seven companies of 180 rank and file in hand he led them forward, received the French fire at 20 to 30 yards and without checking for a moment, fell on with the bayonet; the French column broke and made for their entrenchments leaving a few prisoners and many killed and wounded. Lieut.-Colonel Lockhart had his men so under control that he was able to withdraw them without serious loss from the storm of shot and shell which came from the French batteries when the discomfiture of their infantry was apparent. Within five minutes of the first fire of the French column, Lieut.-Colonel Lockhart was leading his men back to the position amidst the cheers of the rest of our army, but in the two or three minutes which the actual combat lasted we had 29 casualties. The letter already quoted from terminates as follows: "Colonel Lockhart led us on with coolness and bravery that did him honour and gained himself and the regiment the thanks of the Commander-in-Chief."

The fire of the French, advancing as they did in column on a narrow front, was naturally ineffective.

The following is an extract from a despatch of General Hutchinson, commanding in chief:—

"Headquarter camp before Alexandria, September 5th, 1801.

"On the east side of the town two attacks were made to get possession of some heights in front of the entrenched position of the enemy. The action was neither obstinate nor severe and our loss is but small, but it afforded one more opportunity to display the promptness of British officers and the heroism of British soldiers."

The despatch proceeds to give an account of Colonel Lockhart's charge.

Brigadier-General Doyle published the following brigade order on August 20th :—

"Brigadier-General Doyle has been prevented by being on duty from expressing his satisfaction at the steady countenance shown by the brigade in the affair of the 17th and wishes to return his best thanks to Lieut.-Colonel Lockhart and the officers and men of the 30th Regiment for their gallant conduct in charging and putting to rout a superior force of the enemy and he has the pleasure to acquaint that Corps that their spirited behaviour has met the most marked approbation of His Excellency the Commander-in-Chief. It is difficult to particularize individuals where all deserve praise, but Captains Hamilton and Grey of the 30th regiment from their peculiar situation had more the opportunity of distinguishing themselves and the Brigadier-General requests those gentlemen to accept his thanks for their energetic and efficient exertion."

Among those who cheered Lockhart and his men on August 17th, 1801, was Capt.-Lieutenant John Colborne of the 20th who were, on that day, a little to the right of the 30th. Sixty years later, when John Colborne had become Field-Marshal Lord Seaton commanding in Ireland, the exploit of the 30th was still fresh in his memory and he told the story of it to Colonel Mauleverer of the 30th when new colours were presented to the regiment.

The loss of the regiment was: *Killed:* Privates Edward Moriarty, Wm. Johnstone and John Wood. Five privates were seriously wounded, Lieutenant Mansergh and twenty of other ranks were returned wounded.

Besides Lieut.-Colonel Wilkinson, Captain David Maxwell was sick at Rosetta, Lieutenant Hare at Cairo, Lieutenant R. B. Lynch, Rosetta, Lieutenant Murray, sick, present, and 101 non-commissioned officers and men in hospital.

Of the officers wounded in previous actions, Captain Douglas had rejoined.

The regiment maintained an advanced position till the end of the siege and it lost a man, Christopher Dunn, of Smyth's company, killed on the 18th, and another, James Dunn of the same company, killed on the 23rd, and had a few wounded.

On the 25th our batteries on the west side of Alexandria opened fire and on the 26th Menou asked for terms. On September 2nd a capitulation was signed under which he and his army were, like General Belliard, to be sent to France in British vessels.

From the capture of Cairo to that of Alexandria fifteen men had died from disease.

Alexandria had no sooner fallen than General Hutchinson embarked the 30th and eight other battalions with some mounted men and guns under Major-General Craddock with orders to expel the Russians from the Ionian islands. They had been acting as our allies in the Mediterranean since 1798, but the Emperor Paul had been estranged by our refusal to hand the island of Malta over to him when captured and he had become our bitter enemy, hence the expedition under Craddock. The force sailed on September 9th, but was met at sea by Lord William Bentinck with orders to abandon the enterprise. The Emperor Paul, who was quite mad, had contemplated putting to death his own sons as well as many of the nobility, but had been murdered in his palace. Russia, under a new emperor, was again friendly with England.

In consequence of Lord W. Bentinck's orders, the 2nd Battalion Royals, 30th and 89th were landed in Malta on October 11th. Lieut.-Colonel Wilkinson was now in command.

There are two curious entries in the pay lists, September 24th to December 24th. After the name of Alexander Young, the adjutant, is a note in red ink: "Adjutant to 2nd Battalion Royals commencing September 3rd," and after the name of Henry Craig, formerly regimental quartermaster, is a note: "to 89th Foot from September 5th." Both officers rejoined the regiment for duty in December.

In the year 1848 a war medal was granted to the survivors of the army which had fought in the Peninsula and in other campaigns, and in the following year this honour was also conferred on the survivors of the Egyptian campaign of 1801. The medal was issued in 1850. In some cases the war medal had already been granted to the Egyptian veterans for the Peninsula or other service, and to those men a clasp for "Egypt 1801" alone was given. Those without the war medal received the medal and clasp.

Six officers and one drummer and fifteen privates were all that survived to claim the decoration for Egypt.

Brevet-Major Ralph Smith Captain Sir David Maxwell ,, Frederick Holebrooke Paymaster Michael Jones	Received the War Medal with Clasp for Egypt, 1801.

Lieut. Richard Goddard Hare Clarges Clasp for Egypt, 1801.

Lieutenant Hare left the regiment after Egypt and gained the silver medal for the Peninsula with clasps for Corunna, Busaco, Barrosa, Ciudad Rodrigo, Badajoz, Salamanca, Pyrenees, and St. Sebastian. He also received the gold medal for his services as A.A.G. at the battle of the Nivelle. When he received his clasp for Egypt he was a lieutenant-general and Companion of the Bath. This is the Lieutenant Hare who was left sick in Cairo when the regiment left it in July, 1801. The name Clarges was assumed later in life.

Lieutenant Robert Blake Lynch received the clasp only for Egypt. He was in possession of the war medal and a clasp for Fuentes Onoro.

Of the privates, Thomas Murray and James Hall received the clasp only for Egypt as they had received the war medal for the Peninsula. Murray had clasps for Ciudad Rodrigo and Badajoz, and Hall a clasp for Badajoz.

The following received the medal and clasp: Drummer John MacGill,

Privates John Butler, James Barry, Hugh Clarke, John Ivals, John Graves, John Good, William Langston, Joseph Lester, Thomas King, Isaac Marshall, William Misson, Bryan Nugent and James Roche.

Most of the survivors of the Egyptian campaign went to India with the first battalion, so that few served in Egypt and the Peninsula.

Captain Sir David Maxwell when he received the Egyptian medal was already in possession of the Naval war medal with a clasp for March 14th, 1795. This medal was issued in 1849 and he was the only survivor of the six officers of the 30th who had served on the *Princess Royal* and *Terrible*, and taken part in Hotham's victory fifty-four years previously.

The state of the regiment on October 26th, 1801, gives in Malta: 32 officers, 33 sergeants, 20 drummers, 319 rank and file fit for duty, 43 sick present, 25 sick left in Egypt.

Lieutenant Ninian Craig was employed in the adjutant-general's office and Ensign Henry Craig in the quartermaster-general's in Malta. Lieutenant Thomas Woolridge was still brigade-major in Ireland. Captains Christopher Maxwell and R. Montgomery still had their recruiting parties at Ipswich and Marlborough. Lieutenant Hare was still sick in Egypt.

Lieutenant Samuel Bircham, Sergeant-Major Peter Wallace and Sergeant Robert Thompson did not rejoin the regiment for duty, but remained with the Maltese Corps.

The health of the regiment was very good after landing in Malta.

In November the thanks of Parliament were conveyed to Lieut.-General Hutchinson and the officers and men of the army for their zeal, discipline and intrepidity, and in the following month Lieut.-General Hutchinson presented to each of the officers who had served in Egypt a gold medal from Selim III, the reigning Grand Seignor; his Majesty King George III graciously allowed these medals to be worn by a General Order dated October 6th, 1803.

On January 13th, 1802, the regiment embarked for home, Major and Lieut.-Colonel Wm. Lockhart was in command, Lieut.-Colonel Wilkinson being on leave.

As usual, the regiment was dispersed in several transports. The first reached Spithead at the end of March, and was placed in quarantine, as were all the other ships on arrival. Some of them did not reach England till more than two months later. On landing, the companies marched in succession to Winchester.

While the regiment was on passage home, peace had been signed at Amiens with Napoleon Buonaparte, now First Consul and practically ruler of France.

On May 7th orders were issued for the regiment to march from Winchester, five companies to Newcastle and Gateshead and five to Durham.

Major the Hon. John Meade, who had been promoted into the regiment from the 9th Foot in June, 1801, *vice* Reynolds retired, joined at Winchester. Lieut.-Colonel Lockhart had succeeded to a company on the retirement of Major Reynolds.

On May 14th the regiment was at Abingdon, when the strength was 35 sergeants, 20 drummers, 413 rank and file; 6 sergeants and 187 rank and file were wanting to complete.

On reaching Worksop the regiment received orders to proceed to Sunderland and Tynemouth, six companies to Sunderland, four to Tynemouth.

On June 25th the establishment was altered. The officers en second were abolished, and as the regiment had a number of subalterns in excess of the establishment

Lieutenant	Richard Goddard Hare	were placed on ensign's pay.
,,	Robert Blake Lynch	
,,	Christopher Williamson	were placed on half-pay.
,,	James Sullivan	
,,	Thomas James	
,,	James Garnon	
,,	David Crowe	
,,	Edward Jennings	
,,	James Adam Dolling	
,,	Noblet J. Lindsay	

Two captains who had been allowed to purchase the companies of Holbrooke and Lahiff were retired on half-pay the day after their appointment.

The fifty sergeants were reduced to forty. To hasten the reduction eight sergeants were summarily discharged, three without pension, and one was allowed to continue to serve, but on private's pay. No doubt he was promised the first vacancy. The number of privates was raised to 710. This was a hopeless number with voluntary recruiting; the regiment had never in the previous ten years been able to reach an establishment of 550 and recruiting was now worse than ever. It is impossible to divine why the number of sergeants was reduced. Everyone knew that the Peace of Amiens was merely a truce and that we must be at war again in a few months.

The new establishment included an armourer for the first time since the regiment was raised.

CHAPTER VIII

THE REGIMENT DIVIDED INTO TWO BATTALIONS AND SERVES FOR TWO YEARS IN IRELAND, AFTER WHICH THE BATTALIONS SEPARATE. FIRST BATTALION—EXPEDITION TO THE ELBE. EXPEDITIONS TO THE COAST OF JAVA AND MACAO. SERVICE IN SOUTHERN INDIA. SECOND BATTALION—IRISH SERVICE. PORTUGAL. GIBRALTAR. CADIZ. RETURN TO PORTUGAL. THE BATTALION JOINS WELLINGTON IN THE LINES OF TORRES VEDRAS.

1803–October 1810

THE year 1803 opened quietly, though the expectation of war was almost universal, but in April the Admiralty began to press seamen and Colonel Wilkinson was ordered to support the impress officers on the Tyne. There could be no doubt regarding the meaning of this step, and on May 22nd the expected declaration of war against France was issued by the British Government.

In May a War Office circular directed that the regimental field officers should cease to command companies; their total pay was to remain the same as that which they had received as field officers and captains and they were to receive £20 a year as compensation for the loss of non-effective allowance. Under this order the colonel's company was given to Captain Thomas Fleming Roberts from the half-pay of the Coldstream Guards, the lieut.-colonel's to Capt.-Lieutenant George Grey and the major's to Lieutenant Ninian Craig.

On May 25th, Ensign Ben. Nunn was appointed adjutant in place of Lieutenant Alex. Young, transferred to the militia of Renfrewshire. The militia was not formed and Young was placed on half-pay. He had served the campaign in South Carolina as corporal and the siege of Valetta and the campaign in Egypt as adjutant.

In June the regiment embarked at Shields for Harwich, where it landed on the 17th and marched to Ipswich. At Ipswich all the officers were present except the colonel, Major-General Robert Manners, and thirteen officers who were recruiting. The colonel, however, was close at hand, having been appointed to the staff of the Eastern Command on the outbreak of war.

The lieut.-general commanding the Eastern District, Sir James Craig, was also a 30th man, having served in the regiment for the first seven years of his military career.

Among the stations to which officers were sent to recruit were Sleaford, Lincoln, Nottingham, Wakefield and Doncaster, the recruiting ground of the regiment in the time of William III and Queen Anne. The other stations were Cambridge, Alcester, Carlisle, Glasgow, Perth, Armagh, Sligo and Tuam.

The chief inducements offered to recruits were a bounty of £5 16*s*. 9*d*. in cash, a free kit on joining valued at £1 16*s*. 6*d*., a free issue of clothing yearly and a pension after twenty-one years' service.

The pay of the private soldier was 1*s*. a day and the following is the scheme for his messing and daily stoppages :—

Breakfast—milk, tea or saloop 1*d*., butter 1*d*.	2*d*.
Dinner—½ lb. of meat 2½*d*., vegetables and salt 1*d*. . .	3½*d*.
Supper—same as breakfast	2*d*.
Bread—one pound per day	1¼*d*.
Daily stoppage	8¾*d*.

The issue of five pints of small beer had been stopped in 1800 and an allowance of 1*d*. beer money granted in place of it. Saloop was made by mixing sassafras with milk or other liquid.

The ordinary barrack-room accommodated twelve non-commissioned officers and men, and the men, for the most part, slept two in a bed, though single beds had been introduced in a few barracks.

The rations were cooked and eaten in the room in which the men slept. If the men provided any extra diet it was probably oatmeal porridge taken with small beer. Thirty years later we know that this was a common supper in the 30th.

An iron pot with hooks and trivets, a flesh fork and frying pan, two wooden bowls, twelve wooden trenches, porringers and spoons, one large earthen pot for meat, one tin can for beer and two drinking horns were provided.

Besides grown men, it was permissible to enlist growing lads of seventeen to nineteen years of age, and boys under sixteen. These lads and boys were to be enlisted as privates, but on reduced bounty and pay, and no promise was to be made to them of being appointed to the band or drums, but they were to receive men's pay as soon as their commanding officer certified that they were capable of doing men's duty. They were to be "well limbed, open chested, and what is called long in the fork."

Experience in the late war, and the poor success of recruiting in this year, had shown that the inducements, as given above, were not sufficient to raise such great forces as the coming war would evidently demand, and as an increase of pay to the army was considered to be extravagant and uncertain of success in meeting the difficulty, recourse had to be made to compulsion, or rather to compulsion assisted by bounties. The plan which recommended itself to Mr. Addington, the Prime Minister, was to reduce the number of the militia who were already raised by ballot, and to raise, also by ballot, a new army of reserve for service in Great Britain, Ireland and the Channel Islands only. The men were liable to be compulsorily transferred to the regular army, receiving a bounty of £11 7*s*. 6*d*., but, as they could not be sent abroad without their own consent, it is evident that the regiments filled up from the army of reserve would be reduced to the position of fencibles. Mr. Addington hoped, however, that by giving an additional bounty of ten guineas, he would induce a sufficient number to volunteer for service abroad.

A Royal Order directed that the 30th should be formed into two battalions, each of 50 sergeants, 50 corporals, 22 drummers and 1,000 privates, the number of men necessary to complete the regiment to that strength to be taken from the army of reserve of the county of Buckingham. The standard for those subject to ballot was to be lowered to 5 ft. 2 ins.

The regiment moved in the beginning of July into camp at Bromswell, about 9 miles from Ipswich, to await the assembling of the army of reserve.

On July 9th Bt.-Lieut.-Colonel William Lockhart was appointed lieut.-colonel in the regiment to command the second battalion; Bt.-Lieut.-Colonel Robert Piggot from the half-pay of the 130th Foot, Bt.-Majors Philip Vaumorel and Ralph Smyth were appointed majors; 11 captains, 15 lieutenants, 12 ensigns, 1 adjutant, 1 quartermaster, and 1 surgeon were appointed at the same time. Ralph Smyth's promotion appears to be a mistake; he retired as a captain by the sale of his commission to H. C. Briscoe and must have applied for permission to sell before he was gazetted major. Bt.-Lieut.-Colonel Michael Jacob from half-pay was appointed major in the *Gazette* of July 30th, but he exchanged back to half-pay with Major Wright on August 6th.

Besides Smyth the regiment lost in this year Captains David Maxwell, who exchanged to the Guards, John Russel, deceased, and John Douglas, retired. They had all commanded companies in Egypt, where Smyth and Douglas had been wounded. Russel had been regimental sergt.-major in South Carolina in 1781 and adjutant for many years afterwards.

Of the 11 captains appointed only one, Sam. Bircham, belonged to the regiment. The others were brought in from half-pay. Of the 15 lieutenants 8 belonged to the regiment; 7 of them had been placed on half-pay by the reduction of the year before. Two of the 12 ensigns and the adjutant and quartermaster were brought in. The adjutant never joined, and it is surprising how few of the officers brought in played any part in the life of the regiment. Some seized the opportunity to realize their capital by selling out, others returned to half-pay. The fact is that as a class they were too old and their appointment was due to an effort of Government to reduce the non-effective charge.

One officer, Captain Richard Fitzgerald, from the half-pay, was reported "absent without leave" during the year he belonged to the regiment. Like many others he had crossed to France after the peace of 1802 and had been seized by Buonaparte and treated as a prisoner of war on the commencement of hostilities in 1803.

In August, receiving parties of officers and non-commissioned officers were sent to Buckingham, High Wycombe and Aylesbury to conduct to the regiment such men of the army of reserve as were handed over to them by the county authorities. At the same time the regiment was ordered to march to Chelmsford to receive them. Before leaving the camp at Bromswell the 30th and other troops were inspected by H.R.H. the Duke of York, Commander-in-Chief, attended by Sir James Craig and the generals under him, of whom Major-General Robert Manners was one.

The regiment arrived at Chelmsford at the beginning of September and occupied the old barracks. Presently Colonel Lockhart moved to the new barracks, and it was there that the second battalion was formed.

The Government order was that the two battalions were to be formed by October 6th, but a commencement was made on August 25th when Sergeant George Stephenson was appointed sergeant-major of the first battalion and David Glass, who had been regimental sergeant-major since the Egyptian campaign, was appointed to the second. Colonel Lockhart was his old captain, which probably had something to do with the arrangement. The remainder of the staff and the drummers remained with the first battalion, but the companies were divided fairly, five old and five new to each battalion. It was impossible to provide the full complement of sergeants and corporals, but every company had at least two sergeants and a dozen old soldiers in the ranks.

During September and October 1,600 men joined the regiment from the army of reserve and some from the recruiting parties, and they were rapidly told off to companies, clothed and armed. This clothing was the first on which the rank of the non-commissioned officers was indicated by chevrons on the arm instead of by epaulettes and shoulder knots.

There is some uncertainty about the colours. The second battalion without a doubt received its colours in 1803, and perhaps we are safe in assuming that General Manners presented them before quitting his staff appointment in the Eastern district on promotion to lieut.-general, which happened on September 25th. The question is whether the first battalion received new colours. In 1801, in consequence of the union of Great Britain and Ireland, the Cross of St. Patrick had been added to those of St. George and St. Andrew on the national flag, and it has been argued that the colours presented in 1780 would be too far gone to bear the addition of the new cross and of the honours won in Egypt. There is, however, reason to believe that when the staff of the regiment was captured on the unarmed transport *Fidelity* in 1795, the colours were destroyed by Major Lockhart and that new colours were presented in that or the following year, but no trace has been found in official papers of any colours being presented to the first battalion between 1780 and 1831.

From the time the colours lasted it is probable that General Manners presented colours to both battalions in 1803.

In the 30th, as in all regiments, there had been an officers' mess for many years. The difficulty is to determine when the officers ceased to have their mess in hotels or hired rooms and established it in barracks. Here again there is no direct evidence, but at Chelmsford we are certain, for the first time, that a room in barracks with fuel and light was appropriated for the officers' mess.

In October, Colonel Lockhart appointed Sergeant Daniel quartermaster-sergeant. The fact that Daniel and the two sergeant-majors, Stephenson and Glass, all got commissions shows that no mistakes were made in the selections.

In November, Ensign William Stewart was appointed adjutant of the second battalion, Quartermaster Kingsley joined for duty and the staff of the regiment was complete.

In this month routes were issued for the regiment to march to Woodbridge. The first battalion arrived on November 25th and the second on the following day.

The ranks were now more than full and volunteering was opened for other regiments; both battalions gave 150 men. At the same time a large number of the men volunteered for foreign service in the 30th. It looked as if Mr. Addington's plan was a success, but the sickness and desertion were very great. Nearly 80 went sick as soon as they joined. In November 8 died and 11 deserted; in December 6 died and 7 deserted.

In December it was considered that the regiment was fit for duty again and routes were issued for it to proceed to Ireland, by march in several divisions to the neighbourhood of Northampton and thence by canal boats to Stockton Quay and from there by march (21 miles) to Liverpool where transports would await it. The numbers on the routes were, first battalion, 970; second, 933.

The first division of each battalion moved off on December 21st and the first battalion disembarked at Pigeon House, near Dublin, on January 25th, 1804, after two days on board ship. The second were not so lucky; some detachments were four days at sea, but all were landed by the 28th. The regiment was transferred to the Irish Command from the 25th.

The first battalion now marched into Meath, Headquarters and seven companies to Tullamore and three companies under Major Vaumorel to Frankford. The second was quartered in George Street Barracks, Dublin.

The following is a list of the officers on January 24th, 1804, the last muster day in the British Command:—

Colonel—Lieut.-General Robert Manners.

	First Battalion.	Second Battalion.
	1st Lieut.-Colonel, Major-General Sir Charles Green, commanding at Dover	
	2nd Lieut.-Colonel, Wm. Wilkinson	3rd Lieut.-Colonel, Wm. Lockhart
	1st Major, John Meade	3rd Major, Robert Pigot
	2nd Major, Philip Vaumorel	4th Major, Wm. Wright
	Adjt., Ben. Nunn	Wm. Stewart
	Q.M., Arthur Poyntz	John Foster Kingsley
	Surgeon, Robert Pearse	J. A. Kennedy
	Assist.-Surgeons, Vacant	Patrick Timon and Wm. S. Fry

Paymaster, Michael Jones. Acting Paymaster, Elias Malet.

Only one paymaster was allowed, assisted by a company officer for the second battalion.

	First Battalion.	Second Battalion.
1	Capt. Alex. Hamilton (light)	Christopher Maxwell, Bde.-Major Severn District
	Lieut. Wm. Fawcett	Sam. Fox
	Lieut. John Mangin	Ens. John H. Ellis
2	Capt. Geo. Grey	G. D. Robertson, A.D.C. to M.G. Lennox
	Lieut. Robert Blake Lynch	John Hitchen
3	Capt. Thos. Fleming Roberts, A.D.C. to M.G. Smyth	Ninian Craig, Adjt.-General's Dept., Malta
	Lieut. James Garnon	Pat. Crauford
	Ens. Robert Howard	Alex. Fettes
4	Capt. Thos. Hardyman	Geo. Wade (Lieut.-Col.)
	Ens. C. S. Watson	Wm. Wallop
5	Capt. Sir C. W. Burdett, Bart., Bgde.-Major to M.G. Este	Thos. Fleming (Major)
	Lieut. Noblet J. Lindsay	
	Ens. E. F. Champion	Jas. Sullivan

6	Capt. Robert Fitzgerald Abt.	Thos. Leach
	Lieut. John P. Beaumont	Wm. Harpur
	Ens. John Peach	
7	Capt. Peter Ryves Hawker	Stedman Rawlings (light)
	Lieut. David Crowe	Chris. Williamson
	Ens. Henry Cramer	Lieut. Thos. Walker Chambers
8	Capt. T. Daniel	Lord Ashbrook
	Lieut. Thos. Syms	Richard England
	Ens. T. W. Newland	
9	Capt. H. C. Briscoe	Sam. Bircham
	Lieut. Geo. Gordon	Edward McCarthy
	Ens. John Tongue	
10	Capt. Robt. Murray	Thos. Jackson
	Lieut. Pat. O'Brien	Elias Malet

For some time after arrival in Meath the first battalion continued to have a large sick list and in February six men died and five deserted. A decided improvement then took place. Vaccination was introduced in the regiment about this time.

On March 23rd the second battalion was ordered to march to Moate, Kilbeggan and Clara, where it would be only a few miles from the first. It left 150 sick in Dublin and twenty-eight men left in hospital in England had not been able to rejoin. In the months of March, April and May fifty-seven men died. Smallpox was no doubt the chief cause.

Up to this time there had apparently been an intention of forming two efficient battalions, but from now the second became a mere depot to the first, which took the pick of the recruits and the best men from the second, and transferred all its undesirables in exchange, especially the men who did not volunteer for foreign service. The company officers were redistributed; it was decided that in each rank the seniors should belong to the first and the juniors to the second. After creating an immense amount of confusion the unexpected result of this rule was that the junior officers led the second battalion in the Peninsula, while their seniors sweltered with the first in Madras.

Both battalions were moved to the Curragh for drill in July and August and then returned to their former quarters, but the light companies of both remained with the light brigade in Dublin.

The following changes took place among the field officers in 1804: First Lieut.-Colonel, Major-General Sir Charles Green, was promoted colonel of the York Light Infantry on January 16th and struck off the strength of the 30th. He was at the same time given the command of the Leeward Islands and from there he captured the Dutch colony of Surinam. He commanded at Malta in 1807–8 and was promoted lieut.-general in 1809. In 1812 he commanded the Northern District and in the following year was given the London Command. He was appointed commissioner for the Royal Military College and the Royal Military Asylum in 1821. He was made a baronet for his distinguished service.

Lieut.-Colonel (Colonel) Wilkinson was given command of a brigade in September. He remained first lieut.-colonel of the 30th till he became a lieut.-general, but never commanded either battalion again except for a short time in India while awaiting an appointment.

Lieut.-Colonel William Minet, from the York Rangers, was appointed third lieut.-colonel to command the second battalion, Colonel Lockhart as senior taking the first.

On April 9th, Major Robert Pigot had died and Captain Alexander Hamilton had been given the vacancy.

In December Major John Meade had been promoted lieut.-colonel of the 16th garrison battalion and Captain George Grey had got the step. Major Wright as senior had moved to the first battalion and the two juniors, Hamilton and Grey, were majors of the second. Meade, who was third son of the Earl of Clanwilliam, died a lieut.-general and Companion of the Bath.

There were few changes in the early part of 1805. The first battalion continued to gain strength at the expense of the second. The two light companies were still in the light brigade in Dublin ; they may have been under Major Hamilton who was at this time training a light battalion there.

Drafts continued to arrive from the army of defence, as it was now called, but two-thirds of the poor wretches disappeared yearly from desertion, invaliding or death. The following is a specimen of how the system worked. In March a draft of 98 men arrived. Of these 19 were rejected by the commander of the forces in Ireland and 18 of the best were chosen by the first battalion, the remainder going to the second. In August another draft of 85 was, received but, during the year, in the second battalion alone, there were 70 discharges, 22 desertions and 38 deaths.

Both battalions went as usual to the Curragh for July and August, but when the camp broke up the second went into the Royal Barracks, Dublin, and the first to Limerick. It was now a very fine battalion and had been selected for active service.

The British Government was preparing to send a force under Lord Cathcart to North Germany to act in conjunction with the armies of Prussia, Russia and Sweden in an attack on the forces of Napoleon in the north of Europe while the Austrians and Russians attacked in the south.

The two battalions never met again till the end of the war, when the second was absorbed into the first from which it had been formed, but there still remained the constant interchange of officers, and when the second also went on service a depot common to both was established, first at Wakefield and later at Hull.

At the end of October the first battalion marched to Cork, where the battalion companies embarked on November 1st and 2nd, and the light company on the 12th. The battalion was brigaded with the 9th and 89th under Major-General Rowland Hill. The regiment had first seen him as a young captain on Lord Mulgrave's staff at Toulon in 1793.

The strength embarked was :—

Maxwell's company—Lieut.-Colonel Lockhart, Majors Vaumorel and Wright, Captain Chris. Maxwell, Lieutenant J. P. Beaumont, Ensign Denis Wood, Adjutant B. Nunn, Quartermaster A. Poyntz, Surgeon R. Pearse, Assist.-Surgeon Ben. Maxwell, Paymaster Michael Jones, Sergeant-Major G. Stephenson, Quartermaster-Sergt. James Glenn, Paymaster-Sergt. Wm. Gillis.

Roberts' company—Captain Thomas Fleming Roberts, Ensign Hinton East.

Grenadiers—Lieutenant T. G. Richardson, Ensign John Whyte.

Wade's company—Lieutenant John Powell, Ensign John Selwyn.

Fleming's company—Captain Thomas Fleming, Ensign William Sullivan.

Burdett's company—Captain Sir W. C. Burdett, Lieutenant Noblet J. Lindsay.

Light company—Captain Thos. Leach, Lieutenant T. W. Chambers, Lieutenant John Mangin.

Hawker's company—Captain Peter Ryves Hawker, Lieutenant Robert Howard.

Bircham's company—Captain Sam. Bircham, Lieutenant C. S. Watson, Ensign J. Brown.

Murray's company—Captain R. Murray, Lieutenant P. O'Brien, Ensign John Gowan.

A total of thirty-two officers and 1,000 non-commissioned officers and men.

Lord Cathcart's force was to rendezvous in the Downs, but the misfortunes of Hill's brigade began in passing up Channel, when the transport *Ariadne*, with 200 of the 9th on board, was driven ashore in a gale near Calais, and all on board became prisoners of war. The remainder of the brigade reached Yarmouth, where it was landed.

On the weather moderating it re-embarked and the transports assembled in the Downs on December 12th. They were almost immediately dispersed by a heavy gale and the transport *Jenny* (Patterson, Master) with Captain Peter Ryves Hawker's company on board was driven on shore, near Gravelines, to the east of Calais. Some of the 89th were wrecked at the same time on the *Texel.* All became prisoners of war. The 30th were fortunate in getting ashore without loss and in saving their baggage.

The party on the *Jenny* consisted of Captain Thomas Fleming Roberts and 3 privates of his company, evidently his servants, Captain Hawker with Lieutenant Robert Howard, 5 sergeants, 2 drummers and 88 rank and file of his company, Ensign William Sullivan of Captain Fleming's company and 10 men attached. All were sent to Verdun where they arrived on January 5th, 1806, having been well treated by the French. The officers and most of the non-commissioned officers and men remained as prisoners of war at Verdun till the Peace of 1814, but one or two escaped and more died before that time.

Captain Roberts had been serving on the staff in England and only arrived at Cork in time to embark with the expedition. He had been appointed brigade-major to Sir Arthur Wellesley, whose brigade was already in Germany, and had every prospect of rising rapidly in his profession, but was ruined by that unlucky storm.

Of the remaining nine companies of the battalion, five reached Germany, but the transports conveying four companies, and one with seventy-five men of different companies under Ensign Brown, were forced to seek shelter in the harbours of Kent. The whole staff of the battalion except the field officers were with those four companies.

A few men of the 30th seem to have been on the transport *Adventure* which was driven ashore near Yarmouth, but none were lost.

The strength of the five companies which landed and joined Lord Cathcart's army on the Elbe was sixteen officers and 493 other ranks. The officers were: Maxwell's company—Lieut.-Colonel Lockhart, Majors Vaumorel and Wright, Surgeon Pearse, Captain Maxwell, Lieutenant Beaumont, Ensign Wood. Roberts' company—Lieutenant Crauford, Ensign East. Grenadiers —Lieutenants Richardson and Watson. Wade's company—Lieutenant Powell, Ensign Selwyn. Light company—Captain Leach, Lieutenants Chambers and Mangin.

No explanation is forthcoming as to how the staff got separated from Maxwell's company and from the field officers, but the latter may have sailed on the men-of-war of the convoy, as more likely to reach the Elbe.

The stay of the five companies in Germany was not long and they saw no fighting. The main strength of the coalition against Napoleon in the north of Europe lay in the Prussian army, but King Frederick William shamefully fell away from his allies when he heard of Napoleon's victory over the Austrians and Russians at Austerlitz on December 2nd. There was nothing then left for Lord Cathcart to do but to prepare for a return to England. While awaiting instructions from his Government, he placed on board ship the portions of his army which had been most broken up by the storms of December.

The five companies of the 30th embarked on January 28th, 1806, and landed at Margate on February 8th. Their health had been good throughout, only one death having taken place. At Margate they were joined by the remainder of the battalion from Canterbury. The companies which had been driven back by stress of weather had been assembled in that city in January under command of Captain (Brevet-Major) Thomas Fleming.

At the beginning of April the battalion was at Faversham and from there it marched to Portsmouth, arriving on the 17th and 18th. It was quartered in Hilsea Barracks with four companies detached at Fort Cumberland.

On arrival the battalion received orders to embark on board the East India Company's ships fitted up for its conveyance to India.

Meanwhile the second battalion had been ordered to send a reinforcing draft to Portsmouth of seven subalterns and 400 men. The subalterns were Lieutenants Henry Cramer, Alex. Fettes, F. Champion, Wm. L. Cane, Edward McCarthy, Brereton Watson, and James Skirrow. Four officers, Major Grey, Captains Bamford and Fox and Ensign Henry Heathcote, were ordered to conduct the party and return on completion of the duty. Heathcote was the first gentleman cadet from the Royal Military College to be appointed to the 30th.

On arrival at Portsmouth fifty of the draft were rejected and sent back to Ireland. The number transferred to the first battalion was seven officers and 352 other ranks.

There was great bustle at Portsmouth at this time. In addition to the 30th, the 47th and strong drafts were embarking for the East Indies, the 21st Light Dragoons and the skeleton of the 4th/60th were embarking for the Cape and several regiments for the Mediterranean. We had lately taken possession of the Cape for the second time and the 4th/60th was to be filled up from the soldiers of the late Dutch garrison who had transferred their services to the British Crown.

On May 6th, the 1st/30th, under command of Colonel Lockhart, and headed by its own band, marched from Hilsea Barracks to Portsmouth Hard, where it embarked on several Indiamen. The strength was 33 officers, 46 sergeants, 22 drummers, 1,194 rank and file. In addition to those present, 4 officers and 105 other ranks who were prisoners of war were borne on the strength. Included in the rank and file were two volunteers, Arthur and Samuel Robert Poyntz, sons of the quartermaster, who were fourteen and twelve years of age. They both lived to become lieut.-colonels in the army, as did a younger brother, James, who joined the second battalion.

Four-fifths of the men were English and the proportion from Cambridgeshire was larger than at any other time in the regimental history. Two-thirds of the battalion were over 5 ft. 6 ins.

The following was the distribution by companies :—

Maxwell's—Lieut.-Colonel Wm. Lockhart, Majors Philip Vaumorel and Wm. Wright, Captain Chris. Maxwell, Ensign Denis Wood, Adjutant Ben. Nunn, Quartermaster Arthur Poyntz, Surgeon Robert Pearse, Assist.-Surgeons Ben. Maxwell and W. Griffiths, Paymaster M. Jones, Sergt.-Major George Stephenson, Quartermaster-Sergt. James Glenn. No armourer or paymaster-sergeant.

Roberts'—Lieutenants J. P. Beaumont and Wm. Lyon Cane.

Craig's—Lieutenants F. G. Richardson and Alex. Fettes.

Fleming's—Bt.-Major Thos. Fleming, Lieutenant C. F. Champion.

Burdett's—Captain Sir C. W. Burdett, Bart., Lieutenant James Skirrow.

Light—Captain Thos. Leach, Lieutenants T. W. Chambers, John Mangin and W. C. Harpur.

Hawker's—Prisoners of war.

Grenadiers—Captain Sam. Bircham, Lieutenants C. S. Watson, James Browne and Henry Cramer.

Murray's—Captain Robert Murray, Lieutenant Pat. O'Brien, Ensign T. F. O'Brien.

Jackson's—Captain Thos. Jackson, Lieutenants John Powell and J. Selwyn.

The officers absent on duty were Colonel Wm. Wilkinson, commanding a brigade in Ireland, Captain Ninian Craig in the adjt.-general's department at Malta, Lieutenants Noblet J. Lindsay, Pat. Crauford, John Tongue, Wm. Grey and Ed. McCarthy, Ensigns Ed. Crosbie, John Gowan and J. L. White on command, Ensign M. J. Sparks recruiting in Lincolnshire, and Ensign Washington Carden, absent without leave. He seems to have joined before the fleet sailed and to have given a satisfactory explanation.

As each ship was ready it dropped down to St. Helen's and on May 13th a fleet of twenty-nine ships, escorted by H.M.S. *Lion*, 64, Captain Rolles and two frigates, sailed from there for the Cape of Good Hope, India and China.

On August 6th the fleet anchored in St. Simon's Bay, Cape of Good Hope. Assist.-Surgeon Ben. Maxwell and a few rank and file of the 30th had died on the passage.

Although destined for service in Madras, the regiment had been placed at the disposal of Sir Thomas Troubridge, our Naval Commander-in-Chief to the east of a line running due south from Point de Galle in Ceylon ; Sir Edward Pellew commanded to the west of that line. Sir Thomas wanted the 30th to help in the reduction of the Dutch ports, which had become a base for French privateers since the Dutch had lost their independence. The ships carrying the regiment therefore sailed with Captain Rolles when he started on August 15th for Pulo Penang with the portion of the fleet destined for China. Pulo Penang was to be the starting point of Sir Thomas Troubridge's operations. On October 13th the fleet reached Penang, where Sir Thomas was awaiting it with H.M.Ss. *Blenheim, Macassar* and *Greyhound*. All his plans were upset by the despatches brought by Captain Rolles. Sir Thomas found that he was appointed by the Admiralty to command at the Cape, while Sir Edward Pellew was to assume command of all our ships in the East Indies. He sailed as soon as he could for the Cape, but was lost with all his ship's company in a cyclone off Madagascar.

Sir Edward Pellew was able to destroy a great part of the Dutch naval force in Batavia Roads in November, but he postponed operations against their ports till he had been able to make a fresh reconnaissance of them. The 30th had been landed on October 16th, but, not now being wanted in the islands, it re-embarked on November 12th and landed in Madras on January 14th, 1807. Quartermaster-Sergeant James Glenn had died on the voyage.

INDIA

On landing, the regiment marched to Wallahjabad Cantonments, 40 miles west of Madras. The sickness was already excessive.

In February, four companies were detached to Vellore under Major Vaumorel. The other officers were, Bt.-Major Fleming, Captain Jackson, Lieutenants Chambers, Beaumont, Powell, Cramer and Cane, and Assist.-Surgeon Griffin. The strength of other ranks was 374.

During the first nine months in India 163 men of the regiment died. The number of deaths in each month shows the rise and fall of the sickness, which was chiefly dysentery. They were, 11, 28, 36, 30, 22, 15, 4, 8, 9.

In May, Sir Edward Pellew began to prepare for his expedition against the Dutch ports. He asked for men of the 30th to strengthen his ships' companies and a sergeant and fourteen men were sent him to act as marines. In June, Lieutenant C. S. Watson and fifty-four other ranks embarked on the frigate *Psyche*, commanded by Captain Fleetwood Pellew, the admiral's son. The *Psyche* and the frigate *Caroline*, Captain Rainier, were then on the point of sailing to reconnoitre the coast of Java to locate the Dutch men-of-war which had escaped from Sir Edward in Batavia Roads in November.

On August 29th Captains Pellew and Rainier heard from a prize, taken off the eastern extremity of Java, that the ships they were in search of were lying dismantled at Grezzie on the river Sourabaya. The *Psyche* and *Caroline* parted to convey the intelligence to the admiral by different routes.

On September 2nd the *Psyche* observed some vessels in Samarang Roads. The water was too shallow for his ship, but Captain Pellew sent his boats in to attack the enemy's vessels. They brought out, under a heavy fire from the batteries, an armed schooner of eight guns and a large merchant brig which were at once destroyed to enable the *Psyche* to chase three vessels seen in the offing. At half-past three, finding the *Psyche* was closing with them, the enemy ran their vessels aground and opened fire. The water was too shallow to allow the *Psyche* to approach as near as Captain Pellew wished, for his thirty-six guns were all merely 12-pounders, and after being in action for nearly an hour he was upon the point of hoisting out his boats to board when one of the enemy's vessels hauled down her flag, an example which was quickly followed by the others. They proved to be the *Resolutie,* an armed ship of 700 tons having on board the colours and staff of the 23rd Dutch regiment ; the *Ceres,* a twelve-gun brig ; and the corvette *Scipio* of twenty-four guns, whose commander was mortally wounded.

Lieutenant Watson with fifty-four other ranks of the 30th served in these actions on the *Psyche.*

On September 16th a further detachment of fifty non-commissioned officers and men was sent on board ship to serve as marines under Ensign Arthur Poyntz, who had just been promoted from volunteer. Ensign Washington Carden was sent to command this party in November.

By the beginning of October Sir Edward Pellew was ready for his expedition and on the 17th he embarked the wing of the 30th under Colonel Lockhart which was now in garrison at Madras, and sailed to Malacca to await the reports of his son and Captain Rainier. Major Vaumorel brought his wing down from Vellore to hold Fort St. George in Lockhart's absence.

The strength embarked under Colonel Lockhart was—Major Wright, Captains Maxwell, Burdett, Leach, Bircham and Murray, Lieutenants Nunn (adjutant), Richardson, Harpur, Cramer, Cane, Fettes, Owen Wynne Grey, and Selwyn, Ensigns Holt and Todd, Surgeon Pearse, Assist.-Surgeon Sam. Piper, Quartermaster Poyntz and 534 other ranks.

Assist.-Surgeon Sam. Piper is believed to have been the original of Dr. Slammer of the *Pickwick Papers.*

On November 20th Sir Edward Pellew, having received the reports of his son and Captain Rainier, sailed from Malacca with H.M.Ss. *Culloden* and *Powerful* (line-of-battle ships), *Caroline, Fox, Victorie, Samarang, Seaflower, Jaseur* and the *Worcester* transport. He was off Point Panka, in the neighbourhood of Grezzie, on December 5th, and from there in conjunction with Colonel Lockhart he sent Captain Burdett of the 30th and a naval officer under a flag of truce to the Dutch naval commandant with a summons to surrender the ships in that port. The Dutch commodore imprisoned the bearers of the flag of truce and sent no reply. Sir Edward thereupon caused the *Culloden* and *Powerful* to be lightened and sailed up the river, accompanied by the remainder of the squadron. A battery in the island of Madura fired red-hot shot at them, but although the ships were hit no damage was done and the fire of the line-of-battle ships quickly overpowered the enemy's guns.

The Governor and Council of Sourabaya now intervened. They disowned the violent measures of their naval commander and ordered the release of the bearers of the flag of truce and the surrender of the ships.

On taking possession of his prizes, the sixty-eight gun ships *Pluto, Revolutie,* and *Kortinaar* and the *Ruttkoff* pierced for forty guns, Sir Edward Pellew found that the Dutch commodore had scuttled them. He therefore caused them to be burned and, having destroyed the fortifications, stores and ammunition of Grezzie, he sent the 30th back to Madras and broke up his squadron. In his despatch to the Governor of Madras Sir Edward spoke in high terms of the wing of the 30th under Colonel Lockhart and thanked the Madras Government for having given him " that valuable officer " as a colleague.

Colonel Lockhart landed at Madras on February 14th, 1808. Major

Vaumorel had returned with his wing to Vellore in January. In February, Brevet-Major Thomas Fleming died there.

Captain Sir C. Burdett, who had resigned the appointment of brigade-major at Fort St. George in order to accompany the expedition, again took up his staff duty on his return. Lieutenant C. S. Watson, Ensign Washington Carden and sixty-three men remained at sea as marines.

In the regimental returns of this period the name Poonamallee occurs for the first time. It was a depot close to Madras for receiving drafts from England and despatching them to their regiments in the Presidency, and for receiving invalids and time-expired men, and arranging for their passage home.

The regiment was now much healthier than on first arrival in India, but Lieutenant Denis Wood died on June 15th. In this month the number of men under Lieutenant Watson acting as marines had been reduced to 34, but 6 sergeants, 3 drummers and 200 rank and file had been sent on an expedition under Rear-Admiral Drury to take possession of the Portuguese harbour and settlement of Macao in the neighbourhood of Hong Kong. The officers on the expedition were Captain Beaumont, Lieutenants J. G. Richardson, James Skirrow, Alex. Fettes, Washington Carden, Ensign J. L. Perry, Assistant-Surgeon Sam. Piper.

An order came from the Supreme Government at Calcutta after the expedition had sailed, to send in command of the troops an officer of tact and judgment who would be likely to induce the Portuguese to surrender without fighting. He was to explain to them that we had no ill will and no desire to deprive them of their possessions, but that, Napoleon having occupied Portugal and forced it into an alliance with him, we were afraid that he would turn their harbours into nests of privateers to prey upon our commerce as he had done with those of the Dutch. Major Wright, who had just been gazetted brevet-lieut.-colonel, was selected for this delicate mission by the Governor of Madras and despatched to the rendezvous of the expedition at Penang. Immediately after this another communication came from Calcutta to say that it was a pity to break up a fine strong regiment like the 30th and that if possible the companies under Major Wright were to be recalled. A European regiment of the East India Company, the strength of which had fallen to 200 men, would be sent in their place. The effort to recall the detachment was ineffectual. Wright died at Penang on August 16th, but the two companies went on to Macao, under Beaumont, and were present at its surrender.

In August the annual fleet from home arrived and brought Colonel Wilkinson, Lieutenants Hinton East and James Lewin and a draft of sixty-eight volunteers from the Cambridgeshire militia and two men from the second battalion. Colonel Wilkinson had been commanding a brigade in Ireland, but being unable to obtain a command on active service had applied to rejoin the regiment. He took command from August 4th.

In December, Major Vaumorel with the Vellore detachment rejoined Headquarters in Fort St. George.

In January, 1809, Lieut.-Colonel Lockhart was appointed to command the Pondicherry district.

The concentration of the 30th in Fort St. George was a consequence of the rebellion of the Rajahs of Travancore and Cochin. The Madras Government, keeping the 30th to protect the Presidency, hurried south all other troops and applied to the Governor of Ceylon for what European soldiers he could spare.

At the end of January the arrival of reinforcements from Ceylon released the 30th and it embarked for Negapatam, a town with a fairly-sheltered roadstead, 170 miles south of Madras. Here it disembarked at the beginning of February. Colonel Wilkinson had been appointed to command the southern division of the Madras army, with Headquarters at Trichinopoly and the battalion formed part of a field force under him. It saw no fighting but had incessant marching till April 16th when the field force was broken up and the 30th went into garrison at Trichinopoly.

While in the field it had received an order from home abolishing the use of powder and queues and ordering the soldiers' hair to be cut close to the neck.

Trichinopoly is about 70 miles north-west of Negapatam and is rather a picturesque place ; a rock rises within the fort abruptly to a height of 270 feet above the river Cauvery which flows at its foot.

The two companies which had been on the expedition to Macao returned in April under Lieutenant Richardson. Captain Beaumont was detained on a court martial at Penang. When he did rejoin he seems to have brought Lieutenant Watson to Madras. Watson's health had broken down on marine service and he died on June 2nd at S. Thomè, a small bay 5 miles to the south of Madras.

Only one sergeant and nineteen privates of the regiment were now employed as marines.

Colonel Wilkinson made the half-yearly inspection of the battalion in June. The strength was 1,045 privates, all English except 182 Scotch or Irish and four foreigners. The sergeants carried pikes, except the five of the light company, who carried fuzees (light muskets). There were only two buglers, those of the light company. Both drummers and buglers were entered under the head " fifers," showing that there was now a drum and fife band and no longer only two fifes for the grenadier company. Captain Ninian Craig was in the adjutant-general's department in Malta, Captain Sir C. Burdett brigade-major Fort St. George, Lieutenant John Powell brigade-major in the Southern District. The adjutant, Ben Nunn, A.Q.M.-general Southern District. Sergeant-Major George Stephenson was acting adjutant.

In this year the natural discontent of the European officers of the Hon. East India Company's Madras army with their treatment by the Governor, Sir Thomas Barlow, led in many cases to disobedience and, here and there, to open mutiny, in which the officers involved their sepoys. Colonel Wilkinson took stern measures to prevent any outbreak in his command and was completely successful. The Madras officers and their friends have accused him of want of feeling and of uncalled-for violence, and this renders necessary a more extended description of what happened than would otherwise have been thought requisite.

In May the Government of Madras issued a form of declaration of full obedience, according to their commissions, to be signed by all British officers of the East India Company's army in Madras. At this time there were at Trichinopoly, along with the 30th, the 6th Native Cavalry and the 2nd/13th and 2nd/24th Native Infantry, besides the staff of the district and a small body of artillery and engineers.

On July 30th, Colonel Wilkinson assembled the European officers of the native army, and tendered the test to them. Of the officers assembled, twenty accepted and signed it at once, but seven of the 6th Cavalry, one artillery officer, and eight of the 2nd/13th declined.

Colonel Wilkinson appears to have been patient enough and asked the dissatisfied officers to pledge themselves to hold no communication with their sepoys. This, under the circumstances, was an absolutely essential condition. On their refusing even to do this, he handed them over to 200 of the 30th with muskets loaded and bayonets fixed, and gave orders to the officers in command to slay the prisoners at once if there was the least appearance of an attempt at rescue on the part of their sepoys.

The escort marched off with the prisoners for Negapatam, but even then Colonel Wilkinson showed his usual kindness of heart by telling the commanding officer that on reaching Tanjore, he might release any one of his prisoners who would give his parole to complete the journey to Madras peaceably and without further offence.

Leaving Major Vaumorel in command at Trichinopoly, Colonel Wilkinson with a strong detachment visited in August all the garrisons in the Southern District and sent to Madras all officers of the Company's army who refused to sign the test.

He had placed Major Christopher Maxwell in command of the 6th Native Cavalry with Lieutenants Wm. Lyon Cave and James Lewin to help him. Captain Sam. Bircham and Lieutenant John Napper were sent to the 16th Native Infantry.

The conduct of the Governor of Madras had been so provocative that the Governor-General treated the insubordinate officers with great leniency, but it is worthy of record that in Colonel Wilkinson's district alone was there no open mutiny.

In 1810 there was trouble with the native rulers in Travancore and Captain Robert Murray was sent there with seven officers and 343 other ranks. The officers were Lieutenants Tongue, Cramer, Fettes, Selwyn, Nicholson and French and Assist.-Surgeon Griffin. The detachment returned within a month.

In May, Captain Sir Charles Burdett was appointed commandant of the Poonamallee depot, and in October, Lieutenant H. Cramer was appointed adjutant and paymaster under him.

In October, Wilkinson, now a major-general, made the half-yearly inspection of the battalion. He took Captain Wade from the regiment as his A.D.C.

In this month the second battalion which had been left as a depot in Ireland began its glorious career under Wellington.

SECOND BATTALION

When the camp at the Curragh broke up in September, 1805, and the first battalion moved south to embark for active service, the second marched to Dublin and was quartered in the Royal Barracks. Only one man had died in each of the months of August and September at the Curragh, but twenty-one died in the succeeding three months and seven in the first quarter of 1806.

The Royal Barracks were very badly built and began to fall down before they were completed.

In April, 1806, the battalion marched to Strabane, leaving the light company (Malet's) in the light brigade in Dublin. On the 19th of the month, as already recorded, the reinforcing draft of seven lieutenants and 400 other ranks embarked to join the first battalion at Portsmouth.

An increase of pay was granted to all ranks in the regiment from June 25th, 1806, as follows:

	Increase. *s. d.*		*Total.* *s. d.*	
Lieut.-Colonel	4 1	a day	17 0	and 3*s.* command money
Majors, each	1 11	,,	16 0	
Captains, each	1 1	,,	10 6	
Captains, with brevet	—		12 6	
Lieutenants	0 10	,,	6 6	
After 7 years as such	—		7 6	
Ensigns	0 7	,,	5 3	
Adjutant	0 6	,,	8 6	including horse allowance
Quartermaster	0 10	,,	6 6	
Paymaster	—		15 0	
Surgeon	—		11 4	
Assist.-Surgeon	—		7 6	
Sergt.-Major	0 5¼	,,	2 6	
Q.M.-Sergt.	0 5¼		2 6	
Paymaster-Sergt., Armourer and 46 Sergts., each	0 3¼	,,	1 10	
Forty-four Corporals.	0 1¾	,,	1 4	
After 7 years	—		1 5	
After 14 years	—		1 6	

The pay of the privates was to depend on length of service. Instead of being enlisted for life, their first engagement was to be for seven years at 1*s.* a day, with power to re-engage for a second period at 1*s.* 1*d.* and for a third to complete twenty-one years' service at 1*s.* 2*d.*

The two battalions being now separated, Hugh Boyd Wray was appointed paymaster without other duties to the second.

The battalion had been a mere depot to the first, but from now there commenced an effort to fit it for active service. The first step was to get rid of the men of the army of defence who had declined to volunteer for service abroad. There were 234 of them in the battalion and on November 25th they were handed over to the 9th garrison battalion. Inducements also were offered to the men of the Irish militia to volunteer for the line. In spite of the increase of pay the number of recruits was not equal to the great demand for them at this time, and the power to enlist boys and lads to serve on reduced pay till certified as fit for the ranks was exercised more freely than was desirable. In December, 1805, out of 23 Cambridgeshire recruits, 10 were boys. In the first quarter of 1806 the Headquarter recruits were 16 men, 3 lads and 8 boys, and later in the year a party from Scotland was made up of 5 men, 2 lads and 4 boys. Under the new terms of enlistment a number of old soldiers were re-enlisted for a second or third term of service. There were among them representatives of all arms of the service, Life Guards, dragoons, artillery, infantry, and marines. They were good soldiers and a limited number would have been valuable in a young regiment, but there were too many of them and as a rule they were too old. There was nothing between them and the recruits—no backbone in fact to the battalion. Strong parties from the militia might have remedied this, but although Carlow, Mayo, Sligo, Cork, and Longford all gave volunteers, their number and age were not sufficient.

The boys no doubt took to soldiering as a duck takes to water, and they became the men of Badajoz and Waterloo, but for long the battalion was unfit for the field from sheer want of physical strength, and after it joined Lord Wellington, the fevers in the low ground and the icy blasts of the Spanish mountains took a fearful toll of the still growing lads.

On April 2nd, 1807, Bt.-Lieut.-Colonel Wade died when recruiting at Sleaford, and was buried by General Manners in his family vault at Bloxholme, Lincolnshire.

At the end of the month the battalion marched to Londonderry; the number on the route is 457 non-commissioned officers and men. While at Strabane a small detachment had been stationed at Omagh to help in the suppression of illicit whisky stills.

In December the battalion was ordered to march to Longford, where it arrived on January 7th, 1808. Early in June it marched to the Curragh for the drill season, at the end of which it moved to Athlone, arriving on August 3rd.

A draft of 102 non-commissioned officers and men had been sent to embark at Cork and join the first battalion, but it had been turned back at Fermoy and rejoined on August 15th at Athlone.

During the year many recruits had come in, among whom were 133 volunteers from the Irish militia, principally from Tipperary, Meath, Carlow, Kerry, and Leitrim.

On December 24th, Colonel Minet promoted twenty-six boys to men's pay as being fit to do duty in the ranks.

All this of course meant active service, and the time when the battalion was selected is fixed by the recall of the draft for India.

As the battalion was about to proceed to the Peninsula, a few dates of the events of the first phase of the war in Spain and Portugal are here given.

August 1st, 1808. Sir Arthur Wellesley lands in Portugal.

August 21st, 1808. His victory at Vimiera.

August 31st, 1808. Convention of Cintra.

September 9th, 1808. Sir Arthur, with the senior generals, recalled.

October 7th, 1808. Sir John Moore ordered to advance with the British army from Portugal to co-operate with Spanish armies.

November 4th, 1808. Napoleon with strong forces enters Spain and disperses the Spanish armies.

December 21st, 1808. Sir John Moore reaches Sahagun. He finds no Spanish armies to co-operate with.

December 24th, 1808. In danger of being surrounded by overwhelming force, Moore retreats.

January 16th, 1809. His repulse of Soult and heroic death at Corunna.

The British Government had no idea of giving up the struggle. As fast as the Corunna regiments could be re-equipped, they were sent on service again. Unluckily many were sent to Walcheren. The others were ordered to return to Portugal and in place of the corps sent to Walcheren a number of second battalions were ordered to the Peninsula. Sir Arthur Wellesley was appointed to the chief command.

The second battalion of the 30th was one of those selected for service in the Peninsula. At the end of January, 1809, it marched from Athlone to Dublin, and after a week's rest continued the march to Kinsale, where it arrived on March 2nd.

On the 11th of the month the battalion embarked at the Cove of Cork. We have no embarkation returns and the muster is misleading as it does not include the officers of the first battalion serving with the second.

As far as can be ascertained the following was the state of the battalion :—

Lieut.-Colonel Wm. Minet, Majors Alex. Hamilton and George Grey, Captains Thomas Williamson, Elias Malet, Thomas Bradgate Bamford, Rob. Blake Lynch, Sam. Fox and John Hitchen, Lieutenants J. L. White, John Garland and Stephen Masters, Ensigns Michael Archer Eades, Richard Heaviside, Thomas Kettlewell, John Garvey, Wm. Pennefather, James Eager and Andrew Perry, Lieutenant and Adjutant Wm. Stewart, Quartermaster John Foster Kingsley, Surgeon John Hennen, Assist.-Surgeon Edward Brett, Paymaster Hugh Boyd Wray, Sergt.-Major David Glass, Quartermaster-Sergt. Robert Daniel, Paymaster-Sergt. Thomas Cuthbert, Armourer Nathaniel Artis, 38 sergeants, 35 corporals, 17 drummers and 564 privates, of whom 14 were recruits who joined at the port of embarkation.

The officers absent were Captains Lord Forbes, Secretary to our Legation in Brazil, T. W. Chambers and Pat. Beaumont with first battalion in India, R. Machell with 7th Foot and Christopher Williamson with depot at Kinsale. Ensigns John Herring and Thomas Light were also at the depot. Assist.-Surgeon Ed. Purdon was sick in Dublin. He never rejoined. Five lieutenants were with the first battalion in India.

One sergeant and thirty-nine other ranks had been left with the depot, including eighteen recruits who joined at the port of embarkation.

Fifty-three sick had been left in Ireland.

The number of effectives includes thirty-four boys unfit for the ranks and many lads, who though given men's pay were only fit to do ordinary duty in barracks. Of the older men a large number were past their best and should have been discharged.

Ninety-seven women and 209 children had been sent to their homes with a gratuity of £1 2*s.* 8*d.* for each woman and 5*s.* 5*d.* for each child. The heavy baggage was left with the depot.

The battalion landed in Lisbon on April 7th, but a transport with a large number of men on board did not arrive till later.

This is the first occasion on which the men went on service with their hair cut short. An order of the previous year had abolished the foul queue or pigtail. The " flash " worn by the light companies of some regiments disappeared along with the pigtail. The Light Bobs after 1799 used neither to wear the pigtail nor to cut their hair short, but brushed it straight up under the hat. In some corps, to preserve the uniform appearance of the regiment, they wore at the back of the neck, attached to the collar, a bow of black ribbon resembling the tie of the pigtail. This was called the "flash." One breach of uniformity was always sanctioned. It was the custom for smart young men to form the curl of the queue of a lock of a lady's hair. If it differed in colour from that of the wearer, so much the better, as rendering it more conspicuous. It is uncertain if the "flash" was worn in the 30th, but the light company of the 44th, with which that of the 30th was so closely associated, certainly did so. Our information as to the "flash" and the queue comes from the records of the 44th. No doubt the regiment would be clean shaven until it came under Wellington, who never troubled himself about petty details of dress or appearance. In many of his divisions enormous whiskers were the fashion.

On April 27th, Sir Arthur Wellesley wrote to the Secretary of State, Lord Castlereagh, that he had assembled the whole of the British army in Portugal, near Alcobaça, except the second 30th, which was in garrison at Lisbon. He was then preparing to move to the north against the French under Marshal Soult, and proposed to leave behind to watch the frontier and the passages of the Tagus, Mackenzie's brigade and the second battalions of the 24th and 30th, with the 3rd and 4th Dragoons and 7,000 Portuguese. In the succeeding fortnight the battalion had 7 sergeants, 2 drummers and 145 privates detached on escort duty, but the names of the officers are not known.

On May 19th the 30th was ordered to march to Azambuja, to the north-east of Lisbon, but by this time Sir Arthur had made up his mind that the 30th was like other second battalions, and determined to exchange it for an older corps, and he was as usual quite right. The battalion was not physically fit for the field.

The second 30th and 29th were ordered to Gibraltar in exchange for the 48th and 61st and the battalion embarked on the 27th from Lisbon. One ensign, 4 sergeants, 2 drummers, 92 rank and file, were left behind, but joined at Gibraltar. There was little of interest in the time spent in Gibraltar, where Sir John Craddock commanded. Captain Bamford was sent home recruiting, and the quartermaster also went home to bring out the heavy baggage.

In May, two companies were added to the regiment to form a depot. Lieut. Alexander Macnab and Captain James Fullerton from the 2nd Ceylon regiment got the two new companies without purchase, but Macnab remained with the battalion and Bt.-Major James Spawforth, who had exchanged in the previous month from the 8th garrison battalion with Lord Forbes, was ordered to establish the depot at Wakefield, where a company had been raised for the regiment in 1689 and again in 1702. The detachment which had been left under Captain Christopher Williamson at Kinsale, and which had been moved to Cork, was sent to Wakefield. Captain Williamson was in bad health and never returned to duty. He was given command of the army depot in the Isle of Wight.

The first return we have of the depot at Wakefield is signed by Major Spawforth and shows Captain Fullerton detached recruiting at Glasgow, Lieutenant Michael Sparks in Lincolnshire, and Lieutenant Mayne at Cambridge. Quartermaster Kingsley is present, but is about to start to rejoin in Portugal with the heavy baggage of the second battalion and a draft from the depot of thirty-seven men and boys. The non-commissioned officers, rank and file present are 2 sergeants, 1 drummer, 42 men, 18 boys.

In November and December the second battalion sent home 2 lieutenants, 1 ensign, 1 drummer and 7 corporals to join the depot, and also 2 sergeants, 1 drummer, 14 rank and file, who were transferred to the first battalion as its contribution to the depot.

In November, Captain Ninian Craig of the first battalion exchanged with Captain Henry Craig of the Sicilian regiment. It may be assumed that Henry was the son of Ninian, whom he had succeeded as regimental quartermaster in 1794. After the Egyptian campaign of 1801, Henry had been promoted lieutenant in the 89th and attached to the quartermaster-general's office, Malta, where Ninian Craig was in the adjt.-general's. They had continued to serve there together for eight years, but to keep his appointment Henry had been obliged to get transferred to the 35th while that regiment was in garrison in Malta, and eventually to the Sicilian regiment which was commanded by Lieut.-Colonel G. D. Robertson, late 30th regiment.

In December, 1809, the strength of the battalion in the Peninsula was 43 sergeants, 17 drummers, 584 rank and file. This was raised in the following month to 652, sick included, by the arrival of drafts.

In January, 1810, Captain Bamford joined from recruiting, and Captain Thomas Williamson was transferred on account of seniority to the first battalion, being struck off the strength of the second. He remained, however, on the staff in the Peninsula. Captain Machell joined from the West Indies and Captain Macnab was appointed town major of Gibraltar.

In April the French were threatening Tarifa and a force was sent from Gibraltar to drive them off, and it is curious how the accounts disagree about the composition of the force, and as to whether there was any fighting or not.

We can only be sure of the detachment sent by the 30th which consisted of the grenadier company, Captain Hitchen, light company, Captain Malet, and a week later, the companies of Bamford, Lynch and Fox, Lieutenants Garland, Andrews, Heaviside and Eades, Ensigns John Garvey, Pennefather, Freear, John Rumley, Baillie and John Crawford, with 13 sergeants, 3 drummers, 11 corporals, 273 privates.

The detachment crossed the bay to Algeciras by boat and marched to Tarifa. Apparently the French withdrew without fighting. Nothing serious could have happened, for Major Hamilton and the adjutant, Wm. Stewart, went on a month's leave on April 20th. The detachment rejoined Headquarters at the beginning of May. On May 12th orders were received to embark for Cadiz, which was held by a British garrison under Sir Thomas Graham and was now threatened by the French.

The battalion landed at Cadiz on June 2nd with a nominal strength of 704 non-commissioned officers and men, but thirty-four were sick and thirteen were boys. As usual, a number of boys had got men's rating just before the move.

Sir Thomas Graham had divided his force into three brigades in the Isla de Leon and a garrison in the city of Cadiz. The 2nd/30th, 2nd/87th and two companies of Rifles formed the garrison and occupied casemated barracks on the land front on each side of the principal gate. Major-General Hoghton was in command, Captain R. Machell was his brigade-major and Captain E.

Malet was D.A.Q.M.-General of the division. A number of men were lent to the artillery to serve as gunners, artificers and drivers.

We learn from the muster roll that Private William Page was killed in the Portales battery on June 19th. It is not known if the battalion had any wounded during the siege.

In September a draft arrived from the depot under Captain Thomas Walker Chambers, transferred from the first battalion. He brought Captain Henry Craig, Ensign P. P. Neville and eighty-five other ranks, including one of the prisoners of war who had been wrecked on the French coast in 1806. How he managed to escape or get exchanged does not appear.

When the 30th first landed, Sir Thomas Graham was very much dissatisfied with the physique of his troops, with the exception of the Guards and the 79th. He said that the men were either too young or too old for service, which exactly describes the condition of the 30th, a condition, however, which was improving every day.

The arrival of the regiment reminded Sir Thomas of the good service done under him by Lieutenant Michael Sparkes as an acting engineer in Ireland, and he requested the permission of General Manners to employ him at Cadiz. Sparkes, who had been left recruiting in Lincolnshire by the first battalion, on account of his wife and family, joined at Cadiz. He remained with General Graham when the second battalion proceeded to Portugal, and did good service at the battle of Barrosa by putting a house in a state of defence in an advanced position during the action, and on General Graham's recommendation was given, without purchase, a company in the Royal African Corps.

General Graham must have made a more favourable report of the progress of some of his troops, for in September Lord Wellington called for the 2nd/30th and 2nd/44th.

Sir Arthur Wellesley had been created Baron Douro for his defeat of Soult near Oporto, and Viscount Wellington for his victory at Talavera on July 28th, 1809.

On September 25th General Graham wrote to his lordship that the 30th and 44th had embarked. He continues: "I take this opportunity of mentioning two officers of the 30th highly deserving of notice whom I have long known and served with on various occasions; Major Hamilton, who has often distinguished himself by his gallantry, and Captain Malet, a most intelligent and attentive officer."

The battalions disembarked at Lisbon on October 6th, 1810.

We may date from their service together at Cadiz the close friendship which united the 44th and 30th during the war. The older officers were friends long before, for their first battalions had fought alongside each other in Egypt in Doyle's brigade.

CHAPTER IX

SECOND BATTALION. PENINSULA—TORRES VEDRAS, FUENTES ONORO, SABUGAL, CIUDAD RODRIGO, BADAJOZ, SALAMANCA, VILLA MURIEL.

1810–1813

THE 30th suffered in the Peninsula from having too few officers. Colonel Minet was either sick or on detached duty until promoted major-general by seniority on June 4th, 1811. He remained third lieut.-colonel of the regiment till 1815, and died a lieut.-general in 1821.

Major Hamilton commanded the battalion from leaving Cadiz till sent home on promotion in September, 1811. Major Grey served during the whole of 1811, and in 1812 until mortally wounded leading the battalion in April. Hamilton rejoined as lieut.-colonel in June, 1812, and held the command till the battalion was again reunited to the first.

In 1811, there were present four captains commanding companies and one, Bamford, acting as major, but two left at the end of the year and the vacancies were not filled up.

The quartermaster was invalided in December, 1811, and until the battalion returned home it is impossible to say who acted as quartermaster. There is some reason to believe that Bamford did so. He was the only officer above the rank of ensign, except the commanding officer, who had no company to command.

In 1812 Lieutenants White, Rumley, Heaviside, Mayne, Eager, Baillie, Andrews and Garvey permanently commanded companies, and at Salamanca, Chambers and Hitchen being absent, wounded, Neville commanded the light and Brisac the grenadier company. White had joined in 1805, the others in 1808 or later. The senior subalterns were in India with the first battalion.

The captains on the staff in the Peninsula were Brevet-Major T. Williamson, D.A.G., first division; C. Machell, brigade major, fourth brigade; E. Malet, D.A.Q.M.G. with Spanish army; Henry Craig, D.A.A.G., fifth division; William Stewart, brigade major in the second division; Alexander Macnab, commandant of Figueras, the port at the mouth of the Mondego. Lieutenant Daniel was his quartermaster.

With regard to the non-commissioned officers and men, it is a little surprising, after the six years spent in Ireland by the battalion, to find that it was on landing in Portugal a distinctly English corps. In time, the drafts from the Irish militia altered the proportions. The casualties show that in the first eighteen months' service under Wellington the deaths occurred of 62 English, 19 Irish, 1 Scotch and 1 Welsh; but the Irish element was always increasing and, at Badajoz, of the 28 killed 14 were English, 13 Irish and 1 Scotch.

During its service in the Peninsula the establishment of the battalion varied, but as the actual strength never bore any relation to the establishment it is not necessary to trouble about it.

TORRES VEDRAS

On landing at Lisbon, the brigade from Cadiz, consisting of the 2nd/30th and 2nd/44th, marched under Colonel Minet to the camp at Sobral and joined the fifth division, commanded by Lieut.-General Leith, on October 10th. The 1st/4th King's Own, one of the Corunna and Walcheren regiments, was on its way from England to complete the brigade, which was numbered the fourth, and Major-General Dunlop had been appointed to command it but did not arrive for a few weeks.

We have no disembarkation return, but soon after the battalion reached Sobral it had 611 rank and file present and effective. There were also 52 sick in Lisbon and 31 in Cadiz, where Ensign Daniel had been left with the heavy baggage. Two drummers and 11 boys had been left at Lisbon, and as usual a number of servants and grooms had been lent to the staff, among others, one to Captain Hare, D.A.A.G., who had served in the 30th in Egypt. The officers said to be effective with the battalion were Major Hamilton, in command, Major Grey, Captains Bamford, Lynch, Fox, Hitchen, Chambers and Macnab, Lieutenants Garland, George Rumley, Adamson, Mayne, Andrews, Heaviside and Eades, Ensigns Garvey, Pennefather, Eager, Freear, John Rumley, Baillie, Crauford and Neville, Paymaster Wray, Adjutant Stewart, Quartermaster Kingsley, Surgeon Hennen and Assist.-Surgeon Newton. Of those we know from other sources that Captain Macnab, one lieutenant and one ensign were sick in Lisbon.

Owing to Sir Thomas Graham's recommendation, Captain Malet had been appointed D.A.Q.M.G. attached to the Spanish army commanded by Mendizabal.

The other troops in Leith's division consisted of those who had fought under him at Busaco in September, namely, the British brigade under Major-General Hay, containing the third battalion Royal Scots, first battalion 9th, and second battalion 38th; and Major-General Spry's Portuguese brigade, containing the 3rd and 15th line regiments, and the 8th Caçadores. Two companies of the Brunswick Oels Rifles, commonly called the "Owls," were soon added. As the Portuguese brigade contained a battalion of light infantry (Caçadores), Lieut.-General Leith had at his disposal the light companies of the six British and two Portuguese line battalions, two companies of Brunswick Riflemen and twelve companies of Caçadores. This gave him a great superiority in trained light infantry over any French division he was likely to meet.

The light companies and the riflemen attached to brigades always fell in as a battalion on the left of their brigades, under command of the senior captain or of a field officer specially selected to command them, and when opposing the enemy were used by the commander of the division or brigade as he thought fit. At other times, and for discipline, they were entirely under the control of the officer commanding their battalion.

All marching on a road in the Peninsula was done in sections of three to suit the breadth of the ordinary roads.

To the question why Leith's division and other troops were at Sobral, which is about 22 miles from Lisbon, the answer is that it was the very heart and centre of the great lines with which Wellington had covered Lisbon. From the extreme right at Alhandra on the Tagus the line ran north for some 9½ miles to M. Agraca. Part of this line was covered by inundations and it was all strongly fortified. From M. Agraca to the great fortified camp at S. Vincente above Torres Vedras, about 8 miles, there were only two redoubts. From Torres Vedras, north and west to where the river Zizambre falls into the Atlantic, the line was covered by inundations and strong works. The space of 8 miles between M. Agraca and Torres Vedras was therefore the critical place, and here was Sobral with Wellington's Headquarters 4 miles in rear of it at M. Negro, and above Headquarters was the central station of the semaphore telegraph system which was established all along the line. The village of Sobral itself was a little in advance of the true line and was only held by us as an observation post, but in rear of it Wellington had placed the six Anglo-Portuguese divisions of his field army, four divisions in line with a supporting division on either flank. Leith's division was in support of the right. The field army was about 60,000 strong, 42,000 British, of whom 7,000 were detached or sick, 24,000 Portuguese, of whom over 3,000 were sick, and two Spanish divisions which the Marquis de la Romana had brought to Wellington's assistance. The fortified lines were held by 20,000 Portuguese assisted by some British artillery and 2,000 of our marines. The river Tagus was patrolled by our navy.

On October 11th, the day after the 30th joined the fifth division, the French "Army of Portugal" under Marshal Masséna was within a day's march of the lines. On that day his cavalry informed the marshal that their way was barred by a vast line of works. It seems almost incredible that neither Masséna nor his master had ever heard of the preparation of this line.

The "Army of Portugal" consisted of the second corps under General Regnier, the 6th under Marshal Ney, the 8th under Marshal Junot and the cavalry under General Montbrun, and numbered 65,000 men. In quickness of manœuvre the French army was superior to Wellington's.

When retiring before Masséna in September, Lord Wellington had induced the Portuguese Government to order all inhabitants to conceal or destroy what they could not bring off and to retire with what they could drive or carry within the lines of Torres Vedras. This order, which was in a great measure obeyed, not only gave Lord Wellington an ample supply of labour for the construction of his defences, but produced a state of affairs with which a French army was singularly unfitted to cope. The revolutionary and Napoleonic system was to keep supplies for ten days or a fortnight in case of an emergency and on ordinary occasions to live on the country. In a rich district no great harm was done to discipline, but in a poor thinly-populated country the troops had to disperse widely to feed themselves and, control being difficult, the foraging parties soon degenerated into bands of brigands who were guilty of horrible crimes against the defenceless inhabitants.

On the night of October 12th Marshal Junot attacked Sobral and we gave up the village.

Simultaneously with Junot's advance General Regnier felt our line from Monte Agraca to the Tagus, and Masséna on the following morning made a close inspection of the position. From what he saw and from the reports of his corps commanders and the cavalry he recognized that the case was hopeless. He could not manœuvre Wellington out of the position, and a direct attack on the man and the army which had beaten him at Busaco, now that they were strongly reinforced and supported by heavy guns, would be madness. He had no magazines and he therefore determined to fall back on the district about Thomar and Santarem, whose inhabitants had not obeyed the orders of the Portuguese Government for all to bury what food could not be carried and to retire into the lines with their live stock. He could there hope to subsist his army and wait until orders from the Emperor should send Marshal Soult from Andalusia or General Dorsenne with the Army of the North to support him. The cavalry reserves and everything which could be spared were at once sent to the rear and on the 15th the French infantry divisions fell back, and Wellington followed with all his troops except the divisions of Leith, Picton and Cole.

Three days later Wellington called up the divisions of Leith and Cole. Leith's division got the order on the 18th and, leaving its tents standing at Sobral, started on the same day and joined Wellington on the 20th at Cartaxo.

Wellington now discovered that Masséna, finding himself able to subsist his troops on the country, had determined to make a stand where he was. The weather had been very bad, and during the last week the rain had been incessant, the rivers were in flood, and the roads completely destroyed; his Lordship decided therefore not to attack, but to get his men under cover, and wait till famine should once more force the French to retreat. Leith's division was billeted at Alcoentre, and there the first battalion 4th Foot joined Dunlop's brigade.

The state of November 25th shows the effect of a winter campaign in Portugal. In the month, Major Grey, a lieutenant, 9 sergeants, 4 drummers, 99 rank and file had been added to the sick with the battalion, while the sick at Cadiz had been reduced by 10. Although there is no mention in despatches of the regiment having been engaged while at Alcoentre, yet on November 6th Sergeant Charles Watson was killed in action there, and the sick may include a few wounded.

In December, 1810, Wellington exchanged Leith's division for Picton's and the former returned to the lines and was stationed about Torres Vedras.

The returns of December 25th were signed at Torres Vedras by Major Hamilton. The sickness caused by the bad weather when the regiment was on the march had now reached its height, and from this time began to abate. The state shows—Sick at Lisbon, 9 officers and 160 other ranks. Sick present, 1 officer and 31 other ranks.

Allowing for men on command, etc., there were left 20 sergeants, 20 drummers, 512 rank and file, present, fit for duty.

James Poyntz, the son of Arthur Poyntz, quartermaster of the first battalion, was at this time serving with the battalion as a volunteer. He was under sixteen years of age.

By February 25th the sick list had fallen to 2 officers and 98 other ranks at Lisbon, 36 present and 9 at Cadiz.

On February 18th the Spanish army under Mendizabal was routed in the neighbourhood of Badajoz by the French who had invaded Estremadura under

Marshal Soult. Captain Malet, 30th regiment, D.A.Q.M.G., reported to Lord Wellington that it had lost all an army had to lose, and that what remained of it was huddled under the walls of Badajoz.

Wellington now began to close up his rear divisions with the intention of attacking at Santarem.

The following shows the strength of the battalion when it took the field :—

Majors Hamilton and Grey; Captains Bamford, Lynch, Fox, Hitchen, and Chambers; Lieutenants Garland, Mayne, Andrews, Eades, Heaviside, and Garvey; Ensigns Pennefather, Eager, Freear, John Rumley, Baillie, and Neville; Paymaster Wray; Adjutant Stewart; Quartermaster Kingsley; Surgeon John Hennen; Sergt.-Major James Wood; 610 other ranks.

Colonel Minet was president of a court martial in Lisbon; Captain Macnab on duty in Lisbon and Lieutenant Adamson at Belem; Lieutenants George Rumley and Brisac sick at Lisbon; Lieutenant Daniel in charge of heavy baggage at Cadiz; Assist.-Surgeon Brett sick in Ireland; Quartermaster-Sergeant Williamson on duty in Lisbon. He never joined the battalion in the field, but remained at the base to forward supplies to the front. Lieutenant Daniel on arrival from Cadiz was appointed acting quartermaster to Captain Macnab, who was now commandant at Figueras. Neither of them joined the battalion in the field.

MASSÉNA'S RETREAT

On March 5th, a day or two before Wellington was ready to attack, Masséna retreated.

In the past month he had lost 10,000 men from starvation, want of shelter and other causes, and he had exhausted the country which his foraging parties could reach. The allied army when it started in pursuit found terrible evidence of the horrors which attended the French system of licensed plunder. The minds of Masséna's men were embittered by their own sufferings, and by the shame of having to retreat, a thing to which a French army had been long unaccustomed, and they were encouraged to revenge themselves on the inhabitants by Masséna himself, who wished to break the spirit of the Portuguese. As a consequence, frightful atrocities were committed, corpses of the peasantry, male and female, were found along the road or in the houses and, besides those, the mutilated bodies of Frenchmen, for whenever a man fell out sick or weary he was pounced upon by parties of Portuguese and he had no short or easy death. Our advanced parties would drive off groups of inhabitants clustered round some wretched Frenchman who would beg to be killed outright, but when our men had passed, the Portuguese would return to resume their horrible work. The road was encumbered too, not only by dead transport animals, but by those still living which had been cruelly hamstrung and left to die. During the pursuit the division was sheltered as well as possible in the ruined villages and towns, which contained perhaps a tenth of their former inhabitants.

After a little uncertainty it was ascertained that the French were retreating along the paved way which led north, or nearly so, to Coimbra, that is, Masséna, abandoning the valley of the Tagus, was marching for the Mondego. Coimbra is only 20 miles from the sea and was already in our hands, but weakly held. If Wellington could push Masséna past that point to the north-east a new base for us could be established there, all supplies being brought by water up the Mondego. The road from the Tagus to the Mondego on which the armies were marching, although honoured with the name of a paved route, has been described by an English cavalry officer as three feet of mud with here and there a large stone, so the importance of a better route to Coimbra from the sea can be readily appreciated.

The pursuit of Masséna was entrusted to the light and third divisions. They were supported by the first and sixth, and Leith's division was last of all. Whenever the French rear-guard made a stand, orders were sent to our rear divisions to close up rapidly, but Ney, who commanded the French rear, knew his business too well to wait till Wellington could attack with his whole force, and whatever exertions the men of the fifth might make they were never in time to come into action.

On March 12th at Redinha the French held on to their position until the fifth division came into line, but then finding his flanks turned by the light and third divisions, Ney fell back on Condeixa. Here Masséna decided that he could not hold Coimbra but must continue the retreat by the road running north-east by Celorico to Almeida and Rodrigo. On the morning of the 13th the French were on the march for Celorico, leaving as usual the sixth corps under Ney to cover the retreat.

The pursuit continued as before—the light and third divisions leading and the corps in rear endeavouring, in spite of defiles and bad roads, to keep within supporting distance.

On the 16th, Masséna had continued his retreat along the left bank of the Mondego, which receives a number of tributary torrents from the hills to the south. After crossing the Alva, one of these tributaries, by the bridge of Murcella, he halted in a strong position with the torrent and its precipitous valley covering his front. Wellington at once sent the first, third and fifth divisions into the hills to his right to cross the stream nearer its source. This manœuvre caused the French to fall back in haste. On the night of the 18th, Masséna attempted to shake off his pursuers by a march of 20 miles, but on the 19th the British cavalry were again in touch with him and picked up about 600 stragglers. On this day Leith's division crossed the Alva and came in from the mountains on the right on to the main road. On the 20th Wellington made over what supplies were left to the light, third and sixth divisions, which were to continue the pursuit whilst the first and fifth divisions and two Portuguese brigades halted till a fresh supply column should come up. The 30th were quartered in a place called Venda do Valle in regimental returns, somewhere in the neighbourhood of Moita on the main road. The incessant marching and bad weather had increased the number of sick among the rank and file; there were now 157 at Lisbon and 13 at Coimbra, an increase of 44 in the month. The number of sergeants on the sick list had fallen from 7 to 3, and the same two officers as in February, Lieutenants George Rumley and G. A. W. Brisac, were sick at Lisbon.

On March 22nd, Masséna issued orders for a movement to the south on Guarda with a view to marching through the mountains to the valley of the Tagus, getting into communication with Soult in Estremadura and by again threatening Portugal to force Wellington to retire for the protection of that country. His three corps commanders protested strongly that the army was so weakened materially and morally as to be quite unfit for such an operation, and on the order being repeated Ney flatly refused to obey. Masséna appointed Loison to command the sixth corps and sent Ney to France. Junot and Regnier submitted.

On the 26th the eighth corps was at Belmonte, where the ground begins to fall towards the Tagus, which is 60 miles to the south, but Junot had been obliged to leave his artillery at Guarda and his men were starving. Regnier, on his left, with the second corps was in an equally bad plight near Penmacor, the sixth corps and Masséna himself were at Guarda, which is 3,400 feet above sea level. All the troops were high up in the mountains and the ill-fed and ill-clad men suffered severely from cold. On the 28th Masséna gave way and ordered an immediate retreat upon Almeida and Rodrigo.

On the 29th our advanced troops reached Guarda, just as the sixth French corps was starting on the retreat to Almeida. At the first appearance of the British Loison made off, marching as fast as possible for the Coa.

The road from Coimbra to Celorico, owing to the numerous streams running down from the Estrella on the right, had been one of constant ups and downs like a switch-back. From Celorico to Guarda is one long precipitous climb, and from Guarda there is an abrupt descent towards the Coa.

By the 31st, the three French corps, somewhat aided by luck, managed to concentrate on the upper Coa near Sabugal.

Junot's corps, the eighth, which had suffered most, fell back 10 miles behind Regnier who, along with Masséna, held the line of the river for 20 miles. The Coa, rising in the mountains to the south, runs in a fairly straight course past Sabugal to cross the Chaussée from Coimbra to Almeida at the latter town. By April 2nd Wellington had all his troops in hand, including the newly-arrived seventh division, and determined to attack Regnier on the following morning.

The fifth division, which formed the left of the British line, was to attack at Sabugal, crossing by the bridge there. The third division was to cross by fords somewhat higher up, and the light division, making a wide sweep, was to turn the enemy's left. The other divisions were to support the third and fifth. Unluckily, on the morning of the 3rd there was a heavy fog. Picton and Dunlop, who was temporarily in command of the fifth, agreed to delay their advance, but the light division was pushed forward by its commander and but for the admirable conduct of its officers and men might have been destroyed. In the endeavour to overwhelm the light division, Regnier, who was quite ignorant that he had to deal with the whole of Wellington's army, had weakened his right and centre, when suddenly the fog lifted and the third and fifth divisions were seen advancing to the attack. Regnier realized at once that there was no time to be lost; he ordered the brigade in Sabugal to retire in double time to take up a position a mile to the rear, another brigade was ordered to make a front against the light division, and the remainder of the troops were to withdraw with all speed under cover of these two brigades.

Picton's division, after crossing the river, struck the right flank of the French brigade opposed to the light division and routed it; the fifth division occupied Sabugal without losing a man. The manner of it was thus: The French brigade in Sabugal, ignorant of the fact that the head of Dunlop's column was on the other side of the river and only prevented by the fog from crossing, had spent a peaceful morning in cooking their dinners, which had not been a thing of everyday occurrence lately: their interest in this occupation seems to have interfered with their watchfulness, and when the fog lifted the light company of the 44th ran across the bridge at Sabugal, the 30th and the other light companies splashed through the river where they could, and at the same time the order came from Regnier for the French to fall in and double to the rear. There was no help for it; they had to be off and leave their dinners to be eaten by the light companies.

Wellington suspended the pursuit owing to the torrents of rain which

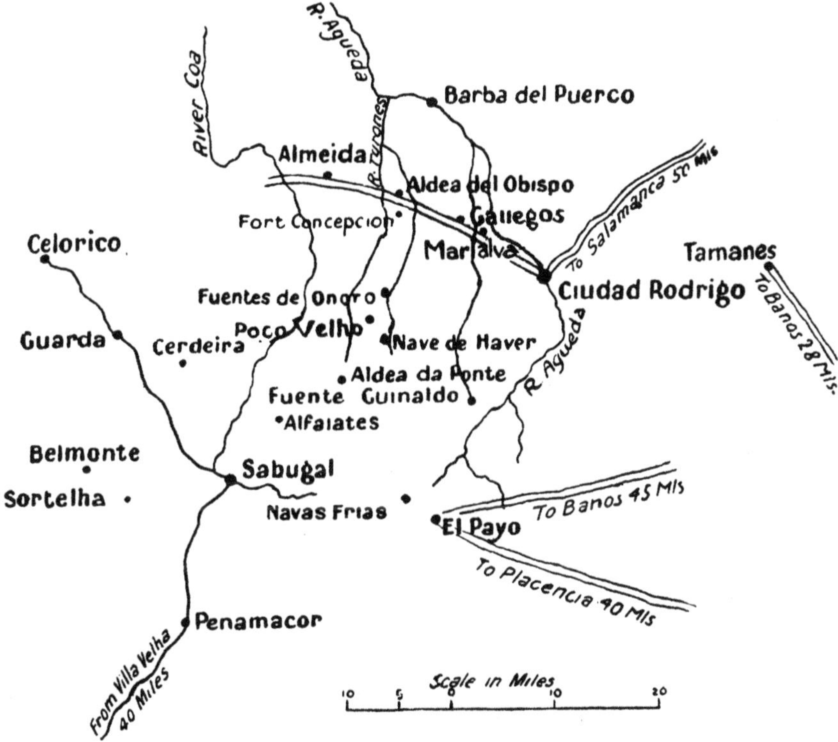

SPANISH AND PORTUGUESE FRONTIER—CIUDAD RODRIGO.

followed the fog, and quartered the army in Sabugal and the neighbouring villages.

On the 4th Masséna fell back 30 miles to the north-east, occupying Fuentes Onoro and other points between Rodrigo and Almeida. Numbers of prisoners were picked up by the British cavalry and Wellington, judging that Masséna must retreat behind the Agueda, resolved to blockade Almeida, for he had no means of besieging it. On April 8th Masséna crossed the Agueda in retreat, and on the following day the 30th and the other regiments of the fifth division occupied Aldea do Obispo, on the main road between Almeida and Rodrigo and close to Fort Concepcion, which the French had blown up. Roughly speaking, the quarters of the division were about 7 miles from Almeida and 16 from Rodrigo. The regiment did not again move till the beginning of May.

The French by April 11th had reached Salamanca. The "Army of Portugal" had lost about 25,000 men since it started in September 1810 to drive the English into the sea, and its artillery, cavalry and transport were mere wrecks.

On April 15th Wellington left his army under command of Sir Brent Spencer (the Colonel Spencer of the Green Hill, Alexandria, 1801), and rode for Estremadura to confer with Beresford as to the recovery of Badajoz. He had not expected Masséna to be able to refit his army and advance again for three weeks, but by May 1st the French army was ready for the field. The infantry was about 42,000 strong and was the most efficient part of it, the 3,000 cavalry were poorly horsed and only twenty guns could be dragged with the army.

The deficiencies of Masséna's army were, however, made good from the army of the north, commanded by Marshal Bessières, who joined and brought 1,700 good cavalry and several gun teams to assist his brother marshal. This gave Masséna a large superiority in cavalry over Wellington and nearly an equality in the number of guns.

Wellington had returned from Estremadura on April 29th.

FUENTES ONORO

The name of the village which gave its name to the battle commencing on May 3rd is variously spelt by English writers. The above is the form used by Sir William Napier. Other forms are Fuentes de Onoro, Fuentes d'Onor and variations upon these. Some assert the true name to be Fuentes de Noria, or the Fountain of the Water-wheel.

Lord Wellington had selected a position to resist an attack from Rodrigo with his left on the ruined Fort Concepcion and covered by the deep ravine of the small river Dos Casas. The fifth division held this part of the line. General Robert Craufurd had returned to command the light division and, General Leith being still on leave, the command fell by seniority to Sir William Erskine. To the right of the fifth division, and higher up the Dos Casas, was the sixth division (Campbell), the first (Spencer), and the third (Picton). The seventh division (Houston) was still higher up the stream at Fuentes Onoro. The light division was at Gallegos in front of the fifth division with pickets at Marialva.

The ravine of the Dos Casas opens out as one ascends it until when Fuentes Onoro is reached the stream is only a slight obstacle.

Wellington's extreme right was covered by the guerillas of Julian Sanchez, posted behind a marsh at Nava de Aver. The position from Fuentes to Concepcion was about 5 miles in length.

On May 2nd the French army issued from Rodrigo, our light division and cavalry retired before it and ultimately formed up behind Gallegos, still covering our army and in touch with the enemy.

On May 3rd Masséna advanced and formed up for battle, the light division and cavalry passing through our army and posting themselves in reserve behind Fuentes Onoro.

Masséna's plan was to make a demonstration with the second corps to hold our fifth division at Fort Concepcion. One division of the eighth corps was to play the same part in the centre against our sixth division, while the bulk of the French army made a direct attack upon Fuentes Onoro. After a sharp struggle this attack was defeated with a loss to the French of 650 men and to us of a little more than one-third of that number. Where the 30th was posted it was soon evident that the French had no real intention of forcing their way across the ravine and the light companies only of the fifth division were sent forward to skirmish.

On the following day both armies held their ground and Masséna, who had not employed his full strength on the 3rd but had suspended the attack when he recognized the strength of the position at Fuentes, made a thorough examination of the ground, with the result that he determined to drive off the guerillas at Nava de Aver, cross the marsh, which was not as impassable as at first reported, and thus turn Wellington's right, while three divisions were to attack Fuentes Onoro. Regnier's corps alone was to hold our fifth and sixth divisions in check and to attack our left if it seemed possible.

Wellington, whose cavalry had given him word of some movement of the French to their left, had placed the seventh division under Houston in Poco Velho and the neighbouring woods to cover his right. It was perhaps too far from support.

The French left attack met with great success. At daybreak their cavalry drove off the guerillas and forced our squadrons back on Poco Velho. Their infantry then came up and stormed Poco Velho, very nearly cut off the seventh division from the rest of the army and threatened Wellington's right flank at Fuentes Onoro, which had meanwhile been attacked by 14,000 men.

Wellington met the French movement by throwing back Spencer's and Picton's divisions and Ashworth's Portuguese on his right at right angles to his original line and to cover the change of front, and to support the seventh division in its retreat to the main body, sent Craufurd's light division down the slopes towards Poco Velho.

These arrangements were successful; the seventh division continued its retreat till in line with Picton's and Spencer's divisions, where it faced about, occupying the heights beyond the Turon. The light division, in actual touch with the enemy and with, at times, its battalion squares surrounded by his cavalry, fell back steadily upon our first division, and passing through it, took up a position in reserve.

The attack on Fuentes had been pushed home according to Masséna's orders and the contest had been fed from time to time from the supports on both sides. About noon the last of the French reserves was thrown in and the village actually carried. The British line seemed for a moment to be broken, but Wellington brought forward the second line of Picton's division which had not yet been in action, and the battle was won.

The French losses amounted to 2,200 and the British to 1,450.

The fifth and sixth divisions can scarcely be held to have been engaged. Their light companies and Caçadores only were employed. The sixth division had only 4 casualties: the seven light companies of the fifth had 27, and of those the 30th had 2 sergeants and 2 rank and file wounded. It was probably in this action that Major Hamilton was wounded. There is no official mention of it, but the friend who wrote the account of Hamilton's life in the *United Service Journal* shortly after his death states that he was wounded in the Peninsula as a major. No date is given, but the author says that among the letters which he had the privilege of reading were some from a general officer to Hamilton with inquiries about his wound. Hamilton was not the man to allow the surgeon to include him among the wounded as long as he was able to serve.

The following was the strength of the battalion at the battle of Fuentes Onoro:—

Majors Hamilton and Grey, Captains Bamford, Lynch, Fox, Hitchen and Stewart; Lieutenants Garland (acting adjutant), Mayne, Andrews, Eades, Heaviside and Garvey; Ensigns Pennefather, Eager, John Rumley and Brisac; Paymaster H. Wray, Quartermaster J. Kingsley, Surgeon Hennen, Sergt.-Major James Wood, 32 sergeants, 14 drummers, 414 rank and file.

Mr. James Poyntz, volunteer, was among the rank and file. No officers were on the sick list, but Lieutenant George Rumley had died in Lisbon.

The 30th does not bear Fuentes Onoro on the colours. That honour was bestowed only on those divisions which were engaged in a combat of musketry.

On the morning of the 6th both armies held the same ground as overnight. There was no movement on either side during the 7th, but at daybreak on the 8th it was discovered that the French army had retreated, covered by Regnier's corps, which stood its ground. When the remainder of the army had crossed the bridge at Rodrigo, Regnier marched to Barba del Puerco and crossed the Agueda there.

The force blockading Almeida had been strengthened after the battle by the sixth division and it was evident that with ordinary care on our part the garrison must surrender shortly. There was a chance that it might, after destroying what it could not carry, break out to the north and try to join the French army by the bridge of Barba del Puerco. Wellington, however, had provided for any such attempt. Sir William Erskine had been ordered to extend the line of the fifth division towards the Agueda, and in particular to occupy the bridge of Barba del Puerco with the 4th King's Own. Erskine had this order by four o'clock of the afternoon of the 10th; what became of it is hard to say, but Colonel Bevan, commanding the 4th, maintained that he only got the order close upon midnight and, the night being dark, delayed his march till daybreak.

At 11.30 p.m. on the 10th General Brenier issued from the north gate of Almeida and marching rapidly in two columns broke through the line of sentries. The nearest British regiment of the sixth division got under arms, but waited for orders. General Pack, with a few Portuguese, followed the French, firing at them and capturing baggage and stragglers, but Brenier hurried on without firing a shot. A troop of the Royal Dragoons delayed him for a time and General Campbell tried to catch him with the 36th regiment. The light companies of the 30th and the other regiments of the fifth division also took up the chase, but they all, along with the 4th King's Own, only arrived in time to see the French dive into the defile of the Agueda. The British, throwing off their knapsacks and pursuing at a run, opened fire and even charged the rear of the French column, taking prisoners and driving some of the enemy over the precipitous banks of the river, but Regnier had posted three battalions of infantry with guns on the opposite bank, and under cover of their fire three-fourths of Brenier's men escaped across the bridge of Barba del Puerco.

Wellington's wrath at the escape of the garrison was great, but whether punishment fell upon the right people was much discussed in the army.

After the victory of Fuentes Onoro, Wellington turned his attention to the south where Beresford with the second and fourth divisions had commenced the siege of Badajoz. Taking with him the third and seventh divisions and leaving the remainder of the army under Sir Brent Spencer to guard the line of the Agueda, he marched for Estremadura on May 14th.

The 30th, with the remainder of the fifth division, changed quarters a few miles to Nava de Aver. Here it remained till June, either quartered in the villages or hutted, as was more usual in summer. The huts were simply shelters of boughs of trees thrown together by the men themselves. By the beginning of June Marshal Marmont, who had succeeded Masséna, had reorganized the Army of Portugal and led it back to Rodrigo. On the 5th he advanced across the Agueda and Sir Brent Spencer fell back towards Sabugal. On the 7th it became apparent that Marmont's main body was no longer in front of Spencer, but that, passing his right, it was in full march to Estremadura to co-operate with Soult in raising the siege of Badajoz. Spencer therefore marched for the bridge over the Tagus at Villa Velha to get in touch with Wellington before Marmont and Soult could unite. The fifth division marched to Castel Branco, where it halted with Slade's cavalry to cover the other divisions while crossing the river. On the 18th and 19th it crossed at Villa Velha and marched to the Caya where it hutted itself near the town of Arouches, on June 23rd.

As Wellington had given up the siege of Badajoz on the 10th and fallen back to meet Spencer, the whole army was united on the Caya.

The front was 12 miles in length from Elvas to Ougula on the Gebora, a tributary of the Guadiana, and was occupied by 54,000 men. The French marshals had 60,000 men with them, but although their cavalry were active in reconnoitring they showed no disposition to make a serious attack. On June 28th, Soult was recalled to Andalusia by the activity of the Spanish armies and guerillas who had been set in motion by Wellington. He left with Marmont the fifth corps and some cavalry, making that marshal's force nearly 50,000 men. Marmont now devoted all his energy to sweeping the country for supplies, both for his own army and to reprovision Badajoz. After having placed in the fortress six months' provisions he found it hard to maintain himself in the exhausted country and fell back to Merida on July 15th.

On July 2nd, there being no longer any need of keeping his army so closely concentrated and the low ground along the river being unhealthy, Lord Wellington had sent the fifth division two marches to the north to Portalegre, where it was quartered for a fortnight in the town and afterwards hutted outside.

During this time there was a melancholy sequel to the escape of the garrison of Almeida in May. Colonel Bevan, of the 4th King's Own, brooding over the fact that he was, as he thought, unjustly blamed for that escape, took his own life. He was buried in the castle yard of Portalegre on July 8th, and every officer of the division made a point of attending his funeral.

On the 20th the division marched to Sierra d'Arouches, where it constructed huts. This place was called by the officers Vauxhall, from a fancied resemblance of the woodland scenery to the artificial glades of the celebrated place of amusement on the Surrey side of the Thames near to where the Oval cricket ground now is. The music of the bands of the different regiments helped the illusion.

The regimental band was now an important feature, and not only played on the march, but in the evening in camp or bivouac when the day's work had not been too hard.

On the 25th the regiment was at Quinta de Almeria. In the monthly return, signed there by Hamilton as major and lieut.-colonel, Lieutenant Garland appears for the first time as adjutant.

No officers were now sick, but 11 sergeants, 6 drummers, and 190 rank and file were in hospital or convalescent, leaving 28 sergeants, 14 drummers and 394 rank and file fit for duty.

Towards the end of July, Lord Wellington, leaving 11,000 men near Badajoz, began to move the remainder of his army back to the country round Ciudad Rodrigo. The siege train he had so long desired had arrived at Oporto and he was determined that if a chance of success offered he would take Rodrigo; the guns and ammunition were to be brought in barges up the Douro and then hauled by bullocks to Almeida.

On the 29th the fifth division marched to Castello de Vide where it was quartered in the town. The road from the Portalegre district had been very hilly and rocky and a day's halt was given.

On the 31st the division marched to Nyssa. After bivouacking in a wood it descended the steep hill to the Tagus on August 1st; the road winds through rocks and precipices, and is often cut out of the side of a hill, and the march of the three brigades and the Portuguese battery attached, afforded a most picturesque sight.

After crossing by the bridge of boats at Villa Velha, the bivouac was formed on the bank of the river.

On the following day the ascent of the mountains of Beira was begun in intense heat and the division bivouacked at Larnades.

On August 3rd Castel Branco was reached; it was seen on its hill top many hours before the wearied troops arrived there. A short march was made next day to Atalaya, woods of oak, chestnut and cork were passed, and the broom and heather were in flower. Quantities of game, including partridges, quails, and hares, were seen. Many officers had their guns and pointers with them; some even a brace of greyhounds, and in a good game country there was an opportunity of a change of diet from the eternal over-driven bullock. A mail from England *only seven weeks* old put everyone in good humour. On the 7th a very hard march of 5 leagues was made and many horses, mules and asses died of fatigue. The mountain road was lined with the offensive corpses of the transport animals of a division in front. Expectations of a battle were raised by the number of inhabitants who were met with, flying from the French. On the 8th the division reached Sabugal and bivouacked; the town had been completely wrecked by the enemy.

On the 11th the division passed the frontier into Spain and reached Navas Frias, a small but clean town at the source of the Agueda. Navas Frias means "cold snows," an appropriate name for one of the highest inhabited spots in the country. Besides the river Agueda, the Coa and the Zezere rise in the district. Ciudad Rodrigo is about 25 miles down the Agueda to the north-east, and Lord Wellington had established his Headquarters at Fuentes Guinaldo, a third of the way to Rodrigo.

The fifth division remained for some weeks in this district watching the passes through the Sierra de Gata, lest Marmont should attempt to advance by them from the Tagus to break the blockade of Ciudad Rodrigo, which had now been formed by the cavalry and the third and light divisions.

The light companies of the fifth division were out in front at St. Payo, 7 miles to the east, and Valverde, a dozen miles to the south of Navas Frias. A battalion, and sometimes two, were hutted in support of the light companies and the rest of the division was kept under cover in Navas Frias and Pena Perda, for although the weather was warm during the day, the cold was intense at night. For the same reason the outposts and supports were relieved weekly and brought into the towns. Small parties of hussars of the King's German Legion were in advance of the light companies. A battery of British artillery was given to the division at this time, in addition to the Portuguese battery.

As the 8th Caçadores and Brunswick Rifle companies were still with the division the relief of the light companies was easy. The Light Bobs always prided themselves on their speed, and on the cold mornings, when proceeding to relieve their comrades in front of Valverde, they usually ran the whole distance, doing it in two hours or about 6 miles an hour. They may have been allowed to have their packs and blankets carried by the regimental mules, which had not much to do, but they certainly carried sixty rounds of ball cartridge, the canteen and haversack, and probably they carried the full kit.

On August 16th there was an alarm that the French were advancing in force through the pass on Payo, but after surprising a post of hussars, who lost twenty men, they fell back. The light companies who had pushed to the front to support the hussars almost caught the French at the village of S. Martino, but they got off. They had wantonly fired on the people coming out of church and killed several, and one light infantry officer, new to the Peninsula, was horrified on entering a house to find the bodies of an old man of eighty and another of fifty, and of a beautiful girl of sixteen all freshly murdered.

When there was not an alarm of a French advance the time hung heavily on the hands of all. One resource was to try to kill or trap the wolves which prowled about the bivouacs and often entered at night.

In September the promotion without purchase of Major and Lieut.-Colonel Hamilton to be Lieut.-Colonel of the Royal West India Rangers *vice* Lieut.-Colonel Charles Turner, appointed lieut.-colonel in the 30th *vice* Minet from July 25th, appeared in orders and Hamilton left for England, where he was granted four months' leave. Major and Brevet Lieut.-Colonel George Grey took command of the 30th. The circumstances under which Hamilton left and returned are rather confusing. Lieut.-Colonel Turner had distinguished himself at the first siege of Badajoz in command of a Portuguese regiment but had been very severely wounded on May 11th. The War Office gave him command of the 30th and promoted Hamilton to command the Royal West India Rangers, but when it was found that his wounds incapacitated Turner for command both appointments were cancelled and Hamilton was given the 30th from July 25th. Turner was awarded £300 for wounds received at Badajoz and is shown in the pension list as belonging to the 30th, which has led to an erroneous belief that the regiment was engaged in the first siege of that fortress.

During this month the weather broke with heavy thunderstorms followed by cold rains and wind, against which huts were no protection. From time to time there was an alarm of a French advance and at last they came in reality, but not from the quarter expected. On September 26th heavy firing was heard to the north-east, and word was passed that the Commander-in-Chief was attacked by superior forces at Guinaldo. General Dunlop at once concentrated the division, which remained under arms all day listening to the firing. At 4 a.m. on the 27th, it marched to join Lord Wellington.

The events which led up to this movement were as follow:—Marshal Marmont in order to save Ciudad Rodrigo determined to unite with the army of the north under General Dorsenne. With this object he marched to his right by Banos, but left General Foy's division at Plasencia with orders to mask the movement by threatening an advance against the fifth division through the passes of the Sierra de Gata. On September 22nd Marmont was at Tamames, to the east of Ciudad Rodrigo, and was joined there by Dorsenne with a large train of supplies for Rodrigo. The combined force was about 60,000 men. Wellington's army only amounted to 48,000 men and they were widely scattered. He determined, however, to hold his ground, believing that the French would attempt no more than the revictualling of Ciudad Rodrigo.

After some skirmishing and a smart combat at El Bodon he realized that if Marmont attacked in earnest the allied army was in great danger, and on the night of the 26th, leaving his fires burning to deceive the enemy, he fell back on Alfayates. On the following day he was joined by the fifth division from the mountains on his right and the fourth, first and sixth from his left.

An officer who was present gives the following account of the movements of the fifth division to join Wellington and in the retreat across the Coa. It should be mentioned that September 28th was the decisive day. Lord Wellington then offered battle, and Marmont in spite of his superiority in numbers declined it. Wellington then continued his march to the Coa and Marmont after victualling Ciudad Rodrigo retreated.

The approximate distances are, Navas Frias to Aldea da Ponte 12 miles; from there to Sabugal 15, and from Sabugal to Guarda 18.

"The division halted to cook at Aldea da Ponte, but when the kettles were on the fires the alarm was given that the French were approaching in force on the other side of the town. This was confirmed by a fire of guns and musketry, and beef and soup were thrown in every direction. A division on our left took off the attention of the enemy and General Dunlop moved to a better position on a hill to our left, where we formed in a close column to support the division in action. Picton's division was likewise in reserve further to the left. We had a bird's-eye view of the action. Our men, though fatigued and hungry, were anxious to engage, hammering their flints and making their usual preparations. On the 28th the division marched towards Sabugal. As bad a march as was ever undertaken, extremely dark and the road broken, craggy and rocky, a perfect valley of stones. The inhabitants on our retreat flying in every direction to the woods and mountains carrying their miserable shreds—a very heavy rain, during which we halted three hours and afterwards slept in a wood. Our baggage had gone to the rear on the 25th, so there were no great coats or blankets.

"On the 29th marched through Sabugal to Villa de Toro. At Sabugal there were two other divisions and we heard our baggage had gone on to Guarda. We had not washed or shaved since the 24th and looked like Portuguese. Chestnuts ripe and good all along the road. On the 30th to Guarda."

Lord Wellington kept his troops on the Coa, leaving the guerillas to watch Ciudad Rodrigo and harass the garrison. The fifth division remained quartered in Guarda, where, in spite of the cold, they seem to have enjoyed life. As the diarist already quoted says, "there was an equal share of duty and merriment." There were officers' guards at each of the five gates of the town and divisional field days twice a week.

Major-General Walker was appointed to command the brigade on October 3rd.

The weather got more severe as time went on and by November 22nd the surrounding hills were white with snow.

On the 24th a movement of the enemy caused the division to advance; one brigade across the Coa and that in which the 30th was to Villa de Toro. It remained there crowded into houses and barns for five days, during which a corporal of the 4th King's Own on bullock guard was frozen to death.

Wellington was now certain that no further attempt was to be expected on the part of the French, and finding great difficulty in feeding his troops, who were suffering severely in the mountains, he broke up his army and sent it by divisions into winter quarters in a richer country.

The fifth division marched back to Guarda and from there down the Celorico road to the Valley of the Mondego. On December 2nd, after an abrupt descent surrounded with precipices, it saw the river foaming through rocky gorges. Although Guarda had been left in the depth of winter, in the valley it was warm as midsummer and all agreed it was like a change to Paradise. The road did not lead down the bank of the river, however, and there were two stiff marches over the spurs of the Sierra Estrella by the switch-back road on which it had advanced in March, before the division reached the head of the navigable water. Here it was quartered by regiments in small towns in the Coimbra district. The 30th were in a place called Medais, which is not in English maps. The weather was cold but clear. Here the battalion rested for nearly a month practising battalion drill, with an occasional field day. The district was not so good for hares, but the little valleys running down from the Sierra Estrella were full of woodcocks.

A surprising thing about this winter campaign is that the commissariat is well spoken of, so it may be concluded that the officer in charge of the divisional supply was an unusually able man.

The prices for everything beyond commissariat supplies were excessive. A dozen sheets of notepaper and a couple of pens cost 4*s.*; cheese, bacon and butter 3*s.* to 3*s.* 6*d.* a pound. The Portuguese did not make butter, using olive oil instead, and the butter was English or Irish. Tea cost 15*s.* a pound. The country people had only heard of tea as a medicine, and when they saw English officers drinking it they thought it was really hot rum and water and not tea. Oranges, olives and chestnuts were cheap and abundant.

Since the return from Estremadura to Beira considerable changes had taken place in the battalion.

Besides Lieut.-Colonel Hamilton, Captains Lynch and Fox had left the Peninsula on transfer to the first battalion.

Surgeon John Hennen, who had been sent down country sick, was transferred to the staff. He ultimately rose to be an inspector-general of hospitals, and died of fever in Gibraltar in 1829. Lieutenant Adamson, who had rejoined from sick leave in July, but who had never been in good health since the first month after landing, was sent down sick in November; he died at Lisbon in the following January.

Lieutenant Michael Archer Eades was sent back to Nyssa with sick in September and did not rejoin. Quartermaster John Foster Kingsley was sent sick to Lisbon in December; from there he was invalided to England. Lieutenant James Eager was sent down to Lisbon to receive the new clothing for the battalion, which was urgently required. Lieutenant William Pennefather, who went sick to Lisbon in September, was invalided to England in December. He did not rejoin in the Peninsula.

Volunteer James Poyntz had quitted the battalion in June, and his connection with the regiment ceased until he was appointed ensign in 1814 from the Royal Military College, Sandhurst.

On September 26th, the day on which Lord Wellington was attacked at Fuentes Guinaldo and the fifth division had concentrated to move to his support, a draft joined from the depot at Wakefield consisting of:—Lieutenant White; Ensigns Smith, Campbell, Carter and Lockwood; 2 sergeants, 1 drummer, and 45 rank and file which brought the effective rank and file up to 373. In the 45 rank and file there were 7 old soldiers and 38 recruits, mostly volunteers from Irish militia.

In the same month Assist.-Surgeon Evans joined for the first time. He served with the regiment till the day of his death many years afterwards.

Immediately after despatching Lieutenant White's draft to the second battalion, Major Spawforth marched the regimental depot to Hull, where it became part of the consolidated depot formed of the depot companies of the 10th, 19th, 30th, 34th, 47th, 65th, and 101st regiments. Assistant-Surgeon Hughes of the 28th regiment, who had been appointed surgeon in place of Hennen, joined at Medais and Mr. Francis Tincombe joined as a volunteer.

While the battalion was in winter quarters Sergeant-Major James Wood was lent as instructor to the Portuguese army.

The sickness in the battalion in the latter part of 1811 had been great—the number of effective rank and file in May had been 414; in September it had fallen to 328, when Lieutenant White's draft brought it up again to 373. At the end of the year the effective strength was 363 rank and file, with 228 sick.

The causes were, first the hot march in June from the Coa to Estremadura followed by some weeks in the feverish valley of the Caya; then the abrupt return to the heights about the source of the Agueda, when the cold weather was coming on with rain and snow.

In this year a man rejoined from prisoner of war in a curious way. In a skirmish in Estremadura a French column was broken by our cavalry and among the prisoners were many English. One was a private in the 30th, the others were sailors. Their story was that they had enlisted in an Irish regiment in the French service with a view to making their escape, but that no opportunity had presented itself till they found themselves in presence of our cavalry, and that they had never fought against their countrymen. Their excuse was accepted.

1812

In the beginning of January Lord Wellington saw that there was an opportunity of snatching Ciudad Rodrigo before the French leaders could combine to protect the fortress.

The artillery and ammunition had been quietly collected at Almeida and its neighbourhood as if for the purpose of re-arming that place.

On January 8th the siege of Ciudad Rodrigo was begun by the light, first, third and fourth divisions with Pack's Portuguese.

In the first week of January Wellington called up the fifth division and drew upon it for working parties, the officers being at the same time invited to volunteer to serve as engineers. Lord Wellington's orders are lost, but Lieutenants A. Baillie, P. P. Neville, F. L. White, Ensign Robert Smith and Privates John Bonsor, Anthony Cullinan, Moses Dyer, John McAndrew, Thomas Murray, Edward Braesland, John Hill (2nd), and Thomas Slavin lived to receive the clasp for Ciudad Rodrigo granted in 1844.

On January 19th the fortress was stormed, and on the following day the fifth division, which had not been employed in the assault, marched in and was quartered in the suburbs and in empty Spanish barracks. When marching in it met the storming parties coming out laden with plunder and fantastically dressed as French officers with cocked hat and jack-boots or any other way which took their fancy.

The exposure in front of Rodrigo had sent the sick list up to 243 rank and file in the battalion.

Lord Wellington had in his possession at the New Year the two principal Portuguese frontier fortresses, Almeida in the north and Elvas in the south, and he had now gained the Spanish fortress of Ciudad Rodrigo opposite to Almeida. He determined to add to these the Spanish fortress of Badajoz, facing Elvas. First of all, Ciudad Rodrigo had to be repaired and armed, and other arrangements made for the safety of the northern frontier in his absence, and before entering on a new campaign it was imperatively necessary that his ragged soldiers should be re-clothed.

The annual clothing had come from England and had been sent up country, some by the Douro and some by the Mondego. As soon as they could be spared he sent the regiments of the other divisions to meet their supplies, keeping the fifth division together at Ciudad Rodrigo to protect them while dispersed. As each regiment got its clothing it marched south to cross the Tagus and join General Hill's army in Estremadura. In the last week of February, feeling that the safety of the northern frontier could now be left to the Spanish troops and the Portuguese militia, backed by the two fortresses, Lord Wellington sent off the regiments of the fifth division to get their clothing and to follow the other divisions to the neighbourhood of Badajoz. The 30th probably re-clothed itself at Penacova on the Mondego, close to where the battalion had passed the previous December.

With improved weather and good quarters at Rodrigo the effective rank and file had risen to 377, and the sick fallen to 1 lieutenant, 11 sergeants, 3 drummers, 193 rank and file.

BADAJOZ

We have no details of the march which followed, but on the 24th of the following month the fifth division was united at Campo Mayor, about 14 miles to the north-east of Badajoz, which lies on the south or left bank of the river Guadiana. The direct route from Penacova to Campo Mayor seems by Miranda de Corvo, Thomar, Abrantes and Portalegre, a new road to the battalion.

Ensign Brooke had died at Castenhiera on the 24th. He had been sent down sick on the 10th.

The siege of Badajoz had already begun. Lord Wellington, after keeping his Headquarters near Ciudad Rodrigo as long as possible to conceal his design, had left on March 5th, and joined his army at Elvas on the 11th. Elvas is about 10 miles from Badajoz, and on the 15th a pontoon bridge was thrown across the Guadiana below its junction with the Caya and about equally distant from the two cities.

On the 16th the light, third and fourth divisions crossed and invested Badajoz. At the same time Sir Thomas Graham marched on Llerena, sixty miles to the south-east, to guard against any interference by Marshal Soult from Andalusia. General Rowland Hill with the second and seventh divisions was already some five and thirty miles to the east at Almendralejo to watch Marshal Marmont and the Army of Portugal.

The fortifications of Badajoz consisted of eight bastions and the castle, all connected by curtains with ravelins, and in some cases the bastions were covered by counter-guards. The castle, the most northerly part of the defences, is on a rocky eminence which rises to 120 feet above the river on its eastern bank, and the ramparts run from the castle with a very irregular semicircular sweep to the south and west, till they touch the river again at S. Vincente. From S. Vincente the wall follows the line of the river, north-east, till it joins the castle once more. The heights of S. Cristobal on the western bank were fortified.

There were two outlying forts, that of Pardaleras on the south and Picurina on the south-east. The eastern side was also covered by the stream Rivillas, which falls into the Guadiana just above the castle, the rivulet is spanned by a bridge which was covered by the S. Roque Lunette, and by damming the bridge an inundation was formed in front of the bastions S. Maria and Trinidad. The glacis was mined on the western front of the fortress. A bridge protected by suitable works led across the river and from it a covered way gave communication with the works of S. Cristobal on a hill opposite the castle.

Wellington's plan was to capture the Picurina and from there breach the bastions S. Maria and Trinidad and the curtain connecting them.

By March 30th he had stormed Picurina and overcome all other obstacles to the formation of his breaching batteries, which opened fire on that day on S. Maria and Trinidad, the two bastions which he had selected for assault.

The fifth division moved on the same day from Campo Mayor by Elvas to Valverde.

On April 4th Wellington called the division into the army before Badajoz to join in the assault, which he hoped to deliver on the following night. The road from Valverde enters Badajoz by the gate of Xeres, and the division bivouacked on the 4th close to the road and 2 miles short of that gate. On the 6th Ensign Smith was sent sick from Valverde to Elvas. Though not mentioned the remainder of the sick must have been sent to the rear at the

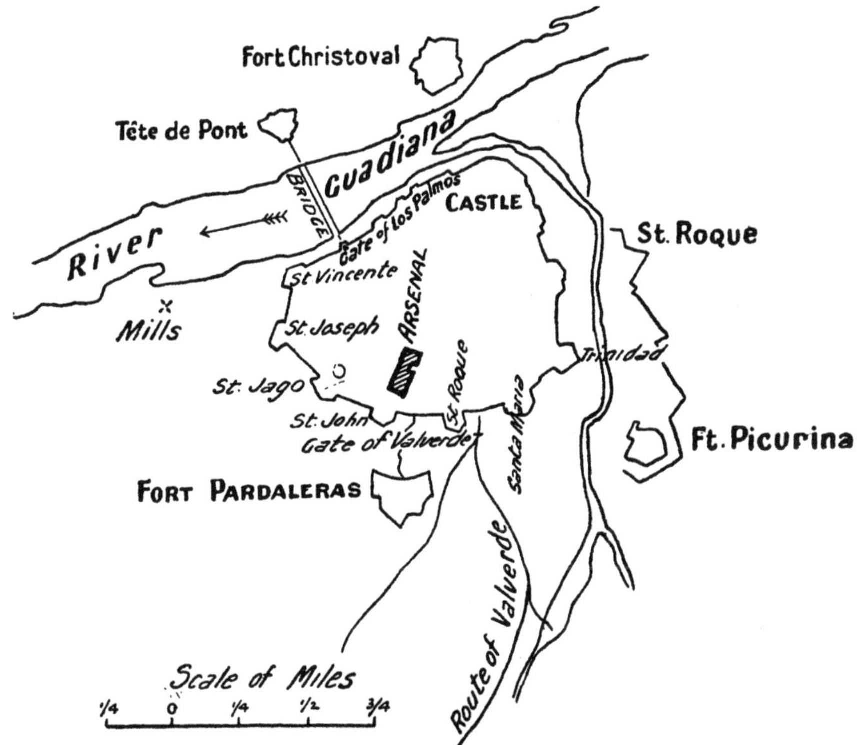

BADAJOZ AND ITS DEFENCES.

same time. Smith was invalided to England. He got the clasp for Badajoz. He did not rejoin in the Peninsula, though he is shown in a War Office list as receiving the clasp for Salamanca.

The breaches in S. Maria and Trinidad had been reported practicable on the 5th, but in spite of the threatening advance of the armies of Soult and Marmont, Wellington determined to delay the assault till the curtain connecting those two bastions had been breached also.

On April 6th all was ready and orders were issued for Badajoz to be stormed that night; the attack to take place at ten o'clock.

The light and fourth divisions were to assault the breaches, the third division under Sir Thomas Picton was to carry the castle by escalade, the fifth division under General Leith was to move to the left, and while Major-General Hays' and the Portuguese brigade made a false attack on the Pardaleras, Major-General Walker with the 4th, 30th and 44th regiments was to move on till the river was reached and then escalade the bastion of S. Vincente which overhung it.

Before dealing with the attack by the fifth division it is necessary to give the result of the other assaults which had an influence on its success.

The third division was early discovered by the enemy and forced to anticipate the time of attack by half an hour; the light and fourth divisions had in consequence to hasten their advance. As is well known, the third division, which seldom failed, carried the castle by escalade after one repulse and thus ensured the capture of Badajoz. The light and fourth divisions met with insuperable obstacles and were called off by Wellington after continuing their heroic efforts for nearly two hours.

On receiving news from General Picton of the capture of the castle, Wellington sent him orders to hold it and to be prepared to make an attack on the town in the morning in concert with the rest of the army. These orders were issued at about half-past eleven o'clock. The attack of the light and fourth divisions had become hopeless by that time and they were recalled by Wellington's order to re-form and prepare for the fresh attack in the morning. Towards midnight then, and before Leith's division came into action, the attack on the centre and right had come to a standstill. How the attack of the fifth division came to be delayed will be dealt with later on, but the thing itself is unquestionable. Colonel Jones, who was brigade major of engineers at Badajoz, says in his *Journal of the Sieges in the Peninsula*, that Walker's brigade did not leave its bivouac till after eleven o'clock. Colonel Lamare, the chief engineer of the garrison, says that S. Vincente was attacked at midnight; that three companies had been withdrawn from the bastion to aid in an effort to recover the castle and that one would have supposed that the English had divined it, their attack was timed so immediately after those companies had left. Finally General Phillipon, Governor of Badajoz, says in his report that after an attempt to regain the castle had failed, he was informed that S. Vincente was attacked.

The bastion S. Vincente, which Walker's brigade was ordered to carry by escalade, is also known as No. 1 Bastion. Four bastions connected by curtains 500 feet in length intervene between it and the front S. Maria–Trinidad, where the three breaches had been made, and the distance in a direct line through the town is 1,300 yards. The distance to the castle is 900 yards. The glacis was mined in front of Nos. 1, 2, 3 and 4 bastions. The town wall running along the river from the right flank of S. Vincente to the castle is broken at 300 feet from the bastion by a tower protecting the gate which gives access to the bridge carrying the road over the Guadiana to S. Cristobal. On this tower were two field-guns flanking S. Vincente. The height of the faces of the bastion is 31 ft. 6 ins. and that of the flanks and curtains 21 ft. This is a point of great importance, for the ladders available were ordinary 30 ft. ladders for one man and, unless it projects well over the top of the wall, it is impossible for men fully armed and carrying sixty ball cartridges, to quit the ladder nimbly and without checking those behind them. The height of the counterscarp was only 11 ft. 9 ins. and the men dropped or were lowered by their comrades into the ditch. In the centre of the ditch was a cunette or trench filled with water, 6 ft. 6 ins. broad and 5 ft. 6 ins. deep.

Outside the ditch was the usual covered way with a line of palisades, which was complete and in good order. At a gate in the palisade near the river was a guard house where an officer's party was stationed to find the sentries beyond the ditch. A ravelin covered the curtain between No. 1 and No. 2 bastion. Immediately in rear of the bastion S. Vincente is a large building marked in some maps as a barrack and in some as a hospital.

From the foregoing it might be thought that it was a physical impossibility to scale the faces of the bastion and that the attack must be confined to the flanks or curtains, which were 10 ft. lower, but there was a weak point in S. Vincente. Nearly fifty years before it had been designed to crown the bastion with a tower and, to bear the increased weight, an additional thickness of 6 ft. 6 ins. had been built on to the scarp. When this strengthening wall had reached 20 ft. 6 ins. high, the plan, according to Spanish custom, was given up. To prevent an enemy making use of the top of the new wall as a step in an escalade a steep slope was ordered to be made from it to the top of the scarp and faced with stone. Had this been done throughout it might have served, but for a short distance near the salient the stone facing had been omitted; there was merely a rough earthen slope on which assailants could assemble and spread themselves out. The French General Belmas, in his history of the siege, calls the place "*une retraite large de deux mètres,*" and says it facilitated the escalade.

The garrison to be encountered by an enemy entering by S. Vincente and forcing his way to S. Maria to assist his comrades attacking the breaches was as follows :—

		Strength.
Nos. 1 and 2 bastions	1 battalion of 9th light infantry	580
" 3 " 4 "	1 battalion of 28th light infantry	590
" 4 " 5 "	1 battalion of 58th line . .	450

Of those two or three companies of the 9th were called away to attempt to recover the castle before Walker's assault was delivered, but his troops were opposed not only by the corps mentioned above, but by a battalion 250 strong formed of French merchants, clerks, canteen-keepers and other followers of the army. It was called the battalion of administration. Walker's men were also in close action during some part of the night with one or both of the two battalions of Hessians serving in the garrison. These battalions had been posted early in the evening as reserve to the troops defending the breaches and must have been moved, after our attack there had ceased, to oppose Walker's brigade. A light company man of the 4th King's Own captured the colour of one battalion, killing the officer who carried it, and two days later presented it to Lord Wellington.

THE ASSAULT

General Walker's brigade broke up from its bivouac at eight o'clock in the evening, but the ladder party had lost its way between the engineer park and the bivouac, and it was past eleven o'clock when the column was able to move off. The light company of the 38th lent by Leith's other British brigade accompanied Walker. General Leith himself followed with the remainder of the 38th and the 15th Portuguese regiment of the line. The night was dark and cloudy and there was a mist hanging over the low ground near the Guadiana. General Walker reached the river somewhat below the town, where there are a number of mills, and formed his brigade behind them. The attacking party, under Lieut.-Colonel Brooke, of the 4th King's Own, consisted of the light companies of the 4th, 30th, 38th, and 44th. The battalion companies of the 4th were in support, and the 30th and 44th in reserve under General Walker. The orders were to enter the ditch at the salient and escalade the flank, and if possible the face of the bastion towards the river ; twelve ladders were carried by Portuguese of the division.

Advancing in silence up the glacis the light companies could hear voices in the guard house and the French officers visiting their sentries. At this moment the moon shone out and a French sentry challenged. The challenge was almost immediately followed by a storm of bullets from the defenders, who were each provided with four loaded muskets. The guiding engineer officer was killed here.

The palisades caused delay and it was with difficulty an opening was made large enough to admit two men abreast. Meanwhile the Portuguese had dropped the ladders and disappeared. The light company men picked them up and entering the ditch were soon trying to rear the ladders against the right flank of the bastion, which seemed the only practicable place, but a mine was sprung beneath their feet and logs of wood and live shells were rolled over on their heads and the field guns in the tower 100 yards off opened with grape. Some of the survivors say that half the light companies went down there. Lieut.-Colonel Brooke saw that success on that face was hopeless and sent an officer back to tell General Walker that he was about to change the point of attack, and to ask that steps might be taken to keep down the fire of the place. General Walker extended two companies of the 44th and advancing them to the edge of the ditch opened fire upon those points of the defences which flanked the left face of S. Vincente, and at the same time Lieut.-Colonel Brooke hit upon the weak spot on the left face close to the salient. Three ladders were raised close together so that the men might support each other and the men thronged up them, but were checked when it was found that their leaders were still about 7 feet below the top of the rampart. Luckily an embrasure was found, closed merely by a gabion, and one daring fellow was hoisted up by his comrades to the sill of the embrasure, and he in turn hoisted up others. Apparently the French fired upon them heavily from all sides, but did not close and thrust them out with the bayonet. The foremost men fell fast, but the number increased till our men were able to attack. All accounts agree that the French fought well, but the bastion was cleared. As Major Piper, who led the 4th in support, was stepping into the bastion he received a contusion from a musket ball which knocked him over on to the heaps of dead and wounded and thereby saved his life, for at that moment the magazine in the bastion blew up, killing many of the French and of our men.

Although the force was divided into a storming party with supports and reserves it must not be supposed that when once the ladders were up there were any gaps or that the supports and reserves waited to see whether those in front could not do it without them. On the contrary, as soon as the ladders were up they were all treading on the heels of those in front to get there. General Walker knew well that it was better to run the risk of loss by crowding the ditch than to leave the least chance of there being a check in the stream of men passing up the ladders.

As soon as he was up General Walker formed his brigade and, perhaps, loaded for the first time, though it is not mentioned in any narrative of the siege.

When the men were formed, one-half of the 4th was sent under Major Piper to dislodge the French who had commenced firing from the large house in rear of the bastion. Major Harding of the 44th accompanied this party, so it is probable that there was some mixing of regiments.

General Walker himself led the rest of the brigade along the ramparts towards the breaches. As soon as he moved, General Leith sent the 38th into the bastion to hold it in case of reverse. It was followed by the 15th Portuguese.

General Walker, whose brilliant leading contributed much to the success of his brigade, carried three bastions in succession by hard fighting, but the force was brought to a standstill on the curtain between Nos. 4 and 5 bastions. According to General Belmas it was the good fighting of the French 28th light infantry and 58th of the line which had this effect, and no doubt they did their duty. They were equal if not superior in numbers to the part of Walker's brigade on the ramparts. There were other contributory causes. General Walker was dashing forward to clear a traverse across the *terre pleine* of the curtain when a field piece behind it was discharged and he fell dreadfully wounded; at the same time one of his followers, seeing the portfire thrown down by the gunner suddenly blaze up, called out "a mine"; the word was re-echoed through the ranks and the whole of those brave fellows, who had performed prodigies of valour, hesitated. General Veillande, Phillipon's second in command, charged at this unlucky moment with men he had brought up from the reserve, most likely the battalion of Hessians, and drove the brigade back, even to S. Vincente, at the point of the bayonet, but here General Leith's forethought availed. The 38th, under Colonel Nugent, were perfectly steady and received the French with such a deadly volley at close quarters that they fell back. Veillande's force was followed and completely broken up. He himself managed to join General Phillipon, the Governor.

General Jones,[1] in his *Journal of the Sieges in the Peninsula*, says that the men who had fled from the imaginary danger of a lighted portfire "now turned round and showed the same intrepidity as previous to their panic, and the whole marching immediately to the breaches the troops defending them dispersed though attack in front had long ceased."

That is in the main what happened, but besides making their way to the breaches, Walker's men occupied the town and so became engaged in a number of street combats of which we know little or nothing. We know from the accounts of Major Piper and Captain Ellers Hopkins some of the adventures of the part of the 4th King's Own which General Walker had detached to attack the French who were firing, from the building in rear, upon his brigade in S. Vincente. This detachment, no doubt by General Walker's orders, did not rejoin him, but kept on his left and passing through the town attempted to attack the defenders of the breaches from the rear. The party, with a bugler

[1] Jones was Captain and Brigade Major of the Royal Engineers at Badajoz. He had the advantage, when preparing his work for publication, of having before him the journal and opinions of Sir John Burgoyne, R.E., the Chief Engineer in the Siege

at its head, marched through the brilliantly lighted but deserted streets without meeting with resistance, though officers and men fell from time to time from shots fired from the houses.

The effort of Major Piper's party to reach the breaches was repulsed and he established himself in the place of S. John (immediately to the north of the arsenal) where the Governor's Headquarters had been earlier in the evening. The troops detailed for the defence of the breaches were still in position, though, as Colonel Jones says, the attack of the light and fourth divisions had long ceased.

Besides the combat caused by the counter-attack of General Veillande, there was a fierce struggle on the other side of the town near the river and the gate of las Palmas. No British account of this has been found, but General Phillipon, the Governor, says in his report that after the loss of the castle and the repulse of two companies of the battalion of the ninth regiment who attempted to retake it, he heard that S. Vincente was attacked. He went towards it and found the battalion of the ninth and the battalion of administration making a vigorous resistance. He ordered the thirty or forty mounted chasseurs and dragoons with him to charge upon the Place de las Palmas, but most of their horses were killed or wounded. The enemy could not be stopped and soon he himself was cut off and could not communicate with his troops. With difficulty he was able to escape by the bridge of las Palmas to S. Cristobal.

General Phillipon was accompanied in his escape by General Veillande and about 100 men. Both the generals were wounded.

When General Phillipon says that he was unable to communicate with his troops he alludes to an attempt he made, during his last stand at the gate of las Palmas, to assemble and carry with him to S. Cristobal as large a number as he could of his garrison. For this purpose he directed Captain de Grasse of his staff to penetrate, if possible, as far as the breaches in Trinidad and S. Maria, where it was understood there was a considerable body of French troops which still kept its formation, and to order all to rally on the Governor at the gate. De Grasse failed, the streets were already blocked by the men of Walker's brigade, reinforced by the 38th and the 15th regiment of Portuguese, and the French were surrendering fast. General Belmas, who gives the narrative of the siege from the French side, fixes the time of Phillipon's escape from the town at one o'clock on the morning of April 7th. General Phillipon thought it was much later.

Soon afterwards Lord Wellington, having heard of General Leith's complete success, ordered the fourth and light divisions to advance again and they entered the town unopposed.

By two o'clock in the morning of the 7th the French garrison had nearly all laid down its arms, though firing was heard for some time longer where parties had thrown themselves into houses which they tried to defend. The conquerors treated their prisoners with the greatest forbearance, and even kindness.

Although the details of the fighting are in part unknown or matter for conjecture, we are on firm ground when we say that the actual capture of Badajoz was achieved by Leith's division, and more especially by Walker's brigade.

The division had carried the point of attack assigned to it, in spite of almost insuperable obstacles and a brave resistance by an enemy who was in superior numbers. It had cleared the ramparts as far as the breaches and forced their defenders to disperse. It had defeated the Governor and his last reserve and had thwarted his attempt to break out with a large part of his garrison, and finally had forced him to fly from the town almost unaccompanied to avoid death or capture. This astonishing feat was undoubtedly aided by the brilliant success of the third division in taking the castle, and the long-continued efforts of the light and fourth divisions at the breaches.

Colonel Jones writes on this point as follows :—

"Badajoz may be said to have been twice carried by escalade this night. First by General Picton with the third division in getting possession of the castle, from which moment further resistance was useless, as from the castle the besiegers could have poured their whole army into the town, and secondly by General Leith with the fifth division, which was the more immediate cause of the fall of the place ; for though General Picton's successful escalade of the castle placed the garrison at the mercy of the besiegers, still the third division remaining formed in the castle without further movement their success produced no instant effect upon the defence and the fifth division met with the same opposition as if the castle had not been escaladed."

The lamentable excesses of the troops when Badajoz was at their mercy have been too often described to need repetition here.

On the 7th Lord Wellington published a General Order as follows :—

"1. The Commander of the Forces returns his thanks to the General Officers, Officers and Soldiers of the third, fourth and Light Divisions, Royal Engineers and Artillery, for the persevering patience, laborious industry and the gallantry which they have uniformly manifested throughout the siege.

"In thanking them for the uncommon gallantry displayed last night in the assault of the place under the most trying circumstances the Commander of the Forces must include among those the General Officers, Officers and Soldiers of the fifth Division.

"2. The Commander of the Forces requests that the men's arms may be immediately got into order again as he hopes that another occasion of meeting the enemy may before long occur.

"3. The Musket ammunition to be completed.

"4. The regiments of the fifth division are to return to their bivouac by regiments as soon as General Leith will think proper except the Royal Scots and the 9th Regiment."

The Royal Scots and the 9th had been in reserve during the assault and only entered on the 7th.

The fifth division had not served in the siege, but was only brought in for the assault, so is not included in the first paragraph.

Lord Wellington in his despatch of the 8th says : "Lieut.-General Leith's arrangements for the false attack upon the Pardalera and for that under Major-General Walker, were likewise most judicious and he availed himself of the circumstances of the moment to push forward and support the attack under Major-General Walker in a manner highly creditable to him. The gallantry and conduct of Major-General Walker who was also wounded and that of the Officers and troops under his command, were highly conspicuous. I must likewise mention Lieut.-Colonel Brooke of the 4th Regiment, the Hon. Lieut.-Colonel Carleton of the 44th and Lieut.-Colonel Grey of the 30th, who was unfortunately killed. The second battalion 38th Regiment and the 15th Portuguese likewise performed their part in a very exemplary manner."

There are some difficulties as to the numbers of the 30th employed in the assault. The ordinary arrangement was to leave in the bivouacs the quartermaster and the officer commanding the guard, but the 30th had only seventeen combatant officers present, so after providing for three staff, Colonel Grey, Bamford, and the adjutant, two carrying the colours, two left in the bivouac, and three with the light company, only seven are available for the remaining nine companies. Two companies might be broken up and the men distributed among the remainder, but it is a curious thing that no officer was hit or man killed in No. 4 or No. 10 company, nor did any man die in either company for a month afterwards. We do not know who was acting quartermaster, though there is some slight reason to believe that Bamford was. He may have been left in the lines with two companies to protect the baggage. In a statement sent to the Royal Calendar six years later, he says that he was present at the assault of Badajoz and succeeded to the command of the battalion when Colonel Grey died, but if Bamford actually shared in the assault he must have taken command when Grey fell, not when he died of wounds, and must therefore have brought the battalion out of action, a very important matter and not likely to be omitted. He was a good officer and eighteen months later was given a non-purchase step in another regiment, but he was awarded nothing for Badajoz.

The pay list, signed by Hamilton in Ireland four years later, does not give the names of the wounded, only of those killed or died of wounds or disease.

In the absence of stronger evidence it is better to assume that all companies shared in the assault. The effective strength on March 25th was 21 officers, including the paymaster, surgeon, and assistant-surgeon, and 423 other ranks. From this must be deducted Ensign Smith and a few sick of other ranks, the men in charge of regimental transport, men lent to staff, commissariat, etc., and the officers' guard, which must have been fairly strong, for everything the battalion possessed from reserve ammunition to the men's packs was in its charge.

The casualties in all ranks were 133, so that if about 400 were engaged in the assault the loss would be one in three. The loss of the 44th was much the same, but the proportion in the leading battalion, the 4th King's Own, was much more, nearly one-half. It not only suffered in the ditch but from the explosion of the magazine in the bastion S. Vincente.

The second 30th was ordered to place Badajoz on its colours and the gold medal was awarded to Lieut.-Colonel Grey. It was delivered to his heirs.

From the disorganization which followed on the success of the assault the returns of casualties furnished to Lord Wellington were faulty and in the brigade from twenty to thirty men were reported as dead who turned out to be only wounded. In his despatch of the 8th, his Lordship shows 36 non-commissioned officers, rank and file killed in the 30th, whereas 28 is the proper number according to the muster rolls. Unluckily we have not even the number of the non-commissioned officers and men wounded in each company, but the following is a roll of the officers and staff sergeants present, showing in what capacity they served and of the non.-commissioned officers and men killed or died of wounds:—

Major (Brevet Lieut.-Colonel) George Grey, died of wounds, April 7th.
Captain Thomas Bradgate Bamford, second in command.
Paymaster Henry Boyd Wray.
Adjutant John Garland.
Surgeon John Hennen.
Assistant-Surgeon John Evans.
Sergt.-Major James Woods.
Paymaster's Clerk T. Cuthbert.
Armourer N. Artis.
Drum-Major George McCann.
Ensigns Campbell and Lockwood carried the colours.
Mr. Francis Tincombe, volunteer, was attached to the light company.

COMPANY.	OFFICERS PRESENT.	NAMES OF KILLED.
1. Capt. Macnab (duty Lisbon).	Lieut. John Rumley.	Pte. John Chamberlain. Pte. Francis Owens, Roscommon.
2. Capt. Machell (Brigade Major: severely wounded).	Lieut. Richard Heaviside.	Sergt. Thomas Aitchison, Londonderry. Sergt. John Tully, Tipperary. Pte. John Clarke, Leicestershire. Pte. Valentine Flanagan, Dungannon.
3. Capt. Craig (D.A.A. Gen., 5th Div.).	Lieut. J. L. White.	Cpl. Robert Langford, Cambridgeshire. Pte. Henry McCann, West Meath. Pte. James Rouston, Yorkshire. Pte. Robert Paul, Lincolnshire.
4. Capt. Stewart (Bde.-Major, 2nd Div.).	Lieut. Mayne.	
5. Capt. Nunn (with 1st Batt., India).	Lieut. Eager.	Pte. Elijah Fletcher, Lincolnshire. Pte. Thomas Thackery, Yorkshire.
6. Capt. Tongue (with 1st Batt. in India).	Lieut. Andrew Baillie (Wounded).	Cpl. William Throstle, Norfolk. Pte. John Rizen, Leicestershire. Pte. John Collins (2), Tipperary.

Company.	Officers Present.	Names of Killed.
7. Fullerton (Depôt), England.	Lieut. M. Andrews.	Pte. William Cook, Cambridgeshire. Pte. William Gunn, Caithness. Pte. James White, Lincoln. Pte. Eugene McDonagh, Londonderry. Pte. Michael Rowe, Kilkenny.
8 Light Company.	Capt. Chambers (Wounded). Lieut. P. P. Neville (Wounded). Ensign John Pratt (Wounded).	Cpl. James McDoole, Granard, Co. Longford. Pte. John Doonican, do. Pte. Richard Kelly, Longford. Pte. Pat Macken, West Meath. Pte. John Hodge, Leicestershire. Pte. Thomas Mobbs, Northampton. Pte. Jos. Tidds, Nuneaton, Leicester.
9. Grenadiers.	Captain John Hitchen (Wounded).	Pte. Samuel McBeth, Strabane.
10. Capt. Spawforth (Depôt, England).	Lieut. John Garvey.	

The wounded numbered 98, and of those there died :—
No. 1 Private Richard Rosengrove, Longford.
,, 3 Private Mat. Devens, Sligo.
,, 5 Corporal James Wadsworth, Cambridge.
Private James Buckingham, Leicester.
,, 8 Private James Cherry, W. Meath.

Two of these men are shown in the pay list as having died a natural death, but we know from correspondence that they died of wounds. Two sergeants of Macnab's company, W. Brassish of Northampton, and George Metcalfe of Oxford, who died on the 10th and 20th of the month were probably also among the wounded.

Two years later the paymaster was asked by the auditor what had been done with the balance due to Private Fletcher when he was killed, and the answer was : "Expended along with a very generous subscription from the officers for the maintenance of his child."

The officers of the 30th serving on the staff at Badajoz were General

Walker's brigade-major, Captain Machell, who, like his general, was wounded, and Captain Craig who was with General Leith as D.A.A.G.

One officer of the regiment missed the assault through no fault of his own. General Sir Robert Wilson had been given a mission to Turkey and Russia and his A.D.C., Lieutenant James N. Charles of the 30th, was awaiting at Cadiz the arrival of his chief from England when he heard of the siege of Badajoz. He hurried from Cadiz, but Wellington had been too quick for him and Badajoz had fallen a week before he arrived. He returned in haste to Cadiz to find that Sir Robert had come and gone. The unlucky A.D.C. only caught up with General Wilson in Russia when the latter was following the retreat of Napoleon's army from Moscow. Sir Robert, who was himself a daring soldier, found no fault with the young man's dash for Badajoz.

The death of Major Grey nearly involved that of his wife, who, like many English ladies, was living in Lisbon. When the news of the capture of Badajoz reached the capital she was sitting upstairs in the verandah of her house, when her attention was attracted by the shouting in the streets, and presently she was aware that some one below was giving important news from the army and her husband's name was mentioned. She at once anticipated the worst and when her fears were confirmed was taken with premature labour and on April 14th gave birth to a boy whom she named after his father. That boy was the future Sir George Grey who was honoured by his Queen and countrymen as one of the greatest of colonial governors and statesmen.

Among those who stormed Badajoz was Private Luke Lydon, in later days Sergeant Lydon and regimental schoolmaster, father of Sergeant Dominick Lydon, wounded in the Crimea, and grandfather of the Regimental Sergeant-Major Lydon who distinguished himself in the South African War.

Those of the wounded who were able to travel were sent down country under Lieutenant J. L. White with Assistant-Surgeon Evans in medical charge.

The siege of Badajoz had been hurried by the approach of Marshal Soult from Andalusia. On the day of the assault he was at Llerena and on the 8th at Fuente del Maestro, only 35 miles distant—the covering armies under Sir Rowland Hill and Sir Thomas Graham had fallen back before him to the Albuera River, a little over 10 miles from Badajoz, and there they were joined by Wellington eager for a battle. He could put 45,000 men in line between Albuera and Talavera Real and he knew that Soult was much weaker. At Fuente del Maestro Soult got the news of the fall of Badajoz and retired again to Andalusia. Wellington on his part was unable to pursue Soult to Seville owing to the advance of Marshal Marmont to the Agueda and the consequent danger to Ciudad Rodrigo and Almeida, which had been utterly neglected by the Spaniards and Portuguese since he left these towns in March. The fifth division therefore again took the familiar route by Portalegre and Nyssa, across the bridge over the Tagus at Villa Velha and through the mountains of Leon to the Agueda. It passed Portalegre on April 15th.

The 4th brigade was now commanded by Major-General Pringle in place of General Walker, whose wounds had compelled him to go to England.

Captain T. B. Bamford commanded the regiment. Captains Hitchen and Chambers and Ensign Pratt had been sent down to Lisbon to recover from their wounds.

Wellington's army began to cross the Tagus on April 16th, and found that

Marmont had passed the Agueda and the Coa and was occupying the hills about Penmacor which look down into the Tagus Valley; the continuous rains had broken the bridge in his rear and Wellington hoped to force him into action. Some days, however, were necessary to close up our army and in that time the floods abated and Marmont was able to recross the Agueda on the 24th.

The 30th and the other regiments of Leith's division reached Sortelha near Sabugal on the 25th, having taken ten days, after leaving Portalegre, to cross the river and climb the mountains. Sortelha is a little walled town of about 900 inhabitants and must have been crowded.

On examination, the state of the fortresses of Ciudad Rodrigo and Almeida was found to be very bad, and as little help could be expected from the Governments of Spain and Portugal Wellington formed the bold resolution of breaking up his army for a time, placing the troops where they could be fed by water carriage and employing the carriages and mules of his army in bringing up provisions and ammunition to the fortresses. In consequence of this decision the fifth division marched to Lamego on the river Douro, about 40 miles above Oporto and 75 from Ciudad Rodrigo. It was an entirely new country to the regiment.

Early in June the fortress having been put in a state of defence and the field magazines replenished Wellington was able to reassemble his divisions. Lieut.-Colonel Hamilton rejoined on June 3rd in time to march up country. He had arrived at Lisbon on April 18th, but as there seemed no chance of active service he remained on leave at the capital. He had married in England and his bride accompanied him to Lisbon.

On the 16th Lord Wellington commenced his advance from Ciudad Rodrigo to Salamanca. On the 17th he forded the river Tormes above and below that town and occupied the heights of S. Cristobal, 5 miles in advance of it, Marmont retiring before him. Ten days were consumed in reducing the forts of Salamanca, but on the 27th they fell into our hands and gave us command of the bridge over the Tormes. Marmont, who had been threatening an attack, retreated on learning of the fall of the forts. On July 2nd he passed the Douro and took up a position on the northern side. His centre was at Tordesillas, where he left the bridge standing, and the fifth division bivouacked in front of it. In consequence of the arrival of the second battalion of the King's Own on May 14th there had been several changes of battalions from one brigade of the division to the other, but in a few days these changes were cancelled and the division stood as before except that Pringle's brigade had two battalions of the 4th instead of one.

In front of Tordesillas the 30th had a happy time ; the weather was fine and both the French and our men spent most of the day bathing and exchanging chaff from their respective banks of the river ; full rations were issued and wine was only too plentiful. On July 9th Marmont commenced a series of manœuvres which caused the 30th many hard marches, but it was scarcely in contact with the enemy till the 18th. The fifth division had been rapidly moved to the right of the army on the previous day to support the light and fourth divisions and had bivouacked at Terracina de la Orden. On the 18th the light and fourth divisions were attacked and their left flank turned. Wellington thereupon ordered all to fall back across the Guarina. The fifth covered the retirement of the other two divisions and then followed, having by its steadiness checked any disposition on the part of the enemy to pursue. A few men are said to have been wounded in the division but are not officially recorded. Marmont now turned the British right and Wellington marched rapidly for his old position on the heights of S. Cristobal.

The morning of the 21st found Wellington's army in the position it had advanced from on June 28th. Marmont had occupied Alba de Tormes, which the Spaniards had abandoned without notice to Wellington, and the French army with the exception of one division was across the Tormes and their leading troops occupied Calvariza Arriba.

Wellington, who was aware that reinforcements, especially in cavalry, were only two days' march behind Marmont, was determined to retreat behind Ciudad Rodrigo and not to fight unless the enemy made some glaring error. Marmont, however, was very close to the road from Salamanca to Ciudad Rodrigo and might force a battle. Towards evening Wellington crossed the Tormes, leaving the third division and D'Urban's cavalry on the right bank.

Picton was absent from ill-health, and the third division was now under Major-General Sir Edward Pakenham, uncle to General Thomas Henry Pakenham, colonel of the 30th regiment in later days.

The army bivouacked with its left near the ford of Santa Marta and the right close to the village of Arapiles. It thus covered the city of Salamanca. The road from Salamanca to Rodrigo ran past the right of our line nearly at right angles to it and about 3 miles from the village of Arapiles. The fourth division was on the right, then came the fifth, sixth, and seventh divisions. The first and light divisions and the bulk of the cavalry were still near the ford of Santa Marta by which they had crossed. The night was one of great discomfort, for as the troops reached their ground a heavy thunderstorm broke over them and many passed the night standing up in the saturated cornfields, where the crops had not yet been cut.

SALAMANCA

At daybreak on the 22nd Marmont strengthened his force in Calvariza Arriba and on some heights in advance of it close to our line. Although the right of our army had extended on the night of the 21st to the neighbourhood of the village of Arapiles we had not occupied it nor two rugged hills in our front called the two Arapiles or Hermanitos. There was now a race for these two hills, and after a skirmish we retained one but the French held the other. Their possession of this strong point was a threat to our line of retreat along the road running south-west to Ciudad Rodrigo ; the road passes at less than 4 miles from the French Hermanito. Wellington therefore extended his right towards the road and occupied the village of Arapiles. He also brought the third division from his left across the river and passing it behind the army placed it on his right close to the Rodrigo road at Aldea Tejada. The first and light divisions also took ground to their right. True to his usual custom, Wellington did not show his hand. The third division was quite out of Marmont's sight, on its left Bradford's Portuguese and the cavalry were hidden by a hill near Las Torres, the fourth division was in the village of Arapiles, to the left of the fourth were the fifth and sixth divisions massed in a hollow on the reverse slope of our Hermanito. The first, seventh, and light divisions with Pack's Portuguese brigade were in reserve. Marmont saw enough to cause him to strengthen his line between his left at the French Hermanito and his right at Calvariza Arriba, but the divisions of Brenier and Sarrut were still on the march through the forest in rear when either his patience gave way or something induced him to believe that Wellington had already begun a retirement on the road to Ciudad Rodrigo. He ordered Maucune to move with his own division and that of Thomières towards the road. Maucune moved off with Thomières' division leading and covered by the fire of fifty pieces of artillery, which took ground to their left by guns in succession to keep up with the movement of the infantry. At the same time Bonet's division drove the allies out of the village of Arapiles, but the troops who should have connected Maucune's divisions with the centre had not yet come up. Maucune too allowed his divisions to move loosely so that it was possible for his leading regiments to be overwhelmed before he could form a line of battle.

Wellington, who had been watching the French closely from the English Hermanito, had sat down to lunch about half-past two o'clock when the officer of the staff at the telescope reported "The French are extending more to their left." "The devil they are!" said Wellington. "Give me a glass quickly." After a long look he said, "I think this will do at last." He had seen that Maucune's divisions were isolated. The necessary orders were given and Pakenham's division struck due south to cross Maucune's path. He marched in open column of battalions ready to form line to his left.

The fifth and sixth divisions ran down the reverse slope of the English Arapiles and the former passing behind the fourth division formed line to its left.

The sixth division remained in rear with the seventh. Bradford's Portuguese and the allied cavalry came up on the right of the fifth division to connect it with Pakenham, and Pringle extended his light companies to the right under Major Alured Faunce of the 4th King's Own with the same object. These formations had of necessity taken time and it was nearly five o'clock when Leith gave the word to advance against Maucune's division. He and his staff were well out in front of the leading brigade and the division, confident in itself and in him, was in the highest spirits and welcomed his order with hearty cheers. The ground was covered with crops of flax, with here and there groups of evergreen oak, but the division had still preserved its order unbroken when General Leith saw that he was close to the enemy, who were still invisible to the men on foot. When the general gave permission, his division opened fire with the same regularity which had characterized its advance.

Meanwhile Pakenham with the third division had headed the French left. Marching straight across its path in column of companies he had made his men bring their right shoulders up and form line without halting. Thomières' division made a feeble effort to withstand him, but the little chance which it had of success was entirely destroyed by the appearance of the British cavalry, for they also were to take a part in Wellington's great combination to crush the French left.

To return to Leith's division—we have seen that, previous to the advance, the light battalion of Pringle's brigade, that is the light companies of the first and second 4th, 30th, and 44th, had been extended to the right to connect with Pakenham. As the division advanced, the light companies closed to their left not only because the interval between the third and fifth divisions was decreasing as they advanced on converging lines, but also to give room to our cavalry to come up between the two divisions. About the time when General Leith gave permission to begin the attack this light battalion suddenly found itself in presence of a French battalion which it had been unable to see from the amount of cover, although the distance is said to have been only about 20 yards. The French were in the act of deploying from column of double companies and both sides fired instinctively. Our men advanced to charge and the French gave way. At this moment Le Marchant's heavy cavalry charged home through the gap between Pakenham and Leith, and Thomières' division was ruined. The battalion opposed to the light companies of Pringle's brigade was struck in flank by two squadrons of the 5th Dragoon Guards and broken up. It was the senior of the four battalions of the 62nd regiment, which were all in Thomières' division. The eagle of the regiment was captured by Lieutenant Pearse of the light company of the 44th and the colours or fanions of two battalions were taken, one by Ensign John Pratt of the light company of the 30th and one by Lieutenant Francis Maguire of the 4th King's Own. The men of Thomières' division threw down their arms and in their panic terror of our horsemen clung for safety to the men of the third and fifth divisions. Two thousand prisoners were taken here.

Leith's division now brought up its right shoulder to align itself with the third and together they resumed their victorious advance with the fourth division on their left, but the battle was not yet won. Marshal Marmont had been struck down by a shell which broke his arm, and General Bonet, who succeeded him, had also been wounded and General Clausel had taken command. He brought up a division from the French right and very opportunely the divisions of Brenier and Sarrut which had been so long expected came up at this moment. Using the French Hermanito as a pivot and guard to his right, Clausel threw back his left and presented a fresh line of four unbroken divisions for his defeated troops to rally upon. As the British advanced, the fourth division found its left flank threatened by the French on their Hermanito, and to take the pressure off the fourth Sir Dennis Pack attacked the height with his Portuguese brigade, but was roughly thrown back. The French now fell upon the fourth in front and flank and defeated it and threw the leading brigade of the fifth into some confusion, so that Pringle's brigade in the second line had to advance to its support. General Leith at this time was severely wounded and Major-General Pringle took command of the division, Lieut.-Colonel Brooke of the 4th King's Own taking his place in command of the brigade. The officers present that day are unanimous in praising the gallantry shown by the French at this point, but Wellington as usual was at hand when wanted and threw the sixth division into the fight and the French were beaten, but only after a hard struggle and heavy losses. Darkness was now coming on, and except by their fire it was hard to define the position of the troops. Lord Wellington, thinking that the Spaniards were still in possession of Alba de Tormes, made arrangements to forestall the retreating enemy at the ford of Huerta, but, covered by Maucune and Foy, who doggedly disputed every inch of ground, the French fell back towards Alba de Tormes. At the point where the road enters the forest a desperate stand was made by the divisions which had kept their order, to enable the broken troops to get away. They were attacked by the fifth and sixth divisions in front, and in flank by the third, and driven off in great confusion, and it was probably at this time that the 22nd regiment of the French line in Brenier's division lost its eagle.

Taking advantage of the darkness the French crossed the river by the bridge at Alba de Tormes and the fords below it while Wellington with the light and first divisions was seeking them fruitlessly at Huerta. Had the Spaniards held Alba, or had they even warned Lord Wellington when they abandoned it, few of Marmont's army would have escaped. As far as the fifth division was concerned there was no pursuit that night, but from early on the 23rd till late on the 24th, the marching was almost incessant. Wellington then granted a halt at Flores de Avila, 40 miles from the field of battle, and there he received the trophies captured by the light companies and others and wrote his despatch.

In a French regiment the senior battalion always carried the eagle. The other battalions carried a fanion or fanon bearing the number of the regiment. The 62nd had all four battalions in action and lost its eagle and two fanions. Wellington used the word colour and others the word standard to describe the fanion, and in the manuscript history of the British Army in possession of the Royal United Service Institution the 30th is said to have captured a standard at Salamanca.

From Flores de Avila Wellington sent home the eagles of the 22nd and 62nd French regiments and six colours. The eagles are now at Chelsea and one standard or fanion of the 62nd is in the museum of the United Service Institution, Whitehall.

Half a dozen years after the battle of Salamanca, Major Crookshank of the 38th, who in the battle commanded the 12th Caçadores in the third division, furnished a statement of his services to the *Royal Calendar*. In this he says that his battalion had captured the eagle of the 22nd French regiment and that he personally had presented it to Sir Edward Pakenham.

The foregoing account of the capture of the eagle and fanions of the 62nd regiment by the light companies of Pringle's brigade is to a large extent taken from the records of the 44th, which were compiled from the statements of Lieut.-Colonel Pearse, K.H., and other officers present at the battle.

In 1844 a discussion was started in the *Naval and Military Gazette* by a writer who asserted that the 44th had captured two eagles at Salamanca. A number of letters followed; among others Lieut.-Colonel John Garland, who had been adjutant of the 30th at Salamanca, wrote to give his recollections. The following is his letter :—

"PARIS,
"June 18th, 1844.

"Seeing in your *Gazette* of the 15th instant a letter from 'One of Greville's Brigade' assuring you that only one eagle was taken or picked up at the Battle of Salamanca by the 44th regiment is quite correct, but I beg also to inform you that (to the best of my recollection, so long a period having now elapsed) another eagle came into possession of an officer in the 30th regiment in the same brigade who accompanied the officer of the 44th regiment to the Headquarters of the British army where they deposited the two eagles. The officers named are Lieut.-Colonel William Pearse, K.H., late lieutenant 44th regiment, and Major John Pratt, now deceased."

Garland was sixty years old when he wrote this letter and he lived seven years longer. We have no knowledge of his state of health or why he distrusted his memory. He may have confused the colour or fanion captured by Pratt with an eagle.

Wellington sent home only two eagles, and of those Pearse of the 44th certainly captured one, and the claim of Crookshank that his Caçadores took that of the 22nd has never been contradicted, though Colonel Hamilton and all the senior officers of the 30th, who contributed their statements of services to the *Royal Calendar*, must have been well acquainted with it.

According to Leith Hay, General Leith's A.D.C., Greville's brigade led and General Pringle's was in support at Salamanca, and his authority is usually accepted, but the records of the 44th say that Greville's brigade was at first in second line and did not pass to the front until the defeat of Thomières' division, when Pringle's brigade was encumbered by the number of prisoners. It really is immaterial for the casualties show that Salamanca was the day of Greville's brigade as Badajoz had been that of Pringle's, then commanded by the gallant Walker. Greville's brigade had 348 casualties and Pringle's 104. Still the fact that Pringle's had so many casualties shows that during the day it was not always in support but at some time took an active share in the fighting—perhaps when General Leith was wounded and the leading brigade checked or in the last struggle in the dark on the edge of the forest. The latter is more likely, as seven men were reported missing from the brigade.

The 30th had :—

KILLED.

Machell's Co.	(Heaviside),	Private	Timothy Meara, Tipperary.
Tongue's	(Baillie),	,,	Thomas Livesay, Northampton.

DIED OF WOUNDS.

Spawforth's	(Garvey),	Private	James Woollen, Lincoln.
,,	,,	,,	Francis Pibble, Lincoln.
Machell's	(Heaviside),	,,	Edward Flanagan, Dungannon.
Chambers'	(Neville),	Sergeant	Jos. Matthews, Hereford.

OTHER WOUNDED.

Lieutenant Garvey and 19 rank and file. One man was a prisoner of war.

The company officers and staff were the same as at Badajoz, except that Hitchens and Chambers were absent on account of wounds, and Lieutenant G. W. Brisac who had been long sick had rejoined and commanded Hitchen's company, and Neville commanded the light company. The colours were carried by Campbell and Lockwood.

The total effective strength was about 330 and about one in twelve became a casualty.

Colonel Hamilton was awarded the gold medal and the battalion was ordered to bear Salamanca on the colours.

At Flores de Avila a draft joined under Lieutenant Freear, who had been on duty at Belem since the battalion landed in 1810. The draft, which was brought from Hull to Lisbon by Ensign Henry Beere, consisted of 15 recruits, 7 volunteers from the militia, and 4 old soldiers. Lieutenant James Eager was sent back sick from here and Mr. Tincombe, who had served so long as a volunteer, was taken on the strength as an ensign. In the War Office list Eager is shown as receiving the clasp for Badajoz, but not for Salamanca. This is apparently a mistake. He rejoined in time to command a company at Villa Muriel.

On the 26th, Wellington's advance was resumed. Napoleon's brother Joseph, whom he had made King of Spain, had left Madrid the day before the battle of Salamanca to join Marshal Marmont with 14,000 men, and he was now close enough to support the beaten army under Clausel. The latter was, however, determined to cross the Douro and retire upon the northern road where lay his magazines and his connection with France, and the King would not give up Madrid; they therefore separated again. Clausel recrossed the Douro and eventually retreated on Burgos, while the King retreated towards the passes of the Guadarama followed by Wellington's right, in which was the fifth division, still commanded by Major-General W. H. Pringle. On August 1st the division was at Cuellar. Wellington had with him 28,000 of his own army and a Spanish division, and Joseph, while making a show of defending the Guadarama, ordered his Court to quit Madrid and move south towards La Mancha. He at the same time summoned Marshal Soult to meet him in that province.

On August 11th, Wellington's cavalry were across the Guadarama and the infantry followed by a finely engineered road, winding through the mountains with easy gradients.

On the following days dreadful scenes took place almost under the eyes of the Allied army when King Joseph's followers and Court were crossing the bridge at Aranjuez to retreat to the south. As all his Spanish adherents who could find carriage followed him for safety, there was a multitude of 20,000 souls. The soldiers escorting them suddenly fell upon the helpless people and plundered and ill-treated them. Lord Wellington could have driven the escort into the river and captured the Court, but perhaps thought his interference would only make matters worse; in any case it was better that the French should be hampered by these people than the Allies and he refused to interfere. At the risk of their lives Marshal Jourdain and other French officers restored order and started their unhappy adherents on a dreadful march of 150 miles to Almanza, during which they were followed by guerillas who killed without mercy those unable to keep up.

The French when evacuating Madrid had left immense stores in the Palace of Buen Retiro which stands in a magnificent park close to the city, and the 2,000 troops left behind in Madrid retired to the Retiro on the 12th when the Allies approached. Immediately the shops which had been closed were re-opened and Madrid became a scene of such joy as had never been witnessed even in the days of its proudest prosperity. Soon after mid-day the Allies began to enter through streets so crowded with gratulatory multitudes that the officers who were on horseback at the head of their men could only make their way with difficulty and could scarcely keep their seats, so eagerly did the Spaniards press to shake hands with them.

The fifth division bivouacked in the park round the Retiro till the French surrendered on the 14th. Two regimental eagles were found in the palace and fort as well as great stores of arms, provisions and ammunition.

After the capture of the Retiro the division returned to higher ground to avoid the great heat and was quartered near the Escorial under the Guadarama. While in the neighbourhood of Madrid Captains Hitchens and Chambers, who had been absent owing to wounds, rejoined, but Lieutenant Freear and Ensign Beere were sent down sick and news came of the death on July 23rd of Ensign Carter at Ciudad Rodrigo. He had been sent in charge of sick to Lisbon after the capture of Rodrigo and must have been on his way up country to rejoin.

The sickness at this time was very great. The battalion, when quartered near the Escorial, had 20 sergeants, 4 drummers, and 340 rank and file sick, which left only 21 sergeants, 13 drummers, and 186 rank and file fit for duty.

Sir William Napier tells us that the discipline of the army was relaxed and the troops "gloomy." An issue of pay would have been of great service, but in spite of Lord Wellington's efforts, both officers and men were many months in arrear and there seemed little prospect of money being soon available. Many officers were obliged to sell their horses to procure means of subsistence.

On the day on which the Retiro surrendered, General Clausel advanced from Burgos with the army which we had beaten at Salamanca, which he had reorganized and made fit for the field again in an incredibly short time.

On August 18th he occupied Valladolid and sent General Foy with a couple of divisions down the north bank of the Douro. On the 25th Foy even threatened Salamanca. On that day Lord Wellington joined Clinton at Arevalo with the fifth division, and on September 1st, the first, fifth and seventh divisions with Pack's and Bradford's Portuguese brigades and two cavalry brigades were united there. The third, fourth and light divisions were left in Madrid and Hill with 20,000 men was summoned from Estremadura.

Clausel fell back before Wellington slowly and showing great skill in retarding our advance without risking a general action, but at length was pushed beyond Burgos. On the 19th, Lord Wellington sent the fifth division and Bradford's Portuguese across the river Arlanzan above the town, and the first with Pack's Portuguese and a cavalry brigade below and at once formed the siege of the castle. The fifth division took no share in the siege, but formed part of the covering force under Lieut.-General Sir Edward Paget who had joined the army. The officers of the regiments not engaged in the siege were as usual invited to volunteer to serve as assistant engineers. How many of the 30th came forward is not known, but from the circumstance of his being wounded we know that Lieutenant P. P. Neville served in the trenches. He is almost certainly the Lieutenant N., assistant engineer, who is mentioned in an officer's diary as volunteering to keep down the enemy's fire with a party of marksmen on the night of September 30th, and who had the top of his ear shot off on that occasion. No other assistant engineer with that initial can be traced. Four days later he was severely wounded. On the evening of October 4th we had exploded a mine successfully under the French defences and formed a lodgment. About midnight the enemy attacked with great vigour, drove out our working party and upset our gabions and sandbags. In the struggle which followed Neville was conspicuous, being seen engaged with two French soldiers, before he fell. He narrowly escaped being completely covered with earth, but managed to grasp the foot of Lieutenant Pitts of the engineers, and make himself known. A sergeant of the 79th Highlanders carried him to his quarters. Neville pressed his watch upon him, but the

Highlander refused it, and when Neville tried to trace him later, he found that he had been killed in the succeeding assault.

Lord Wellington had from the first great doubts as to whether with his inadequate means, especially in artillery and trained engineers, he could take Burgos, but he hoped that, as in former sieges, he might make the bayonet supply the place of the means his Government refused him.

After five assaults the place was still untaken when the approach of a relieving force of superior strength made him abandon the siege. This French army, which was under General Souham, numbered 44,000 men. Lord Wellington had 33,000, but only 21,000 were of his own tried Anglo-Portuguese army, the remainder were Spanish troops and guerillas.

Having made up his mind to retreat Lord Wellington carried out his intentions with his usual boldness and skill. By crossing the Arlanzan by the bridge of Burgos he could gain a march from the French, but the bridge and the road to it were commanded by the Castle. Nevertheless, during the night of October 21st, he filed his army across with slight loss. There should have been none. The artillery moved with muffled wheels and the disciplined Anglo-Portuguese army kept perfect silence and order, but the mounted guerillas lost their nerve when approaching the bridge and began to gallop, and the castle guns opened a fire which was at first accurate, but the range was soon lost and the casualties were trifling; two men of the 30th were reported missing this night. On the night of the 22nd the infantry had reached Hormillas and the cavalry rear-guard the bridge of Baniel. General Souham heard of Wellington's retreat on the 22nd and by making a long march was able to attack and beat our cavalry at Baniel on the 23rd. Two more men of the 30th were reported missing during the action on the 22nd, and two on the 23rd; the infantry had crossed the Pisuerga in two bodies, one at Cordovilla above its junction with the Arlanzan and the other below it at Torquemada; the fifth division and the Spaniards crossed at Cordovilla. Torquemada was unfortunately a centre of the wine trade, and the great vaults being broken open there was a disgraceful scene of drunkenness, 12,000 men being said to have been helpless at one time. Whether the disorder spread to Cordovilla we do not know, but the 30th left no man behind them there.

Wellington, who was not a man to allow himself to be chased, determined that it was time to give the enemy a check, to enable his own convoys of sick and other trains to get well to the rear. With this object he fell back a short march across the Carion on the 24th, and took up a strong position on the right bank. In this retreat the light companies of the fifth division formed the rear-guard and we are indebted to the history of the 44th for a very characteristic story of Lord Wellington. The light companies had halted to protect the printing press of the Spanish army, which had stuck in a small stream, when Lord Wellington as usual turned up to see what was the matter. At this moment up jumped a hare, and a brace of greyhounds belonging to an officer of the 44th got away after it. Away went Lord Wellington at full speed, followed by his orderly, and when the hare was killed after a sharp course he sent the orderly back with it and his compliments to the owner of the greyhounds.

Lord Wellington had intended that all his troops should have a short march so as to be fresh for the fight on the morrow, but through the carelessness of the staff, the fifth division followed the remainder of the Anglo-Portuguese troops down the river Pisuerga for two leagues past the point where it should have turned off to the right for the village of Villa Muriel on the right bank of the Carion, which it was to hold. The division had to retrace its steps and arrived at its bivouac after dark with the men much fatigued. No rations were served and neither officers nor men had anything to eat or drink. Another consequence of this blunder was that the baggage and women of the division were captured by the French and the men on guard became prisoners. The 30th had two wounded on this day.

VILLA MURIEL

Our information about this action, in which the fifth division gained great honour, is very scanty. Lord Wellington and Major-General Oswald, commanding the division, have given us an outline of the engagement, and but little more has been gleaned from private accounts. Sir William Napier's description is forcible and picturesque, but he was not near the scene of action, and has not sifted with his usual care the evidence on which his narrative is based. He imagines that the fifth division was driven back and the Carion crossed in force by the French under Maucune and gives as one of the causes of this discomfiture Oswald's neglect to occupy the village of Villa Muriel in sufficient strength or to make use of a dry canal running parallel to the river. This is an entirely false view of what happened. The way in which the village and canal were occupied seems to have been excellent, but when the Spaniards on the left gave way and Wellington marched with the British troops of the fifth division to support them, the Portuguese of the division, who were left to hold the village and fords, certainly lost some ground. On the return of the British brigades the ground was regained and the French driven across the river. The division suffered no reverse. Napier, therefore, must be discarded. His wish to do justice to his comrades in arms is beyond doubt, but if the light division had fought a successful rear-guard action in which 34 officers had been killed and wounded we should have had a better and fuller narrative.

When Wellington turned to fight on October 25th, his line ran from Duenas slightly below the junction of the rivers Pisuerga and Carion, for ten miles up the latter river to Palencia. Villa Muriel is about half-way between the two places. In Napier's phrase his position was a range of hills, lofty, but descending with an easy sweep to the Carion in front. There were bridges at Duenas and Banos close to it on the Pisuerga, and at S. Isidro, Villa Muriel and Palencia on the Carion. Orders had been given to mine them all and blow them up in case of attack. The points in the line of defence which concern us are the village and bridge of Villa Muriel held by the fifth division, and the town and bridges of Palencia. Palencia was held by the Spaniards, who might be about 10,000 strong. They watched the river for the greater part of its course between Palencia and Villa Muriel, but, as Wellington said, "they were not in a state of discipline which could gain them either the confidence of their allies or of themselves," and the 3rd Royal Scots was detached by the fifth division to Palencia to make sure that the bridges would be destroyed in case of need.

General Oswald, who had arrived from England that morning, was placed in command of the fifth division. He uses the words first and second in talking of his brigades. The first (Royal Scots, 9th and 38th), was commanded by

Major-General Barnes; the second (1st and 2nd 4th, 30th and 44th), was under Major-General Pringle, who had up to that time commanded the division.

Major-General Oswald describes the position about 8 a.m. as follows:—"Major-General Pringle had already posted the troops and the greater part of the division were admirably disposed of about the village, as also in the dry bed of a canal running in its rear, in some places parallel to the Carion. Certain of the troops were formed in column of attack supported by reserves ready to fall upon the enemy if, in consequence of the mine failing, he should venture to push a column along the narrow bridge. As the enemy closed towards the bridge he opened a heavy fire of artillery upon the village. At that moment Lord Wellington entered it and passed the formed columns well sheltered both from fire and observation. His lordship approved of the manner the post was occupied and of the advantage taken of the canal and village to mask the troops."

General Oswald does not mention his artillery, but we know that one battery, Lawson's, was on some rising ground in rear of the village. The column of attack at the bridge was the 2nd 4th, and its first battalion was in support of it. The 44th would be on the left of the 4th, and the 30th on the left of all, perhaps in the canal bed.

The French brought eighteen guns into action and quickly dismounted two of Lawson's guns and forced him to withdraw the remainder higher up the hill. They managed also to get some guns to bear on the troops in the canal and inflicted some loss. Among others the British major of a Portuguese regiment was killed, his head being carried off by a ball when he was sitting reading a London paper which had arrived that morning.

General Oswald continues:—"The French, supported by a heavy and superior fire, rushed gallantly on the bridge, the mine not exploding and destroying the arch till the leading section had almost reached the spot. Shortly afterwards the main body retired, leaving apparently only a few light troops."

Immediately before this repulse of the enemy a staff officer had arrived from Palencia to inform Wellington that the Spaniards had been attacked at Palencia and had given way so hastily that there had been no time to destroy the bridges. The fugitives were making for the heights in rear and the French cavalry was in pursuit. There was now no pressing danger at Villa Muriel, for the river was in flood and the fords scarcely practicable for infantry and not at all for artillery. Wellington therefore ordered the British brigades of the fifth division to leave the Portuguese and some light troops and detachments to watch the fords and to withdraw and to move along the hills towards Palencia ready to attack in flank any advance of the French from that place. In the letter to his brother Sir Henry Wellesley, already quoted, he says: "If I had not ordered a movement on the enemy's flank by the British troops, not only the enemy would not have been driven across the Carion, but they would have carried the heights above Villa Muriel on which the Spanish troops were posted." This masterly movement entirely checked the French advance and the Spaniards were given time to re-organize. Our men not knowing by whose orders they were moved were very angry at being called on to quit Villa Muriel and very unfairly blamed their new commander, Oswald, for what they considered a retreat, saying: "If General Leith were here we could beat the French easily." Later in the day the division was moved again and formed on the right of the Spaniards on the heights. Villa Muriel was now a trifle to the left front of the division.

Immediately after Lord Wellington had moved the division from the village in the morning the French had concentrated their fire upon the Portuguese light troops at the ford and obliged them to seek cover in broken ground to the rear. A body of French cavalry had taken advantage of this to cross the ford and to make the broken arch of the bridge passable for single men by ladders and other means, and by degrees a considerable number of infantry crossed and occupied the village and the canal. They made no further progress, but skirmished with our men and the Spaniards on the heights above them; their artillery had been increased by another ten pieces, and some of them at least were able to reach the reserves from across the river. These guns may have belonged to the Young Guard who were engaged. There was no move of importance in this part of the field on either side for the greater part of the day, but Wellington, who had heard that the bridge of Banos near the junction of the rivers had been lost, had already made preparations for retreat. His first step was to throw back across the river Carion the French who had crossed at Villa Muriel.

Towards evening he rode up to General Oswald and said that all was quiet at Palencia and that the fifth division must now reoccupy the line of the Carion and hold it till ordered to retire, which would probably be on the following morning. He ordered a Spanish brigade to support the fifth division on its left.

General Oswald says of his advance that he directed "the first brigade to attack the enemy's left flank; the second to advance in support extending to the left to succour the Spaniards who were unsuccessfully contending with the enemy in their front. The rapid advance of the first brigade cut off the enemy from the bridge and forced them into the river, where many were drowned. Our casualties were not heavy considering the fire to which our men were exposed. The enemy suffered more severely."

General Oswald's very modest account does not do justice to his own services and those of his division. On receiving Wellington's order to recover the line of the river, he advanced with Barnes' brigade in line of columns. Pringle's on the left was deployed into line, perhaps on account of the fire of the French pieces on the further bank of the river. All went well till the Spanish brigade on the left came under the close fire of this battery when they broke badly. Encouraged by the steadiness of Pringle's brigade on their right they were rallied and re-formed and the advance was resumed. The first brigade struck the river about the bridge and ford, cutting off the French retreat, and Pringle cleared the line of the canal with the bayonet. The Spaniards, now full of spirit, pursued the French mid deep in the river and killed many. The friend who wrote the sketch of Hamilton's career for the *United Service Journal*, after the colonel's decease, uses what seem to be Hamilton's own words in describing the decisive phase of the action. "The 30th and 44th did rather a dashing thing when they advanced in line against seventeen pieces of cannon, cleared the adjoining village and took more prisoners than their own force was composed of." The qualifying word "rather" must surely have come from Hamilton himself. Both artillery and musketry fire went on long after dark across the Carion. During this long day's fighting our men's pouches had to be refilled more than once.

Lord Wellington's baggage had been moving to the rear all day and he now issued orders for it not to stop before Valladolid was reached and for the army to retire during the night across the Pisuerga by the bridge at Cabezon, covered by the fifth division which was to hold its ground at Villa Muriel.

At daybreak on the 26th, the rest of the army having got well away, the fifth division started on its fifteen-mile march to Cabezon, which it reached in the evening without having been attacked. As soon as it had crossed, the bridge was destroyed and the bivouac established near the bank of the river.

The French have been blamed for not having made a more strenuous effort to improve their success when they had got the Spaniards on the run at Palencia, but their leading divisions had fought at Salamanca three months before and had learned not to be too venturesome when dealing with Wellington. It is said, too, that when their army passed through Torquemada it had found the stores of wine far from exhausted and had got even more drunk than the British before it.

The want of discipline and cohesion on the part of the Spaniards was caused by their corrupt Government. They often showed as individuals the courage which formerly had made their infantry the terror of Europe.

The 30th had engaged on the 25th :—Lieut.-Colonel Hamilton; Captains Bamford, Hitchens (wounded), and Chambers; Lieutenants White, Mayne, Andrews (wounded), Heaviside, Garvey, Freear, Rumley (severely wounded), Brisac (wounded), Baillie and Campbell; Ensigns Beere (wounded), Madden (wounded), and Tincombe (wounded); Paymaster Wray and Surgeon Hughes.

Beere, Madden and Tincombe may have been wounded carrying the colours.

Garland, the adjutant, and Lockwood had been sent down country sick. Pratt was in charge of a convoy of sick. Neville had not recovered from the wounds received at Burgos. Assistant-Surgeon Evans was in charge of sick at Salamanca.

Mr. Hughes serving as a volunteer, Sergt.-Major Woods, 21 sergeants, 11 drummers, 176 rank and file were present. Seventeen sergeants, 4 drummers and 310 rank and file were absent sick or wounded.

The casualties were :—

Killed—Two sergeants and 2 privates.—Sergeants James Keith, of Paisley, and John Nightingale, of London; Privates Patrick Shearin, of Longford, and Morgan Blakeney, of Athy.

Wounded—Seven officers and 25 other ranks carried off the field, and 9 rank and file left in the enemy's hands. Private Tanzy is known to have died of his wounds.

One private is reported missing, unwounded.

Mr. Hughes serving as a volunteer was reported missing. Seven men in addition to the above were reported missing at this time and are supposed to have been taken with the divisional baggage. All became prisoners of war and nearly all fought at Waterloo. The nine wounded prisoners were taken by the French during their temporary occupation of Villa Muriel.

The regiment had 230 officers and men engaged at Villa Muriel, of whom more than one in five were hit.

Very soon after the action a draft arrived from the depot. The date is not known but on October 25th it is shown as marching up country. Captain Richardson from the first battalion was in command, and the other officers are given as Lieutenant O'Halloran, Ensigns Sam Robert Poyntz, Madden and Kelly. The strength of non-commissioned officers and men was ninety. The above officers arrived about the same time as the draft, but they did not all belong to it. Some came independently, others, like O'Halloran, with a party of convalescents of different corps from the base hospital. Madden must have preceded them all, for he fought and was wounded at Villa Muriel.

On the 27th the French were opposite Cabezon but made no attempt to force a passage. Wellington continued his retreat down the left of the Pisuerga and they tried to out-march him on the right bank. On the 29th our army reached the point where the Pisuerga falls into the Douro and crossed to the left or southern bank of the last named river. On November 9th, Wellington was back in his old position of S. Cristobal covering Salamanca. On the following day he was joined by General Hill with the divisions left behind at Madrid in August. Since the morning after the action of Villa Muriel the 30th had no men reported missing, but the sufferings of the army had been very great and so had been the losses from straggling.

The French armies of the centre and the south under King Joseph and Marshal Soult now joined that of the north under General Souham, and Wellington hoped that they would attack him. Acting on the advice of Marshal Soult, King Joseph attempted to seize Wellington's line of retreat to Ciudad Rodrigo, at the same time keeping carefully out of reach of such a blow as Marmont had received, but on the 15th Lord Wellington marched across the front of the French army in order of battle and regained his communication with Rodrigo.

On the 16th he continued his retreat, the light division and the cavalry forming the rear-guard. The march was through a forest practicable in all directions and filled with great herds of swine fattening on the sweet acorns. The alarmed herds galloped between the columns and the men, some of whom had been obliged, like the pigs, to feed on acorns, opened fire and killed many. Wellington, thinking from the volume of fire that the French had broken through his flank guards, hurried to the scene of action, but it was impossible in some divisions to make the hungry soldiers keep their ranks and the French captured 2,000 of them. The 30th had one man missing.

The night was passed in great discomfort. There were no rations except a pound of flour, which was useless, for the wood of the forest was too green to burn, and there was no other fuel. The rain was ceaseless and cold.

The march on the 17th was even more uncomfortable. The weather did not improve and the constant rain had flooded the numerous rivulets crossing the road and the men had frequently to wade deep. Our cavalry also seem to have been badly handled. The enemy rode along the flanks of our columns and carried off stragglers and baggage with impunity, and even made prize of Sir Edward Paget. That great soldier had noticed that a gap had been allowed between the fifth and seventh divisions, and was riding to put it right when a squadron of Frenchmen carried him off. Wellington might easily have met with the same fate for he was, as usual, riding from point to point where he thought his presence was most required. An attempt of the French to cut off his rear-guard at the fords of the Huebra was frustrated by his presence. When the army halted, the word was given for all baggage and stores to push on and not stop till Ciudad Rodrigo was reached.

The 30th had another man reported missing this day.

The 18th was the last march in retreat and was perhaps the worst, although the French, who were themselves starving and exhausted, no longer pressed the army, a few cavalry merely following to glean the stragglers and baggage.

The marshy plains exhausted the men, who frequently lost their shoes in the adhesive soil, and there was as usual constant wading across the gullies which carried off the water. Hundreds of men sank exhausted, unable to keep up with their comrades.

In the afternoon, however, the weather cleared and the sun came out for the first time for many days. Sloping ground was reached which gave dry bivouacs, and fuel and rations were issued. Sir William Cope's history of the Rifle Brigade says that at the first issue, men with fixed bayonets had to be posted to protect the bags of biscuits from the starving soldiers.

On the following day the army was billeted in the villages round Ciudad Rodrigo and the cavalry and mounted guerillas were sent back to bring in the men who had not been able to keep up. A thousand or 1,500 were thus rescued. As the French had retired, these men must have perished miserably where they were but for the mounted troops. There appear to have been five of the 30th among the men so brought in.

Napier estimates the total loss of the army in the double retreat from Burgos and Madrid to have been not less than 9,000, including those lost in the siege of Burgos. Probably three-fourths of this loss was due to straggling and want of discipline and this led to the following circular letter being addressed by Lord Wellington to officers commanding divisions and brigades :—

" The discipline of every army after a long and active campaign becomes in some degree relaxed—but I am concerned here to observe that the army under my command has fallen off in this respect to a greater degree than any army with which I have ever served or of which I have ever read. Yet this army has met with no disaster. It has suffered no privations which but trifling attention on the part of the officers could not have prevented, nor has it suffered any hardships except those resulting from the necessity of being exposed to the inclemencies of the weather when they are most severe.

" It must, however, be obvious to every officer that the moment the troops commenced their retreat from the neighbourhood of Burgos on the one hand and from Madrid on the other the officers lost all command over their men. Irregularities and outrages of all descriptions were committed with impunity and losses have been sustained which ought never to have occurred."

How far is this just to the army and in particular how far can it be said to apply to the 30th ?

In the first few days of the retreat (from October 21st to the 26th) the regiment had 2 drummers and 23 rank and file taken prisoners, of whom nine are known to have been wounded and seven we have reason to believe were captured through no fault of their own with the divisional baggage. During those days the division was in the rear-guard and hard pressed by the enemy.

From October 26th to November 15th, no men were reported missing. On the last three days of the march when the French horsemen were circling round our columns carrying off baggage and stragglers by hundreds, only three prisoners were lost, and for three half-starved exhausted men to fall out and get snapped up is no great matter.

After the French had abandoned the pursuit when approaching Ciudad Rodrigo five men had to be helped in, unable to walk, but that has no bearing on the question of discipline. The fact remains that from October 26th to November 19th, when the army was at its worst, the 30th only lost three prisoners. It should be added that General Oswald prided himself on the way the fifth division held together and the absence of straggling.

There is another side to this question of discipline. The commissariat claimed to have issued three days' bread to the army before leaving Salamanca, but so far no battalion has been found which acknowledges to having received this issue. In fact what Lord Wellington said of the officers of his army was almost mild when compared with what those officers said of the staff and commissariat who were, as a body, accused of attending to their own comfort and safety and letting the army get along as best it could. As everyone knows, when an army is on the move, especially in retreat, the supply officer must be a very able man whose whole heart is in his work if he is to do any good at all. It cannot be said that this description fits the civilian commissaries in the English army in the Peninsular or Waterloo campaigns. Wellington could do little with them; if he sent one home he might have a worse one sent out to him.

It is to be feared, therefore, that the supplies which the Commander-in-Chief believed to be issued to his army did not reach the regiments owing to neglect on the part of subordinate officials. He was at this time without the services of Sir George Murray, his trusted quartermaster-general, and of Sir Richard Kennedy, the very efficient head of the commissariat.

Before leaving the subject of discipline, two things should be mentioned which must have had a powerful effect upon it one way and the other. The first is that no officer or man had been paid since April 25th. The second that Lord Wellington throughout the campaign really surpassed himself. He seemed to know, as if by instinct, where things might go wrong and his presence be of use, and his composure was always unruffled and his judgment quick and sure. The army soon recovered its strength and discipline in billets.

For a time Lord Wellington kept his men together behind the Agueda, but when the French drew back and went into winter quarters, he broke up his army and sent the fifth division to the neighbourhood of Lamego, whence it had started for the Salamanca campaign in June. The 30th was stationed at Villa Nova del Rey which does not appear on ordinary maps. They reached it about December 4th.

On the second of that month when passing near Celerico, where there was a base hospital, Ensign Kelly, who had come out with Captain Richardson's draft in October, went sick on the march and died in the hospital wagon before reaching the town. The men were in rags and mostly barefooted, but the new clothing was on its way up from Lisbon and Lord Wellington authorized the free issue of a pair of shoes. The paymaster-general, too, was now in a position to issue pay to and for August 24th. The Headquarters of the army were at Frenada near Celerico.

On December 1st the following order appeared :—

"The Commander of the Forces desires that all effective privates of the second battalions of the 24th, 30th, 44th, 53rd and 58th shall be transferred into four companies of those battalions.

"The four companies of the second 30th and four companies of the second 44th, are to be the 4th provisional battalion under Lieut.-Colonel Hamilton of the 30th.

"The Staff of the second 30th are to perform the duties of the provisional battalion to which they are attached.

"The 4th to be in Major-General Walker's brigade, fifth division."

The reason for this order is given in a long correspondence between the Duke of York, Commander-in-Chief of the army, and Lord Wellington. His Royal Highness had since 1803 been striving for a permanent establishment of the army into a service and a home battalion. The loss of the pick of the army in the Corunna campaign and in the disastrous expedition to Walcheren had forced the Government to send the home or second battalions to the Peninsula before they were fit, and the Duke's scheme seemed to have come to an end, but in 1812 he felt that he was in a position to give Wellington other troops and get the second battalions home again. Wellington might at one time have been glad of the exchange, but now that the lads of 1809 were grown into men and had been three years in the field he objected strongly, protesting that the second battalions were the best troops he had. His Royal Highness was perfectly courteous, but quite resolved. He pointed out that he had to think not only of the service in the Peninsula, but of the efficiency of the army all over the world. Lord Wellington had to give way, but resolved to keep what he could. He therefore broke up all his second battalions as described in the order of December 1st, and formed a number of provisional battalions, by which means he hoped to keep all the effective men of the second battalions.

December 25th is the last day on which the battalion was mustered before being broken up and the following was its state :—

Present, effective—Lieut.-Colonel Hamilton ; Captains Bamford, Hitchen, Chambers and Richardson ; Lieutenants White, Mayne, Andrews, Heaviside, Garvey, Eager, Freear, Rumley, Baillie and Campbell ; Ensigns Lockwood, Pratt and S. Poyntz ; Adjutant Garland ; Surgeon Hughes ; Assist.-Surgeons Evans and Clarke ; Sergt.-Major Woods ; Paymaster's Clerk Thos. Cuthbert ; Armourer Nat. Artis ; Drum-Major Thos. Vipond ; 24 sergeants, 10 drummers, 206 rank and file.

Major Leach, Captains Nunn and Tongue were with the first battalion in India. Captains Malet, Craig and Stewart were on the staff in the Peninsula, Macnab was commandant at Figueras, Machell sick, Spawforth and Fullerton at the depot at Hull, Lieutenants Neville, Brisac and Madden wounded, and Lieutenant O'Halloran sick, Daniel acting quartermaster at Figueras, Harrison on the way to join, Charles A.D.C. to Sir Robert Wilson, Hughes attached to Portuguese army, Ensign Tincombe wounded, Flude and Beere sick, Travers attached to 71st regiment, Paymaster Wray "gone for money." Q.M.-Sergt. J. Williamson, 3 sergeants and 22 rank and file were on command ; 10 sergeants, 5 drummers, 317 rank and file were sick or wounded.

The strength of the depot at the same time was : Sergeants, 17 ; drummers, 8 ; rank and file, 167.

The skeletons of the six companies which had been deprived of their effectives embarked on February 13th, 1813, at Lisbon and reached Portsmouth on the 24th.

Captain Bamford was in command and the strength was 6 sergeants, 4 drummers, and 60 privates. Many of the officers, especially those who were sick or wounded, no doubt went home independently. We have no embarkation or disembarkation returns. The six companies joined the depot at Hull in the beginning of March, 1813.

We know a little more about the four companies kept by Lieut.-Colonel Hamilton. Their strength was Lieut.-Colonel Hamilton; Captains Hitchen, Chambers and Richardson; Lieutenants Garvey, Rumley, Freear, Eager, Harrison and Campbell; Ensigns Pratt and Lockwood; Adjutant J. Garland, Paymaster Wray, Assistant-Surgeons Evans and Clarke, Sergt.-Major James Wood, Armourer-Sergeant Nat. Artis, Drum-Major Thomas Vipond, 474 other ranks.

During its service in the Peninsula the 4th battalion remained quartered at Villa Nova del Rey. We have no records of it, but it must, like the remainder of the army, have greatly improved in health with the prolonged rest and been busy preparing for the great advance of 1813, but it was not allowed to share in it. The Duke of York insisted that it was ruin to a battalion to break it up and that he must have the second battalion complete at home.

The result was the following order by Lord Wellington, dated Frenada, May 10th, 1813:

"1. The 4th Provisional Battalion is to march to Lisbon and the detachments which form it are to proceed to England by the first opportunity.

"2. The Commander of the Forces returns his thanks to the commanding and other officers and the soldiers of the 2nd battalion 30th and 2nd battalion 44th, for their services during the time they have belonged to this army."

To this Major-General Robinson, who had taken command of the brigade in March, added an order complimenting Lieut.-Colonel Hamilton on the efficient state to which he had brought his battalion.

Only one change of importance had taken place since the battalion was formed. Captain John Hitchen, who had distinguished himself at Badajoz and Villa Muriel, being twice wounded in six months, was appointed to the 12th Royal Garrison Battalion at Lisbon, and Lieutenant James Eager accompanied him.

Lieut.-Colonel Hamilton embarked with his four companies at Lisbon on June 18th, and landed at Cowes on the 29th. From Cowes the wing marched to Sandown. The strength was 377 non-com. officers and men. The remainder of the four companies, consisting of from 60 to 70 men on command or too sick to be moved, had been left in Portugal under Lieutenant Daniel.

Colonel Hamilton at once asked permission to give furloughs to his veterans, which was granted without restriction.

While the provisional battalion was serving in the Peninsula, the six skeleton companies which returned home under Captain Bamford at the end of February had gone some way towards forming a battalion. They had been stationed for a month at Hull, but on April 14th they moved under

Major Bailey to Berwick, taking with them the two depot companies; the strength was already 245 non-commissioned officers and men.

When the Headquarters and four service companies were expected to return home, Major Bailey was ordered to take his half battalion to Jersey. It had left Hull with eight companies, but the depot companies were now absorbed. The half battalion embarked at Holy Island on June 26th, and on July 16th landed at St. Aubin's in Jersey and marched to Grouville Barracks.

The strength was:—Major N. W. Bailey; Lieutenants R. C. Elliott, A. Baillie, J. Gowan, William Pennefather, Brisac, T. O'Halloran, J. Roe (1), J. Roe (2), R. Smith; Ensigns G. Madden, H. Beere, S. Poyntz, D. Macdonald, E. Prendergast; Quartermaster J. Kingsley, Surgeon J. C. Elkington, 29 sergeants, 353 drummers, rank and file.

Assistant-Surgeon J. G. Elkington of the 24th Foot had been appointed surgeon in March, *vice* Hughes.

On September 20th, Lieut.-Colonel Hamilton and his four companies also landed at St. Aubin's after being ten days in crossing from the Isle of Wight. He brought with him the same officers who had embarked at Lisbon with the addition of Ensign La Touche. The battalion was now complete, with the exception of the men in Portugal under Lieutenant Daniel, and was strong enough to take the field if called upon.

Several changes of importance took place among the officers about this time.

Lieut.-Colonel (Brevet-Colonel) William Lockhart, who had been commanding a district in India almost continuously since 1808, was promoted to major-general on June 4th (the King's Birthday). This gave a lieutenant-colonelcy without purchase to Major Christopher Maxwell, and the majority was given to Brevet-Major Charles Albert Vigoureux of the 38th, then serving in the Peninsula. On June 21st Major Vigoureux greatly distinguished himself leading a light battalion at the battle of Vitoria, and on receiving news of his promotion he returned home, reporting his arrival on September 24th. For his services in the Peninsula he received the brevet of Lieut.-Colonel, and was made a Companion of the Order of the Bath.

On September 1st, Captain T. B. Bamford had received a well-earned promotion, being given a non-purchase majority in the 6th West Indian Regiment.

On September 30th, Captain Elias Malet, who had served as D.A. Quartermaster-General attached to the Spanish army from the time the battalion joined Lord Wellington at Torres Vedras, was killed in action at the passage of the Bidasoa (Bridge of Vera).

On September 2nd, Bt.-Major Spawforth was given a majority in the 96th regiment.

The month of September was marked by the first appointment of colour-sergeants.

The rank had been created by an order of July 6th to show the appreciation by His Royal Highness the Prince Regent of the meritorious services of the non-commissioned officers of the army during the war, but the execution of the order had been postponed in the battalion pending the return of Lieut.-Colonel Hamilton from the Peninsula.

The colour-sergeants were to be men of approved valour, to whom the duty of attending the colours in the field was to be entrusted. They were to wear above the chevron the honourable badge of a regimental colour supported by two cross swords and were to draw a special rate of pay of 2*s*. 4*d*. a day.

The first colour-sergeants were Thomas Connelly, Mathew Donellan, George MacCann (who had been drum-major at Badajoz), John Ward, William Brien, Joseph Berrige, Joshua Scotton, Christopher Barnewell, and Joshua Harrison. Although the order said that there was to be a colour-sergeant in each company, Colonel Hamilton did not transfer any of the men appointed, so that some companies had three colour-sergeants and some had none. Probably he thought that discipline would suffer if he thrust men into a new company over the heads of non-commissioned officers who were doing their work to the satisfaction of their captains.

Colonel Hamilton had scarcely got his battalion together again when he was warned to prepare a draft for the first battalion in India. Luckily it was a small one of only twenty men, for the second battalion was soon wanted for active service again.

In November the battalion was inspected at Grouville by General Don, commanding in the island, when it was able to show under arms 37 sergeants, 30 corporals, 13 drummers, and 537 privates.

General Don said of the non-commissioned officers that " they were very young, but performed their duties with promptitude and energy, the privates, with the exception of about 100, who were much too young for any service, were a good body of men with the appearance of health and cleanliness, their conduct was soldier-like on duty and in quarters.

" The regiment is in the habit of practising with ball ammunition."

In spite of continual efforts on Colonel Hamilton's part nearly all the captains were still absent, either with the first battalion in India, or on the staff in the Peninsula. Lord Wellington always strongly resented any attempt to meddle with his staff. In addition to those on the staff in the Peninsula Bt. Lieut-Colonel Williamson, who had been on the staff of the first division in Spain, had been appointed superintendent of the military hospital at Deal.

Lieutenant Daniel was still in Portugal in command of the 60 or 70 men left there. Lieutenant Hughes was still serving with the 9th Portuguese regiment; Ensign Rogers, who had lately been gazetted to the 30th from volunteer with the 77th regiment, was allowed to continue to serve in Spain with the hope of obtaining a vacancy in that corps in which his father was a captain.

Immediately after this inspection Lieutenant and Adjutant John Garland was given a company in the 73rd as a reward for his distinguished services. He was succeeded as Adjutant by Lieutenant M. Andrews, who had commanded a company at Fuentes Onoro, Badajoz, Salamanca, and Villa Muriel, where he had been wounded.

On December 19th H.R.H. the Commander-in-Chief ordered the second battalions of the 30th and 81st to be ready to embark for Holland.

The causes which led to the battalion being again sent on active service are narrated in the following chapter.

CHAPTER X

SECOND BATTALION. SIEGE OF ANTWERP. REGIMENTAL DEPOT RE-FORMED AT COLCHESTER. RETURN OF PRISONERS OF WAR.

ON October 19th, 1813, Napoleon suffered a disastrous defeat at Leipsic at the hands of the Russians, Prussians, Austrians and Swedes, and, as he retired towards France, the nations subject to him saw an opportunity of regaining their freedom. In November the Dutch rose and appointed a provisional government which sent delegates to London to ask for help and to invite the Prince of Orange to return to his country.

His son, the Hereditary Prince, had during his years of exile served with the British army in the Peninsula, and was a general officer in our service.

The result of the Dutch appeal was the despatch of a British force to Holland, and Sir Thomas Graham was selected to command it. He was given two divisions; the first, commanded by Major-General Cooke, included a brigade of Guards, the 4th Royal Scots, a wing of the 21st Fusiliers, the first battalions of the 33rd and 58th, and the second of the 37th, 38th, 44th, 69th, and 91st. The second division, commanded by Major-General Mackenzie, consisted of the second battalions of the 25th, 52nd, 56th, 73rd, 78th, with four companies of the 95th (Rifles). To this division the second battalions of the 30th and 81st were to be posted on arrival. It will be seen that the Government in its desire to do something had again disregarded the organization of the army in regiments with a home and a foreign battalion. His Royal Highness the Commander-in-Chief in writing to Sir Thomas Graham expressed his regret that he had to send him the second battalions, containing as they did so large a number of old men and boys. It appears therefore that most of the second battalions were still in the state from which the 30th had emerged three years before.

At the end of December the transports arrived at Jersey and on January 1st, 1814, the draft of twenty men for the first battalion in India embarked for conveyance to the Isle of Wight, where the East India fleet was to assemble. Lieutenants W. Pennefather, S. R. Poyntz, W. Campbell, and D. Macdonald, all of them Peninsula officers, were ordered to join the draft in the Isle of Wight.

On January 2nd, the battalion embarked in three transports; Lieut.-Colonel Hamilton with the Headquarter wing in the *Union*, Lieut.-Colonel Vigoureux with the other wing on the *Saragossa* and Lieutenant B. W. Nicholson with the men to form the depot on the *Earl of Cathcart*. Portsmouth was reached on the 19th, and there Lieutenant Nicholson, Ensigns Tincombe and Drake, Quartermaster Kingsley, 11 sergeants, 7 corporals, 2 drummers, 69 privates fit for duty, 37 recruits, 16 lads and boys, and 20 invalids for discharge were landed to march to Winchester and form a depot. Among the sergeants was Luke Lydon, the acting schoolmaster. The depot also took on its strength 48 men on furlough, 53 in Portugal under Lieutenant Daniel, 68 recruiting, and 33 prisoners of war lost in Spain.

The property of the officers' mess was left in charge of the depot. While in the field the officers were to mess by companies as they had done in the Peninsula.

On the morning of the 22nd, the fleet of transports with the 30th and 81st on board, sailed from Spithead, convoyed by H.M.S. *Rinaldo*. After six days of baffling winds the majority came to an anchor in the Downs, but the drunken master of the *Union*, against orders, anchored off Dover on the evening of the 27th. Here he was caught by a heavy gale and the ship was expected to part from her anchors. A regiment of the garrison was ordered to line the shore to save life, in case she should be cast away, and prayers were offered for her in church. On the evening of the 29th the wind came off shore, and on the following morning a pilot managed to get on board. By his order the cables were cut and he got the ship safely round the Foreland and into Ramsgate. She was found to be so badly strained that Colonel Hamilton and his wing were transferred to the transport *Sophia* in which they joined the *Rinaldo* on the 31st in the Downs, where the convoy was awaiting the arrival of more transports. The *Saragossa* was not there, having lost the convoy at the start.

On February 2nd, the fleet sailed, but did not pass Schouwen Light till the evening of the 6th. After that a quantity of ice was met with, but Helvoet Sluys was reached on the 10th. The *Saragossa* with Major Vigoureux and his wing arrived on the 14th, none of the men and scarcely any of the officers having put foot on shore for forty-four days. The battalion was assembled at Willemstad and marched to join Sir Thomas Graham at Loenhout, which it reached on March 2nd.

The following are the officers who landed in Holland: Lieut.-Colonel Hamilton; Major (Lieut.-Colonel) Vigoureux; Captains T. W. Chambers, R. Machell; Lieutenants J. Gowan, J. L. White, R. Mayne, R. Heaviside, R. C. Elliott, J. Rumley, A. Baillie, P. P. Neville, John Roe (1), John Roe (2), T. O'Halloran, R. Harrison, P. Lockwood, J. Pratt, J. Flude; Ensigns H. Beere, E. Prendergast, W. Warren, T. Moneypenny, and D. La Touche, Adjutant M. Andrews, Surgeon J. Elkington, Assist.-Surgeons J. Evans and Patrick Clarke, Paymaster H. B. Wray. In addition to the above Lieutenant A. W. Freear and Ensign J. James joined in a transport from Harwich on the 23rd, and Major N. W. Bailey by mail packet on the 25th.

No quartermaster accompanied the battalion; Kingsley had been ordered to India to take the place of Wilson deceased, and Quartermaster-Sergeant J. Williamson who was to succeed had not yet returned from Portugal.

Sergeant-Major James Wood and the other staff sergeants who had served in the Peninsula were present. The number of non-commissioned officers and men disembarking was about 480.

When the 30th disembarked, the French had quitted the Netherlands, partly under pressure from Sir Thomas Graham's force and a Prussian corps under General Bulow, partly to concentrate for the defence of France against the main armies of the Allies under the Emperors of Austria and Russia and the King of Prussia. They had left garrisons in Antwerp and Bergen-op-Zoom and a few lesser places.

Carnot, the great War Minister of the Republic, had offered his services to Napoleon and been placed in command of Antwerp. He had under him 12,000 men, and there were several men-of-war in the harbour. The Prussian army had now followed the French to the main scene of action and Sir Thomas Graham was left to his own resources, which consisted of little but infantry and might number 7,000 effectives.

The French men-of-war in the Scheldt caused him uneasiness and ten days before the 30th joined him, he had made an unsuccessful effort to burn them by long range fire from mortars. He now determined to carry the fortress of Bergen-op-Zoom by storm and to follow this up by the assault of a smaller place, Fort Lillo, which is still nearer Antwerp. He would then effectually command the river Scheldt below Antwerp. The 30th had no share in the attack on Bergen-op-Zoom, which failed, but they were detailed for the attack on Lillo which was to have followed had that on Bergen been successful.

From this point the part which the regiment played in the campaign under Sir Thomas Graham and in the following year under Wellington is described by Macready in his inimitable diary.

Edward Neville Macready, the younger brother of the great actor, proceeded to the Netherlands at the age of sixteen in hope of being allowed to serve as a volunteer in one of the British regiments under Sir Thomas Graham, to whom he had a letter of introduction. When he presented his letter, Sir Thomas decided to place the lad with his own old friend Colonel Hamilton of the 30th.

Macready joined at Loenhout on February 28th, 1814, and like all volunteers was attached to the light company, then commanded by Chambers.

His narrative may be accepted with trust and his judgments are fair, except for a bias against all in authority over him, especially when they call him out for drill. That always puts him into paroxysms of rage. As is well known, the Duke of Wellington's army consisted in great part of raw levies, and in the judgment of the more experienced officers was not, as a whole, fit to manœuvre against the French, however bravely it might fight in a fixed position ; but the boy of sixteen shows his rage and contempt in a most amusing manner when there is the least attempt at instruction. None of his would-be teachers escapes. Colonel Hamilton, the Prince of Orange, Sir Thomas Bradford, Sir Thomas Belson, all are equally incompetent and foolish. One feels that it can only have been by a strong effort that Macready avoided saying that " The Duke " could not handle troops.

The early part of the diary is concerned with Sir Thomas Graham's efforts to shut up the French fleet in Antwerp. After the failure of the attack on Bergen, Sir Thomas effected his object by the erection of batteries on the right bank of the Scheldt near Fort Frederick, and it was to assist in this that the 30th was brought on March 13th, from Brasschaet to Putte.

MACREADY'S " JOURNAL "

" The morning after I arrived at Loenhout (Feb. 28th, 1814), the regiment marched to relieve the 35th, who were erecting a battery near the village of Brasschaet to command the high road from Breda to Antwerp. Our men appeared fine young fellows, remarkably clean in appearance, and I soon discovered from their swarthy cheeks and frequent exclamations of 'Jesu Maria!' and other less holy expressions, that many had shared in the Peninsula campaigns. We arrived at our destination about six o'clock and half the men were set to work with pickaxes and spades on ground which turned the iron at every stroke.

"Our post was composed of two or three barns well loop-holed and connected by stockades. Our videttes were about a mile in advance. In twenty-four hours we were relieved and returned to our hovels at Loenhout. The snow was at this time deep on the ground and continued so till late in April. This, with thick fogs, frequent sleet showers and chilling breezes, rendered our working day very harassing.

"The villages we occupied had been abandoned by their inhabitants and destroyed by the Prussians. Not an article of furniture nor a pane of glass was to be found in them; they were mere sheds, and when we woke in the morning we were petrified with cold and frequently covered with snow. Indeed, but for the accomplishments of smoking and gin-drinking, which I speedily acquired and most strongly recommend to all campaigners in Holland, I think this unusual freezing would have proved as injurious as it was unpleasant.

"After one day of rest we marched back again to our battery and I was sent on an outlying picket with Lieutenant Rumley and a sub-division of the light company. The nights we passed on this duty—for from this day we were invariably detached—were I think the most tedious in my life. The cold was intense. After our twenty-four hours' work we returned to Loenhout and next day marched back to Brasschaet.

"On the following morning, as we were entering Wustwezel, to which we were now directed to change our cantonments, a staff officer galloped up to the head of the column and ordered us to march to Calmhout as Bergen-op-Zoom had been stormed on the preceding evening and we, with some other corps on the following night, were to attack Lillo; anxiety of some kind beamed in every face and we stepped out famously. The spire of Calmhout was already in sight when our brigade-major met us with a horribly long face, and told us to counter-march, for that Bergen was lost, and with it 2,000 men of Cooke's division. The road, which seemed to fly from under us on our advance, appeared to lengthen as we returned, and it was night before we entered the ruins of Wustwezel. Our men had gone 30 miles after twenty-four hours' work, preceded by a march, and were growling like bears.

"'Why weren't we sent there?' was indignantly murmured from the ranks. For my part, I felt, like all young fellows on these occasions, an apprehensive sort of anxiety as we pushed on, and abused Cooke's people as roundly as the rest of them, when we retired. But notwithstanding the fretful indignation which I heard vent itself so generally around me, I remarked that our sorrow for this imagined deprivation of honour by no means influenced our appetites; like Ancient Pistol, we could eat and swear.

"Our regiment continued to take the Brasschaet duty as usual, and the outposts were ordered to be doubly vigilant. Our situation was rather critical as the garrison of Antwerp were equal to us in number, and Maison could have thrown a few thousand men over the river whenever he pleased. He attacked some Saxons a few miles to our left one day, and completely routed them;

we remaining under arms during the firing, which was very brisk. About March 13th, the regiment changed its quarters to Putte, a neat village in tolerable repair, between Bergen-op-Zoom and Antwerp; and as we marched thither from Brasschaet, our stupid guide led us close up to the French videttes, who sat on their horses like statues, and we did not molest them.

"We now ceased our Brasschaet duty, and, in exchange, were to go twice a week to occupy a battery erected on the Scheldt, to prevent any communication by the river between Antwerp and Bergen-op-Zoom. It was erected near an old fort called Frederick Hendrick, and was about 10 miles from our cantonment. We started for it the first time at twelve o'clock on the night of the 19th. It was dark as pitch, and our men were incessantly slipping over the icy roughness of the road, and occasionally some unfortunate fellow would plump through the thin ice of the running stream.

"The only guide we had to keep the path were the curses of those who were falling off it, and just as light dawned, and we congratulating each other on the conclusion of our annoyance, we found ourselves checked by a branch of the inundation which crossed the road. It was about up to our middles, and through it we dashed. The very recollection of it chills me. We reached our post in a dense fog, and I was told we had relieved the 78th Highlanders. About eight o'clock a breeze sprung up, and, carrying off the grey mantle night had left behind, showed us a 6-pounder pointed down the dyke, and a howitzer in battery towards the river. The remains of the old fort and two or three houses were close to us, and about a mile lower down the dyke was our battery of six long 24-pounders, with its furnace for heating shot. In a contrary direction, and at about double the distance, was Lillo, and Liefenschoeck immediately opposite; between them were a French line-of-battle ship and some craft at anchor. The river was here at least a mile in breadth, and the enemy having opened the sluices, the whole country was under water. The forts and houses looked like islands, which the roads and dykes connected with each other. Our advance posts, composed of riflemen, were on the bank towards Lillo, nearly a mile from us, and about 300 yards from the French sentinel. Just before the fog had cleared up, we were alarmed by a heavy fire from our lower battery, which was thundering hot shot at an enormous lump of ice, which sustained the cannonade with exemplary *sang froid*. The officer of artillery had mistaken it for a boat.

"The day passed quietly enough, with the exception of a few shots exchanged at the advanced posts. I was down with the riflemen and admired them very much. They hit every time I saw them fire. We were on the alert all night, and twice got under arms, anticipating a sally from Lillo, and conjecturing, from the noise on board the ships, that they were moving down to co-operate with the garrison. Morning broke, and exhibited *L'Anversois* (seventy-four) laid broadside opposite our howitzer, with her stern-chasers pointed at our heavy battery. She was about half a mile from us, and behind her were the small craft. 'This looked rebellion.' However, she did not open, so we went to breakfast, and I was just discussing an egg, when off went a broadside, down came the chimney in a shower of brickbats, and our poor egg-woman was cut in two by a round shot. We hurried under the dyke, and formed regularly into companies, while they kept up their infernal din, and slashed mud and ice about our ears in bushels full.

"The cannonade was kept up till near eleven o'clock when some Congreve rockets arrived and, though badly thrown, alarmed the enemy so much that they cut their cables and stood off in confusion. When we returned to Putte we buried our dead with military honours."

Macready does not give the following incident of the day. Colonel Hamilton had sat down on a bank to breakfast close to the colours when the enemy in Fort Lillo managed to get some guns to bear on that part of the line. Drum-major Vipond and Private Joseph Gabett of No. 4 were killed, two men had each a leg carried off and two more were slightly wounded. One of the colour staves was cut in two and Hamilton was upset and covered with mud and blood and fragments of flesh. He had his knees drawn up to support his plate and the officers and men present averred that a round shot had struck the ground beneath them. He was wounded three times during his service, but his escapes were so numerous that the regiment used to say "Nothing will hit him."

MACREADY'S "JOURNAL" (*continued*)

"The ships whose intention was to pass Bergen, returned to Antwerp and the force at Fort Frederick was reduced to a wing, and afterwards to a captain, two subalterns and 100 men. I went there successively with Majors Bailey and Vigoureux and Captains Machell and Chambers. The inundation had entirely flooded the road and we were obliged to march on the dyke around by Santoliet, about 16 miles. Our bivouacs were shockingly cold.

"The day Chambers commanded our detachment a small party of the enemy advanced to annoy our riflemen and with ten men I went with him to drive them in. They retired immediately, one of them was wounded.

"The weather became milder in April, but hardly more pleasant, as the thaws rendered the roads (which were principally on the top of the dykes) deep in mud and so slippery as to be difficult to walk on, and as a volunteer I was out with every party that marched.

"Soon after this, we marched from Putte, and with the 52nd, 73rd, and 81st, occupied Brasschaet where a dozen officers were stored in every room, and our soldiers lodged in barns and stables. In consequence of the progress of the Allies in France, Sir T. Graham and Carnot concluded [1] an armistice. We were now idlers, and each one followed his business or desires. Some of our moneyed men took trips to Brussels and other towns in possession of the Allies, and as my inclination was as strong, though my purse was weaker than theirs, I resolved to have a look at Antwerp and its French garrison. I borrowed a broad-brimmed Dutch hat and a white handkerchief, and thinking my grey great-coat, waistcoat, and trousers sufficiently bourgeois, mounted the diligence as it passed through Brasschaet and in less than two hours rattled over the drawbridge of the Porte de Beigerout. Here we were stopped, and a French corporal stepping up, demanded our passports. 'Je n'en ai pas,' said I. 'Eh bien, Monsieur,' said the rascal with a smile (for he knew I was an Englishman), 'descendez, s'il vous plait, vous trouverez des camarades, je crois.' I looked about the hole into which the ruffian was so politely ushering

[1] The Allied Armies entered Paris on March 31st; Sir Thomas Graham heard of it on April 4th. Napoleon abdicated and was sent to Elba in April.

me, and discovered Gowan, of ours, in similar durance. He did not understand French, so I was spokesman ; and hearing that our appearance before Monsieur Fauconnet, the Lieutenant-Governor, was indispensable, I requested as early a presentation as possible. We were accordingly paraded between the bayonets of the party, and marched up to our destination.

" Fauconnet was a gentleman, and learning that curiosity alone was the motive of our trip, after chiding us for our indiscretion, and reprimanding the corporal for marching us as prisoners, gave us leave to remain in the town till seven o'clock, after which hour, if found, we should be detained. We amused ourselves till nearly six, looking at the dockyards and other public buildings. The garrison were a set of ragged, slovenly-looking fellows, but I've no doubt effective enough on the working-day. A great number of them were our own countrymen, and were inclined to be very civil ; but I avoided them. There is something loathsome in the very idea of such traitors. We dined at a *cabaret* where, notwithstanding all the Abbé Barnel says against masons, Gowan being one saved us from being insulted by a large body of French officers. On our return from Antwerp, we found an order had arrived for march towards Brussels."

On May 1st the battalion moved into the suburbs of Antwerp, and on the following day marched to Malines—strength 30 officers, 473 other ranks.

MACREADY'S " JOURNAL " (*continued*)

" I was billeted on a doctor. Vive le Docteur Jansens ! He was an excellent antithesis to Sangrado ; his wine was excellent ; his cook inimitable ; the beds were incomparable. Since February 23rd I had not been in one. I never thought Charles XII a man of taste for taking his long snooze at Demotica till I revelled between Madame Jansens' white sheets. In a few days our corps marched to Vilvorde. I was sorry to leave Jansens' good cheer, although lately we had diminished our good understanding by repeated arguments about the battle of Trafalgar. The obstinate old fool would assert that we were entirely beaten, and to all my proofs to the contrary he replied by flourishing a poker and repeating—' Mais, monsieur, un moment—votre fameux amiral n'etait-il pas tué ? '—' Eh bien, oui ! '—' Ah ! ah ! c'est cela que je dis—il etait tuê et la flotte ecrasêe.' Nothing would convince him to the contrary, so I dropped all discussion and forswore disputation with my future landlords.[1] On May 19th our brigade assembled at Malines, and next day marched into Brussels, to the mutual delight of ourselves and the inhabitants. Our billets were excellent people of the first rank, who welcomed us to their houses, and entertained us most hospitably, and in a few days we became *enfants de la famille*.

" In the middle of June it was whispered that our regiment was to move towards the frontier, and though I was anxious to traverse a country in which every field is hallowed by some gallant action, yet I dreaded the confirmation of the rumour. I was young in pleasure, and Brussels was the scene of my

[1] On May 6th our Government had ordered all hostilities to cease. Sir Thomas Graham, now Lord Lynedoch, had returned home and the army was commanded by the Prince of Orange.

initiation. My entrance upon life was cheerless, and my welcome but a rough one for a green and bashful boy; but here the friendliness of home and the fascination of new delights alternately awaited me. At length the dreaded order appeared, and our destination was Tournay. On June 27th we left Brussels. On our march to Ath, we fell in with the 33rd French regiment of infantry returning from the north of Germany, where they had formed a part of Davoust's *corps d'armée*. Several of our fellows conversed with them in Spanish, and we had some difficulty in preventing a little conversation *à la Belcher* between the parties: swords and bayonets were unsheathed more than once. The soldiers all retained the eagle and 'N' on their chacos, and some who had lost these insignia had imprinted their resemblance in chalk.

"The Duchess of Oldenburg was at Ath when we arrived, and our flank companies remained behind the regiment as her guard-of-honour. After her departure [1] we followed the corps, and, as we marched left in front, the grenadiers swore they would run over us. The challenge was no sooner given than the lights shouted 'double quick,' and away we went—a musket on my shoulder and sixty rounds in my cartouche-box. We stopped at Leuza, about 7 miles from Ath, when only three bacon bolters (and these were overgrown light bobs) were visible. I was in high favour with the greasy rogues for keeping up, and received three or four pats on the shoulders which nearly shook out the little wind I had left, accompanied with the assurance that I 'was of the right stuff.' A Roman general, on being hailed 'Imperator,' could not have felt more grateful for his exaltation. On this day, 30th, after a long march, we entered Tournay.

"At the beginning of August we were much surprised and chagrined to receive an order to make forced marches to Antwerp, in consequence of the turbulent behaviour of the French Marine. Our friends were determined to give us some mark of their esteem, and on the evening before we marched we were invited to a ball at the Stadthouse. Our efforts at gaiety were useless; we lingered out the hours in the expression of fruitless regret and when the bugle sounded we left the ball-room. I took an affectionate farewell of my friends and shouldering Brown Bess, trudged away to Avelghem.

"Our second march was to Oudenarde,[2] the scene of Marlborough's defeat of Vendôme, and Ghent was our next halting-place. We halted one day and made a long march to St. Nicholas. As we were moving the following morning towards Locheren, a sergeant of the 44th regiment, which had served with ours in Egypt and Spain, happened to be passing in a cart. As soon as his button was recognized, a scream of congratulation was heard through the column, and every canteen was unstrapped in an instant. He was dragged from his seat and shoved from rank to rank, every fellow stopping his mouth with his canteen and shouting, 'Good luck to the old boys, and how are they?' till the worthy non-commissioned officer was replaced in his seat speechless and motionless. This is certainly an instance of killing kindness, but as an affectionate remembrance of auld lang syne even the most starched discip-

[1] The Duchess of Oldenburg, sister of the Czar of Russia, was on her way to England as guest of the Prince Regent. She was joined in England by her brother.

[2] On this march to Oudenarde the battalion passed the scene of the action of July 8th, 1693, when Lord Castleton's Foot (30th regt.) led the way at the storming of the French lines.

linarians must forgive it. For my part I felt a thrill of joy and pride at this rough exhibition of feeling in our fellows. We crossed the Scheldt on a flying bridge, and on August 27th entered Antwerp, which was garrisoned by the 21st, 25th, 30th, 33rd, 37th, 54th, 56th, 69th, 78th, 81st and 95th regiments. Major-General Halkett was governor, and Colonel Crawford commanded in the citadel.

" Shortly after our arrival, the shipping and stores were divided according to decision of the Vienna Congress, and the French with their share sailed away. In consequence, the force assembled here was broken up, and only our regiment, the 25th, 37th and 81st remained, but we were soon reinforced by numerous bodies of Hanoverians.

" As I was parading for muster on September 24th I learned that on the 8th of the month I had been appointed an ensign in the 30th regiment. It may easily be imagined how welcome was this intelligence, as I had been above six months a volunteer.

" General Halkett, who was much beloved, was ordered to give up the command to General Mackenzie, in consequence of his disagreement with the civil authorities. They would not keep the town clean, and I believe he sent fatigue parties, who collected the filth and deposited it at their doors. Whatever was the real cause he was ordered away, and the officers of the garrison resolved to give him a dinner. We sat down 300 red jackets, English and German, neither nation remarkable for temperance. All were soon drunk, and wine having got the better of our manners, we attacked the head of the mayor with champagne corks. Never was the *caput* of an Antwerp magistrate in such danger since the days of Alva. Poor man, he underwent the operation like a philosopher, ' he stirred not, he spoke not, he looked not around.' The concussive corks started the powder from his jazey, which, settling on the perspiration that poured down his cheeks, gave to his physiognomy an expression ' horrible, most horrible.' To describe the confusion of this party would be impossible—it may be fancied when I mention that 700 bottles of champagne alone were consumed."

Before Major-General Colin Halkett returned home he inspected the 30th in Antwerp. There was no change of importance among the officers except the death of Lieutenant Flude. He had suffered from a good deal of sickness in the Peninsula. Captain Machell had been appointed town major of Antwerp by the Prince of Orange. General Halkett says of Colonel Hamilton that he is " an old and zealous officer whose exertions have been unremitting on behalf of his corps." Hamilton was certainly an old officer, having joined at twelve, but he was only 40 years of age.

The report says that the privates are a fine body of men with a healthy and clean appearance, well drilled and conducted—regular and orderly in quarters.

" The officers mess together—unanimity prevails."

The mess had been formed as soon as the regiment had gone into garrison at Antwerp.

The non-commissioned officers, rank and file present at inspection numbered 5 staff-sergeants, including Sergeant Poole appointed drum-major *vice* Vipond killed in action, 26 sergeants, 443 rank and file, but there were only 3 fusils, 22 pikes and 417 muskets.

No doubt the battalion got along by taking the arms of the sick and orderlies and giving them to the duty men, but the deficiency of arms draws attention to the frightful risks the British Government sometimes runs for the sake of a petty economy, for the shortage arose not from the neglect of the regiment, but from a Government order that no steps were to be taken to complete the arms of the 30th and several other regiments. By this order the depot, which was rapidly increasing in numbers, was harder hit than the service companies.

THE DEPOT

About this time the battalion began to be filled up to the strength of officers and men which carried it through the Waterloo campaign, and therefore this opportunity is taken to show how the depot was able to supply so many trained soldiers to the battalion.

As already narrated the second battalion had put ashore at Portsmouth on January 19th, 1814, Lieutenant Nicholson, 2 ensigns, 1 quartermaster, and 159 non-commissioned officers and men with orders to march to Winchester and form a depot. After a few days' halt at Winchester the detachment marched to Chelmsford, where it remained for a fortnight and then marched to Colchester.

The non-commissioned officers selected for the depot were generally senior to those with the service companies and the rank and file included men whom General Don had pronounced too young for service, men who had not recovered from the hardships of the Peninsula and a large proportion of married men, for, up to this time, no woman with children had been allowed to accompany the service companies to the Netherlands. As the depot served for both battalions, all the non-commissioned officers and men on recruiting duty were taken on its strength.

Captain William Stewart was appointed to command on February 22nd. After his promotion from adjutant to captain he had served as brigade major in the second division till very severely wounded at the siege of Pampeluna in July, 1813. His wounds obliged him to return home, and when convalescent the Commander-in-Chief had selected him for the depot.

In April, 1814, James Poyntz, who when a boy had served as a volunteer with the second battalion at Torres Vedras and Fuentes Onoro, was appointed ensign, but was given leave to remain at the Royal Military College to complete his studies.

In this month the abdication of Napoleon and the subsequent peace had a great effect on the depot through the release of all prisoners of war on both sides.

During the war the 30th had lost to the enemy 4 officers and 103 non-commissioned officers and men captured when the transport *Jenny* was wrecked in 1806, 1 man taken at Salamanca and 31 taken at Villa Muriel and during the retreat from Burgos. Of the men captured on the *Jenny* two had managed to escape and 4 officers and 80 men now rejoined. Of the prisoners lost in the Peninsula 30 came back. The officers, Captains Thomas Fleming Roberts and Peter Ryves Hawker, Lieutenant Robert Howard and Ensign William Sullivan all got a step without purchase; the other ranks received a gratuity of a sovereign apiece.

During their long captivity at Verdun in France, the prisoners of war were fed on horseflesh, and they were in the habit of using as small change among themselves, the teeth of the horses killed for their rations. Captain Howard brought home with him on release a bag of these tokens, and some were kindly presented to the officers' mess of the regiment in a handsome ebony box, by Captain Howard's grand-nephew, Colonel Howard of Wigfair.

In addition to the released prisoners of war, the 48 men on furlough and the men left in the Peninsula, sick and on command under Lieutenant Daniel, rejoined in the early part of 1814. Of these a party of 28 were on the transport *Queen*, which brought from the Peninsula 325 men, mostly sick and invalids of the artillery, with 63 women and 58 children ; the crew numbered 21. On January 11th, 1814, she anchored in Carrick Roads near Falmouth and was there caught by the great gale. Her cables parted and she went ashore on Trefusis Point, and was beaten to pieces by the sea. One hundred men with four women with great difficulty got ashore, 363 perished. By some extraordinary chance the 30th men were saved except Privates Philip Grant, Philip Garrett, Josh. Hunt, George Rowdon, and James Waters.

The gale was accompanied and followed by snow, which buried wagons and mail coaches, and it was not till May that a route was issued for the party to march to Colchester from Exeter.

By this time the depot had 341 rank and file, but it had only 194 muskets and was forbidden to ask for more. Things were not so bad as if the newly joined men were raw recruits, but trained soldiers though they were, they needed a course of drill and ball practice. Sixty-three of the men who had been prisoners of war since 1806 were to fight at Waterloo a year later.

Captain Stewart may have made shift to give every man some ball practice, but he could not enforce the constant platoon (firing) exercise which was a necessity with the muzzle loader, for all soldiers trained or untrained. After a soldier discharged his piece he was for an appreciable time disarmed until he had reloaded and returned his ramrod. To reduce this interval of time, he was practised almost daily in the motions of loading and firing, till he could fire with extreme rapidity, but without hurry or slurring over the motions, and he learned to level justly, that is to hold the musket at the time of discharge parallel to the ground ; neither allowing the muzzle to droop nor firing in the air ; with an enemy in close order not a hundred yards off this sufficed. The constant use of the word "level," instead of "aim," shows to which our ancestors attached most importance. All our commanders, from Marlborough downwards, had insisted upon the constant and careful practice of the platoon exercise, and this was one cause of the extreme deadliness of the British fire.

Besides the return of the prisoners of war from France there were now many changes among the officers.

Captain A. Gore from the half-pay of the late Loyal Cheshire Regiment was gazetted in February as a consequence of the retirement of Captain Richardson. Ensign Madden, who had been wounded at Villa Muriel, was promoted into the 89th in April. Captain and Brevet-Major Matthew Ryan from the half-pay of the 85th exchanged with Captain and Brevet-Lieutenant-Colonel Thomas Williamson. Lieutenant Owen Wynne Grey of the first battalion, a cousin of Lieut.-Colonel George Grey who was killed at Badajoz, got Hawker's company on the latter's promotion, and Lieutenant Donald

Sinclair got his company in consequence of the promotion of Roberts. Ensigns H. Beere and Francis Tincombe, both Peninsular men, got their lieutenancies without purchase in September, and volunteer Edward McCready was gazetted to the ensigncy vacated by Beere; this, of course, is intended for Edward Neville Macready, the author of the Journal. Lieutenant J. L. White, who had commanded a company at Badajoz, Salamanca and Villa Muriel, was promoted into the 14th Foot in November, Ensign Prendergast, who was to fall at Waterloo, got White's step and James Bullen, who was also to fall in the great battle, succeeded Prendergast as ensign. In October, Quartermaster-Sergeant John Williamson was gazetted as quartermaster in succession to Wilson, who had died in India. Kingsley on his own application was allowed to go to India, in place of Wilson, and Williamson remained with the battalion in which he had served so long. He had lately returned with Daniel from the Peninsula. Colour-Sergeant Joshua Harrison succeeded as quartermaster-sergeant.

In September, 1814, the Government saw cause to increase its force in the Netherlands and Captain Stewart was ordered to prepare a draft for embarkation consisting of Brevet-Major Matthew Ryan, Captains Donald Sinclair, Lieutenants B. W. Nicholson and Robert Daniel, Ensign R. N. Rogers and Ed. Drake, 5 sergeants, 210 rank and file with 22 women and 12 children.

No man unfit to take the field was to be sent and the draft was to be completed with arms at Harwich, the port of embarkation. The quartermaster was to go with the draft. This means Williamson, although he was not yet gazetted.

The draft embarked early in October. It marched to Harwich from Colchester, to which the depot had returned after spending two months at Weeley Barracks, 7 miles from the town.

In that month, on Colonel Hamilton's application, Captain Macnab, who had returned from duty in Portugal, was ordered to join his battalion and a month later Captain Howard was sent to Antwerp at his own request.

On December 15th orders were received at the depot for Majors Bailey, Vigoureux and Roberts, one captain and two subalterns and Quartermaster Kingsley to be ready to sail for India with a draft of 3 sergeants and 79 rank and file to join the first battalion. In consequence of an appeal from Colonel Hamilton, Bailey and Vigoureux were allowed to stay with the second battalion though transferred to the first.

On December 31st Major Thomas Fleming Roberts was allowed to purchase a lieut.-colonelcy in the Regiment of Roll. Captain T. W. Chambers of the light company purchased the vacant majority and Lieutenant R. Heaviside the company. The latter was allowed to purchase over the heads of several lieutenants senior to him, but while Chambers was promoted major on February 16th, 1815, the promotion of Heaviside was delayed till June 15th, with the evident intention that he might remain junior to Easte, who got the non-purchase company vacant by Jackson's death and whose promotion was dated June 14th. A very just act on the part of H.R.H. the Commander-in-Chief.

Colonel Hamilton applied for Captain Henry Craig to be sent to him on return from staff duty in Spain; but an exchange was arranged with Captain James Finucane of the 102nd Foot, who was ordered to join. The connection of the Craigs with the regiment, which had lasted for forty years, now came to an end.

On April 10th, 1815, Lieutenant Tincombe with a small draft embarked at Harwich and on the 13th landed at Ostend and marched to join the second battalion, now at Brussels.

The depot affairs have been dwelt on at some length partly to show how, owing to Colonel Hamilton's unceasing efforts, the regiment which had suffered from want of officers in the Peninsula, entered on the Waterloo Campaign splendidly officered with four field officers in the battalion, and either three or four officers with each company, and secondly to dispel an idea that the second battalions which fought at Waterloo were necessarily composed of young soldiers. We have all heard that the battle was won by lads who in many cases were wearing the uniform of the militia regiments from which they had recently been transferred, and many fallacious theories have been founded on that supposed fact. The history of the depot of the 30th gives no countenance to such ideas. In 1814 the depot starting with 69 men fit for duty, 16 sick, 37 recruits and 16 boys and lads, received up to June 47 recruits and volunteers from the militia and during the remainder of the year only 3, who joined in October. It received, however, 197 old soldiers, that is 110 released prisoners of war, 48 men returned from furlough and 39 from Portugal. In October it gave a draft of 210 rank and file to the second battalion, and in April, 1815, another of 29. The three recruits who joined in October may have been in this draft and perhaps two or three who joined in January and February 1815, but with these exceptions the men who carried the regiment through the awful day of Waterloo had served in it for at least twelve months and most of them had served under Hamilton in the Peninsula. The non-commissioned officers, though still young men, were veteran soldiers.

CHAPTER XI

QUATRE BRAS AND WATERLOO. THE END OF THE SECOND BATTALION

1815–1818

THE opening of the year 1815 found the second battalion at Antwerp in daily expectation of an order for its reduction: already twenty-four battalions had been disbanded and their officers placed on half-pay. The officers of the 30th may be pardoned, therefore, if they heard without regret that Napoleon had escaped from Elba, had landed near Cannes on March 1st, and had been received as Emperor at Paris three weeks later.

Great Britain and her Allies refused to recognize Napoleon and prepared for a new invasion of France. Here Macready takes up the story again.

MACREADY'S "DIARY"

"Our only wish now was to be ordered forward, which we constantly expected, as the regiment had appeared in orders as a part of Sir C. Halkett's brigade, in the Walloon country. The Duke of Cambridge passed through Antwerp, and reviewed the garrison, and soon after our corps was minutely inspected by Sir H. Clinton, whose report, together with the repeated applications of Halkett, caused the wished-for order to appear, and we marched on April 8th. Our comrades of the 81st gave us three cheers as we defiled over the drawbridge. We marched to Malines, and next day to Brussels, through Vilvorde, where our old friends received us with shouts of congratulation. We halted at Brussels for three weeks, our worthy old hosts received us most kindly, and with a melancholy grace, which (though they would not shock our vanity by the confession) plainly proved they thought we marched but as sacrifices to the fire-eyed maid of smoky war.

"I was now appointed to the light company, an honour seldom conferred on an ensign, and rendered doubly flattering to me, as it was offered at the moment of danger. Captain Chambers had purchased his majority, and John Rumley commanded the Caçadores. Pratt was my brother subaltern. A detachment arrived from England under Tincombe, and we completed the company to sixty, as handsome and active fellows as ever stepped. As we marched them to and from parade, every one stopped to admire them, and there was quite a murmur of—'Ils sont des chasseurs, jolis garçons, nets, sveltes.' They were the prettiest company I ever looked at. 'Ashes to ashes; dust to dust'—poor fellows! a few years have left little of this gallant band, but recollections so dear and deep, that time alone renders them more indelible. Towards the end of April we left Brussels, and marched to Hal to join our brigade, whose Headquarters were at Chaussée de Notre Dame; we occupied two villages, Petite Rue and Steinkirk."

On July 24th, 1692, the morning of the battle of Steinkirk, the regiment (Castleton's Foot) had marched down the same road from Hal to that place.

With the arrival at Brussels on April 13th, 1815, of Tincombe's draft, consisting besides himself of Ensigns James and Bullen, 2 drummers, 2 corporals, and 27 privates, the battalion was completed to the strength with which it was to enter the Waterloo campaign. One officer alone was absent. Captain James Finucane had visited France during the peace and had nearly become a prisoner of war on the return of Napoleon; it was impossible to reach the regiment by crossing the Belgian frontier, but he was lucky enough to find a ship at Bordeaux sailing for Ostend, and thus was able to join on May 17th. Of the officers in the country but not with the battalion, Captain Machell was town major at Antwerp, where Lieutenant Andrew Baillie was also on the staff, Lieutenant P. P. Neville was with the engineer department and Ensign Berrige was on duty at Ostend, now our chief port of embarkation.

In the whole history of the second battalion it had never been so well prepared for war. The name of Colonel Hamilton has been handed down by tradition as that of a model commanding officer, as well as a distinguished soldier. Of the other three field officers, Vigoureux and Chambers had proved themselves soldiers of a very high order, and Bailey although he had not had their opportunities of distinction was to show on June 18th that he was worthy to command such a corps. Instead of the companies, as at Salamanca, being led in action by one very junior subaltern apiece, six companies had captains and all had three or four subalterns who in many cases had themselves commanded companies in the field. The staff sergeants had all served several campaigns and only eight of the company sergeants had not been in the Peninsula. The rank and file were young, strong, healthy men of whom 300 had served in the Peninsula, and sixty had belonged to Captain Hawker's company which had been shipwrecked in the *Jenny* and remained so long prisoners of war in France. Three-fourths of the men killed in the campaign were entitled to prize money won in the Peninsula but not yet distributed.

FORMATION OF THE OPPOSING ARMIES

On March 30th, Napoleon stopped all intercourse between France and Belgium.

On April 5th, the Duke of Wellington arrived at Brussels, and on the 11th relieved the Prince of Orange in command, and commenced the organization of the Allied army into corps and divisions, the constitution of which will be given later.

As usual, he looked into the actual details of the soldier's life and issued a series of orders regarding an increase in the number of camp kettles, the issue of tents, the fitting of the blankets with loops to serve in place of tents in case the latter did not come up, the reduction of the weight the men had to carry by returning the great coats into store. He fixed the daily ration at :—2 lb. of bread, ½ lb. of meat, ¼ lb. of vegetables, and a dram and a quarter of brandy. The daily stoppage for rations was 5½*d.*

He restored daily payments in the following order :—" So much benefit was felt during the late war, particularly by the soldiers of the army, from the system then adopted of paying them every day, that the Commander of the

Forces has determined to adopt it again." Daily payments were continued in the 30th for fifty years after this, and long after weekly payments had been adopted by other regiments.

By the end of May the organization and distribution of the army were complete and were as follows :—

The Allied Army commanded by the Duke of Wellington :—

1st Army Corps	The Prince of Orange.
2nd Army Corps	Lord Hill.
Reserve	Sir Thomas Picton. The Duke of Brunswick.
Cavalry	The Earl of Uxbridge.
Total strength	82,000 Infantry. 14,000 Cavalry. 204 Guns.

The first army corps consisted of the :—
First division (Guards) under Major-General Cooke.
Third division under Lieut.-General Sir Charles Baron Alten.
Second Dutch-Belgian division, Lieut.-General de Perponcher.
Third Dutch-Belgian division under General Baron Chassé.

The third division was composed of :—

5th British Brigade under Major-General Sir Colin Halkett. Brigade Major Capt. W. Crofton, 54th Foot	2nd 30th 1st 33rd 2nd 69th 2nd 73rd	615 rank and file 561 rank and file 516 rank and file 562 rank and file 2,254
2nd Brigade King's German Legion under Colonel Von Ompteda	1st Light Battalion 2nd Light Battalion 5th Line Battalion 8th Line Battalion	The Light Battalions carried rifles. 1,527 rank and file
1st Hanoverian Brigade under Major-General Count Kielmansegge	1st Field Battalions of the regiments Bremen, Verden, Duke of York, Luneburg, Grubenhagen, Jaëger Corps	3,189
	Total Infantry strength of division	6,970
Artillery under Lieut.-Colonel Williamson	Major Lloyd's British Battery Captain Cleves, K.G.L.	6 9-pounders 6 9-pounders

The numbers given above include all men in the country, but the number present and fit for duty in the 30th on the 25th of May was 34 sergeants, 14 drummers, 570 rank and file. On taking the field an officers' guard under John Roe (1) was left to look after regimental property, and the men in charge of transport, ammunition, regimental stores, etc., cannot have been in the ranks. The drummers under age, too, would be left behind. The regiment, therefore, cannot have taken the field with more than 30 sergeants, 10 drummers and 540 rank and file. According to Macready the number was about 510 rank and file, and he must have had some reason for saying so.

The King's German Legion was raised in Hanover, so that the third division was composed entirely of subjects of King George though of different nationalities. The Hanoverian field battalions were newly raised. They had been stiffened by having officers and non-commissioned officers of the King's German Legion attached to them and at Waterloo were placed between the British brigade and the magnificent soldiers of the Legion. In that position they fought heroically on the whole, though they showed their want of training on one or two occasions.

All the general officers except General Halkett were Hanoverian, although Sir Charles Alten had been so long in the British army as to be looked upon as a countryman.

Colin Halkett and his brother Hew on the other hand, from their long connection with the Legion and the Hanoverian army, are frequently spoken of as Germans, but they were the sons of a British general officer and belonged to a well-known Scottish family.

The first corps was posted as follows :—The Headquarters of the third division were at Soignies, having the Guards at Enghien on its right and Chassé's division on its left. Beyond Chassé on the extreme left was Perponcher's strong division at Nivelle, Frasne and Quatre Bras, where he communicated with the Prussian right under Ziethen.

The second corps under Lord Hill was about Ath and Audenarde on the right of the Guards. The reserve, consisting of the fifth division under Sir Thomas Picton and the Brunswickers under their Duke, were at Brussels. The British cavalry and that of the King's German Legion were chiefly on the Dender in rear of the right of the army, and the Dutch-Belgian in rear of the divisions of Chassé and Perponcher.

On the left of the Duke of Wellington's army were the Prussians under Field-Marshal Blücher. He had a total of—99,715 infantry, 11,875 cavalry, and 312 guns. It is evident that if the Duke of Wellington was suddenly attacked on his left, the Dutch-Belgians would have to bear the brunt of the fighting till the army concentrated. Their first support would be the Brunswickers and Sir Thomas Picton's division from Brussels, the next Sir Charles Alten's division and then the Guards ; Lord Hill's corps and the British cavalry would take a day longer to reach Quatre Bras.

Napoleon left Paris on June 12th, and on the 14th joined his army, which had been concentrated behind the French frontier opposite the right of the Prussian army and the left of Wellington's.

On the 15th, the French, under Napoleon himself, attacked and drove back Ziethen's Prussian corps, and Marshal Blücher ordered his army to concentrate about Ligny, 6 or 7 miles from Quatre Bras, to which Wellington's left extended.

The French army under Napoleon was thus constituted :—

The five corps of d'Erlon, Reille, Vandamme, Gerard, and de Lobau numbered 94,569 men with 206 guns.

The four corps of reserve cavalry under Pajol, Excelmans, Kellerman and Milhaud had 13,327 men and 48 guns.

The four regiments of Grenadiers, four of Chasseurs, two of Tirailleurs, and two of Voltigeurs of the Imperial Guard numbered 12,555 ; the two regiments of heavy and three of light cavalry of the Guard had 3,488 men.

The artillery of the Guard had 96 guns. They were chiefly 12-pounders, which outclassed the 9-pounders of our field artillery. The grand total of the troops under Napoleon was 83,753 infantry, 20,959 cavalry, 19,224 artillery with 350 guns.

QUATRE BRAS

Macready's narrative is, and always must be, the real account of the part played by the 30th regiment in the Waterloo campaign, but it has been thought best to preface it by a sketch of the operations in order to supply, as it were, a framework to his picture and also to give some details which are not in his diary.

At 4 p.m. on June 15th, Wellington at Brussels received information of the general advance of the French and issued orders for the concentration of all divisions at their Headquarters, and later for a general movement to the left.

The 30th with the rest of Alten's division marched at 2 a.m. on the 16th (leaving the light company on picket), through Braine le Comte and Nivelles on Quatre Bras, which it reached shortly after 5 p.m., having marched 22 miles. Major Bailey was on short leave to Brussels and the movement to the left was so sudden that he was unable to find the battalion till the 17th.

The light company by a rapid march was able to join the battalion soon after it entered the field of battle. The story of the light company will be told by Macready.

Early on the 16th the Prince of Orange had taken up a position with the Dutch-Belgian troops on the Nivelles–Namur road with his right on Quatre Bras. The wood of Bossu was in front of his right, the farm of Gemioncourt on front of his right centre and Piermont in front of the left. He was able to occupy the wood of Bossu, Gemioncourt and Piermont, but had not enough troops up to occupy the wood of Delhutte further to his left front, and it was seized by the French light troops.

The importance of the woods and farms lies in the fact that except for them the ground was without cover. There were no hedges or enclosures except round the farms. A stream issuing from the wood of Bossu flows past Gemioncourt and in the valley of this stream the French, after their capture of the farm, were able to keep their reserves and form their men for attack out of sight of our people.

Shortly before two o'clock in the afternoon Napoleon attacked Marshal Blücher, and about the same time Marshal Ney with 18,000 men and 36 guns fell upon the Prince of Orange who had less than 7,000 infantry and 16 guns. The Prince's position was almost desperate when, shortly before three o'clock, Picton's division (Kempt's and Pack's British and Best's Hanoverian brigades) came on the ground and drew up along the Nivelles–Namur road. Just as they appeared Ney captured Gemioncourt.

The Duke of Wellington now took command but could not recover Gemioncourt, and the enemy quickly occupied the southern part of the wood of Bossu. The Duke of Brunswick's corps now came up and the greater part was placed with its left facing Gemioncourt and its right touching the wood of Bossu.

Brunswick was at once so hard pressed that Wellington ordered Picton with his two British brigades to advance on his left. They drove the enemy back and nearly reached Gemioncourt, but the Duke of Brunswick fell and his men gave ground. The resistance of the Dutch-Belgian troops had by this time nearly come to an end. Picton's British and Hanoverians, and the greater part of the Brunswickers fought heroically, but their losses were great. The French cavalry attacks forced them into squares which were a target for guns and light infantry and towards five o'clock the ammunition was running short. About that hour two brigades of Sir Charles Alten's division arrived with two batteries ; Halkett's brigade took up a position on Picton's right, Kielmansegge and his brigade passed in rear of Picton and formed up on his left.

Halkett's brigade had marched from Nivelles right in front with the 30th leading.

General Halkett in his Waterloo letter says that " as soon as he had cleared the wood of Bossu," he received an order from Sir Thomas Picton to cause his troops to bring their right shoulders up and drive the French from the wood of Bossu. Perhaps he wheeled up his battalions from column of route to the right, for the line was formed in inverted order, the 30th on the left near Quatre Bras, and the 33rd on the right.

When on the point of advancing General Halkett received an urgent appeal from General Pack, of Picton's division, who said that he could no longer hold his ground without assistance. Halkett ordered Colonel Morrice to take his regiment, the 69th, to the help of General Pack, and to act under that officer's orders.

General Halkett now advanced his three remaining battalions in line of columns at deploying distance, covered by the flank battalion under Lieut.-Colonel Vigoureux, C.B., 30th regiment. No. 9 company of the 30th acted with the flankers instead of the light company, which was detached. If the brigade had 1,900 bayonets in the field, as is usually allowed, then after deducting the 69th, the flank battalion, and one company of the 30th, the line would be about 530 yards long, so that if the 30th, as its officers assert, advanced in the open ground between the wood and the Charleroi road, the 73rd and 33rd entered the wood of Bossu. At this time the resistance of the Dutch-Belgians in the wood was thought not to have entirely ceased, but General Halkett riding in front of his brigade met the troops of the Duke of Brunswick falling back owing to the giving way of the Dutch-Belgians in the wood on their right. On Halkett's appeal and the promised support of his brigade the Brunswickers halted, and as soon as he was satisfied with their steadiness, Halkett rode far to the front towards Gemioncourt. What he saw there caused him to send an A.D.C. to order the 69th to form square and to warn Pack of an impending cavalry attack. He had already informed General Picton that the support of his brigade was so necessary to the Brunswickers that he could not carry out the original order to clear the wood and act against the French left.

The position of the troops likely to form the object of the French cavalry attack was as follows from left to right :—

The 42nd and 44th of Pack's brigade were on a ridge looking down on Gemioncourt near a cross-road which falls into the Charleroi road from the east. The 69th was in close column about 200 yards in rear of them. All three were to the east of the Charleroi road.

The 30th, in column, was in echelon about 200 yards further back on the other side of the road. The 73rd and 33rd, which had just come out of the wood, were in line of columns at deploying distance to the right of the 30th.

The Brunswickers to the right front of the 33rd were on the edge of the wood.

The French attack was made with splendid boldness and vigour, and met with a large measure of success, partly owing to the action of the Prince of Orange who ordered the 30th and 69th to deploy into line although Colonel Morrice told him that General Halkett had ordered him to form square. The Prince said that there was no danger of a cavalry attack. At that moment

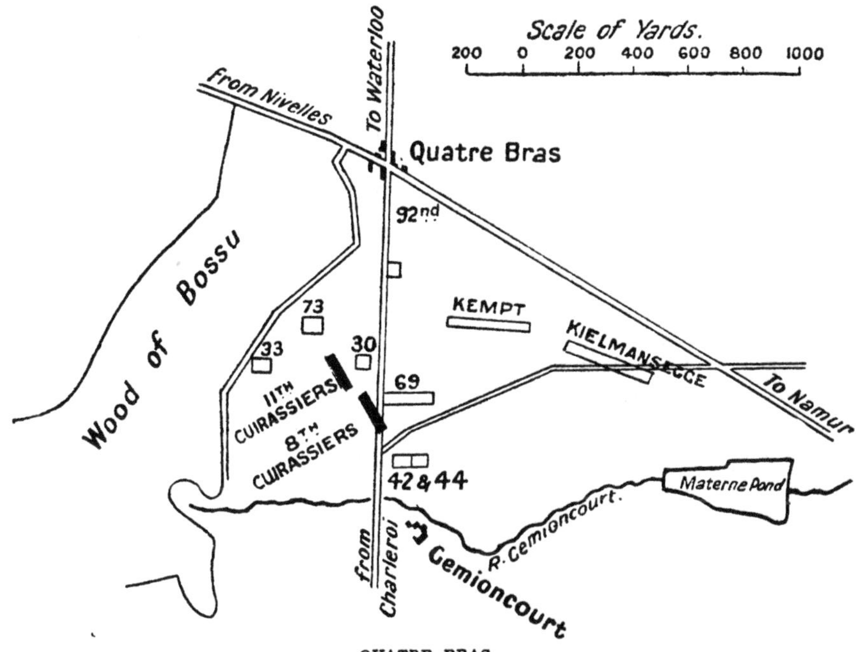

QUATRE BRAS.
Gitaut's Cuirassiers advance from Gemioncourt against Halkett's Brigade and suddenly change front to the right and fall on 30th and 69th.

Gitaut's brigade of cuirassiers of Kellerman's corps was crossing the bridge of Gemioncourt led by Kellerman himself.

The brigade came up from the valley at a trot, formed line of squadrons to the west of the road and advanced so as to threaten the whole of Halkett's brigade. The 33rd formed square and stood firm, the 73rd retired into the wood which it had lately quitted, but the French catching sight of the 30th and 69th in line, changed front to the right and charged them both. The right wing of the 69th was caught before completing the formation of square on the centre, and 150 men were killed and wounded and a colour lost. The left wing rallied on the square formed by the 42nd and 44th. The 30th were formed in square by Colonel Hamilton with great rapidity, and the flank battalion rallied on it. The French were beaten off and pursued by the volleys of the 30th and flank battalion.

Colonel Morrice told Captain Rudyard, R.A., that the survivors of the 69th were saved by the destructive volleys of a battalion, which he believed to belong to the Guards. There were no Guards near and the volleys were those fired by the 30th and flank battalion. The mistake was natural, for Morrice had reason to believe that Halkett's brigade was fighting in the wood of Bossu and the Guards were expected on the field. Morrice was killed at Waterloo.

Some of the cuirassiers galloped up to Quatre Bras, where they were met by the fire of the 92nd lining the banks of the enclosures and of Cleeve's battery. The survivors fled to the valley of Gemioncourt, where they were re-formed and reinforced by lancers. Picton, who had watched the repulse of the cavalry with delight from his own side of the Charleroi road, galloped over as soon as he could and called out, "Who commands here?" and upon Colonel Hamilton answering, "It is the 30th," Picton said with one of his customary oaths, "That was not my question." Finally he said he would make the highest report to the Duke of the commanding officer *and* the regiment. Owing to his death at Waterloo this report was not made.

Colonel Hamilton, like most old campaigners, chewed tobacco, and tradition says that in action it was his custom when he thought he saw any sign of nervousness or impatience in the ranks to ask audibly if any one had any tobacco. Half a dozen brass boxes were of course offered to the great man, who, with all eyes fixed on him, would select one and cutting a convenient quid, insert it in his cheek, nor until he had it comfortably under weigh would he deign to notice the enemy. He thought a 30-yard range quite long enough. This tradition may be somewhat wide of the truth, as family and regimental traditions often are, but it shows that he was regarded by his men as an exceptionally cool and resolute officer.

Halkett's advance was resumed, but he was soon called to the right. The ground here slopes downwards from right to left, and in addition to that the 33rd in advancing rose higher with every step, while the 30th descended a little, so that the 33rd were soon about 40 feet above the 30th and formed a splendid target for the French guns in the wood of Bossu, from whose fire the 30th were shielded. After suffering heavily the 33rd formed line obliquely to its right, facing the guns, and advanced to join a battalion of Brunswickers fighting on the edge of the wood. At this time an alarm of cavalry in rear was raised, and fearing to be caught in line like the 69th the regiment entered the wood.

There *was* a renewed cavalry attack but chiefly on the corps close to the Charleroi road. Still the left of the cavalry extended so far as to force into square though it did not attack the battalion of the 73rd which was now advancing again, covered by one of its battalion companies extended. The cavalry would therefore pass in rear of the centre of the 33rd when the latter was in line facing the guns.

The 30th were forewarned of this renewed cavalry attack. After being complimented by Picton for the repulse of the cuirassier brigade, Colonel Hamilton had ridden to the front to watch the enemy, but on nearing the enclosures of Gemioncourt he was fired at by two tirailleurs hidden in a tree and shot through the left leg. He was able to ride back to his battalion and warn it to form square; he then rode to the rear to look for the regimental surgeon who had established his dressing station near Quatre Bras.

Before leaving the battlefield he met his light company hurrying up to rejoin from detachment and said, "They have tickled me again and now one leg cannot laugh at the other." (At Toulon he had been severely wounded in the right leg.) A little further on he came upon Surgeon Elkington, who did what he could for him and sent him on without dismounting to the base hospital, where it was found that he was severely wounded and the leg in danger.

The light company had a smart run up to the battalion and they only arrived in time to form upon it when the square was charged on two faces by cuirassiers and lancers. The attack was repulsed and the tremendous fire from the six-deep square (30th and flankers combined) and from other troops threw

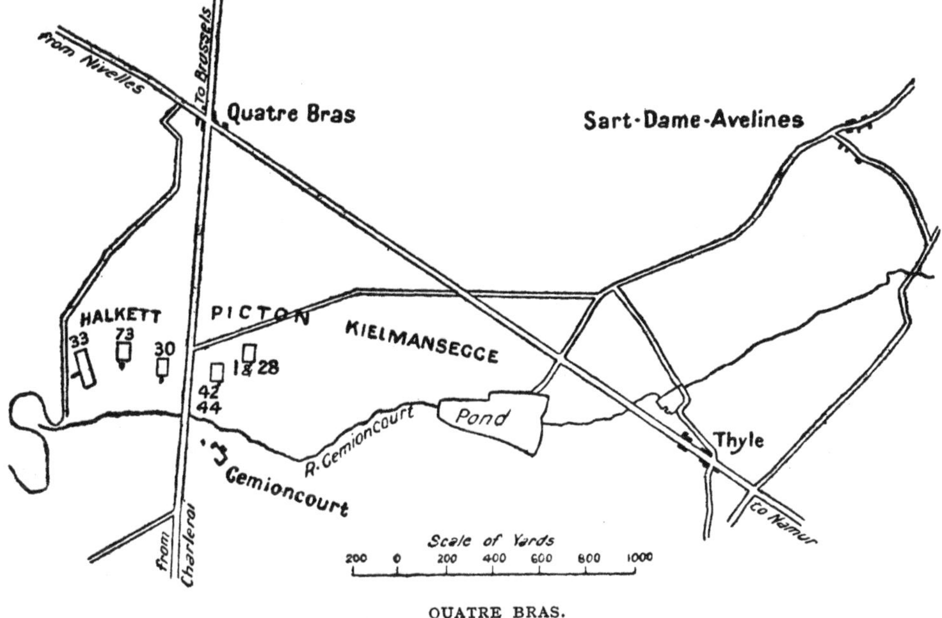

QUATRE BRAS.

Position of Halkett's Brigade and the right of Picton's Division when Gitaut's Cuirassiers delivered second charge on 30th and Picton's right.

Line taken by Brigade through Sart-Dame-Avelines on 17th covering retreat of Army from Quatre Bras.

the French cavalry into such disorder that they could not be rallied, as before, at Gemioncourt but fled for some way after passing the stream, carried away with them Bauduin's infantry brigade and spread a panic as far as Charleroi.

Picton paid the regiment another visit and said they were a set of noble fellows but immediately afterwards damned them for making a noise like devils. There was a good deal of laughter and handshaking in the square over the second repulse of the cavalry, and the praises the regiment had received and congratulations were shouted by the other light companies to that of the 30th for being up in time to take its share, but that was going on when he arrived and he had not taken any notice of it. It looks as if some of those in the square had allowed themselves to give the old hero a cheer, a proceeding which would call for the strongest language.

The 42nd and 44th had shared in this defeat of the French cavalry. Whether a wing of the 3rd Royal Scots and the 28th did so is a question. Field-Marshal Sir William Gomm, who at Quatre Bras was a lieut.-colonel and A.-Q.M.G. of Picton's division, sent in later years to Siborne, the historian, an extract from his diary to the effect that Sir Thomas Picton had in person led those two battalions in one square to the support of the 42nd and 44th, and that the 30th advanced at the same time and that its square flanked the one above mentioned and received with equal firmness the charge of French cavalry. This explains how Picton came to be so far forward.

Siborne, who for some reason had made up his mind that the Royal Scots and 28th had advanced to the support of the 42nd and 44th, and had been withdrawn again earlier in the day, persuaded Sir William Gomm that the weight of evidence was against him. It is difficult to see what better evidence there can be than the diary of the chief staff officer of the division engaged. In any case the two advanced battalions of Pack's brigade had at last got the support which their general had so urgently demanded; and the Royal Scots must have taken part in the movement, for in the subsequent advance they were on the immediate left of the 30th.

The manner of the advance of Halkett's brigade and its position when attacked by Gitaut's cuirassiers has been so much disputed that it is as well to see what has been said by the regiments engaged in this part of the field.

The *History of the Black Watch*, by Archibald Forbes, states that that regiment and the 44th were in advance on the ridge looking down on Gemioncourt. The 69th was in a hollow behind them. They saw the cuirassier brigade after crossing the stream form line and advance on the west side of the road, avoiding them. Presently the 8th cuirassiers, the right regiment of the brigade, caught sight of the 69th in line and changing front to its right charged and broke that regiment.

The Journal of Occurrences Concerning the 69th (Butler) says that after being detached by Halkett that regiment crossed to the east of the Charleroi road and advanced under Colonel Morrice in close column to a hollow in rear of the 42nd and 44th. Here it was forming square, as ordered by General Halkett, when the Prince of Orange came up and forced it to deploy. It saw the cuirassiers and attempted to form square (on the centre). The grenadiers and Nos. 1, 2 and 3 companies were in movement to complete the square, but the captain of No. 1 turned those companies about to fight. The right wing was sabred and the colour lost, but the left wing rallied on the 42nd and 44th.

Captain Harty, 33rd regiment (Waterloo Letter 141) says that throughout the battle of Quatre Bras he was serving with the flank battalion under Lieut.-Colonel Vigoureux, which acted with and in vicinity of the 30th regiment. He was in the square when the cuirassiers were repulsed, and when Picton rode up and calling for the commanding officer told Colonel Hamilton that he would report the gallant conduct of the square to the Duke.

Lieutenant Pattison, 33rd regiment (W.L. 142), says that Halkett's brigade on arrival at Quatre Bras was ordered forward to the right to support Picton, that it advanced in open column at deploying distance, that on reaching a prominent part of the field the 33rd met the French cavalry and formed square on the leading company. The cavalry then moved to the left and broke in on the 69th, which was in a low part of the field of battle. The 33rd then became a target for the French guns. It formed line in an angular direction and moved towards a Brunswick battalion fiercely engaged at the corner of the wood. An alarm of cavalry in the rear caused the battalion to enter the wood.

Sergeant Morris, 73rd regiment, in his *Recollections*, says that his regiment, marching through high crops, did not see the French cavalry till they were so close that there was no time to form square. The battalion therefore retired to the wood through which it had advanced. It quickly resumed the advance, covered by his company extended. The French cavalry was again met with but seeing the battalion well prepared for it, went off to the left where it attacked the 42nd regiment.

No further evidence is necessary of the position of the 30th when attacked by cavalry.

After the repulse of the last real effort of the French cavalry the battle turned in favour of the British. The Guards and several batteries arrived on the field, and Halkett was able to advance his skirmishers to the enclosures of Gemioncourt. The Guards after heavy fighting cleared the wood of Bossu and showed themselves at the southern end, a trifle to the right front of Halkett's brigade.

Macready describes the close of the battle and the doings of the light company in his diary.

The casualties at Quatre Bras were—British, 2,275; Hanoverians, 369; Brunswickers, 879; Dutch-Belgians, about 1,000.

The 30th escaped with the loss of Colonel Hamilton and Lieutenant Lockwood severely wounded, Lieutenants Harrison and Roe (2) slightly wounded but able to remain with the regiment, 1 sergeant and 4 rank and file killed, 2 sergeants and 27 rank and file wounded, and 5 missing, 2 of whom seem to have rejoined on the following day. The small loss is owing to the fact that except the light company the battalion was not engaged in a serious musketry conflict and never came under really effective artillery fire. Its main duty was the repulse of the magnificent French cavalry, and at one time, owing to the Prince of Orange, that was not an easy task.

The steadiness and gallantry of the regiment were conspicuous and it received the thanks and praises of the Prince of Orange, Sir Thomas Picton, Sir Charles Alten, Count Kielmansegge and General Halkett.

MACREADY'S "DIARY"

"On May 20th our regiment and the 33rd occupied Soignies, a neat little town, the headquarters of our division. Here we were drilled out of all patience, and, like the soldiers of ancient Rome, I longed for war as a respite from fatigue.

"About June 12th all intercourse between France and Belgium was suspended, we consequently expected soon to march, and were all in a charming state of anxiety. As a group of us were standing in the square of Soignies discussing the probable events of the approaching campaign about four o'clock on the evening of June 15th, a rumour got afloat that the French had crossed the frontier. We were looking at each other with a half-incredulous and half-apprehensive sort of smile, when General Halkett galloped up, and called out,

'Are any light infantry officers among you?' 'Yes, sir,' said I. 'Parade your company in ten minutes' time on this spot,' was his reply, and I went away to arouse the men. I ordered my servant to put the baggage, and a box of light infantry appointments that had just arrived, on my pony, and by the time specified, Rumley, Pratt, and I, were on the ground with the company. Lieut.-Colonel Vigoureux was ordered to plant us on picket at a village called Naast, about a league from Soignies. As we marched to our post, we met several regiments of our division hurrying to the town; it was evident that the game was a-foot. We placed our company in a large barn, and threw out advanced pickets on the Rouelx and Nivelles roads, communicating by patrols with our old friends the Jägers von Kielmansegg. A corporal came in from Head-quarters about nine o'clock, and told us the French had certainly advanced, and were pressing the Prussians considerably, also that our whole division had entered the town of Soignies, and occupied the church and other buildings. The night passed quietly.

"About eight o'clock next morning, it was agreed that I should ride over to the regiment to order out a few necessaries; and accordingly away I went. I cantered on unconsciously and pulling up in the market-place, was thunder-struck. Not a soul was stirring. The silence of the tomb reigned where I expected to have met 10,000 men. The breath left my body as if extracted by an air pump. I ran into a house and asked, 'Where are the troops?' 'They marched at two this morning,' was the chilling reply. 'By what road?' 'Towards Braine le Compte,' was all I heard, when, jumping on my pony's back, I endeavoured by sympathetic heel to convey the rapidity of my ideas into his carcase. But vain were my efforts; Soignies was his home, and his obstinacy invincible. So getting off, I made the best of my way on foot, lugging him by the bridle. On reaching Naast, I found Rumley and Pratt in deep consultation with the burgomaster, who had informed them that the French had passed the Sambre, and occupied Charleroi; that the Prussians were falling back and our troops hastening up to support them. Our patrols had examined the country on every side, and not a soldier was to be seen.

"We were most unpleasantly situated; ignorant whether we were left here by mistake or design, and dreading equally the consequences of quitting our post without orders, or the division being engaged during our absence. Our commissions were safe by remaining where we were; but we were determined to risk them, and all the hopes of young ambition, rather than be absent from the field of glory. Away we marched towards Braine le Compte. There we learned that the troops had struck off to Nivelles and we followed their route. We soon got among the baggage of the army, passed it, and quickened our pace on hearing a noise like distant peals of thunder. 'The Dutch artillery are practising,' said a young soldier, in a tremulously inquisitive tone. 'They've redder targets than your cheeks, my boy, that fire those guns,' replied a swarthy veteran, who had learned this music in Spain. In a few minutes, regular discharges could be distinguished. They came from the Prussians and French at Ligny. We redoubled our speed, and entering Nivelles found considerable difficulty in forcing our way through the crowds of baggage animals, commissaries, quartermasters, and women, who thronged the streets. Some of our regimental women came up, blessed us, and kissed their husbands—many for the last time.

"They told us the division had halted at Nivelles, and marched again, about an hour before, towards Quatre Bras, where the enemy were said to be. We met our stores, and, seizing them, took out an allowance of spirits for each man. The repeated 'God in heaven bless you, my dear child,' of a poor woman who was choking in a ditch, and who shared my gin, was a better renovator than the spirits. Thus re-inspired, our boys started double quick, for the firing increased. It was now past three o'clock. We passed the division of Guards; Rumley had some words with a staff officer about crossing their line of march, and our fellows began to laugh and jeer them. They had some cause, for I never saw such a number of men knocked up in my life. 'Shall I carry your honour on my back?' said one of ours to a grenadier guardsman, as he was sitting down. 'Haven't you some gruel for that young gentleman?' shouted another, and continued, 'It's a cruel shame to send gentlemen's sons on such business; you see they don't like it; they had quite enough at Bergy-my-Zoon.' High words arose; and we stopped our men, who, however, took leave of them, saying, 'Good-bye to ye, young gentlemen; pray don't hurry yourselves—we'll do your work, never fear.'[1]

"We continued our double-quick and struck across some fields. Our men had now marched about 18 miles, and run 6; they began to faint very fast; but not a soldier fell out till he dropped, black in face and senseless. Many of them reeled while replying to our encouragements, 'Never fear, Sir, I'll keep up.' We cut their pack straps, took off their stocks, and left them gasping. At length we had a confused view of the field, with our troops and the enemy firing away, under their sulphurous canopy. Clouds of birds were flying and squealing above the smoke. We loosened our ammunition, and pushed on for it. Hedges, trees, and ditches were passed like thought. After scrambling through a thick thorny plantation, we found ourselves close to a body of men with whose uniform we were not acquainted. Not above twenty of our company were present. We advanced to these people, and found them to belong to the Nassau Usingen contingent. They had just been driven from the wood of Bossu, which was in front of us and between ourselves and the army. We made for it, and came up with Sir George Berkeley, adjutant-general to the Prince's corps, who had just escaped from the lancers. He told us our regiment had entered the field about a quarter of an hour before, and that they were at the other side of the wood, which we must pass on the left, near Quatre Bras, as the enemy occupied the whole of it. This we were convinced of by the numerous round shot coming from it, one of which slashed dirt and mud over the whole company. 'Close your files, and hould up your heads, my lads,' roared an old campaigner, named Terry O'Niel. One feels a thrill at these moments. We soon reached Quatre Bras, where the Brunswickers and some of Picton's people were in square; and on turning the end of the wood, found ourselves in the hurly-burly of the battle. The roaring of great guns and musketry, the bursting of shells, and shouts of combatants, raised an infernal, an indescribable din; while the galloping of horses, the mingled crowds of wounded and fugitives (Belgians), the volume of smoke, the flashing of fire, struck out a scene which accorded admirably with the

[1] This, of course, is only soldiers' chaff. For Major Macready's own opinion of these regiments, see p. 340. The Guards had already marched 4 miles further than the light company of the 30th.

music. As we passed a spot where the 44th, our old chums, had suffered considerably, the poor wounded fellows raised themselves up and welcomed us with faint shouts, 'Push on, old 30th—pay 'em off for the 44th. You're much wanted, my boys. Success to ye, my darlings.' Here we met our old Colonel riding out of the field, shot through the leg.

"Hamilton showed us our regiment and we reached it just as a body of lancers and cuirassiers had enveloped two faces of its square. We formed up to the left and fired away. The tremendous volley our square (which in the hurry of formation was six deep on the two sides attacked) gave them, sent off these fellows with the loss of a number of men, and their commanding officer. He was a gallant soul—he fell while crying to his men, 'Avancez, mes enfans! Courage! Encore une fois, Français!' I don't know what might have been my sensations on entering this field coolly, but as it was, I was so fagged and choked with running, and was pressed so suddenly into the very thick of the business, that I can't recollect thinking at all, except that the Highlanders (over whom I stumbled at every step), were most provokingly distributed. On our repulse of the cavalry, Sir Thomas Picton rode up and thanked us warmly, as this body had cut up two or three regiments. I think the men I saw dead belonged chiefly to the 11th Cuirassiers. They were savage-looking fellows—fine subjects for Bonaparte or Salvator. The light bobs were ordered to pursue the rest of them, so we dashed out and followed firing until we were brought up by a line of tirailleurs, with whom we kept up a brisk fire. Lockwood, of our grenadiers, came with us, and dropped with a shot in his head as he was speaking to me. He was a noble fellow; his last words were an exhortation to his company to do their duty.

"The cannonade and skirmishing were lively on both sides. Our broken columns and their cavalry were re-forming, while the heavy fire from the wood in our rear showed that the Guards and the enemy were nobly disputing it. On the left of our company were some Hanoverian Jägers, one of whom covered a soldier of ours named Tracy; a tirailleur dashed out from their line and shot the German through the head; upon which Tracy ran over to him, and before he could get off blew his skull to pieces. This enraptured their officers; who, to say the truth, were marvellously distempered with drink or choler, and they were loading us with praises, such as 'Engleesh and Hanover viell goot for the Franzosen.'

"When the advance of the enemy's cavalry obliged us all to retire to our columns, they made a faint charge on ours and some other battalions, but being uniformly repulsed, retired, and we occupied our former ground. We now descended a slope towards our right in the direction of a deep ravine across which the Royal Scots and ourselves drove a heavy body of infantry after a severe fire. The enemy were retiring from the wood, and the Guards pressed them very closely. A retrograde movement was perceptible along their whole line, and it was performed in beautiful style; their skirmishers and columns kept their alignment and distance as if on parade; the dimness of the evening made the firing doubly vivid, and above its roar, one occasionally heard German bugles sounding the 'Advance and fire.' We had no cavalry to spoil the spectacle, but the light troops pushed rapidly from hedge to hedge. Major Chambers of ours was pushing on with two companies towards a house in our front, and I joined with as many of the light infantry as I could collect. We rushed into the court-yard, but were repulsed; he re-formed us in the orchards, directed the men how to attack, and it was carried in an instant, by battering open the doors and ramming the muskets into the windows. We found 140 wounded and some excellent beer in the house.

"The Belgians behaved vilely; if a man was wounded, he generally left the field accompanied by his whole company. As our company entered the field, an order was brought to Hamilton to leave his light infantry to stop these rascals; Poor H. in the warmth of his feelings, shouted, 'My light company is detached, so I can't leave it, but, d——n them, let them run, we want no cowards here.' This was carried to the Prince of Orange, and Hamilton lost the Order of William of the Netherlands.

"Our regiment piled arms about ten o'clock at night, and lay down to sleep, covered by the ravine in front. The dead and dying were around us, but no one slept the worse. Military men know this, but it appears incredible to the uninitiated that a few hours of glory should give the heart such a stoical insensibility. Warren and I pigged together under a cuirassier's cloak, and John Rumley made a pillow of my body. I was ordered up in the middle of the night to reconnoitre some figures moving in our front; they were Brunswickers. About two in the morning I was awoke by a heavy fire, and seeing flashes all around me I started up and found that some one had raised a false report of the enemy's advance; it was a mere joke, but no joke to us poor caçadores, who were ordered to the front to support the pickets in case of an attack. We extended between the captured farmhouse and the wood, and at daybreak the enemy showed a line of tirailleurs opposite to us."

The brigade bivouacked about the captured farm and stood to its arms at daybreak, when the Duke of Wellington arrived from Genappe, where he had slept. He was dressed as usual when in the field, in a short blue frock coat and shorter blue cloak with a white stock, leather breeches and Hessian boots, a low cocked hat without a feather, but with a black cockade on which were three small ones each about an inch in diameter, being those of Spain, Portugal, and the Netherlands, in token of his holding military rank in the three countries.

The Duke rode to the front on arrival and inspected Ney's position. After visiting the front, he proceeded to the left to await reports from the Prussians, to whom he had sent his aide-de-camp, Lieut.-Colonel Sir Alexander Gordon, with a cavalry escort. Gordon was able to find General Ziethen covering the retreat of the Prussians upon Wavre and by half-past seven o'clock returned with his news which Wellington summed up in his own pithy style, "Old Blücher has got a damned good licking and has fallen back, we must do so too; I suppose in England they will say that we have been licked." He sent a message to Prince Blücher that he was falling back in line with him and would accept battle in front of Waterloo if assured of the support of two Prussian corps.

Orders were issued for the reserve ammunition to be parked forthwith behind Genappe and for all other baggage on the roads in rear to proceed to Brussels. The troops were ordered to fall back at 10 a.m. Meanwhile the dragoons were sent to search the battlefield and bring in the wounded who were supported on the dragoons' horses or carried in blankets to Quatre Bras, whence they were sent by wheeled carriage to Brussels. The rest of the troops piled arms and slept, and Wellington sat down to read an English paper which had just arrived; presently he put the paper over his head and lay down too.

Throughout the night the cavalry and Hill's corps had been arriving and he had his whole army in hand.

The army in its retreat was ordered to use the three main roads, Nivelles, Charleroi and Braine le Comte to Brussels. Alten's division, strengthened by some light troops, was to hold the ground and follow as rear-guard, the whole to be covered by Lord Uxbridge and the cavalry. Alten's division moved by Thyle to Sart-dame-Avelines and thence by country roads to Genappe where it halted to cover the passage of the defile by the main body.

The retreat is described by Macready.

MACREADY'S "DIARY"

"We were every moment expecting an attack, when, about ten o'clock, we were ordered to withdraw as quietly as possible, and our place was occupied by Jägers. We joined the regiment, and, while looking for something eatable. I fell in with Major Watson, 69th, who appeared dreadfully chagrined. He told me that his corps had been dreadfully cut up, and had lost their King's colour, and then devoutly d——d the Prince of Orange! Before I could comfort my grumbling intestines with anything masticable, my division stood to its arms, and moved to its left. A shot was heard in the middle of the column; we turned round—poor Strachan, 73rd, was dead. A piece accidentally went off at the long trail, and the ball went through his body. We passed over the ground that Picton's division had so gallantly defended, and, crossing one road, came to a second, down which we proceeded for half a mile, and halted at a village. We thought we were going to join the Prussians in forcing the enemy's right. I slept here nearly an hour on a fine soft dunghill, though the firing between the Jägers and the French was very heavy. Numbers of our wounded were carried through the yards we were in, and, before we marched, balls repeatedly whizzed through it. This showed our friends were losing ground. We stood up, and turned into a narrow road which branched off to the left. On entering a wood we found parties of hussars scattered here and there to cover the retreat. These groups—some of whom were sleeping, bridles in hand, others smoking their German pipes, a few cleaning their horses, and the rest irregularly idling about amongst trees—had a picturesque appearance.

"About three or four o'clock a most furious storm occurred. The rain came down in torrents, and in a moment we were drenched to the skin. The thunder rolled awfully above our heads, and the lightning glistened among the bayonets. The enemy's artillery, pushing on closer every minute, mingled its roar with this hubbub of the elements. These things look and feel ominous to a retreating army. As we descended a steep declivity our men rolled head over heels from top to bottom, and the road in the low ground was knee-deep for a quarter of a mile. About half-past five we came on the Charleroi chaussée as the covering division. Our Jägers and Brunswickers were busy on the flanks, the cannonade was brisk, and report said that the enemy had captured two of our guns. At this time the 7th Hussars charged some Red Lancers near Genappe, and were sadly beaten. The Life Guards then came up and fully revenged them. I saw a regiment of Heavy Dragoons deploy to the support of the Household troops if necessary. They had on their red cloaks and looked like giants. Numbers of the hussars galloped by us so covered with mud that their uniforms could not be distinguished ; from counter to tail, and from spur to plume, horse and man were one cake of dirt. They must have had pretty rolling or running. The rain continued unabated, and the night drew on. It must have been near eight o'clock when we formed in contiguous columns on the position of Waterloo.

" Fortunately we occupied ground from which the Guards had been ordered rather hastily, and they had left a good deal of wood and some biscuits. Rumley, Pratt, and I shared a fowl, which we roasted, or rather warmed on a ram-rod, and the third of this animal with an onion at Genappe, some beer at the captured farmhouse, and a couple of biscuits, was all that exhilarated my inward man on June 16th, 17th and 18th. We lay down on the mud around our fires, and the rain continued pouring on us all night. In the morning we were almost petrified with cold, many could not stand, and some were quite stupefied. Poor Pratt, who had fainted the day before at Genappe, set off (at our earnest entreaty, and promise to call him when things looked serious) towards Mount St. Jean, and shortly after we found him at our fire unconscious of where he had been or what he was about. It was a miserable night ; however, motion brought us about in some degree, and we began to gape and stroll about the field, and, the rain having ceased, our soldiers were busily employed in firing off and cleaning their pieces in readiness for action."

Macready does not mention one incident of the day's march on the 17th. While the division retreated by the country roads Surgeon Elkington with his field panniers followed the main road, keeping as nearly as possible abreast of the regiment which he could see from time to time and recognize by the adjutant's white horse. Here he met Major Bailey, who up till now had not been able to find the regiment or his baggage, and was still in plain clothes, but luckily he was mounted. Elkington was in red and Bailey exchanged coats with him and borrowing a cocked hat from an assistant surgeon rode off to join and take command of the battalion.

The loss in the retreat from Quatre Bras was 1 rank and file killed, 3 wounded, 9 missing.

THE WATERLOO POSITION

The great road running south from Brussels as it approaches the village of Mt. St. Jean, about a dozen miles from the capital, divides into two branches ; one inclining slightly to the west goes to Nivelles and the other with a somewhat slighter inclination to the east to Charleroi. The roads are good and broad with a paved centre. At a trifle over half a mile south of the fork, both roads are crossed by a country road running in a general line from west to east as far as Wavre, to which place Marshal Blücher had retired after the battle of Ligny. Where this road crosses the Nivelles and the Charleroi roads and for a mile to the east it follows the ridge of a low hill, and it was on this ridge that Wellington drew up his army on June 18th to bar the road to Brussels and to await the arrival of his colleague, Marshal Blücher.

Wavre is only 10 miles off, but the roads from it were unpaved and nearly impassable in wet weather.

The downward slope to the north of the ridge is considerable, and Wellington's second line and reserves were covered from view and artillery fire.

The French army under Napoleon occupied a ridge parallel to the British, and about three-quarters of a mile to the south.

The third division was in the centre of our army with its left on the Charleroi road.

On the right of the third division were the Guards. On the left, beyond the Charleroi road, was the fifth division under Picton. The Duke of Wellington had occupied in this part of his line the farm of La Haie Sainte in front of the centre on the Charleroi road and the chateau and grounds of Hougomont, 420 yards in front of his right. The defence of La Haie Sainte was entrusted to the third division and directly concerns us. From the point where the Charleroi road is cut by that to Wavre the ground slopes to the south. At 200 yards from the crest and on the left side of the Charleroi road there is a knoll and gravel pit which was held by a battalion of the 95th (Rifles) of Picton's division; 40 yards further down and on the right side of the road are the farm buildings of La Haie Sainte, and for another 240 yards the orchard attached to the farm borders the road. Both orchard and buildings were held by the second light battalion of the King's German Legion. Soon after passing the orchard the bottom of the valley is reached and the road begins to rise again. It enters a cutting which marks a swell of the ground about 750 yards from our position. On this site the French established a battery of seventy-four guns. A hundred yards further on there is another cutting which reaches to La Belle Alliance, the centre of the French position; on the ridge through which this cutting passes, the artillery of the French guard was placed. The distance from the left of the third division to La Belle Alliance was 1,450 yards and the road falls about 80 feet and rises again about 100 to La Belle Alliance.

The following was the disposition of the third division. The first light battalion of the King's German Legion was in support of the second which held La Haie Sainte. The first, like the remaining battalions of the division, was in quarter column. To the right of the first light was the fifth line battalion and the eighth line battalion was in second line. The front line of the division was at deploying interval from the left and the columns in second line covered the intervals. To the right of Ompteda's brigade was Kielmansegge's with the battalions Luneburg, Verden and Bremen in front (the two last acting as one corps) and Duke of York and Grubenhagen (also acting as one corps) in rear. Halkett's brigade, on the right of Kielmansegge's, had the 30th and 73rd (acting as one corps) in front and the 33rd and 69th in rear: they also acted as one corps. The distance between first and second line was about 80 yards.

Starting from the Charleroi road the ridge which the third division defended ran west, and the road from Wavre followed the crest. At first it was a hollow road with banks about 8 feet high, but after 400 yards the road had reached the level of the ground alongside. The ridge was always rising and at 500 yards from the Charleroi road it had gained about 50 feet. Throughout this distance there was good cover from fire close behind the road. The road now forms a fork, one branch going north-east to Braine l'Alleud, the other inclining to the south of west. Close to this fork were the battalions Bremen and Verden, and 150 yards further on were the 30th and 73rd. At the fork the ridge juts boldly out to the south like a bastion and the ground rises into a down which dominates the whole field, and the crest line is about 200 yards in front of the road. During the battle the 30th and 73rd were frequently obliged to advance and hold the crest, sometimes for considerable periods, but when it was possible they were withdrawn under cover.

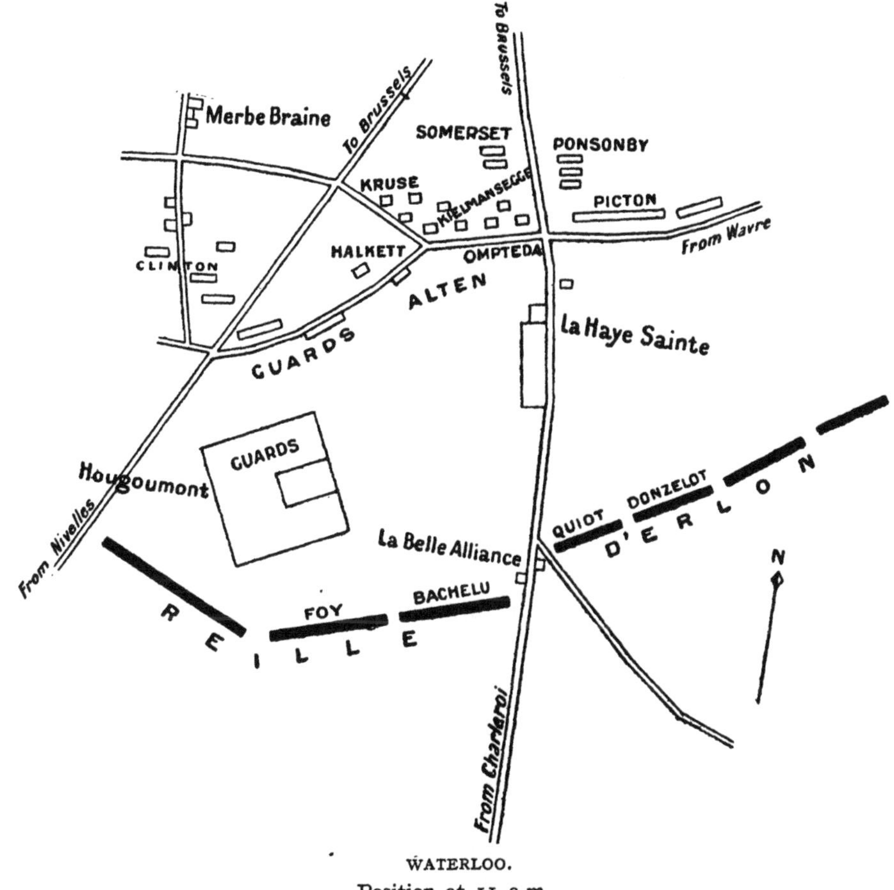

WATERLOO.
Position at 11 a.m.

To obtain this they had to go well back, especially after the French got guns up to the north-west corner of the enclosures at La Haie Sainte.

A tongue of elevated land not much lower than the down defended by the two battalions ran from it to La Belle Alliance where the ground rose again. The third battalion of the 1st Guards on the right of the 30th had also to fight in advance of the road at a distance from cover.

Kruse's brigade of Nassauers supported the third division and advanced one of its three battalions to cover the deploying interval between Bremen and Verden and the 30th and 73rd.

Lloyd's and Cleeve's batteries of 9-pounders were placed a hundred yards in front of our line, Lloyd to the left and Cleeve to the right of the 30th.

In rear of the third division were the light cavalry brigades of Arenschildt and Dornberg. Behind Ompteda and Picton were the heavy brigades of Lord Edward Somerset and Sir William Ponsonby.

In the forenoon the softness of the ground lessened the effectiveness of artillery fire, and the rye crops, as high as a man's head, concealed the movements of troops, but they were soon trodden down.

After providing for a detached force at Hal, and other services, Wellington was able to place in the field at Waterloo, 49,608 infantry, 12,402 cavalry, and 5,645 artillery with 156 guns—a total of 67,655 men. Of these the British and King's German Legion furnished a trifle under 30,000 men with 96 guns.

Napoleon had in the morning 47,579 infantry, 13,792 cavalry, and 7,529 artillery, a total of 68,900 men with 245 guns. He had however to employ during the day the sixth corps, the Young Guard and three battalions of the Old Guard against the Prussians—that is about the fifth of his army.

About a fifth of Wellington's army also was of no use to him on that day. In the heat of the moment the British and Germans who fought at Waterloo used the word cowardice in describing the failure of a considerable part of the Dutch-Belgian army to support them, but all now recognize that from political reasons the hearts of the men were not in the cause. Regardless of their feelings the Congress of Vienna had formed the Dutch and Belgians into one kingdom and one army. Had the Dutch stood alone or if the Belgians had been on the French side and pitted against the Prussians, the fighting of these two brave nations would have been very different.

The field of battle on June 18th was only 1½ miles long from right to left and 2½ miles in depth from the rear of the reserves on either side to the rear of the reserves on the other, and this concentration made the struggle all the more deadly.

As squares played such an important part in the fighting, a word about their formation is necessary. Halkett's brigade stood two deep and the squares were formed of the sixteen battalion companies of two regiments. The battalions having closed on each other and opened to column or half distance, upon the order to form square No. 3 company of the 73rd and No. 8 of the 30th closed on the front companies, the remaining companies as they got their distance went " sub-divisions right " and " sub-divisions left " according to which battalion they belonged to, except Nos. 8 and 9 of the 73rd, and Nos. 3 and 2 of the 30th which closed up and turned about to form the rear face. The square was thus formed four deep, the grenadiers (No. 1), and light (No. 10) company, when not skirmishing, were inside the square ready if necessary to fill up gaps or sally out if ordered.

The rank and file killed, wounded and missing, on the 16th and 17th, numbered 324 ; the brigade should therefore have had under arms at Waterloo from 1,550 to 1,600, the 30th, 494, and the 73rd, 444 rank and file. This gives these two battalions 200 for the flank companies and 738 for the square. After two days' marching and fighting, with little to eat, there must have been some sickness, and the number of bayonets in the front rank can scarcely have exceeded 45 on each face.

Siborne's Field State for June 18th is valueless as regards the 30th. He includes among the present men whom we know to have been absent sick or on command.

In the absence of Colonel Hamilton (wounded), the 30th were commanded by Major Bailey; Lieut.-Colonel Vigoureux commanded the light companies of the brigade with Lieutenant John Pratt as his adjutant. Colonel G. Harris, afterwards Lord Harris, of the 73rd commanded the two battalions. Captain Crofton, 54th regiment, the brigade-major, though wounded at Quatre Bras, was present.

WATERLOO

At daybreak on the 18th the pickets were drawn closer to our line. As the morning advanced the rain ceased and the clouds rose, but they still hung overhead. Soon there was a rattle of musketry caused by the troops on both sides discharging and cleaning their pieces.

The quartermasters of Halkett's brigade arrived with their carts and distributed the ration of biscuit, meat and brandy, but there was no time to cook, and this brings us to the question, what had the men to eat from the mid-day meal on the 15th to the morning of the 19th. From Macready's account all that he had beyond a couple of biscuits was due to his success in foraging, but the conditions with regard to the light company differed from those of the rest of the battalion. A full ration was issued to the latter for the 16th, and an attempt was made to cook during the halt at Braine le Comte, but the march was resumed before the cooking was finished, and the soup and meat poured out on the road side. The men therefore had bread or biscuit for the 16th, and we know that on the evening of the 17th they came upon a small fresh supply of biscuit left behind by the Guards near La Haie Sainte. No other issue was made on the 17th. To sum up, between the mid-day meal on the 15th and the morning of the 19th the men received somewhat over two days' bread rations and two days' meat but had no time to cook the latter. In the Duke's General Orders two days after the battle the Dep.-Assist. Commissary General of the third division was removed from the army for being absent and neglecting his division. He had been amusing himself in Brussels while the division was marching and fighting.

Sir Charles Alten seems to have delayed taking up his position as long as possible in order to let the men breakfast, for it is on record that the quartermasters were still there when the first gun was fired about half-past eleven. It carried off the head of a pioneer of the 69th at the ration carts, and Mathew Stephens, the quartermaster of the corps, one of the heroes of the battle of St. Vincent, was heard to say, "It is time for all peaceable non-combatants to be off." The carts then started for Brussels. Alten's division then retired a little behind the crest of the hill to the position it was to occupy. When the battle was on the point of joining, Wellington rode down the front of the line from the Nivelles road as far as La Haie Sainte and returning placed himself about 250 yards to the right front of the 30th to watch the French advance. His staff was in blue like himself, except the adjutant-general, Sir Edward Barnes, who was in red.

As usual the front was covered by light troops and Lieutenant John Pratt, acting as adjutant to Lieut.-Colonel Vigoureux, received Major-General Halkett's orders for the conduct of the flank battalion of the brigade. They were as follows:—

"To protect and cover guns. To establish themselves as near to the enemy as is compatible with prudence, and to have a good interval between files, to be obstinate in resisting infantry but to attempt no formation or offer useless resistance to cavalry, but to retire in time upon the squares in rear, no matter of what nation, and to resume their position at once when the cavalry retired."

Colonel Vigoureux extended his men at twelve paces between files and advanced.

The first French attack was on Hougoumont. The light troops only of Halkett's brigade were seriously engaged; the remainder were lying in line behind the ridge.

The way in which Lieut.-Colonel Vigoureux handled his flank battalion was much admired. A brigade of Bachelu's division marching straight towards our position with the intention of taking Hougoumont in reverse was encountered by him, and when it wheeled to its left to get in rear of the Chateau he hung upon its flank, and plied it with musketry. Cleeve's battery joined in, and the brigade was driven in disorder to the position from which it had started.

At half-past one Napoleon was ready for the real attack by which he hoped to destroy the English left and left centre. The movement was made by General d'Erlon's corps, and the chief effort was against Picton's division, but along with this there was an attack on La Haie Sainte which concerns us.

The left brigade of Quiot's division of d'Erlon's corps on the east of the Charleroi road, supported by Bachelu's division of Reille's Corps and a strong body of cuirassiers on the west, advanced against the farm while the remainder of d'Erlon's corps attacked Picton. The defenders were driven from the orchard and garden but managed to hold the buildings. The 95th (Rifle Brigade) in the sand pit were driven back to the crest of the hill. The first light battalion of the King's German Legion which was in support had crossed the road to help Picton, and in its place the Duke of Wellington sent the Luneburg battalion down the hill in quarter column to reinforce La Haie.

It was at this time that the French cuirassiers rode to the front. Macready describes how, according to orders, our skirmishers cleared the front and took refuge in different squares. The Luneburg battalion was newly raised and had not learned that in its close formation it could laugh at the best cavalry in the world. It attempted, rather hurriedly, to regain the heights, was caught and sabred. Only a few survivors were rallied during the day. The Hanoverian Jägers from too much confidence and want of military experience met with a like fate.

The cuirassiers then continued their advance to the French left of La Haie Sainte and mounted the hill, but Lord Uxbridge saw his opportunity and met them with Lord Edward Somerset's brigade. The 1st Life Guards swept past the 30th and 73rd, who had formed square and overthrew the French just as they had topped the crest. To the left the hollow road embarrassed both sides but ultimately the cuirassiers fled in a body past La Haie Sainte, and were pursued by Lord Edward Somerset through their own batteries. Sir William Ponsonby further to the left had routed the infantry opposed to Picton, taken 3,000 prisoners and like Lord Edward Somerset had ridden through the French batteries, cutting down gunners and over-turning guns. Two eagles were captured, that of the 45th regiment by the Greys, and that of the 105th by the Royal Dragoons. The 105th is the regiment with which the 30th was in conflict six hours later at the moment of victory.

We had been brilliantly successful, but it had cost us dear; we had almost expended our heavy cavalry, or at least that part of it which would fight; we had lost the enclosures round La Haie Sainte, where the defenders were confined to the buildings, and the third division had been seriously weakened. Picton too had been killed at the head of his division.

Napoleon now determined to take advantage of his superiority in cavalry to force our position to the west of the Charleroi road. To prepare for this the 12-pounder guns of his guard were deployed on the heights near La Belle Alliance and lighter batteries were pushed forward closer to our line under cover of a dense line of skirmishers. Our light companies were now recalled by Halkett for fear they should be destroyed, but not before Lieut.-Colonel Vigoureux, Lieutenants Rumley and Pratt, and many men of the 30th company had been hit.

As a preliminary to the grand attack a determined effort was made to carry La Haie Sainte. Two columns of Donzelot's and Quiot's divisions surrounded the farm and passing beyond it engaged the troops of the third division on the heights. The attack was unsuccessful but our line from the left of Halkett's brigade to La Haie Sainte was still further weakened.

During this combat, Marshal Ney had assembled in the hollow ground between La Haie Sainte and La Belle Alliance 43 squadrons, the cuirassiers of Milhaud in the front, the red lancers of the Guard in support and the chasseurs of the Guard in reserve. At about four o'clock all was ready for the attack. Mindful of the hollow road in rear of La Haie Sainte, Ney directed the right of his force on the squares of Verden and Bremen, and of the 30th and 73rd; the left attacked Maitland's Guards' brigade. Sir Charles Alten thus sums up the onslaught upon his division and its result. "The cannonade by this time on the part of the enemy was most destructive to our infantry squares, yet none showed even a disposition to give way but filled up the space over the bodies of their brave comrades as they fell. The enemy's cavalry now appeared in crowds upon the position, charged the square of the 30th and 73rd, the one of the Grubenhagen and Osnabruck and that of the Bremen and Verden field battalions five or six times, but were as often repulsed by the coolness of our troops who reserved their fire until they approached within twenty paces."

After each attack the French were charged by our cavalry while shaken by the fire of the squares and driven down the hill, and our artillerymen who, when the enemy's cavalry advanced, had by the Duke's order abandoned their pieces and taken shelter in the nearest square, ran out and followed the retreating horsemen with their fire.

In one of the earliest of those attacks the adjutant of the 6th cuirassiers had his horse shot and fell in front of the side of the square formed by the 30th. Lieutenant Hughes saved him from the bayonets raised to kill him, and taking him by the hand drew him into the square and for further safety made the young Frenchman take his arm. They stood thus until the Imperial Guard formed up for attack on the heights near La Belle Alliance, nearly two hours later, when the adjutant implored Hughes to send him to the rear that he might not be killed by his friends. Hughes pointed out to him that his life would not be worth a minute's purchase if he left the square, but in a moment of confusion they were separated and in all likelihood the young Frenchman was killed in attempting to escape. This confusion was that caused by the giving way of some foreign troops on the left in rallying whom Hughes is said to have been wounded.

Ney was not discouraged by the failure of his first onslaught, but obtained from the Emperor leave to strengthen his force with Kellerman's two divisions of reserve cavalry and the heavy cavalry division of the Guard.

After each attack as soon as the French cavalry had retired the artillery fire had recommenced on our squares and Ney, before making his second great advance, allowed some time to elapse for it to take effect. When he thought the artillery fire had done its work the enormous mass of cavalry advanced again and covered the ridge and its reverse slope upon which our squares were posted, but although the French charged again and again it was in vain, not a square gave way. The cavalry attack lasted altogether about two hours, during which the square composed of the 30th and 73rd was charged eleven times. Gradually the charges lost their vigour and the French horsemen opened a sharp carbine fire on the immovable squares, but this died down, many horsemen left the ridge and finally an advance of our cavalry swept the whole body into the hollow in front of La Belle Alliance. All the heavy cavalry on both sides had now been engaged.

The third division got no rest, however, for before the cavalry attack had ceased, Donzelot had renewed the attack on La Haie Sainte and Bachelu had moved against the Hanoverians on Halkett's left. The Prince of Orange deployed and moved forward the 5th and 8th line battalions of the Legion to charge Bachelu's columns, but some French cavalry falling back from the right saw their opportunity and charged the 8th in flank, cutting down most of its right wing and capturing a colour.

About 6 p.m. Major Baring, who had in vain sent repeatedly to the rear for rifle ammunition, had fired his last cartridge and withdrawn his men from La Haie Sainte. The French occupation of La Haie Sainte was hailed with loud cheering by their army and put fresh spirit into their men, upon whom the long and apparently unsuccessful combat had told heavily. A swarm of skirmishers was pushed forward against the third division, supported by a mass of cavalry rallied by Marshal Ney from different corps. Ompteda commanding the brigade of the Legion was forced by the Prince of Orange to deploy the fifth battalion again and advance against the skirmishers. Irritated by the Prince's language the gallant Ompteda put himself at the head of the battalion and advanced rapidly against the enemy who fell back and on approaching La Haie Sainte, as Ompteda had foreseen, saved themselves in the enclosures, thus uncovering their cavalry who charged and overthrew the battalion, killing Ompteda and most of his men. The French infantry were now actually on the ridge held by the third division, and it was not long before they had guns up.

In his great history of the campaign Siborne, after describing these events, says :—" Of all the troops comprising the Anglo-allied army the most exposed to the fierce onslaught of the French cavalry and to the continuous cannonade of their artillery were the two British squares posted during a very great portion of the battle in advance—at times considerably so—of the narrow road which ran along the crest of the Duke's position. They consisted of the third battalion of the 1st Guards, and of the 30th and the 73rd, acting together as one corps. It was upon these troops that fell the first burst of the grand cavalry attacks and it was upon these troops also that the French gunners seldom neglected to pour their destructive missiles.

"It was not long after this that the square of the 30th and 73rd was attacked by some French artillery, which trotted boldly up the slope in front of these regiments and, having approached within a fearfully short distance, unlimbered two of its guns, from which several rounds of grape were discharged into the very heart of the square."

At this time also guns were brought up against the Hanoverian square formed by the battalions Bremen and Verden 160 yards on the left of the 30th and 73rd. One side of the square was completely blown away and the fire continued till the square was reduced to a clump of men. The square formed by the battalions Grubenhagen and York also suffered severely, and as has been told, the brigade of the King's German Legion near La Haie Sainte had been already very severely handled, the two line battalions being nearly destroyed.

Alten had been wounded and Kielmansegge now commanded the division. The Prince of Orange, whose gallantry and dogged perseverance command respect in spite of his blunders, put himself at the head of Kruse's Nassau brigade and brought them up to recover the position, but he was badly wounded and the Nassauers fell back. At this moment the Duke of Wellington in person led into action five battalions of Brunswickers, whom he had brought from the right, and placed them on Halkett's left. The fire was so awful that the Brunswickers fell back, carrying with them Kruse's brigade and the remains of Kilmansegge's and Ompteda's men. There was now a gap in the line from the square formed of the 30th and the 73rd for the 650 yards to Picton's division, and the position was very serious. The Duke, assisted by a number of staff and other officers, rallied the Brunswickers, and Kielmansegge, after giving them a few minutes' breathing time, was able to lead back into line the remains of the two noble brigades of the Legion and Hanoverians. It was in helping to rally the Brunswickers that Hughes is said to have been wounded.

In describing the loss of La Haie Sainte and the events which followed, no mention has been made of a movement to the point of danger made by the 30th and 73rd. Sir Colin Halkett merely tells us that when the farm fell he moved the two regiments to their left, but suffered severely and did not persist in the movement. It may be assumed that he moved to his left simultaneously with Ompteda's advance and halted when he found that through the failure and death of that gallant officer he was left alone with two weak battalions in face of overwhelming odds. Perhaps it is at this time we should place an apparently well authenticated story of Sir Alexander Gordon, the Duke's A.D.C. Peering through the smoke which hung low upon the battlefield he asked "What is that square lying down in front?" The answer was "That is the position from which the 30th and the 73rd have just moved." Their loss had been nearly 300.

After his movement to the left General Halkett did not resume his original position but placed his brigade at the junction of the Wavre road, and that to Merbe Braine, about 100 yards nearer to La Haie Sainte. The Guards also moved to their left to keep their distance from Halkett. It was evident that the battle, as far as our army was concerned, was to be fought out on the ground between the Charleroi and Nivelles roads, and the Duke massed all his available men there. On the front originally occupied by the third division, the remains of Ompteda's and Kielmansegge's brigades were on the left touching the Charleroi road. On their right was Kruse's Nassau brigade, then the Brunswickers, and then Halkett's brigade, and on its right was Maitland's brigade of Guards. To the right of the Guards stood the Light Brigade, under General Adams, which had been brought from the right across the Nivelles road. It was composed of the 52nd (two battalions), the 71st and the second battalion of the 95th (Rifle Brigade), with some companies of the third battalion. The remainder of Clinton's division, that is a Hanoverian brigade under Hew Halkett and a brigade of the King's German Legion, was also moved to its left. In second line, in rear of Ompteda's and Kielmansegge's brigades was a brigade of Hanoverians brought across the Charleroi road from Picton's division and they were strengthened by Sir John Lambert's British brigade posted on the eastern side of the road. Vivian's light cavalry, released from duty on the left of our army by the approach of the Prussians, formed up behind the Nassauers and Brunswickers. Chassé's Dutch-Belgian division was brought up from Braine l'Alleud and one brigade of it was posted behind our Guards and the other behind Halkett. Sir John Vandeleur's light cavalry brigade which had followed Vivian's from the left, took post on the Nivelles road in rear of the Dutch-Belgians. Lloyd's battery, in spite of the exhaustion of the few gunners left, was still able to keep up a fire, but it was very slow. Cleeve had run out of ammunition and taken his battery to the rear. Luckily his place was filled by a Dutch-Belgian horse battery of eight guns brought up by Major Van der Smissen, which formed on Lloyd's right.

More than one-third of the casualties among the officers of the 30th took place during and after the attack of the French guard. If we allow one-fourth of the casualties among the men for this last phase, the brigade should have had over 1,100 rank and file, and the 30th and 73rd, 630, to meet the attack.

Napoleon on his part was preparing for the last struggle. He had brought forward to La Belle Alliance seven battalions of his Guard, of which two were to remain in reserve and the other five to head the last great attack, which was to be joined in by every corps which retained the power to act. By his orders no rest was given to the third division, fresh swarms of tirailleurs ascended the heights and their fire at close quarters shook the Nassauers and Brunswickers who began to fall back. At this moment Count Kielmansegge with the remains of the Legion and Hanoverians dashed boldly forward with drums beating and charged the enemy. They were cheered on by Sir Hussey Vivian and his officers. The Nassauers and Brunswickers took up the movement and the line from Halkett's left to the Charleroi road was re-established where it had stood in the morning.

The French Guard was now formed up for attack and the Duke ordered his men to deploy into line four deep. Halkett kept his brigade behind the crest of the hill during the tremendous artillery fire with which Napoleon prepared the way for his Guard. When the attack came it was at first in two columns in echelon from the right and directed against the British Guards and Adams' brigade. As the right attacking column advanced obliquely across the front of the third division Halkett led the 33rd and 69th forward from second line on the left of Maitland's Guards' brigade and was able by firing to his right front to assist in the defeat of this column of the Imperial Guard by Maitland. Van der Smissen's Dutch-Belgian battery came into action on the left of the 69th and helped materially. When advancing to the crest of the hill, the 33rd and 69th had met the full force of the French artillery fire and in the opinion of the officers of the 30th, whom they passed on their way, lost more severely than any troops that day, in an equally short space of time. This fire perforce ceased as the French mounted the hill.

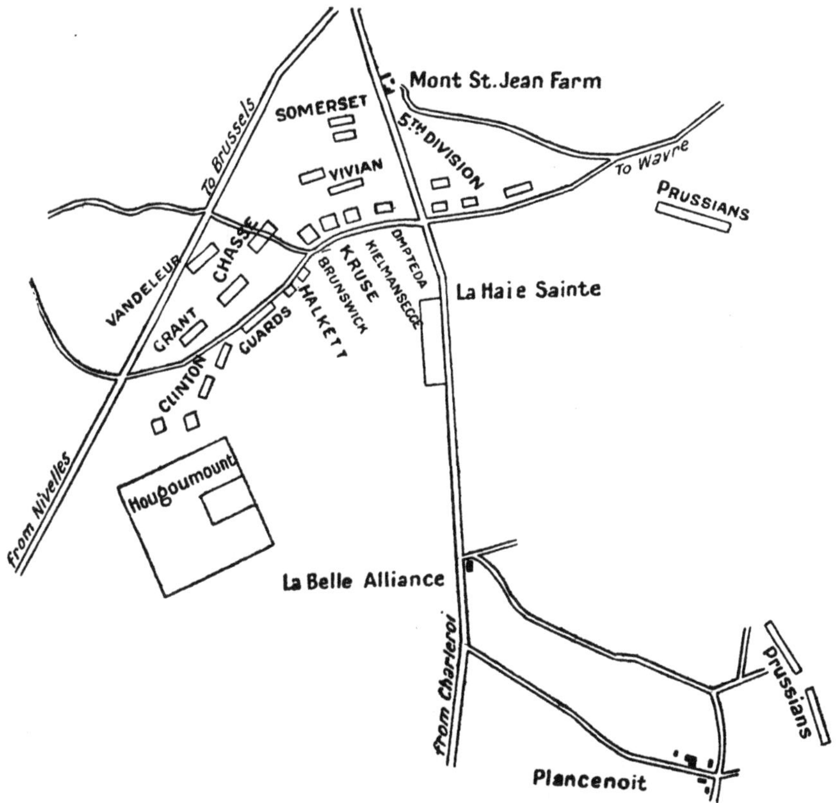

WELLINGTON'S DISPOSITIONS TO MEET THE ATTACK OF THE IMPERIAL GUARD.

The left column of the Imperial Guard which had appeared to be directed against Adams' brigade swerved to its right and attacked Byng's British Guards. Colbourne saw his opportunity, wheeled up the 52nd, fired into the left of the column and charged. This was a fatal stroke, and the Imperial Guard fell back in confusion. Colbourne had only anticipated the orders of Wellington, who was present. No time was given to the French to regain their order. The remainder of Adams' brigade formed line on the 52nd, Hew Halkett's brigade was brought up in quarter column on the right and as a further protection against the French cavalry, Vivian's light cavalry brigade from behind the third division was ordered to the front. Adams was then able to push the

Imperial Guard across the front of our army to La Belle Alliance, where it was rallied by Napoleon.

When the Imperial Guard advanced one or two battalions had been thrown out to the right as a flank guard. Something had delayed the advance of this body, which appeared in front of Halkett's left after the defeat of the first column by Maitland's brigade. After assisting in this defeat Halkett was able to hasten to his left to call upon the 30th and 73rd to resist this new attack. Under him the two battalions advanced in line four deep from the hollow. On the summit of the ridge there is here a space of 100 yards of nearly level ground and the French had gained this little plateau and were crossing it when the 30th and 73rd appeared on its northern edge; the advance of the two battalions had been sheltered from artillery fire by their opponents. When the column and the line were about 80 yards apart,[1] the French, who had attempted no deployment, halted and fired. Their volley from, perhaps, 150 muskets, was answered by one twice as heavy from the two front ranks of Halkett's battalions, and the French did not wait for the British charge. The Imperial Guard scarcely lived up to its reputation, but it was the old story of column against line, and in this case the line was backed by a formidable artillery.

At the moment of success, Halkett was called away to take command of the division in place of Kielmansegge, wounded. This was a misfortune for the brigade, which, owing to the retreat of the enemy, found itself exposed, on the crest line, to the full fury of the French artillery fire. To remain there was destruction; no one in Wellington's absence could sanction an advance, so of necessity the word was given to retire. Halkett may have given it before he left. Unfortunately, the officer in command seems to have thought it necessary to form squares before moving, and the losses were very heavy. In the 30th Macnab was killed by a grape shot; Prendergast blown to pieces by a shell; James and Bullen had their legs smashed by a round shot. Bailey was wounded by a musket ball and forty men are said to have fallen. During the retreat one square ran into the other, the officers and men who tried to check the rush were literally swept off their feet and the brigade did not halt till it came in line with the Brunswickers and was out of fire. According to a letter from Macready to Siborne the men of the 30th never lost their calmness. Some were laughing and saying to their officers, "By God, Sir, I'd halt, but I cannot get my feet under me."

When the brigade halted Halkett came up and was restoring order when Wellington returned from the right where he had directed the defeat of the two columns of the Imperial Guard and had started Adams on his victorious march. He treated the mishap of the fifth brigade lightly and not as a matter calling for his personal attention.

General Halkett was severely wounded at this time. He had before this been slightly wounded twice and had four horses killed under him.

The Prince of Orange commanding the corps, Sir Charles Alten and his three brigadiers were all down and it is hard to say who commanded the division. The 30th were commanded by Major Chambers. He had shown his usual energy in re-forming the brigade but soon afterwards fell, shot through

[1] This is Captain Howard's estimate. On the following morning he paced the distance between the lines of dead and wounded of the French Guard and of our own men.

the heart by a musket ball. He had just said in jest that he was too small to be hit. Captain Gore, who himself had been wounded shortly before, says, in praising the brave men who gave their lives for their country on that day, " Of that number there is no one more worthy of the voice of praise than my friend and companion in arms, Major T. W. Chambers. He was an active, zealous and intelligent officer and a great loss to his regiment both as a soldier and a gentleman." Captain Robert Howard succeeded to the command.

Major Kelly, of the 73rd, an officer on the staff of the quartermaster-general, now took command of the two battalions, and soon afterwards they were attacked by a body of which the brigade of Bourgeois formed part. The brigade was composed of the 28th and 105th regiments and was brought from a considerable distance beyond the Charleroi road. Its attack should have been simultaneous with that of the Imperial Guard.

The following is Major Kelly's narrative. He is quite an hour out in his time, for it now must have been half-past seven or later.

" About half-past six when passing with the Duke and other officers in rear of the fifth brigade there appeared to be some little confusion, when the Duke without addressing himself to any one in particular said : 'See what's wrong there.' I rode up to the brigade and, while addressing myself to Sir Colin Halkett, he at that instant received a wound in the face, the ball passing through his mouth, and he had to retire. Colonel Elphinstone of the 33rd ran up and asked if I had any orders. I replied none except to inquire into the cause of the confusion. He stated that they were much pressed and the men exhausted, Harris of the 73rd had been severely wounded and he commanded the brigade and he added : 'What is to be done ; what would you do ?' At this period the attacking column was again retiring. I advised him to direct the officers and men to resume their respective stations (they had got intermixed from the frequent formation of square) to form as extended a front as possible, directing them to cover themselves as well as they could by lying down, to renew or check their flints and to fresh prime so as to be ready for the next attack.

" Thus situated we remained for a short time inactive when at last the attacking column made its appearance through the fog and smoke which throughout the day lay thick on the ground. Their advance was, as usual with the French, very noisy, and evidently reluctant, the officers being in advance some yards cheering their men on. They, however, kept up a confused and running fire, which we did not reply to till they reached nearly a level with us (that is they topped the ridge), when a well-directed volley put them in a confusion from which they did not appear to recover, but after a short interval of musketry on both sides they turned about to a man and fled. After a short period the order to advance came down the line from the right, which we obeyed as soon as the brigade was formed."

A memento of the last struggle of the 30th with the enemy on this day remains in the shape of the drum of the 105th French regiment now in the officers' mess of the regiment.

In a further letter Major Kelly says of the flight of the French : " I presume it might have been about this time that some of our troops had got to the rear or flank of the enemy and caused this sudden retreat."

Major Kelly's narrative has thus brought us to the time when Adams' Light Brigade, following the defeated Imperial Guard across the front of our position, had halted near the corner of the orchard of La Haie Sainte. The farm was at once re-occupied by the Allies and the Duke of Wellington riding down past it joined Adams. After a look at the Imperial Guard rallied by Napoleon in front of La Belle Alliance, the Duke said to Adams, " They won't stand, better go on." Then turning to Sir John Colbourne he gave the order, " Go on, Colbourne ! Go on ! " The whole Allied army was ordered to advance in support of Adams and of Vivian and Vandeleur who pressed forward on the right. The Duke himself, facing his army from near the position to the right front of the 30th, from which he had watched the French advance in the morning, gave the signal for the line to advance with his cocked hat held high above his head. At this moment the rays of the setting sun broke through the clouds which had hung low on the battle-field and fixed the hour as about eight o'clock.

The 30th and 73rd halted when they came to the ridge which they had so long and so well defended and allowed the Dutch-Belgians to pass from the rear into second line. The latter were for the most part in a state of high excitement and cheering wildly, with their shakos on the points of their bayonets, and they received a good deal of rough chaff from our men as they passed to the right of the 30th.

The resistance of the French was feeble and their exhausted army practically dissolved. Adams and the two light cavalry brigades pursued as far as Rosomme whence the pursuit was taken up by the Prussians who had captured Planchenoit and broken in on the French line of retreat.

Throughout the battle the Duke had never been more than a few hundred yards from the 30th and during the earlier charges of the French cavalry he took shelter in the square formed by that regiment and the 73rd. The following is his account of the day to his friend, Marshal Beresford :—

" You will have heard of our battle of the 18th. Never did I see such a pounding match. Both sides were what the boxers call gluttons. I had the infantry for some time in squares and we had the French cavalry walking about us as if they had been our own. I never saw the British infantry behave so well."

When the Duke's A.D.C. and friend, Colonel the Hon. Sir Arthur Gordon, fell with his leg smashed by a shot just in rear of the 30th, Sergeant-Major James Woods was ordered to take a party and carry him at once to Mont St. Jean in the hope that immediate amputation might save so valuable a life, and when another A.D.C., Lieut.-Colonel Canning, fell on almost the same spot shot through the body, he was carried into the square where he died, holding Captain Gore's hand and propped up by knapsacks, while the Earl of March, a brother A.D.C., listened to his last wishes.

MACREADY'S "DIARY"

" The line of hill we occupied descended in a very gentle slope, and was covered with grain higher than our heads. At about half-past ten o'clock the enemy began moving his forces, and displayed strong columns of infantry and cavalry opposite every part of our position. A superb line of Red Lancers stretched from their left, far beyond our right flank, but, from the nature of the ground and the disposition of Clinton's division, they were not much feared.

"Thus, at near eleven o'clock, stood the contending armies. Our army might amount to more than 60,000 men; and the enemy's probably exceeded 80,000. We (I mean the multitude) were not aware that Blücher could afford us any assistance, as we heard that he was completely beaten and hotly pursued; but no British soldier could dread the result when Wellington commanded. Our poor fellows looked wretchedly, but the joke and laugh were bandied between them heartily and thoughtlessly, as in their happiest hours. About eleven o'clock some rations and spirits came up; the latter were immediately served out to the men, but I dared not drink on my empty stomach. I had just stuck a ramrod through a noble piece of beef, and was fixing it on the fire, when an A.D.C. galloped up, and roared out, 'Stand to your arms.' We were in line in an instant. Considerable movements were perceptible among the enemy's columns, and from the number of mounted officers riding to and from one group of horsemen, I should think Napoleon was there. Our artillery arrived at full gallop, and the guns were disposed on the most favourable ground in front of their respective divisions. The regiments formed column and marched a little to the rear, under cover of the brow of the hill; our company and 73rd Grenadiers protecting Cleve's and Lloyd's brigade of guns. The men were in a great measure covered by the crest of the hill, but the whole French army, with the exception of its reserve, was exposed to our artillery. There was a pause for some minutes, and I imagined there were few of the many thousands assembled that did not experience a sort of chill and rising sensations in their breasts. It was indeed a spirit-stirring sight—the chivalry of two mighty nations in grand and deadly rivalry.

"At length the enemy's left appeared in motion towards Hougoumont, and old Cleve slapped away at them. When the first shot was fired I threw off a wet blanket I had wrapt round me, gave myself a shake, and, like Joe Miller's soldier, considered all was clear gain that I might bring out of the battle. Cleve's guns, which told most gloriously on the columns as they approached the orchard, were unanswered for some minutes, but we soon saw the enemy's artillery trotting down the hill, and at once they opened from 200 pieces. The cannonade extended along the whole line, and the musketry commenced in thundering volleys at Hougoumont. The skirmishers were soon ordered to extend twelve paces each file, and to descend the slope in order to protect the guns. Jerome Bonaparte's corps attacked the orchard and chateau, defended by some companies of our Guards under Lord Saltoun. The obstinacy of the assailants was only exceeded by the gallantry of their opponents. For an hour and a half they were muzzle to muzzle and bayonet to bayonet; fresh bodies were poured in incessantly by the enemy, and the Guards repeatedly reinforced their comrades. I saw them amid the flames of the trees and out-houses, to which the French had set fire, alternately advancing and retiring, first the red and then the blue jackets prevailing. Around single trees whole sections lay dead. At length the overwhelming force of the enemy enabled him to establish himself in the orchard and gardens, and the building itself became the point of attack. From its doors and windows our gallant

Guardsmen poured an unceasing shower of bullets, and the enemy fell dead in heaps around them; repeated and successful sallies astonished the Frenchmen and convinced them of the inutility of their perseverance. After two hours of most determined exertion they retired from this spot, leaving it covered with the bodies of their countrymen.

"The conduct of the Guards was most glorious. On the retreat of the enemy, the firing still continued at this point; but it was no longer considered as an attack, merely occupation for both parties. During the contest for the chateau our columns were lying at length under the hill to shelter themselves as much as possible from the showers of shot and shell which were tearing up every part of the field. The Light Dragoons, to the right of Hougoumont, were skirmishing with the Lancers in beautiful but not very effective style, for they seemed to think their broad-swords no match for the lance; it was all pistolling, and at a distance which would have satisfied even Bob Acres. The artillery on both sides, covered by their respective light troops, who kept up a brisk fire, were dealing destruction around them; and the only bodies in motion were the groups of staff officers who attracted the fire of the enemy and the curses of their friends wherever they appeared. Our company and 73rd Grenadiers, after a pretty long skirmish, had pushed the French tirailleurs close under their guns, and our shot began to whistle among the artillerymen, when we perceived a body of cavalry coming down on us at a gallop. We were too far extended to effect any formation and the ground was quite open, so Colonel Vigoureux gave the word to us to make off, and away we went at score. Pratt with some men reached a Hanoverian square; Rumley, one of the Nassau's, and I, with about a dozen men, made our own. The rest of our men were dispersed into La Haie Sainte and various squares, and some of them cut down. Our rapid retreat was peculiarly dangerous, as we had to run through high corn towards our guns, which opened with grape on the enemy's cavalry. Kielmansegge's Jägers, who were on our left, trusting to their numbers and the nature of the ground, stood, and were annihilated. After cutting them to pieces, the cavalry galloped up the slope, sabred the greater part of Lloyd's artillerymen, and charged a Hanoverian square. They were repulsed, and, before they could effect their retreat, were destroyed by a squadron of our Life Guards. These ruffians laughed at us as we scudded from their uplifted sabres, but as their own proverb says, 'Il rit bien qui rit le dernier.' I could not help grinning at some of les bons sabreurs, though certainly they made noble-looking corpses. Their charge was a gallant piece of service—of course, as they were destroyed, it will be called a rash one; but had they been satisfied with the destruction of a regiment of Jägers and a brigade of artillery, they might have returned to their comrades covered with success and glory. Our company re-assembled at Cleve's brigade, and lay down among the guns until the advance of the enemy's tirailleurs, when we proceeded once more down the slope."

. . . .

"When Napoleon saw his columns irretrievably routed on the left, he appears to have determined on a grand and desperate push upon our centre; infantry had alone advanced against Hougoumont and Picton's line, but they were now to be supported by the whole of his cavalry, and accompanied by a formidable force of artillery. Before the commencement of this attack our company and Grenadiers of the 73rd were skirmishing briskly in the low ground, covering our guns and annoying those of the enemy. The line of tirailleurs opposed to us was not stronger than our own, but on a sudden they were reinforced by numerous bodies, and several guns began playing on us with canister. Our poor fellows dropped very fast, and Colonel Vigoureux, Rumley and Pratt were carried off badly wounded in about two minutes. I was now commander of our company. We stood under this hurricane of small shot till Halkett sent to order us in, and I brought away about a third of the light bobs, the rest were killed or wounded; and I really wonder how one of them escaped. As our bugler was killed, I shouted and made signals to move by the left, in order to avoid the fire of our guns, and to put as good a face upon the business as possible.

"When I reached Lloyd's abandoned guns, I stood near them for about a minute to contemplate the scene; it was grand beyond description. Hougoumont and its woods sent up a broad flame through the dark masses of smoke that overhung the field; beneath this cloud the French were indistinctly visible. Here a waving mass of long red feathers could be seen; there gleams as from a sheet of steel, showed that the cuirassiers were moving; 400 cannon were belching forth fire and death on every side; the roaring and shouting were indistinguishably commixed—together they gave me the idea of a labouring volcano. Bodies of infantry and cavalry were pouring down on us, and it was time to leave contemplation, so I moved towards our columns, which were standing up in square. Our regiment and 73rd formed one, and 33rd and 69th another; to our right beyond them were the Guards, and on our left the Hanoverians and German Legion of our division. As I entered the rear face of our square I had to step over a body, and, looking down, recognized Harry Beere, an officer of our Grenadiers, who about an hour before shook hands with me, laughing, as I left the columns. I was on the usual terms of military intimacy with poor Harry—i.e., if either of us had died a natural death, the other would have pitied him as a good fellow, and smiled at his neighbour as he congratulated him on the step; but seeing his herculean frame and animated countenance thus suddenly stiff and motionless before me (I know not whence the feeling could originate, for I had just seen my dearest friend drop almost with indifference), the tears started to my eyes as I sighed out, 'Poor Harry!' The tear was not dry on my cheek, when poor Harry was no longer thought of. In a few minutes after, the enemy's cavalry galloped up and crowned the crest of our position (our guns were abandoned), and formed between the two brigades (batteries) about 100 paces in our front. Their first charge was magnificent. As soon as they quickened their trot into a gallop the cuirassiers bent their heads, so that the peaks of their helmets looked like visors, and they seemed cased in armour from the plume to the saddle. Not a shot was fired till they were within 30 yards, when the word was given, and our men fired away at them. The effect was magical. Through the smoke we could see helmets falling, cavaliers starting from their seats with convulsive springs as they received our balls, horses plunging and rearing in the agonies of fright and pain, and crowds of the soldiery dismounted, part of the squadron in retreat, but the more daring remainder backing their horses to force them on our bayonets. Our fire soon disposed of these gentlemen. The main body re-formed in our front, and rapidly and gallantly repeated their attacks. In fact from this time (about four o'clock) till near six we had a constant repetition of these brave but unavailing charges. There was no difficulty in repulsing them, but our ammunition decreased alarmingly. At length an artillery wagon galloped up, emptied two or three casks of cartridges into the square, and we were all comfortable.

"The best cavalry is contemptible to a steady and well supplied infantry regiment; even our men saw this, and began to pity the useless perseverance of their assailants, and, as they advanced, would growl out, 'Here comes these d——d fools again!' One of their superior officers tried a *ruse de guerre* by advancing and dropping his sword, as though he surrendered; some of us were deceived by him, but Halkett ordered the men to fire, and he coolly retired, saluting us. Their devotion was invincible. One officer whom we had taken prisoner, was asked what force Napoleon might have in the field, and replied, with a smile of mingled derision and threatening, 'Vous verrez bientôt sa force, Messieurs.' A private cuirassier was wounded and dragged into the square; his only cry, 'Tuez, donc, tuez, tuez, moi, soldats!' and as one of our men dropped dead close to him, he seized his bayonet, and forced it into his own neck; but this not despatching him, he raised up his cuirass, and plunging the bayonet into his stomach, kept working it about till he ceased to breathe.

"Though we constantly thrashed our steel-clad opponents, we found more troublesome customers in the round shot and grape which all this time played on us with terrible effect, and fully avenged the Cuirassiers. As often as the volleys created openings in our square, the cavalry dashed on, but they were uniformly unsuccessful. A regiment on our right seemed disconcerted, and at one moment was in considerable confusion. Halkett rode out and seizing their snow-white colour, waved it over his head, and restored them to something like order, though not before his horse was shot under him. At the height of their unsteadiness we got the order to 'right face' to move to their assistance; some of the men mistook it for 'right about face,' and faced accordingly, when old Major McLain, 73rd, called out, 'No, my boys, it's "right face"; you'll never hear the "right about" as long as a French bayonet is in front of you.' In a few minutes he was mortally wounded. A regiment of Light Dragoons, by their facings either the 16th or 23rd, came up to our left and charged the cuirassiers. We cheered each other as they passed us; they did all they could, but were obliged to retire after a few minutes at the sabre. A body of Belgian cavalry advanced for the same purpose, but on passing our square they stopped short. Our noble Halkett rode out to them, and offered to charge at their head; it was of no use; the Prince of Orange came and exhorted them to do their duty, but in vain; they hesitated till a few shots whizzed through them, when they turned about, and galloped like fury, or, rather, like fear. As they passed the right face of our square, the men, irritated by their rascally conduct, unanimously took up their pieces and fired a volley into them, and many 'a good fellow was destroyed so cowardly.'

"The enemy's cavalry were by this time nearly disposed of, and, as they had discovered the inutility of their charges they commenced annoying us by a spirited and well-directed carbine fire. While we were employed in this manner it was impossible to see further than the columns on our right and left, but I imagine most of the army was similarly situated; all the British and Germans were doing their duty. About six o'clock I perceived some artillery trotting up our hill, which I knew by their caps to belong to the Imperial Guard. I had hardly mentioned this to a brother officer when two guns unlimbered within 70 paces of us, and by their first discharge of grape, blew seven men into the centre of the square. They immediately reloaded and kept up a constant and destructive fire. It was noble to see our fellows fill up the gaps after every discharge. I was much distressed at this moment; having ordered up three of my light bobs, they had hardly taken their station when two of them fell, horribly lacerated. One of them looked up in my face and uttered a sort of reproachful groan, and I involuntarily exclaimed, 'By G——, I couldn't help it!' We would willingly have charged these guns, but, had we deployed, the cavalry that flanked them would have made an example of us.

"The *vivida vis animi*—the glow which fires one upon entering into action—had ceased. It was now to be seen which side had most bottom, and would stand killing longest. The Duke visited us frequently at this momentous period; he was coolness personified. As he crossed the rear face of our square a shell fell amongst our grenadiers, and he checked his horse to see its effect. Some men were blown to pieces by the explosion, and he merely stirred the rein of his charger, apparently as little concerned at their fate as at his own danger. No leader ever possessed so fully the confidence of his soldiery, 'but none did love him';—wherever he appeared, a murmur of 'Silence—stand to your front—here's the Duke!' was heard through the columns, and then all was steady as on a parade. His aides-de-camp, Colonels Canning and Gordon, fell near our square, and the former died within it. As he came near us late in the evening, Halkett rode out to him and represented our weak state, begging his Grace to afford us a little support. 'It's impossible, Halkett,' said he. And our General replied, 'If so, sir, you may depend on the brigade to a man!' Our colours were ordered to the rear. This measure has been reprobated by many, but I know I never in my life felt such joy, or looked on danger with so light a heart, as when I saw our dear old rags in safety. Our brigade did not stand 800 men, and how could they be expected to protect four stands of colours from the most dreaded troops in Europe, approaching with an awful superiority of numbers?[1]

"It was near seven o'clock, and our front had sustained three attacks from fresh troops, when the Imperial Guard were seen ascending our position in as correct order as at a review. As they rose step by step before us, and crossed the ridge, their red epaulettes and cross-belts put on over their blue great-coats, gave them a gigantic appearance, which was increased by their high hairy caps and long red feathers, which waved with the nod of their heads as they kept time to a drum in the centre of their column. 'Now for a clawing,' I muttered, and I confess, when I saw the imposing advance of these men, and thought of the character they had gained, I looked for nothing but a bayonet in my body, and I half breathed a confident sort of wish that it might not touch my vitals.

"While we were moving up the slope, Halkett, as well as the noise permitted us to hear him, addressed us and said, 'My boys, you have done every-

[1] Can Macready's memory, for once, have played him a trick and caused him to insert the numbers left after the close of the action?

thing I could have wished, and more than I could expect, but much remains to be done ; at this moment we have nothing for it but a charge.' Our brave fellows replied by three cheers. The enemy halted, carried arms about 40 paces from us, and fired a volley. We returned it, and giving our 'Hurrah !' brought down the bayonets. Our surprise was inexpressible, when, pushing through the clearing smoke, we saw the back of the Imperial Grenadiers ; we halted and stared at each other as if mistrusting our eyesight. Some 9-pounders from the rear of our right poured in the grape amongst them, and the slaughter was dreadful. In no part of the field did I see carcases so heaped upon each other. I never could account for their flight, nor did I ever hear an admissible reason assigned for it. It was a most providential panic. We could not pursue on account of their cavalry, and their artillery was still shockingly destructive.[1]

"About this time Baron Alten was wounded, and General Halkett went to take the command of the division. There was a hedge[2] in our rear, to which we were ordered to move, as some cover from the fire. As we descended the declivity, the enemy thought we were flying, and, according to their invariable custom, turned a trebly furious cannonade upon us. Shot, shell, and grape came like a hurricane through the square and the hurly-burly of these moments can never be effaced from memory. A shriek from forty or fifty men burst forth amid the thunder and hissing of the shots. I was knocked off my legs by the fall of a brother officer, and just as I recovered my feet, an intimate friend, in the delirium of agony, occasioned by five wounds, seized me by the collar, screaming, 'Is it deep, Mac ; is it deep ?' Another officer was seen to halt, as if paralysed, and stare upon a burning fuse, till it fired the powder and shattered him to pieces. At this instant the regiments on our right rushed amongst us in frightful confusion, and our men passed the hedge at an accelerated pace. The exertions of the officers were rendered of no avail by the irresistible pressure, and, crying with rage and shame, they seized individuals to halt them, they were themselves hurried on by the current. At this moment, some one huzza'd, we all joined, and the men halted. Major Chambers ordered me to dash out with our light bobs and grenadiers, whilst the regiment marched up to the hedge, and re-formed. The whole brigade was within an ace of ruin. Our men were as steady as rocks till the others came amongst them, when the disorder was extreme. The officers did wonders, but the shout alone saved us. I never could discover who raised it, nor can I conceive what the enemy was about during our confusion. Fifty cuirassiers would have annihilated our brigade. Some of them advanced, when everything was remedied, and forced my party to retire ; but as they did not appear inclined to charge, I was reinforced, and we continued to amuse them, while the 33rd and 69th, having formed four deep, went to occupy their position in the line. Some Brunswickers had formed on our left as a support ; they gave way once, but were rallied, and now stood their ground famously. Cook's and Clinton's divisions had also to repulse attacks of the Imperial Guard.

[1] He had passed Lloyd's nine-pounders in the advance, but apparently did not see the eight Dutch-Belgian guns on Lloyd's right.

[2] A twig from this hedge, presented by a grand nephew of Macready, is preserved in the officers' mess, 30th Regiment. It is beech.

"The ground between Hougoumont and the hill was now occupied by the second division, which, on the advance of the lancers, had moved up, and altered the original convex of the division to a concave, thus raking the advance of the French columns. There was severe fighting on this point, and the Welsh Fusiliers suffered terribly. The Prussians had ere this begun to push the enemy's right, and it was evident, from the lull which took place near us—for cannonading and close skirmishing with columns in grey great-coats was now all our work—that affairs were altering. We were in line four deep, and the enemy's column within 150 yards of us, and yet neither party advanced. I lost some men while covering the regiment, but the dead horses and soldiers formed capital shelter for both sides. I was wondering at the apathetic listlessness that seemed to possess us all, when suddenly the enemy's columns fired away with considerable effect. Major Chambers dropped dead, General Halkett of ours was shot through the face, and the casualties were again numerous. The fire toward the right of the French became tremendous, our opponents rapidly and unexpectedly disappeared, and the regiment of German Hussars galloped past to our right, cheering us (this is the moment mentioned in the despatch as the general charge), and swearing they'd pay 'em off for us. I believe the Guards, Adams' brigade, and some other corps followed the cavalry, but we did not attempt it. We marched to the crest of the hill, and the noise moved rapidly from us. The enemy must have defended some of their guns well, as long after the dragoons had passed, a solitary shot whizzed through us, and carried off the four men it had encountered.

"This must have been near eight o'clock. Soon after we piled our arms and lay down to rest. I remember as long as I remained awake I was thinking on the work, and considering whether it would be called an action or a battle. I certainly considered we had, 'spilt blood enough to make our title good' to the latter honour; but I fancied that, so far as we were concerned, some grand bayoneting charge, some concluding *coup de théâtre*, or rather *coup de grace*, was wanting to entitle us to it. I had no idea, till I awoke in the morning, that the victory was so complete. I congratulated myself in having had the honour of serving on this memorable day with the 30th regiment. Its conduct was repeatedly noticed and warmly thanked by the Prince of Orange, Picton, Alten, and Halkett; the latter was unceasing in his praises. Its loss was severe. From the number of sick, and on detached duties, it did not enter the field above 460 bayonets, of whom only 160 were in line at night; 279 men killed or wounded, and 21 men away, with disabled officers or soldiers. Our light company marched into the field, 3 officers and 51 men; of these 2 officers, 1 sergeant, 1 bugler, and 37 rank and file were killed or wounded; 6 more were away assisting them, and we stood at night, 1 commissioned, 2 non-commissioned officers, and 8 privates. When we formed four deep and the poor light bobs could only muster a front of two men, I really did not know whether I should laugh or cry. Our officers killed were Major Chambers, Captain McNabb, Lieutenants Beere and Prendergast, and Ensigns James and Bullen. Poor young Bullen was much regretted; he had left his home contrary to the wishes of a fond mother, and had only been with us three weeks. His legs were both terribly shattered. Just before the amputation of one of them, he was smiling, and saying he must now return to his mamma, and he thought £150 per annum (his half-pay and two pensions), would make her more comfortable. He bore the operation nobly, but as soon as it was ended, exclaimed, 'Gentlemen, you have done for me!' and breathed his last. When Chambers fell, his friend Nicholson threw himself on the body and sobbed aloud, 'My friend—my friend!' Harrison was standing near me in our square; a poor fellow, his servant, came up and said, 'My dear master, I am wounded and must away; but I wish to say good-bye to you for I know I shall never see you again.' The words were hardly out of his mouth, when a round shot dashed his head to pieces, and covered us with his blood and brains. Two of our officers were not on terms; the one saw the other behaving gallantly, he ran up to him and cried, 'Shake hands, and forgive all that has passed; you're a noble fellow.' The field in the morning presented a most distressing spectacle. It was covered with lacerated, mangled carcases, caps, cartridge-boxes, guns, tumbrels, books, and arms of all kinds; the poor wounded chargers looking patience in their misery, were nibbling the trampled grain round the spot they lay upon, while our wounded were bitterly reviling us and calling for assistance, which we had not in our power to give. I spoke to numbers of Frenchmen; they were not very communicative, but a common phrase among them was, 'Monsieur, nous sommes joliment foulés.' I went to look for our poor fellows who had fallen while skirmishing, but every one was dispatched—the sabre had settled their worldly affairs.

"My remaining eight lights stole me a capital breakfast, after which, about ten o'clock, we left this glorious spot encumbered with thousands of the dead and dying. Our acquaintance with the enemy had been but short, and we had some reason to complain of a few atrocities on their part; but while valour and heroic devotion to a cause are commendable, their praise as soldiers cannot be refused them. Our own countrymen excited a softer feeling—they were our friends and fellow soldiers, but they died the death that every soldier looks for, and they fell by gallant foemen. 'Peace to the souls of the heroes, their deeds were great in battle!'"

It is difficult to arrive at the exact number of casualties in the Waterloo campaign. The number of killed was at first given officially as 6 officers and 51 other ranks. In an amended return Colonel Hamilton gives the number as 36 other ranks, and this is no doubt correct. With regard to the wounded, Surgeon J. G. Elkington, in his diary, puts the number as 15 officers and 208 other ranks. A list compiled from the Medal Roll and Pay List gives 15 officers and 202 other ranks. Macready says that 279 men were killed or wounded. Deducting 51, the number he supposed to be killed, this gives us 228 wounded, or 20 more than Elkington's numbers. The Medal Roll and Pay List are probably both wrong. They do not even agree with each other. Elkington's figures may be accepted as right. He was a man of ability and experience in war and he remained on the field for two days, till all the wounded had been removed to hospital. Macready's figures are very likely right too and may have been derived from Elkington himself when he rejoined at Bavai. Both diaries may have been written up at the halt there. The numbers given by Elkington in his diary are the numbers he sent to general hospital; but there may easily have been an additional twenty men whose wounds were dressed in the regimental hospital and who accompanied the advance. We know from Macready that two subalterns, John Roe (2) and Harrison, were wounded at Quatre Bras, but not included in the returns. Roe (2) is the only one whom we know to have been wounded both at Quatre Bras and at Waterloo.

One reason for any inaccuracies there may be in the pay list is that the paymaster's clerk, like the other staff-sergeants, fought in the battle and was wounded. Another is that the authorities at home cleared the hospitals after Wellington advanced by taking many wounded to England. Neither the Duke nor the men's regiments had even a roll of the men so taken for some months. The pay list was not completed for five months after the battle when most of the wounded were back at duty. Perhaps it is creditable that the lists are so nearly correct.

The following is the distribution of the casualties compiled from the medal roll and pay list. Unless otherwise stated, the casualty took place on the 18th.

OFFICERS KILLED—Major Thomas Walker Chambers, Captain Alexander Macnab, Lieutenants Henry Beere and Edward Prendergast, Ensigns John James and James Bullen.

OFFICERS WOUNDED—Lieut.-Colonel Alexander Hamilton (16th), Major Morris William Bailey, Br.-Lieut.-Colonel Charles Albert Vigoureux, Captain Arthur Gore, Lieutenants Richard Mayne, R. C. Elliott, John Rumley, Robert Daniel, Richard Harrison (16th), Robert Hughes, John Roe (2) (16th and 18th), Purefoy Lockwood (16th), John Pratt, Charles Ousely Warren, Thomas Moneypenny, Matthias Andrews (adjutant).

Of these, Hamilton, Bailey, Vigoureux, Rumley, Lockwood, Pratt and Andrews were severely and Warren dangerously wounded.

No. 1 Grenadier Company: Officers present: Captain A. Gore (w.), Lieutenants Lockwood (w.), and Beere (k.).

Killed: Private Garrett Vizer, Meath.

Died of wounds: Colour-Sergeant Christopher Barnewell, Longford; Privates Thomas Jones, Yorkshire; Thomas Harker, Lincoln; Michael Lahay.

Other wounded: Q.M.-Sergeant Joshua Harrington, Paymaster-Sergeant Thomas Cuthbert, Drum-Major Thomas Poole, Corporal Hugh Macdonald, Drummer John Brice and 18 privates.

No. 2.—Officers: Captain Donald Sinclair, Lieutenants Mayne (w.), and Prendergast (k.).

Killed: Sergeant Thomas Catlin, Bedford; Privates Wm. Brewitt, Lincoln; Wm. Eaden, Oxford; John Grady, Tullamore; John Goodchild, Suffolk; Wm. Hopefield, Leicester.

Died of wounds: John Fletcher, Leicester; John Neill, Carlow.

Other wounded: Sergeant John Smith, Corporal John Flinn and 10 privates.

No. 3.—Officers: Lieutenants Elliott (w.), and Hughes (w.), Ensign Robert Naylor Rogers.

Killed: Privates Thomas Miskill, Galway; John Parker, Dublin; Robert Sparrow, Cambridge.

Died of wounds: Corporal R. Brown, Northampton; Privates Joseph Jordan, Leicester; Wm. Cranaway, Cambridge; Thomas Jones; R. Morrisey; Wm. Prior, Cambridge.

Other wounded: Sergeants Jonathan Cart and George Sheridan, and 15 privates.

No. 4.—Officers: Captain MacNab (k.), Lieutenants F. K. Tincombe and Moneypenny (w.).

Killed: Private John Lister, Middlesex.

Died of wounds: John Addy and John Towey, Sligo.

Other wounded: Sergeant Ben Detheridge, Corporal Pat Murphy and 10 privates.

No. 5.—Officers: Brevet-Major Mat. Ryan, Lieutenants Robert Daniel (w.), and David Latouche; Ensign James Bullen (k.).

Killed: Privates Francis McGlim (16th), Longford; John O'Brien, (16th), Tipperary; George Webb, Cambridge.

Died of wounds: Thomas Freeman and John Hilton.

Other wounded: Sergeant John Carroll, Corporal Denis Moran and 17 privates.

No. 6.—Officers: Captain Robert Howard, Lieutenants John Gowan, John Roe (2) (w.), (16th and 18th), Edward Drake.

Killed: Sergeant John Wilkinson, Leeds; Corporal John Gadborough, Nottingham; Privates James Campbell, Edinburgh; George Smith, Worcester (16th).

Died of wounds: Sergeant Charles Kilmartin, Sligo.

Other wounded: Corporals Charles Cooke and Joseph Leville, and 9 privates.

No. 7.—Officers: Captain James Finucane, Lieutenants A. W. Freear and John Roe (1) (detached with baggage guard).

Killed: Sergeant Pat Gunning, Sligo.

Wounded: 20 privates.

No. 8.—Officers: Lieutenants B. W. Nicholson and Theophilus O'Halloran, Ensign James (k.).

Killed: Timothy Maher, Templemore; Francis Smith, Leicester.

Died of wounds: Nicholas Lumsden, Wexford; Wm. Moran, Carlow; James Pick; Wm. Perchender, Cambridge; Charles Storer.

Other wounded: 11 privates.

No. 9.—Officers: Lieutenants R. Heaviside, R. Harrison (w.), (16th), W. O. Warren (w.).

Killed: Sergeant John Suttle, London; Privates Wm. Brien, Roscommon; Henry Lynch, Carlow; Henry Millar, Cambridge; Pat McGrath, Down; John Raysdale, Leicester.

Died of wounds: Thomas Bailey and John Sharpe.

Other wounded: Colour-Sergeant Joseph Scotton, Corporals Joseph Andrews and Thomas Dobbs and 26 privates.

No. 10 (Light).—Officers: Lieutenants Rumley (w.) and Pratt (w.), Ensign Edward Neville Macready.

Killed: Drummer Thomas Elliott, Meath; Privates John Ashby, Suffolk; Ben Colley; Pat Morton, Tipperary; Charles Norfolk, Yorkshire; Ben Siddons, Leicester; Thomas Freeney.

Died of wounds: Henry Day; Richard Snoden; Richard King, Louth.

Other wounded: Sergeants Wm. Frohock and Isaac Burrows, Corporal Sam Lumsden and 15 privates.

RECAPITULATION

	Officers Killed and Wounded.	Other Ranks. Killed.	Died of Wounds.	Other Wounded.	Total other Ranks.	Total.
Staff	5					5
Cos.						
1	3	3	4	23	30	33
2	2	6	2	12	20	22
3	2	3	6	17	26	28
4	2	1	2	14	17	19
5	2	3	2	19	24	26
6	1	4	1	11	16	17
7	—	1	—	20	21	21
8	1	2	5	11	18	19
9	2	6	2	29	37	39
10	2	7	3	18	28	30
	22	36	27	174	237	259
Correct number for the three days as given by Surgeon Elkington .						266

We have given reasons for believing that the strength of the 30th at Waterloo could not exceed 34 officers, 29 sergeants, 10 drummers and 494 rank and file, a total of 567, and of these 227 (somewhat more than two out of five) were killed or wounded. If we accept Macready's figures the proportion would be nearly a half, and he may be right.

The number of missing reported in the campaign was 5 at Quatre Bras, of whom 2 rejoined at once, 9 on the following day, and 14 at Waterloo. A note attached to the list of casualties furnished by Wellington on June 30th, says, "The men returned missing had gone to the rear with wounded officers and soldiers and the greater number have since rejoined." The nine reported absent on the 17th were probably employed loading the wounded at Quatre Bras on carts and accompanied them to Brussels. The adjutant found a party of nine of the regiment employed in a base hospital there after Waterloo. The detaining of effective men in hospital always made Wellington very angry, but after a battle the surgeons were in sore need of help.

The rewards for Waterloo were the thanks of the Prince Regent and both Houses of Parliament, the grant of a medal to all ranks in the regiment, whether alive or dead, the Companionship of the Bath to the three surviving field officers (Lieut.-Colonel Vigoureux had already been made a Companion for his share in the victory of Vitoria), a Brevet Majority to Captain Robert Howard, who brought the regiment out of action, prize money for artillery captured according to the following scale :—

			£	s.	d.
5	Field Officers	each	433	2	4
9	Captains	,,	90	7	3
31	Subalterns	,,	34	14	9
42	Sergeants	,,	19	4	4
585	Drummers, Rank and File . .	,,	2	11	4

All subalterns present were granted two years' service towards increased pay, that is, they were to receive the additional shilling a day after five instead of seven years' service as lieutenants.

The non-commissioned officers and men were to be entered in musters and pay lists on a special roll under the heading "Waterloo Men," and to be granted two years' service towards increase of pay and pension.

In October the King of the Netherlands conferred the honour of Wilhelm's Order upon 8 general officers, 4 colonels, and 13 lieut.-colonels. Among others, Lieut.-General Sir Charles Alten and Major-General Sir Colin Halkett received the first class and the commanding officers of the 33rd, 69th and 73rd, the fourth class. The reason why it was not conferred upon the commanding officer of the 30th is given by Macready. Colonel Hamilton was repeatedly urged to ask for the Order, but he was wont to reply that he valued the honours he had received solely on the ground that they were given without any solicitation on his part. It might have occurred to the King of the Netherlands that by including the name of a hero like Hamilton he was doing honour to his Order as well as to the man to whom he gave it.

ADVANCE ON PARIS

During the advance on Paris the battalion, which was never in touch with the enemy, was commanded by Brevet-Major Howard until June 30th, when Brevet Lieut.-Colonel Bailey re-joined. It formed part of a column of which the usual order of march was—a brigade of Belgian cavalry, the first division, third division, and a Belgian infantry brigade. The route was through Nivelles, Binche, Bavai, Le Cateau, Caulaincourt, Vermand, Roye, Pont Maxence, and Senlis. From Le Cateau to the Somme, the regiment retreated on the same road in 1914. In passing the fortified town of Mons, Captain James Finucane was left behind to act as commandant and a day's march further on Surgeon Elkington joined the battalion which had halted for a day in bivouac near Bavai. On the 23rd the division was at Le Cateau, where nearly a hundred years later the regiment was to take part in the battle fought by General Smith-Dorrien to check the German invasion of France in 1914. After the division had passed on, the inhabitants of Le Cateau petitioned that they might have English troops stationed in the town. They implored the Duke to let them have a guard "of brave Englishmen even if it was only 100 men" to protect them from our allies—no small compliment. On June 25th, Sergeant-Major Woods was appointed assistant baggage master to the division. The army had outmarched its supplies and Macready gives an amusing account of the shifts it was reduced to :—

"We bivouacked near Le Cateau on the 24th, and at Caulaincourt on the day following. We had very strict orders against plundering of any description, and all marauders taken by the provosts-marshal were to be immediately executed. Two of our men had been taken, but the Duke released them, as his note expressed, 'in consequence of the excellent behaviour of our corps.' However, 'neither compliments nor commands can fill empty stomachs,' and as our commissariat only served out three rations in twenty days, our men were obliged to forage or starve. I had myself mounted Ryan's horse, and was reconnoitring the neighbourhood of Caulaincourt, when I heard a confused noise of pigs squeaking and men swearing, behind a hedge very near me. I rode up and discovered four soldiers of the 33rd cutting the throats of an old sow and her litter. 'Hallo!' said I, 'what are you gentlemen about; don't you know that if the provost sees you, you'll swing for this?'

'Yes, sir,' answered the spokesman, 'we do, but then there's no rations—and perhaps your honour would like a pig.' 'Why, as the pigs are killed, I have no objection; but I assure you, my good fellows, this is a very serious business, and I wish I could have prevented it. I'll thank you for that pig to the left. Good morning.' I galloped off with my prize and the ram-rod was through him immediately!"

Captain Richard Heaviside was now in command of the light company, news having arrived of his promotion on the 15th of the month. Andrews and Pratt re-joined about the same time as Lieut.-Colonel Bailey.

On July 1st and 2nd the fifth brigade, which was on the left of the allied army, halted at Aunoy, 7 miles from Paris, and the heights of Montmartre could be seen from its position. The Prussian artillery was heard far to the right and it was expected that on the following day an assault would be delivered by the Duke on the French position, on the northern side of Paris. This waste of life was avoided by a capitulation, the French leaders agreeing, on conditions, to abandon Paris and withdraw their army behind the Loire. On the 5th the Duke's army closed in on Paris and on the 6th the 30th occupied the suburb of La Chapelle close to the fortifications of the city; it was still on the north road by which it had advanced. On the 9th the army moved to its right and bivouacked in the Bois de Boulogne. The 30th was close to Passy. Here the men soon hutted themselves and tents were issued as soon as they could be provided; forty-seven were allowed. They were the ordinary bell tents such as we have now, but in those days they were made to hold fifteen men.

On July 18th, Major-General Sir Thomas Bradford and Colonel Sir C. Belson, who had arrived from England, were appointed to command the third division and the fifth brigade respectively. Sir Thomas was the future colonel of the 30th.

Macready gives a characteristic account of the life in camp:—

"General Halkett was still laid up and Lockwood of ours had gone home with a silver plate in his skull, on which was engraved 'bomb proof.' In consequence of the increase in our numbers, we began to entertain the Sovereigns with field days. We generally manœuvred on the Plain of St. Denis. When the Duke did not require us, Sir T. Bradford and Sir P. Belson had us out and when their fantastic tricks were finished, old Hamilton paraded us to caricature light infantry movements.

"Until late in September the camp was pleasant and some parts of it remarkably picturesque. Heaviside, Harrison and myself had our tent in a very fine grove of old trees beneath which we had built huts and enclosed the whole with a staked hedge. Numbers daily stopped to admire our little establishment, and among the rest Lady Castlereagh. The comfort of the interior was not, however, strictly correspondent with the elegance of the outside appearance. We had not an article of furniture, but an old camp stool; the ground was our bedstead, a bundle of straw our mattress, and H——'s blanket, for I had none, our covering. I always turned in completely dressed."

On July 22nd, the anniversary of the victory of Salamanca, the Duke announced to his army that the Emperor of Russia would review it on the following Monday, the 24th instant. On that day the Duke's army entered Paris in triumph, with the men's caps decorated with laurel leaves. The line when formed reached from the bridge over the Seine at Neuilly, under the triumphal arch in the Place de l'Etoile, and down the Champs Elysées to near the Place de la Concorde (Place Louis XV). The Emperors of Austria and Russia and the King of Prussia, after riding down the line, stationed themselves in the Place and the Duke of Wellington led his army past them in open column at half distance. After passing, the troops wheeled in succession to their left and passed through the Place Vendôme, then to their right and after passing through the Place des Victoires gained the Gate of St. Denis whence they returned to the Bois de Boulogne. The march past took from 11 a.m. to 3 p.m.

In the beginning of August it was known how the Waterloo vacancies were to be filled up. Captain Samuel Bircham got the Majority *vice* Chambers, dated July 20th, and Lieutenant Benjamin Walter Nicholson got Bircham's company. Ensign Edward Windus was promoted *vice* Beere, July 17th; Ensign John Donnolly *vice* Prendergast, July 18th; Ensign James Poyntz *vice* Lewin, July 19th; and Ensign Edward Macready *vice* Nicholson, July 20th. These promotions were of course without purchase. Owing to Hamilton's strong recommendation, Macready passed over an officer senior to him, who had not been at Waterloo.

On August 24th the composition of the fifth brigade was altered, and it now consisted of the 3rd Buffs, 12th, 2nd/30th and 2nd/33rd.

In September the Duke ordered the free issue of a pair of shoes to each man, which must have been very welcome as their kits were in a very bad condition; a pair of shoes cost 6*s.* 2*d.*

On October 24th Colonel (Brigadier) C. P. Belson, inspected the battalion when there were present :—

Lieut.-Colonel Hamilton, Lieut.-Colonel Bailey, Major Howard; Captains Gore, Ryan (Br.-Major), Sinclair, Heaviside, and Nicholson; Lieutenants Gowan, Mayne, Elliott, Freear, Daniel, Neville, Roe (1), Harrison, Roe (2), Hughes, Pratt, Tincombe, Warren, Moneypenny, La Touche, Rogers, and Macready; Ensigns Warren, Backhouse, Lardner, Stephens and Gregg; Paymaster Wray; Adjutant Andrews; (Lieut.) Quartermaster Williamson; Surgeon Elkington and Assistant-Surgeons Evans and Clarke; 36 sergeants, 12 drummers and 427 rank and file. Besides recovered men, a draft of twenty had joined from the depot. Captain Machell was still town major at Antwerp, and Captain Finucane commandant at Mons. Lieut.-Colonel Vigoureux, Lieutenants Rumley and Lockwood were still incapacitated by wounds. Lieutenant O'Halloran was in command of wounded and sick at Brussels.

Brigadier Belson's report was very favourable, but he says that the colours "which have been carried for twelve years, are so bad that they are not capable of receiving the badges and names of the actions in which the regiment has been engaged and as a substitute Lieut.-Colonel Hamilton has annexed silk streamers on which such names, etc., are placed."

The regiment had been very healthy in tents and huts during the preceding three months, but the weather was now broken and at the end of the month the battalion moved into billets, at first in Clichy and a couple of days later at Clermont, and on November 3rd at Vanves on the left bank of the Seine;

a week later it moved to Mont Rouge, a mile from the Barrier d'Enfer on the Orleans road.

It was now decided that the battalion should return home to resume its proper work of training recruits and furnishing drafts to the first battalion in India. The depot was ordered to be in readiness to join the second battalion, into which it was to be absorbed, as soon as the latter had reached England.

In a General Order Sir P. Belson thanked the battalion for its good conduct while under his command.

For the march to the coast the 30th was brigaded with the 2nd, 12th and 33rd under Colonel Stirke of the 12th. On December 4th the battalion marched through the suburbs of Paris and struck the great north road at St. Denis. It experienced foul weather as usual in crossing the Channel and Private Thomas Roycroft of No. 7 company was drowned in landing at Ramsgate on December 26th.

The following is Macready's account of the catastrophe:—

"On Christmas Day we marched our 300 gallant bloods on board miserable smacks of from 60 to 100 tons burthen. The Government had contracted with a house at Dover for the transport of the troops at a very cheap rate, and consequently lost two-thirds of the cavalry horses for want of accommodation. Our men were so crowded on and below decks, that there was hardly room, if there had been hands, to work the vessel properly. The master of the one I went in assured me that if we encountered such a breeze as the 12th regiment had weathered a few days before, we should founder, and added that he would give £100 to be off his agreement. We started from Calais with a fair breeze which chopped round, when we were about mid-Channel, and freshened into a gale. Smack went our main-brace, and roll came the seas over our weather quarter. In a few minutes we were in imminent danger; but, providentially, the wind lulled, and we repaired our rigging. We tried in vain to make Dover, and towards night changed our course, and stood up for Ramsgate. When we reached its offing, the gale was stiff and momentarily increasing, so that, though the high-water flag was not hoisted, the masters of our vessels, trusting to their trifling draught of water, dashed for the harbour. The first vessel came crash aground between the two piers, the next ran her bowsprit into her stern rigging and swung broadside on to the sea—all followed, and all got foul of each other. The piers were covered with crowds of all sorts of people, running here and there with lights, and appearing by their gesticulations (for the roaring of the elements drowned their voices) to recommend us to take to our boats. We were twelve crank ships locked together, with our gangways, bowsprits, and timbers crashing with every heave of the water; the spanker booms were lashing right and left with each kick of the vessels, and the uproar of the wind, rain, waves, speaking-trumpets, ropes and sails cracking, and crews shouting was indescribable. At length part of us were landed in boats, and the remainder ascended the pier by rope-ladders thrown down. Two men fell, and were crushed to pieces between the vessels. It was about ten o'clock when we landed, and were ordered to march off to Margate, though it rained terribly, and our arms were still on board the ships. We reached the warm welcome of an inn about twelve o'clock, and at seven next morning were ordered back to Ramsgate.

The vessels were all ruined, and the horses of the Enniskillen Dragoons were piled up on the quay in heaps opposite every ship. As soon as we had received our goods, we continued our march to Sandwich, and next day to Dover, where we took up our quarters in the Castle."

The officers landing were the same as were present at Colonel Belson's inspection, except that Lieut.-Colonel Bailey, Ensigns Warren, Backhouse, Lardner and Assistant-Surgeon Evans, who were ordered to join the first battalion with the draft preparing to sail in the January fleet, were already in England. Neville had been left at Paris sick, John Roe (2) was with the heavy baggage at Ostend. Drake had joined from sick leave and Ensigns C. Dean and Lewin on appointment.

The battalion remained for three days in the Castle at Dover, during which time it was inspected by Major-General Sir George Cooke. The depot, numbering 7 drummers and 116 rank and file, were taken on the strength and arrangements were completed for sending out the draft to the first battalion. Lieut.-Colonels Bailey and Vigoureux, Captain Machell, Ensigns Backhouse, Frizell, Charles Warren, Joseph Berrige, Charles Lardner, Frederick Liardet and Assistant-Surgeon John Evans were ordered to sail with the draft. The two field officers obtained leave of absence, but the remaining officers embarked. The men of the depot were mostly sick and convalescents from Belgium, or were employed on recruiting, and it would have been hard to have sent abroad the Waterloo men who had just returned home, so there was a difficulty about finding sufficient numbers for the draft. Ultimately about thirty-eight were got together, of whom ten were Waterloo men, several of whom had been wounded. The resources of the War Office, however, were not exhausted. From the prison ships lying off the Isle of Wight 200 young men and lads convicted of desertion, not aggravated by any other offence, were selected and embarked on the East India Fleet; of these fifty were assigned to the 30th and the regiment was ordered to send clothing to Tilbury at once for them. In July the two drafts joined the first battalion at Fort St. George, Madras.

The fifty prisoners were promised that if at the end of seven years their commanding officer testified that they had done good and faithful service they would then be put on the same footing as other soldiers and the restriction on their return to England withdrawn, otherwise they would be expended in India.

The battalion embarked for Cork on December 31st on His Majesty's transport *Alexander* with Brevet-Major Howard in command. In spite of contrary winds, the *Alexander* managed to get to Portland on January 6th, 1816. From there it was blown back to Spithead, where it lay at anchor wind-bound for nearly three weeks. On the 23rd it sailed again and in spite of a violent gale managed to get into Cork Harbour just before dark on the 29th. It was on this day that the 59th and 82nd were wrecked in Tramore Bay, when hundreds perished.

An attempt was made to land the men on the 30th, but was given up as too dangerous even in the sheltered water at Cove. At last on February 1st they got ashore and marched to Cork on the way to Limerick, where the battalion was to be quartered. What followed is like a page out of *Harry Lorrequer*. The left wing marched to Mallow on the 2nd and its doings are not recorded, but on the evening of that day the officers of the Headquarter wing supped at an inn at Cork and celebrated their home-coming with such a prolonged and joyous sitting that the wing had to march in the morning with only one officer. About ten o'clock the officers left behind realized the position and called for post-chaises in which they managed to catch up their wing before it arrived at Mallow. At Buttevant the battalion halted from February 5th to the 19th, when the march was resumed and Limerick was reached on the 23rd. The garrison of Limerick was commanded by Major-General Barry, an especial friend of the regiment.

The following pictures of regimental life are from Macready's "Journal."

"On March 17th St. Patrick was propitiated with due libations and the immortal Shamrock was swallowed I know not how often. As Rumley and I were returning from the Mess, our light bobs seized us and carried us back to their barracks, swearing we must see how happy they were. Such days are the saturnalia of our poor fellows, so we could not refuse them. They brought us to our room on their shoulders, preceded by a blind fiddler, torturing his catgut into 'St. Patrick's Day in the Morning.' Our strict disciplinarians may frown at this, but heaven knows I care not for their wrinkles. If I can contribute to the happiness of the brave fellows under me, by now and then running with them a few yards out of the highway of regulation, I'll do it, though Sir David and the old school stood before me and thundered 'Hold, hold.'"

Sir David is of course General Dundas, sometimes known as "Old Pivot" who had been instrumental in enforcing uniformity of drill and discipline in the army, and was regarded by young officers as the personification of all that was stiff and unbending.

On May 1st medals were presented with an appropriate speech by Colonel Hamilton to the officers and men who had served in the Waterloo campaign.

The anniversary of the victory was marked in a whole-hearted way. The day commenced with the chairing of their officers by the men, followed by a parade at which new colours were presented, then a dinner to the men on the barrack square and, to wind up, a ball given by the officers to 400 guests.

Macready gives a lively account of the chairing :—

"The soldiers assembled at an early hour in the morning, neatly dressed, with white trousers, uncovered caps and side arms, bearing numerous chairs, lavishly and really tastefully ornamented with flowers, ribbons and laurels and declared their intention of carrying their officers round the town. Entreaties were of no avail, resistance had a worse effect ; accordingly the band drew up, the men of our respective companies, after fixing a leaf of laurel in our caps, hoisted us up, and away we went to the quick-steps of 'Waterloo,' 'The Downfall of Paris,' 'Garryowen,' and 'The White Cockade.' Thousands of people joined our fellows, and every five minutes greeted us with thundering cheers. Women would dash from their houses, and try to push through the crowd to shake hands with us, or give us an audible 'Arrah, God bless your good-looking face, honey ! I'm sure, ye're a brave one.' We were halted opposite General Barry's door. He came out, and bowed to us all, and giving a hip, hip, set three such cheers a-going as I never heard before or since. After this we were carried to the barracks. The day was delightful; its brilliancy, the scene before us, and its various associations, were really intoxicating."

On September 11th and 12th, the battalion marched to Tralee, and the following out-stations, Bantry, Beer Island, Clonakilty, Castle Island, Widdy Island, Dingle, Mill Street, Ross Castle, Skibbereen and Kenmare. A strength of about four companies was at Headquarters, one at Bantry, two at Beer Island and small detachments at the other places.

From Tralee a draft consisting of Lieutenant Sullivan, Ensign Deane and fifty-seven non-commissioned officers and men was despatched in December to join the first battalion. In the beginning of 1817 the reduction of the army was being rapidly carried out. In January the 30th was hit by an order for one lieutenant per company to be put on half-pay. Accordingly on March 25th Lieutenants Philip McPherson, Ben Ottley, Meyrick Shaw, Thomas Moneypenny, Edward Windus, Edward Macready, and John Stewart were placed on the British half-pay and James Battersby and Henry Robson on the Irish half-pay.

In March, General Robert Manners was informed that as soon as a few more regiments returned from France the second battalion of the 30th would be disbanded. This was followed by an order fixing the establishment of the regiment at :—1 colonel, 2 lieut.-colonels, 2 majors, 10 captains, 22 lieutenants, 8 ensigns, 1 paymaster, 1 adjutant, 1 quartermaster, 1 surgeon, 2 assistant-surgeons, and a recruiting company of 1 captain and 2 lieutenants.

The second battalion was to be disbanded on April 24th, and the supernumerary officers placed on half-pay from June 25th. This order affected the first battalion as well as the second, for as usual a large number of officers nominally belonging to the second battalion were serving with the first in India, whilst Lieut.-Colonels Bailey and Vigoureux and a dozen other officers who fought at Waterloo nominally belonged to the first battalion.

Of the non-commissioned officers and men, the staff-sergeants, and others entitled to it, were to be pensioned, half of the younger men to be selected to remain in the regiment and all others to be discharged as " unfit for the active military service of the line." This appears, and no doubt was, ungenerous treatment, but the men discharged were not all Waterloo men by any means. The second battalion in the last two years had given nearly 100 men to the first, and the first had sent home in three years about 250 invalids, and those of them who did not pass the invaliding board were sent to the second battalion. Probably all fitted for service in India were retained. Recruiting had been stopped.

On April 8th routes were issued for the battalion to march from Tralee and the out-stations to Fermoy, and there it was paid to the 24th and broken up, the officers receiving full pay to June 24th.

The numbers transferred to the first battalion and the depot were 17 officers, 16 sergeants and 289 drummers, rank and file. The numbers struck off on reduction were :—29 officers transferred to half-pay, 19 sergeants and 235 rank and file discharged.

The officers retained were Lieut.-Colonel Hamilton, Captains Fullerton and Howard (Bt. Major), Lieutenants Gowan, Mayne, Andrews, Freear,

Rumley, Bailey, Daniell, Neville, Roe (1), O'Halloran, Harrison, Roe (2), Masters, Ensign Gregg.

The officers struck off were :—Majors Maxwell (Bt. Lieut.-Colonel) and Bircham ; Captains Ryan (Bt. Major), Sinclair, Grey, Easte, Heaviside, Lewin, Nicholson and Teulon ; Lieutenants Ness, Campbell, Hughes, Pratt, J. Poyntz, Atkinson, Darling, Tincombe, Macdonald and Warren ; Ensigns Robson, Elliott, Macdougall, Lewin and Boyes ; Quartermaster John Williamson ; Surgeon Elkington ; Assistant-Surgeons Huggins and Palmer.

Of the assistant surgeons who had been so long with the second battalion, John Evans had gone to the first battalion and Pat. Clarke had exchanged to half-pay.

The second battalion had now ceased to exist and Macready in his farewell gave expression to feelings which were common to all the officers and men who had the honour and happiness of serving in it. He says :—

"Thus after a companionship of three years, of mingled hardship and happiness, I had to bid adieu to my dear second battalion. This brave corps —which will be remembered as long as the names of Fuentes d'Onoro, Badajos, Salamanca, Muriel, Quatre Bras, and Waterloo are emblazoned in the highest pages of British achievement—was not more distinguished by its professional exertions, than by the cordial and brotherly unanimity which pervaded its internal regulations. The men were devoted to their colours and their officers ; never, while the regiment existed, had they been known to shrink from either ; the officers, scrupulously attentive to their soldiers, entered with feeling into their wants and wishes, and received a pleasing return when circumstances threw the power of obliging into the hands of the private. The thanks of Generals Leith, Hay, Oswald, Picton, Alten, Halkett, hot from the heart in the full fury of the raging battle, speak their praises as mere soldiers, while the declarations of Generals Barry, Buller, and Gordon, in Ireland, that 'the internal economy of this regiment had been seldom equalled, but never surpassed by any in the service ;' and that 'this gallant corps substantiates its claim to its country's gratitude, not more by its exertions in the field, than by its uniformly exemplary conduct in quarters,' indicated that it was not a brutal fierceness, but a truly noble feeling for the honour of their country and corps, that excited their energies on the day of action. During the year we were in Ireland but two men were brought before courts-martial, and during the existence of the battalion, from 1803 to 1817, not an officer was cashiered."

After the break-up of the second battalion the detachment of the regiment under Colonel Hamilton moved to Cork and on June 24th embarked on the transport *Regulus*, which landed it at Chatham on July 8th. At Chatham it was joined by Lieut.-Colonels Bailey and Vigoureux who had up till now been excused embarkation for India on account of their wounds. Lieutenants Rogers and James Poyntz also rejoined from half-pay, exchanging with Baillie and Masters and paying the difference.

On August 13th the detachment marched to Sheerness. Here four boys joined from the Royal Military Asylum, better known as the Duke of York's School. They were the first of the many excellent soldiers the regiment was to receive from that institution.

In December the detachment was placed under orders to sail to join Headquarters in Madras. Major Bailey at this time arranged an exchange with Major Dalrymple of the 80th Foot and Lieutenant Macready exchanged back to the regiment from half-pay with Lieutenant Freear. With his usual liberality Colonel Hamilton had written that if there was any difficulty about finding the money to pay the difference, Macready might draw upon him. Luckily his assistance was not wanted.

In February, 1818, the detachment embarked on the East India Fleet for Madras, leaving a depot or recruiting company, as it was called, at Sheerness.

The following is the strength of the detachment :—

Colonel Hamilton, C.B., Captain Fullerton, Lieutenants Mayne, Rumley, Harrison, Neville, Roe (2), J. Poyntz and Macready, with 203 other ranks.

The depot was constituted as follows :—

Brevet Major Howard was in command with Lieutenant Robert Daniell and Ensign J. N. Gregg at duty; Lieutenant Matthias Andrews, formerly adjutant of the second battalion, was superintending officer of recruiting at Northampton but borne on the strength of the depot, as were all the officers of the regiment on leave in England. Among these was Lieut.-Colonel Vigoureux on sick leave; his Waterloo wound caused him great suffering for the remainder of his life. A bullet had lodged near the spine and could not be extracted.

The strength of the depot in non-commissioned officers and men was :— 8 sergeants, 3 drummers, 57 rank and file, of whom 22 were sick.

After Colonel Hamilton and his detachment had sailed, the depot marched to Canterbury, where it was quartered. Major Dalrymple, who had exchanged with Lieut.-Colonel Bailey and Lieutenant J. Stewart, who had exchanged with Rogers, joined and sailed for India in March with Lieutenant John Roe (1), and Ensign Gregg to join Headquarters.

In June one sergeant and 20 men were embarked on board the *Lord Melville* convict ship bound for New South Wales. They were to join Headquarters at Madras and were made use of as a convict guard as far as New South Wales. They reached Headquarters.

As all recruiting had long stopped, the despatch of this draft brought the effective strength of the depot down to one captain, 5 subalterns, 7 sergeants, 3 drummers, 8 corporals, and the officers' servants.

CHAPTER XII

INDIA, THE MAHRATTA WAR. THE SECOND BATTALION ABSORBED BY THE FIRST. ASSEERGHUR. SEVEN YEARS IN THE NIZAM'S TERRITORY. RETURN TO THE COAST AND EMBARKATION FOR HOME. PEACE SERVICE AT HOME, IN THE COLONIES AND IN THE IONIAN ISLANDS UP TO THE OUTBREAK OF THE CRIMEAN WAR.

1811–1853

WE left the first battalion at Trichinopoly in October, 1810, at the end of Colonel Wilkinson's long command of fourteen years.

On August 20th, 1811, Colonel Lockhart inspected the battalion. He tells us that it practised ball firing before him and that the mark was hit at 100 paces at the rate of once for every twenty-five shots—the first intimation of musketry results in the regiment. The regimental way was to take out the battalion early in the week with an ample supply of blank cartridge, perhaps thirty rounds, and practise deployments, forming line, etc., the company of formation and each company in succession, as soon as formed, firing one or two volleys, followed perhaps by sub-division volleys from flanks to centre, or file firing from the right or left. A few days later the battalion paraded with fifteen rounds of ball, and went through the movements and firing previously rehearsed; this may have been the course followed at the inspection. The firing was probably at chatties. The target for instruction was of canvas five feet by two, with a six-inch bull's-eye.

In November, 1811, the battalion proceeded to Cannanore, a seaport on the West Coast. Colonel Lockhart rejoined in November, but on January 1st, 1812, took command of the district of Malabar and Kanara. Colonel Vaumorel succeeded to the command of the battalion. Colonel Lockhart never rejoined for duty, though, from his commanding the Malabar district, he was scarcely separated from the regiment in which he had served for twenty-seven years. He had distinguished himself leading it in Egypt in 1801, when Colonel Wilkinson was on the sick list, and it may be said that from 1796 to 1811, owing to Colonel Wilkinson being frequently selected to command and train other troops, these two very capable officers had shared the command of the regiment.

Another veteran officer, Arthur Poyntz, was lost to the regiment at the beginning of 1812. As a non-commissioned officer he had shared in the fighting in South Carolina in 1781–2. He was promoted quartermaster in 1796, and had served in that rank in Corsica and in Sicily, at the occupation of Malta, and the capture of Valetta, in the Egyptian expedition under Sir Ralph Abercromby in 1801, and the expedition to the Elbe in 1805–6. He died at Madras. He was succeeded in his appointment by Quartermaster-

Sergeant Nicholas Wilson. In January, 1812, Captain Sir C. W. Burdett was struck off the strength in consequence of his promotion to a majority in the 56th. As a major in the 56th, he commanded a column in some operations against the Mahrattas, and ultimately died in Ceylon in 1831 as a Lieutenant-Colonel (retired).

Early in March a rebellion broke out in the Wynaad district, and a couple of posts held by Native Infantry were surrounded and in danger of capture. Two columns were dispatched to their relief, one from Cannanore under Major Webber, 3rd Native Infantry, consisting of 50 men of the 30th under Lieutenant Washington Carden and Ensign J. P. Perry, the second battalion of the 3rd Native Infantry and a detachment of the 5th Native Infantry. The other column started from Seringapatam.

Major Webber marched to relieve Manantoddy, where two companies of the first battalion of his own regiment were besieged. When ascending the Cotiaddy Pass, he was attacked by the rebels on April 9th. He succeeded in dispersing them and relieving the garrison with the loss of 1 private of the 30th killed and 2 privates wounded, and 2 officers and 18 men of the 3rd Native Infantry wounded. The other column was equally successful, and on the 23rd, Lieutenant Carden started on the march back to Cannanore.

In July, Captain Fox, who had been transferred from the second battalion on account of seniority, arrived with a draft of two sergeants and sixty-six privates from England. The other officers with the draft were Lieutenant Walter Ross, Ensigns Edward Parry, R. Wedge and William Atkinson.

In September, Captain Robert Blake Lynch, who had been obliged, like Captain Fox, to leave the Peninsula on transfer from the second battalion, arrived in India. On October 6th, Major-General William Wilkinson was appointed commander of the forces in the Bombay Presidency; he took Lieutenant John Wade with him as aide-de-camp.

On January 18th, 1813, the battalion, commanded by Lieut.-Colonel Vaumorel, was inspected by Major-General Wetherall, and some portions of his report are here quoted for the following reasons. It is the first independent report we have after the regiment landed in India; the reports for 1809–10–11–12 were made by Major-General Wilkinson and Colonel Lockhart on the work of their own hands. A second reason is that the report, being made only twelve months after those officers had completed their long service with the regiment, enables us to do justice to the success of their methods.

After praising the zeal and ability of the officers and touching upon the excellence of the mess and band, the *esprit de corps* of the rank and file and the steadfastness and fidelity of the non-commissioned officers in discharge of their duty, the General goes on to say :—

" INTERIOR ECONOMY.—Every branch under this head is in a most perfect state, and the result of an old and well regulated system which has been strictly attended to by successive officers upon whom the command of the regiment has devolved. Such exact uniformity has been established that any individual, either officer, non-com. officer or soldier, armed and appointed, stamps the character and appearance of the whole corps.

" GENERAL OBSERVATIONS.—I have great pleasure in reporting the high state of order and discipline which distinguishes this corps. The most perfect subordination exists in the several ranks and the system of responsibility throughout each is completely established.

"I have the satisfaction of stating that this corps is equal to the most active field service."

This report not only enables us to judge of the first battalion, but it lets us into the secret of the remarkable success of the second, for the men who formed and led the second battalion were all trained under Wilkinson and Lockhart and took with them to the second battalion the traditional discipline of the regiment. The chief of them, Lieut.-Colonel Hamilton, was Wilkinson's subaltern, and had seventeen years' service in the regiment before he was promoted to a majority in the newly raised second battalion.

There are some other points of interest in this inspection report. One is an illustration of the demoralization produced by high bounties and rewards. Three men were brought before the general by Colonel Vaumorel with an assurance that he had investigated their cases and was satisfied of the truth of their stories, which were, that they had never belonged to the 30th, but had been arrested as deserters from it, falsely sworn to by crimps, who pocketed the reward, and shipped off to India. The case of one of them, Patrick Ahern, was peculiarly distressing. He was falsely sworn to as a deserter by a man who owed him money and was in consequence taken from his wife and three children, without either being permitted to see them, or make any provision for their support. The men were ordered by the Commander-in-Chief to be sent home at once. Another point is the first appearance of an attempt to find a substitute for corporal punishment. Up to this time, regimental courts martial could only award sentences of corporal punishment. General courts martial in India frequently transported offenders to New South Wales. The officer commanding a battalion had then, as now, the power of confining to barracks. There were two degrees of this punishment: a man awarded the second degree, "confinement with disgrace," had to turn his coat inside out and take it to the master tailor who sewed a large C on the back, and charged him *2d.* For the time of his sentence the culprit had to wear his coat reversed. The commanding officer had also at his disposal the black hole, about 20 feet square, but more as a place of safety for drunken and violent men than as a means of punishment.

Colonel Vaumorel, urged on by Major-General Wetherall, seems to have resorted to some expedients of doubtful legality to avoid the infliction of the corporal punishment awarded by regimental courts martial. In many cases 150 or 200 lashes were commuted to confinement to the black hole in irons, or tying to a log. Major-General Wetherall and the Commander-in-Chief strongly urged on Government the provision of cells and the reduction, at least, of corporal punishment. It was hard to get money for building, and up to the time of the regiment's leaving Madras, only a few cells were provided. Still a beginning was made and the cat was no longer the one cure-all.

The Commander-in-Chief at Madras at this time was Lieut.-General Sir Thomas Hislop, who had been D.A.Q.M.G. at Toulon in 1793, and was a very good friend to the 30th.

Several officers died in 1813. The first to go was Captain Benjamin Nunn, who died on April 17th at Cannanore. He had done good service as adjutant.

Major and Brevet Lieut.-Colonel Thomas Leach died on September 2nd, Major John Curry on October 8th, and Lieutenant John Napper on the 18th. None of these officers had war service in the 30th. Besides the losses by death Assistant-Surgeon Patrick Griffin, who had been so long with the battalion, left it on promotion to the 85th Foot.

When a year or more might elapse before the reply to a letter was received from England, it was obvious that the Commander-in-Chief in India must have some power to fill up vacancies and so maintain the efficiency of the regiments serving under him. The extent of that power was always in dispute between him and the Secretary of State representing the Home Government.

On the death of so many officers in the 30th, the Governor-General, the Earl of Moira (the Lord Rawdon of 1781 in S. Carolina, and afterwards Marquis of Hastings), proceeded to fill up the vacancies. The series of appointments by his Lordship extend over a period of fourteen months, but as they were disallowed by the Home Government, it is as well to take them all together.

The first appointment was that of his first A.D.C., Major the Hon. Leicester Stanhope to be a major in the regiment *vice* Leach deceased.

He subsequently appointed his second A.D.C., Captain Philip Stanhope, R.A., and his extra A.D.C., Lieutenant Gerard Moore, to be captains in the regiment, but he transferred them again to other corps. In December, 1813, he appointed Captain Samuel Bircham to be major *vice* Curry deceased. Bircham's company he gave to Lieutenant W. C. Harpur to date from October 9th.

In February, 1814, he promoted Lieutenant Hinton East to a company in the 17th Foot, and brought Lieutenant William Croudace from the second Ceylon regiment into the 30th in his place. In the same month he promoted Sergt.-Major William Ledge to be quartermaster *vice* Nicholas Wilson deceased.

In November, 1814, news arrived in India that Lieutenants W. C. Harpur and Henry Cramer had been already promoted at home, so his Lordship gave Bircham's company to Lieutenant Alexander Fettes, and he at the same time promoted Lieutenant Donald Sinclair to a company rendered vacant by the re-transfer of his extra A.D.C.

These appointments were all successively disallowed by the Home Government, but each officer had served for eighteen months in his higher appointment.

Fettes, disappointed in getting his company, exchanged to the half-pay of the 14th Light Dragoons. He was now forty years of age.

There were several other changes among the officers in 1814. In April, Lieutenant John Wade, A.D.C. to Major-General Wilkinson, went home on leave in bad health. He died on May 24th in the following year. John Powell, now a captain, succeeded him as A.D.C. Captain and Brevet-Major Thomas Jackson went home along with Wade, and died on June 14th, 1815. On July 6th, 1814, Lieutenant R. Wedge died. In the same month the small draft arrived which had been despatched by the second battalion when it embarked at Jersey for Holland. Lieutenant Edward Parry died on December 29th.

Major-General Lockhart, who had been promoted on June 4th in the previous year, was sent home in August as his command was not tenable by an officer above the rank of colonel, but he was still borne on the strength of the regiment as second lieut.-colonel.

An order had been issued on December 20th to start a canteen in the regiment, but some months probably elapsed before it was acted upon, for in February, 1815, the battalion left Cannanore for Madras, but first took part in a concentration of troops caused by a disputed succession to the throne of the tributary state of Karnul. A large force was assembled at Gooty, about 300 miles to the north of Cannanore. The 30th reached Gooty at the end of March and were placed in the first brigade along with a flank battalion, and the second battalion of the 4th Native Infantry.

In June the need of concentration was over and the 30th and other troops for Vellore and the Presidency marched down country under command of Colonel Vaumorel. A halt was made for about six weeks at Vellore and on August 12th the regiment marched into Fort St. George at Madras.

In July, a draft of 91 non-commissioned officers and men had arrived from the depot at Colchester. It included seven men of the company which had been wrecked at Gravelines in 1806 and who had remained prisoners of war till released at the peace of 1814. The officers with the draft were Lieutenant John Garvey, Ensign Windus and Quartermaster Kingsley. Ledge, who was acting as quartermaster, received an ensign's commission.

On January 13th, 1815, Lieut.-General Wilkinson (he had been promoted in the previous year) was ordered home as his command was not tenable by any one above the rank of Major-General. He was still, however, kept on the strength of the regiment as first lieut.-colonel. On resigning command of the troops in the Bombay Presidency, Lieut.-General Wilkinson was thanked for his services by the Governor in Council.

The canteen system was started now, if not before. Colonel Vaumorel, after some experience of its working, decided to forbid the sale of spirits and to allow British-brewed beer alone to be sold. The trouble was that there was no check upon the sale of native spirits outside.

In his final report before the battalion left Cannanore Major-General Wetherall had noted that the men were somewhat addicted to drink, though he finishes, " In summing up the character of the battalion, I have no hesitation in saying that I consider the general state of it calculated for the most active and enterprising service."

In the depressing climate of Madras, where the regiment spent four years, both officers and men drank heavily, as did the whole European population. The only thing which could have arrested the decline would have been the prospect of active service or a return to England. The courts martial rose to an excessive number, but there was a remarkable absence of serious crime; drunk or not the men always respected their uniform. The inspecting officer writes :—" I have no hesitation in saying that this is the best dressed corps I have met with in my service." In another place he says that " whatever the men may do amiss they keep it to themselves, for to see a man of the 30th drunk or improperly dressed is a very rare occurrence, which makes the number of courts martial all the more deplorable ! "

In December, 1815, colour-sergeants appear for the first time in the battalion, but they are far from being restricted to company duty. Of the half-dozen first appointed, one is regimental schoolmaster, another is sergt.-major at Poonamallee, and another store sergeant.

THE PINDAREE WAR

The peaceful inhabitants of India had suffered for a long time from the depredations of bands of marauders called Pindarries, who found shelter in the country of the Mahratta confederacy, whose chiefs shared their spoils. On March 10th, 1816, a strong body of these robbers, after a secret march of some hundreds of miles, surprised and sacked the town of Guntur, only 48 miles from Masulipatam on the Bay of Bengal.

Three companies of the 30th were at once despatched by sea to Masulipatam and marched to Guntur, which is in the hills and difficult of access. Lieut.-Colonel C. Maxwell was in command and had with him Bt.-Major R. B. Lynch, Captains John Powell and James Skirrow, Lieutenants J. L. Light, William Campbell, D. McDonald, R. Frazer, Ben. Ottley, Ed. Windus, and W. Ledge.

Lieut.-Colonel Maxwell was too late to bring the Pindarries to action and the companies returned to Madras in June. This was Colonel Maxwell's last service. He went home on leave and was placed on half-pay on the reduction in 1817, and retired in 1826. His father, Sir David, died in 1831, after being on the half-pay of the 30th for 68 years, and the long connection between the regiment and the family of Cardoness came to an end. Five members of the family had served in the corps.

The atrocities which accompanied the sack of Guntur were so inhuman that the Governor-General determined to risk a Mahratta war and have done with the Pindarries once for all, and he began to prepare for a campaign in the year following.

In July the draft sent by the second battalion on landing in England after Waterloo reached Madras under Captain R. Machell.

News arrived also that Lieut.-General Wilkinson and Major-General Lockhart had been placed on the retired list and they were finally struck off the strength. Major-General Lockhart died in 1818. His life was in all probability shortened by an attack of cholera during his service in India. Lieut.-General Wilkinson lived till after the accession of H.M. Queen Victoria, when he was given the Grand Cross of St. Michael and St. George for his distinguished services.

In July the reduction consequent upon the general peace began to take effect and orders were received to reduce the establishment by 1 captain, 2 lieutenants, 8 sergeants, and 8 corporals.

In 1817 the Viceroy's preparations were complete and as the Mahratta chiefs took the part of the Pindarries, seven divisions were mobilized. The Mahrattas were beaten by Sir Thomas Munro at Kirkee, and by Sir Thomas Hislop and Sir John Malcolm at Mahidpur, and the Confederacy submitted in 1818.

The 30th did not take the field as a battalion, but Captain Machell was appointed to a flank battalion formed for service from native regiments in 1817, and at the beginning of 1818 Lieutenant Richard Frazer was sent to Nagpore to act under the British Resident. Nagpore was a centre of disturbance. In February, 1818, the flank companies were sent again to Masulipatam and Guntur, where for seven months they were employed in protecting the district from broken bands of Pindarries who had escaped the grand army. The officers with the flank companies were Captains John Powell (light),

Henry Cramer (grenadiers), Lieutenants Thomas Jones, John Garvey, J. H. Light, and Assistant-Surgeon Sam Piper.

Lieutenant P. P. Neville, who had come out with Machell's draft, volunteered for the appointment of acting engineer, which he had held in the Peninsula and in the Low Countries. He was employed throughout the war and gained another bar to his war medal.

On March 21st, 1818, Lieut. and Adjutant George Stephenson died.

Lieutenant William Atkinson, who had been placed on half-pay in 1817, was now recalled and appointed adjutant.

On June 6th, 1818, Colonel Hamilton marched into Fort George with the detachment of the second battalion which he had brought from home, and with the colours which had been presented to it on the first anniversary of Waterloo. The detachment was played in by the bands and drums of both battalions.

The two battalions were now amalgamated, and the list of officers given below shows how strongly the second battalion was represented in the regiment. The letters T. E. P. W. show the officers who served at Toulon, in Egypt, the Peninsula, including the defence of Cadiz, and at Waterloo, including the siege of Antwerp.

T. E.	Lieut.-Col.	Philip Vaumorel.
T. E. P. W.	,,	Alexander Hamilton, served also in Hotham's naval victory in 1794, and at Bastia.
P. W.	Major	C. A. Vigoureux.
	,,	J. Dalrymple.
T.	,,	Sam. Bircham, served also in Hotham's victory and at Bastia.
E.	Captain	Robert Murray.
E. P.	,,	Robert Blake Lynch.
P.	,,	Sam. Fox.
P. W.	,,	R. Machell, Siege of Antwerp. On Staff there in 1815.
	,,	J. Fullerton, second battalion (recruiting 1809–17).
	,,	John Tongue.
	,,	John Powell.
W.	,,	Robert Howard.
	,,	James Skirrow.
	,,	Henry Cramer.
W.	,,	Mat. Ryan.
	,,	O. W. Grey.
	,,	Robert Baker.
	Lieutenant	Thos. Jones.
	,,	Will. Sullivan.
	,,	J. T. Barlow.
	,,	Wash. Carden, active service as marine officer and in Wynaad.
	,,	J. P. Perry, active service in Wynaad.
P. W.	,,	Richard Mayne.
P. W.	,,	M. Andrews.
P.	,,	T. Kettlewell.

P.	Lieutenant	J. Garvey.
	,,	J. Light.
P. W.	,,	J. Rumley.
P. W.	,,	R. Daniell.
P. W.	,,	P. P. Neville.
P. W.	,,	John Roe (1)
P. W.	,,	Theo. O'Halloran.
P. W.	,,	R. Harrison.
	,,	Walter Ness.
P. W.	,,	John Roe (2).
P. W.	,,	S. R. Poyntz.
P.	,,	J. W. Poyntz.
	,,	R. Frazer.
W.	,,	E. N. Macready.
	,,	J. Stewart.
	,,	A. C. MacDougall.
P.	,,	W. Campbell.
	,,	W. Stanhope.
	,,	W. B. Frizell.
	,,	Charles Jago.
	,,	J. Battersby.
	Ensign	Charles Warren, joined second battalion in France after Waterloo.
	,,	J. L. Backhouse, do.
P.	,,	J. Berridge, duty at Ostend in 1815.
	,,	J. [illegible] Gregg, joined second battalion in France after Waterloo.
	,,	C. Deane, do.
	,,	W. Ledge.
	,,	C. Rumley.
	,,	H. H. Lewis.
	Adjutant	W. Atkinson (Lieut.).
P.	Q.-Master	J. F. Kingsley.
	Surgeon	R. Pearse.
	A.-Surgeon	S. Piper.
P. W.	A.-Surgeon	J. Evans.

Of the officers in the above list, Major Bircham, Brevet-Major Ryan, Lieutenants Campbell, Stanhope, Jago and Battersby were already placed on half-pay, on reduction, but authority to strike them off had not been received in India.

Bircham had scarcely been absent from the regiment since he was born at Gibraltar in 1769, the son of a corporal in the grenadier company who had been one of the daring men who had scaled the rocks of Locmaria when General Hodgson forced a landing in Belleisle. At twelve years of age young Samuel Bircham was appointed a drummer and nearly twelve years later was promoted ensign from sergeant-major for conspicuous courage and conduct at Cap Brun on October 10th, 1793. Major Bircham was brought on full pay again in the 1st Royal Veteran Battalion on January 30th, 1823. When the Veteran Battalion was disbanded in 1827, he was posted as senior major to the Ceylon

Rifle Regiment. On July 22nd, 1830, he was promoted Brevet Lieut.-Colonel and died in that rank in 1834.

Lieut.-Colonel Vaumorel was granted two years' leave on sick certificate to Europe from October 5th, 1818. He never rejoined, but died at home on September 7th, 1820. He had a very honourable record of service in the regiment. In December, 1793, the two companies under his command had held their post at Fort Mulgrave, near Toulon, after being abandoned by other troops, and thus enabled the allied forces and the thousands of civilians under their protection to embark in safety. He had served in the occupation of Malta and capture of Valetta, and in the Egyptian campaign, where he was present in every action. On Lieut.-Colonel Vaumorel's departure, Lieut.-Colonel Alexander Hamilton, C.B., took command.

In September Lieutenant Sam Robert Poyntz exchanged to the 2nd Ceylon Regiment with Lieutenant William Boyton. Besides Lieut.-Colonel Vaumorel and Lieutenant Poyntz, the regiment lost Lieutenant Walter Ness, who died at Pondicherry in September. On the 6th of the following month Captain Hinton East retired on half-pay.

The Viceroy by this time was satisfied that the power of the native chiefs who supported the Pindarries was broken and that the greater part of the troops in the field could be safely sent to quarters, leaving a force under Sir John Malcolm at Mhow, and another under Brigadier-General Doveton based on Jalna and Ellichpore to deal with Appa Sahib, the ex-Rajah of Nagpore, who had escaped from confinement in May and started a guerilla warfare in the hills to the north of those places.

Orders were given for the 30th to relieve the Royal Scots, who had their Headquarters at Secunderabad, close to the Nizam's capital of Haidarabad, and a wing at Jalna with Brigadier-General Doveton. The flank companies of the 30th which had been at Masulipatam since February, 1818, were to march from there and join the battalion at Secunderabad. The eight companies with Headquarters left Madras on October 8th, when the Governor bade them farewell in a very flattering manner. They crossed the Kistna on November 24th, and reached Secunderabad on December 10th. They had a very rough march owing to the north-east monsoon and twenty-seven men died in the two months, some of them from cholera. Lieutenant Lardner had died the day before the regiment marched, and Lieutenant Winrow, who had either been left at, or sent back to Madras, died there on November 11th. The flank companies reached Secunderabad a fortnight before Headquarters.

Macready's account of marching in the monsoon is interesting. The "general" was usually beat at 3 a.m., when tents were struck and loaded on elephants. Strings of pack bullocks were then brought forward to receive the officers' baggage, the men's knapsacks, etc. At 3.30 the battalion fell in and moved off "at attention" with the band playing and bayonets fixed. After a short distance the music ceased, bayonets were unfixed, the men marched "at ease" and the officers mounted their horses. The men sang on the march and covered over 3 miles an hour, halts included.

No uniformity of trousers was required; pantaloons of chintz of many different colours were seen in the same section. Some men walked in shoes, some in sandals and a great number barefoot. The wearing of fancy clothes may have been carried too far, but there was sense behind it. Marching in the rainy season involved among other discomforts constant wading, and it was absolutely necessary that the men should have dry clothing to put on after the day's march. The regimental trousers once thoroughly wet would stay wet, while the water could be wrung out of the light chintz and a very little fire would dry it. In dry great-coats, trousers, socks and shoes the men had a good chance of escaping chills and fevers. The 30th great-coats, made out of native blanket stuff, were very good.

Halts for several days at a time frequently occurred when the regiment was encamped on some rising ground forming an island in the midst of a submerged district. The larger rivers were crossed in circular wicker boats which were swept down by the current and might ground on a mud bank far below the landing place aimed at. There was little of the country which was not mud. The absence of wheeled transport tells a tale.

On arrival at Secunderabad, orders were received for the left wing to be ready to proceed on field service, and here Lieutenant John Rumley, who had been ill when leaving Madras, but had kept up in the hope of active service, broke down altogether and had to be ordered back to the coast. Macready writes in his diary :—" His friends and many of the old light company which he had so often led to honour, came to say farewell, and this seemed to wound him to the very heart. They were going to the front and he not with them." He managed to reach the coast, but died March 16th, 1819, at Goomrapoondy, near Madras, and a monument was erected over his grave by the regiment.

On November 22nd, 1818, the left wing marched for Ellichpore to join General Doveton ; the strength was five companies of 439 officers and men. The officers were Major Dalrymple, in command, Captains John Tongue and John Powell, Lieutenants Thomas Kettlewell, J. H. Light, P. P. Neville, R. Harrison, E. Macready, and John Stewart ; Ensigns E. N. Gregg and William Ledge ; Assistant-Surgeon John Evans.

On January 27th, 1819, after a march of 390 miles, the wing joined the general at Wackara, at the foot of the Calligong range of hills, and relieved the wing of the Royal Scots who started to march down country. General Doveton distributed his force with a view to block the issues from the Calligong hills where Appa Sahib was lurking, but the ex-rajah broke through the line and took refuge in the fortress of Asseerghur, the governor of which was closely connected with the great Mahratta chief, Scindia.

Owing to the reputed strength of the fortress and the danger of interference by the Mahrattas, it was thought necessary to assemble for the siege of Asseerghur an army of 12,000 men, of whom nearly a fourth were Europeans. General Doveton called up his siege train from Jalna and recalled the wing of the Royal Scots. On February 14th he was at Kandeish and on March 3rd in front of Asseerghur, where he was joined by Sir John Malcolm with his field force from Mhow. The King's troops were 1 troop horse and 3 companies foot artillery, 5 companies each of the Royal Scots and 30th, and 8 companies of the 67th. The E.I. Company's troops were the Madras European regiment and 10 battalions of native infantry with pioneers and sappers and a few cavalry, regular and irregular. The great majority were from Madras, but both Bombay and Bengal were represented.

The fortress of Asseerghur stands on an isolated rock, 700 feet in height. The town at the foot of the rock on the north side was slightly fortified and the track from the town to the fortress was covered by a work called the lower fort. To attempt to carry the fortress by assault was a very hazardous undertaking, and it was difficult to bring effective artillery fire to bear upon it ; the weakness of the place lay in the facts that approach under cover through nullahs and broken ground was easy, and that the defenders lacked heavy artillery and ammunition.

Our plan of attack was to storm the town and to establish a battery in it to breach the lower fort. The town was merely covered by a wall with a fortified gate in the centre but open at both ends.

At midnight on March 17th a column of attack was formed. The assaulting party under Lieut.-Colonel Frazer, Royal Scots, consisted of five companies of his own regiment, the light company of the 30th under Captain Powell, the flank companies of the 67th and Madras Europeans, five companies of the 12th Madras native regiment and a detachment of sappers and miners. The reserve under Major Dalrymple, 30th, was formed of the remaining companies of his wing, one company of the 67th, one of the Madras Europeans, nine companies of native infantry, a detachment of cavalry and four horse artillery guns.

At 1 a.m. on the 18th the assaulting troops rushed the wall at each end and converged on the gate. The town when entered afforded good cover from the fire of the fort and the casualties were few. A howitzer battery was established and the greater part of the troops withdrawn. Firing continued all day between our men and the enemy in the fort. Among other incidents is one worth recording. A man of the light company of the 30th was shot through the right shoulder. His comrades picked him up and were going to send him to the rear when he besought his officer (Macready), to allow him to stay on the ground, *as he always fired left-handed.*

Macready nearly lost his life on that day through his devotion to duty. He had been suffering from dysentery for some time before, but insisted on being allowed to accompany his beloved light company in the assault, and in his weak condition narrowly escaped falling a victim to the heat of the sun.

As soon as we were thoroughly established in the town, the companies of the 67th which had not been engaged relieved the storming party. On the 19th, the wing of the 30th relieved the 67th. On this day the garrison made a bold sally, in repelling which Lieut.-Colonel Frazer fell. On the 21st, before daybreak, the enemy evacuated the lower fort. At 7 a.m. our magazine in rear of the breaching battery blew up and killed a native officer and 34 rank and file ; 66 were wounded. Upon this the garrison re-occupied the lower fort. It was again evacuated when we were able to re-open fire.

This concludes the active participation of the 30th in the siege. On the 24th, the brigade to which the regiment belonged marched to a point 4 miles to the north-east of the fort to protect the battering train which we proposed to employ in that quarter Soon the greater part of our artillery was massed there. It now included Sir John Malcolm's siege train from Saugor. We did not succeed in getting good gun positions and the enemy did not suffer much, but, as they could make no effective reply to our fire, they soon thought that they had had enough and on April 9th the governor of the fortress made an unconditional surrender. The losses on both sides were small ; the 30th had nine men wounded and nine died from the climate during the siege.

A memento of the siege of Asseerghur still exists in the regiment in the shape of a metal ghurrie or gong, captured on this occasion.

On the return of the troops to quarters the Commander-in-Chief at Madras issued the following complimentary order :—

"HEADQUARTER, MADRAS,
"*April 28th*, 1819.

"His Excellency, Lieut.-General Sir Thomas Hislop, Bart., G.C.B., publishes with great satisfaction the following extract of a report from Brigadier-General Doveton, C.B., commanding the Hyderabad subsidiary force. 'The conduct of the detachment of H.M. Royal Scots, under the command of Captain Wetherall, and of H.M. 30th, under the command of Major Dalrymple, during the siege of Asseerghur, has been most exemplary, as to reflect the most distinguished credit on their several commanding officers, as well as the whole of the officers and men composing those detachments.'"

Many years later the East India Company petitioned H.M. Government to have the capture of Asseerghur added to the list of seventeen actions in India between 1803 and 1826, for which a medal was granted, but met with a refusal on the ground that the application was made too late, the list was closed.

The wing marched from Asseerghur on April 15th, 1819, and arrived at Jalna on May 3rd.

The paymaster, James Crookshank, had died on March 25th, and, to the joy of the regiment, the vacancy had been filled by the appointment of Hugh Boyd Wray, who had served with the second battalion throughout the campaigns in the Peninsula and the Low Countries.

The little excitement of the march up country and the siege of Asseerghur had exercised a most beneficial effect upon the regiment. The number of courts martial during a period of six months fell from 120 to 13: 8 at Headquarters and 5 in the wing under Major Dalrymple, which General Doveton always speaks of as "this exemplary detachment." In the inspection reports, notice is taken of the efforts of Colonel Hamilton and his officers to prevent the commission of crime.

There was constant ball firing. We only hear now and then of results, through the reports of inspecting generals, and an average of one hit in ten rounds was the best which the regiment ever made, but that was at field firing and the mark was probably a line of chatties. The ranges were 100 and 200 yards. To hit a chattie with Brown Bess at 200 yards demanded a good deal of luck as well as skill.

So far nothing has been said on the subject of duelling, which played so prominent a part in the life of our ancestors. It went on in the 30th as in other corps, but during the early service of the regiment in India we have no information except in a case where a duel was prevented, a misfortune which always caused an infinity of trouble. Two subalterns were charged before a court martial with sending challenges. One was acquitted and the other was sentenced to a nominal punishment and recommended to mercy.

After entering the Nizam's territory, whether it was the stimulating air or that it was difficult to settle down quietly immediately after the siege of Asseerghur, there was quite an epidemic of duels. The left wing at Jalna, with only twelve officers present, managed to fight six in as many months, while the older and less lively people at Headquarters only brought off four. These encounters were nearly always harmless and caused no interruption in the friendship between the combatants. In fact, General Doveton wrote of the wing at Jalna, "perfect unanimity prevails among the officers of this exemplary detachment."

Macready and his dear friend Neville exchanged a couple of shots apiece at eleven paces one morning without damage; they even went so far as to separate (they had shared a bungalow together), but Macready did not at first take kindly to his new chum, and we find him entering in his diary that he wishes he were back with Neville even if it entailed fighting him once a year.

Lieutenant Neville in this year received a decoration of honour from the Nizam for the capture of a band of Pindarries. Beyond the granting of the honour we know nothing of this exploit.

On June 19th, 1820, the left wing marched in to Secunderabad from Jalna. A small draft from the depot had arrived six weeks before under Lieut. Henry Ramus. It had acted as guard on a convict ship to New South Wales and been sent on from there. In September another draft arrived of forty-seven rank and file under Lieutenant Andrews (adjutant of the second battalion at Waterloo), and Lieutenant Peter Cheape. This was the last draft received from Canterbury, for in May the depot had moved to Albany Barracks, near Cowes, in the Isle of Wight. It had re-opened recruiting, which had been stopped in 1817, and had parties out on the old recruiting grounds in Lincolnshire, Yorkshire and Cambridgeshire, but the bulk of the recruits were now coming from Armagh, Monaghan and Ballinasloe. With very few exceptions the men were enlisted for unlimited service.

There were at this time a good many officers on command and on the Staff.

Captain Machell was acting paymaster to a flank battalion, Captain Owen Wynne Grey in command of a battalion of the Rajah of Travancore's army, Lieutenant Duncan Henry Kennedy doing duty with Russell's brigade, which consisted of two regiments of the Nizam's troops officered by Europeans and called after Mr. Russell, the Resident, who had organized it. Lieutenant Kennedy had with him a sergeant, a corporal, and a dozen men from the regiment. Lieutenant C. R. Macleod, A.D.C. to the Viceroy. Ensign Charles Rumley, A.D.C. to Major-General Rumley in the northern district of Madras Presidency.

In August, Lieutenant Robert Daniel was struck off the strength on exchange with Lieutenant James Ralph of the 59th Foot. Lieutenant Daniel had been promoted ensign from quartermaster-sergeant at the siege of Cadiz in 1810. When the regiment joined Lord Wellington, he had been left at the base and acted in 1810–11–12–13 as quartermaster to Captain MacNab, commandant at Figueras. Lieutenant Daniel served with the regiment at the siege of Antwerp, and the battles of Quatre Bras and Waterloo, where he was wounded.

Lieut.-Colonel Charles Albert Vigoureux, C.B., who had been for years incapacitated by the wound received at Waterloo, arrived about this time in the fleet from England.

The desirability of forming a sporting club had been for some time in the minds of many officers and the Cambridge Club came into existence in 1821.

It started of course with a dinner, which in honour of the "Old Egyptians" was held on March 21st, the day of the battle of Alexandria. Membership of the club was restricted to officers of the regiment.

The club gave a cup worth 1,000 rupees, to be run for annually at the Secunderabad Races, and organized numerous meetings and expeditions of its own for tiger and deer shooting and pig-sticking. In this year too the regiment sent 1,000 rupees for the relief of distress, which was acute in some parts of Ireland.

Fifty-four non-commissioned officers and men joined the regiment during the year, of whom three-fourths were volunteers from regiments serving in India, and the remainder a small draft from the depot. The losses by death were 3 officers and 60 non-commissioned officers and men.

In February, 1822, Lieutenant Sutherland Hall Sutherland joined with 1 sergeant and 40 men from the depot, and in May Lieutenant J. W. Poyntz brought a draft of 1 sergeant and 60 rank and file.

There must have been some fresh disturbance in Nagpore, for Lieutenants John Stewart, C. R. Macleod, R. Barlow, and Joseph Berrige, were placed under the orders of the British Resident there.

In September Lieutenant Blackall brought out a draft of 30 men from the depot and in November 22 volunteers were received from other regiments, and John Rogerson Gillespie joined on appointment as assistant-surgeon in place of Evans deceased.

On June 12th, 1823, Lieut.-General James Montgomerie, from the 74th Foot, was appointed colonel of the regiment *vice* General Robert Manners, deceased.

In July, Brevet-Major Howard joined with a draft from the depot of 2 lieutenants, 2 ensigns, and 146 other ranks. In November, Assist.-Surgeon Sam Piper left the regiment on promotion into the 83rd Foot; he was succeeded by Assist.-Surgeon J. Campbell, M.D.

On March 9th Lieutenant James Nesbitt Gregg went home on sick leave. He died in Europe.

In November two small drafts joined from the depot under Lieutenants W. B. Frizell and S. Tressider.

In July, 1825, official intimation was received of the death of Quartermaster Kingsley at sea, on his way home, and of the appointment of Sergeant John Ward to succeed him from October 21st, 1824.

Sergeant Ward had served with the second battalion at Sabugal, Fuentes Onoro, Badajoz, Villa Muriel and Waterloo, and was one of the original colour-sergeants made by Colonel Hamilton on the return of Headquarters from the Peninsula.

On August 20th authority was received from the Commander-in-Chief for the words "Badajoz" and "Salamanca" to be borne on the colours of the regiment.

On December 2nd, Lieutenant Baxter, Ensign Staff, and Quartermaster J. Ward, arrived from the depot with 146 other ranks; Captain Henry Cramer who had commanded the depot for two years arrived in India a month before the draft. Twenty-one transfers from the 67th were also received about this time.

In August, 1826, the long stay of the regiment at Secunderabad came to an end and on the 25th of the month eight companies marched for Madras under Major Dalrymple, Colonel Hamilton being left in command of the Haidarabad Division, and Lieut.-Colonel Vigoureux of the Secunderabad Brigade. Two companies also were left to hold the cantonments till relieved.

After leaving Secunderabad, the regiment erected a monument in the cemetery there to the memory of 17 officers, 61 non-commissioned officers, 526 privates, 39 European women and 55 children of the 30th, who had died during its service in the Nizam's dominions, that is between October 8th, 1818, and March 1st, 1827, the latter being the date upon which the last man died. He was evidently one of the sick left behind. We have no list of the officers to whose memory the monument was raised, but the following are the names of those who died within that period :—

Lieutenant John Rumley :—Served in every action fought by the 30th in the Peninsula, severely wounded at Villa Muriel and at Waterloo, where he commanded the light company. "He was so mild, so good, so gallant a fellow that the whole regiment loved him" (Macready). Died at Goomrapoondy, March 16th, 1819.

Lieutenant Richard Harrison :—Served in the Peninsula, siege of Antwerp, Quatre Bras (wounded), Waterloo. Died May 19th, 1819. A man of distinguished courage.

Lieutenant William Ledge :—Died June 21st, 1819. Promoted from regimental sergeant-major in 1815.

Lieutenant William Stanhope :—Died in September, 1819.

Lieutenant Archibald Campbell McDougall :—Killed accidentally when riding at Secunderabad, August 26th, 1820, very much regretted by his brother officers.

Assist.-Surgeon John Evans :—Died at Secunderabad, July 16th, 1821. The wounded of Badajoz, Salamanca, Villa Muriel, Antwerp, Quatre Bras and Waterloo, had all been carefully tended by him.

Lieutenant John Garvey :—Died at Poonamallee, August 31st, 1821. Commanded a company at Badajoz and at Salamanca (wounded).

Lieutenant John Roe (2) :—Died December 30th, 1821. Had served at siege of Antwerp and had the distinction of having been wounded both at Quatre Bras and Waterloo.

Lieutenant Duncan Henry Kennedy :—Died October 15th, 1822. Joined 30th from 86th, March 1st, 1819. Served in Russell's brigade. Was twice wounded in actions about Nagpore in 1818.

Captain Richard Machell :—Died November 17th, 1822, at Pulo Penang. Served at Badajoz (severely wounded), Salamanca and siege of Antwerp. Town major of Antwerp at time of Waterloo.

Lieutenant Daniel Vanderzee :—Died at Secunderabad, December 18th, 1822.

Quartermaster John Foster Kingsley :—Died at sea, October 21st, 1824. Served with second battalion in Peninsula till invalided in 1811.

Lieutenant Samuel Tressider :—Died at Secunderabad, December 5th, 1824.

Ensign Charles Battley :—Died at Kamptee, January 15th, 1826.

Ensign Richard Wilson :—Died at Secunderabad, August 15th, 1825.

Lieutenant George William Thompson :—Died at Secunderabad, October

24th, 1825. One of the officers left behind when the regiment marched for Madras.

Lieutenant Henry Hartley Lewis :—Died at Ongole, December 14th, 1826. Had served at Asseerghur. Ongole is about half-way to Madras. Lewis was on the march with the two companies left behind at Secunderabad.

Lieut.-General Robert Manners, colonel of the regiment, died on June 13th, 1823, and Colonel Philip Vaumorel on September 7th, 1820, but they did not serve in the Nizam's territory.

On October 15th the regiment arrived at Fort St. George and Colonel Hamilton took command. In November the two companies left at Secunderabad marched towards the coast under Lieut.-Colonel Vigoureux. They joined Headquarters on January 5th, 1827. A detachment of 1 sergeant, 2 corporals, and 21 privates remained at Secunderabad, chiefly men unable to march. The march of the headquarters and eight companies to Madras had been in the rainy season and naturally was accompanied by much suffering and considerable loss of life. The two companies which moved in November had better weather, but they had more weakly men in the ranks and they also suffered heavily. The total loss of the regiment by death from the beginning of August, 1826, to March 1st, 1827, was 2 officers, 4 sergeants, 84 rank and file.

From the time the 30th left Secunderabad, the officers who still remained to connect the regiment with the heroic times of the Republican and Napoleonic wars began to disappear rapidly.

In the beginning of 1827, notification was received that Brevet-Majors Sam. Fox and Robert Howard had been promoted to half-pay majorities in the previous May.

In February, Lieut.-Colonel Charles Albert Vigoureux, C.B., was struck off the strength on his exchange to the 45th Foot with Lieut.-Colonel Stacpoole. Stacpoole was merely brought in to sell, and never joined. Lieut.-Colonel Vigoureux was equally beloved and respected. After holding a commission in a Corsican regiment he was appointed captain in the 42nd Highlanders. In May, 1805, he was transferred to the 6th Foot, and in July, 1807, to the 38th. In April, 1808, he received the brevet of major. When the Peninsular War commenced, he went to Portugal with the 38th. At Fuentes Onoro and Vitoria he commanded the flank battalion of his brigade, and for the latter action he was made a Companion of the Bath and a brevet lieut.-colonel. Although he was unaware of it, he was at this time junior major of the 30th, having been gazetted a little over a fortnight earlier. Lieut.-Colonel Vigoureux served in the 30th at the siege of Antwerp and throughout the Waterloo Campaign, commanding the flank battalion of Halkett's brigade at Quatre Bras and Waterloo, where he was severely wounded. He was a second time gazetted a Companion of the Bath for his services at Waterloo. He had always thoroughly identified himself with the 30th, and was a leading spirit of the Cambridge Sporting Club, and rarely absent from its meets. He died in February, 1841, his death being beyond doubt accelerated by the sufferings from the wound received at Waterloo ; a bullet had lodged near the spine and could not be extracted.

In February, Lieutenant Pepper Parke Neville was struck off in consequence of his exchange to the 13th Light Dragoons. Lieutenant Neville had served with the second battalion in the Peninsula. He was in action with the light company at Fuentes Onoro and had served as assistant engineer at the siege of Ciudad Rodrigo. He was with the light company in the assault at Badajoz, where he was severely wounded. He recovered in time to command the light company at Salamanca. When the siege of Burgos was formed, he again volunteered as assistant engineer and was severely wounded. He served in Holland in 1814 with the regiment, but missed Waterloo, having again volunteered for the engineer department and having been sent on special duty to the French frontier. During the Mahratta War he again volunteered for service as an engineer and received a medal. He also received a decoration from the Nizam. After leaving the regiment, he had the singular fortune to receive the Legion of Honour from the French Government for having by encouragement and example restored order in the panic-stricken crew of a French ship on which he was returning to Europe, and thus saved all hands. In later life his tall soldierly figure was conspicuous at the Court of Queen Victoria, in whose bodyguard he was an officer. He died a Knight of Windsor.

On March 9th, 1827, Captain William Sullivan died at sea. He had been unlucky all his life. In 1806 he had been wrecked as an ensign with Hawker's company on the French coast and remained a prisoner of war for eight years. He became a lieutenant without purchase on his release in 1814, but had to serve eight years more as lieutenant and brevet captain before he got his company.

In May, 1827, Colonel Hamilton was placed in command of the troops in garrison at Fort St. George, the command to date from October 15th, 1826; Major Dalrymple assumed command of the regiment from the same date.

In July, Lieutenant Edward Neville Macready embarked for England on two years' sick leave. He with Mansel and Tobin had been on a shooting trip in April, and all returned with fever. Mansel's attack was slight but Tobin died on May 26th and Macready only recovered after being at death's door. When he was at the worst Colonel Hamilton kindly had him taken into his own cooler quarters. During Macready's illness a great friend of his, Lieutenant George Lynch Backhouse, had died on May 15th.

On August 4th, Captain M. Young died at Ellichpore. He was then, as he had been for some years, acting under the orders of the Resident of Haidarabad. His death gave a step without purchase to Lieutenant Richard Mayne, who had commanded a company in the Peninsula and served at the siege of Antwerp and battles of Quatre Bras and Waterloo—Mayne was at this time on sick leave in England. He never rejoined, but died on December 3rd.

On August 5th the regiment marched for Trichinopoly, where it arrived on the 27th.

On August 27th, Lieutenant Thomas Kettlewell accepted a half-pay company. He had served in the Peninsula in 1809.

In October Lieut. and Brevet Captain Matthias Andrews was appointed captain in the 44th regiment without purchase. He had commanded a company in every action with the second battalion in the Peninsula, and had served as adjutant at the siege of Antwerp and at Quatre Bras and Waterloo, where he was wounded.

On October 31st a draft of 1 corporal and 29 privates arrived from the depot.

When the regiment left Madras Colonel Hamilton had remained there in command of Fort St. George. In December he sailed for England on two years' sick leave.

Surgeon Robert Pearce, who had scarcely been absent from the regiment since appointed to it in September, 1803, was granted two years' sick leave at the same time as Colonel Hamilton. He never rejoined, but went on half-pay on September 11th, 1828.

In April, 1828, a draft arrived from the depot consisting of Captain John Gordon Geddes, Lieutenants C. H. Marechaux and J. P. Meike, 4 corporals and 53 privates.

The regiment was now ordered to prepare to embark for England, and in September the non-commissioned officers and men were invited to volunteer for other regiments serving in India and 4 sergeants, 15 corporals, 7 drummers, and 320 privates left the regiment. In this month notification was received of the promotion of Major Dalrymple to be second lieut.-colonel *vice* Stacpoole, and of Captain J. Powell to be Major *vice* Dalrymple, to date from October 18th, 1827. Powell purchased over R. B. Lynch and John Tongue.

Lynch had been over twenty-two years a captain and had been under fire when Powell was at school; he seems to have lost hope of promotion, for he accepted a half-pay majority to date from February 26th, 1828. He had served at the siege of Valetta, in the Egyptian campaign and in the Peninsula.

Further orders were received for the regiment to return to the Presidency with a view to embarkation, and it left Trichinopoly under command of Lieut.-Colonel Dalrymple on October 7th. On November 4th it reached St. Thomas's Mount, a few miles from Madras, and there volunteering was opened again; 2 sergeants, 4 corporals, 3 drummers, and 120 privates left for other regiments.

On November 14th the regiment marched from St. Thomas's Mount for Wallajabad, and arrived on the 16th. Lieut.-Colonel Dalrymple took command of the troops in cantonments, and Major Powell of the 30th. Volunteering was again opened and 30 men left the regiment.

In view of an early embarkation everything had been returned to store except arms and accoutrements for 400 men, but military training was not suspended. On December 15th, and on the three following days, the regiment paraded as strong as possible for field days. On the first two days, each man fired 20 rounds of blunt (blank) cartridge, and on the third, 9 rounds of ball. On the fourth day, 20 rounds of blunt were fired. The number in the ranks averaged 275, and 292 flints had to be replaced after the four days' firing, so a flint had to be replaced after 65 discharges or thereabouts.

On January 13th, 1829, the 30th marched for St. Thomas's Mount again and there it encamped, awaiting embarkation. Cholera had been hanging about the regiment ever since the march from Trichinopoly and the death rate was high. In the three months ending on January 31st, 1829, there died 1 lieut.-colonel, 1 captain, 4 sergeants, and 28 rank and file. Captain Mann died on December 26th, 1828, and Lieut.-Colonel Dalrymple on January 9th, 1829. The latter was an excellent officer who had commanded the left wing with great credit at Asseerghur.

Captain William Mann had no war service in the 30th, but he was one of the "Die Hards" of Albuera. When the French began to fall back on that day, Colonel Inglis of the 57th, who was lying wounded near the colours, urged his regiment to advance, and the senior surviving officer, Lieutenant and

Adjutant William Mann, called upon the 150 men left standing to follow him. The movement was stopped by Marshal Beresford, who thought that the 57th had done more than enough. As soon as Mann got his company, he was given the gold medal for his service at Albuera. The cholera did not follow the regiment to St. Thomas's Mount.

On February 14th, 1829, the regiment embarked for England on three ships, after nearly twenty-three years' service in India.

On the *Alfred*, Headquarters and four companies—Major Powell, Captain J. H. Light, Lieutenants J. W. Poyntz, George Mansel, R. A. Andrews, C. W. Barrow, J. M. T. Borton, and C. J. Boyes, Paymaster H. B. Wray, Adjutant William Atkinson, Quartermaster J. Ward, Assistant-Surgeon J. K. Adams, and 153 other ranks.

On the *James Pattison*, two companies—Captain J. G. Geddes, Lieutenant H. M. Dickson, Ensigns H. Magee, F. C. Waldron, W. H. Heard, and 60 other ranks.

On the *Barrett Junior*, four companies—Captain T. Jones, Lieutenants William Baxter, C. H. Marechaux, J. R. Burrows, and E. K. Gregg, with 139 other ranks.

Total, 1 field officer, 3 captains, 11 lieutenants, 3 ensigns, 4 staff, 30 sergeants, 6 drummers and 316 rank and file.

Leave had been given to 5 sergeants and 17 men, time expired, to remain and draw their pensions in India. As all the arms and accoutrements belonged to the East India Company, they had been returned to store before embarkation.

HOME SERVICE

On June 15th, 1829, Headquarters disembarked at Gravesend, and on the 27th the service companies were united under Major Powell at Chatham.

While the 30th was on passage home Lieut.-General James Montgomerie, colonel of the regiment, had died on April 18th, 1829, and had been succeeded by Lieut.-General Sir Thomas Bradford, G.C.B., who was not unknown to the 30th. Bradford's Portuguese brigade had been a few hundred yards to the right of the 30th at Salamanca, and the regiment had been in his division at Paris in 1816.

On July 1st, the service companies started to march to Portsmouth, and thence to join the depot at Albany Barracks in the Isle of Wight.

A correspondent of the *Morning Post* gives us a glimpse of the regiment as it passed through Brighton on the 6th. " The Marine Hotel was rendered very gay yesterday; the 30th Infantry having marched into the town, commanded by Major Powell. The officers, about 18 in number, breakfasted and took their mess there, the band playing without and amusing all there assembled. The men had a veteran appearance; some of them were very much the worse for wear, but the greater proportion looked healthy. They left us this morning as they had arrived, without muskets or side arms, for the Isle of Wight, where their depot is and where the aged and infirm in their ranks will receive their discharge."

On the 11th the regiment crossed from Havant to the Isle of Wight, and marched in to Albany Barracks where the depot, 370 strong, was awaiting it under Colonel Alexander Hamilton, C.B.

The following had been the history of the depot since 1822. In 1823 Captain Cramer took it from the Isle of Wight to Portsmouth. In February, 1825, Brevet Major Samuel Fox relieved Captain Cramer in command.

In May the depot returned to Albany Barracks, but in September it proceeded to Chatham. On April 25th, 1826, it was at Colchester with a detachment at Chatham, and here Brevet-Major Fox went on half-pay and Captain John Proctor from the half-pay of the 43rd Foot joined and took command. In July the depot returned to Chatham, and from there moved in October to Canterbury. In June, 1828, Captain Proctor marched it to Dover Castle. Captain John Tongue joined and took command in October. In December the depot returned to Albany Barracks in the Isle of Wight and on January 1st, 1829, Major Robert Murray, who had been on leave from India, joined and took command. In this month, to compensate the 30th for the volunteers it had given the regiments serving in India, the depots of the Indian regiments were called upon for volunteers for the 30th. To cope with the increased numbers the depot was formed into four companies, of which the officers were:—Captains John Tongue, John Proctor, Washington Carden, and C. H. Roberts; Lieutenants Jos. Berrige, Jos. Tompson, and A. G. Wright; Assist.-Surgeon J. R. Gillespie.

In June, Colonel Hamilton and all officers on leave in England, joined the depot.

On August 30th, Colonel Alexander Hamilton, C.B., went on leave pending retirement. No man had ever exercised a stronger influence in a regiment. He commanded the second battalion, except for a few months, from the time it left Cadiz in October, 1810, till it was absorbed by the first in 1818, and he commanded the regiment in India, from October 1819, till he went sick in September, 1827. He was as kindhearted and generous as he was brave, and under him the regiment knew that it was efficient and capable of great things, but during the last year in Madras the climate and age had begun to tell and he was no longer the same man. He died at Woolwich on June 4th, 1838, aged 63. The physician attending him gave the cause of death as "decay of nature."

Major Robert Murray was gazetted out on August 6th, Captain Henry Cramer having purchased the majority. Major Murray had joined in 1795, and had served in Sicily and at the capture of Valetta. During the Egyptian campaign of 1801 he was present as a subaltern with the light company in every action.

News was received in August that Edward Neville Macready had got his company by purchase on July 16th, through the retirement of Captain Owen Wynne Grey, who had been left in India commanding a battalion of the Rajah of Travancore's army.

In August 8 subalterns and 2 assistant-surgeons were placed on half-pay and struck off the strength, being supernumerary to the establishment, which was:—1 colonel, 1 lieut.-colonel, 2 majors, 8 captains, 10 lieutenants, 5 ensigns, 1 surgeon, 1 assistant-surgeon, 1 paymaster, 1 adjutant, 1 quartermaster, 6 staff sergeants, 34 sergeants, 14 drummers, 36 corporals, 705 privates.

The following are the names of the officers reduced:—

Lieutenants Thomas Robson Kirkstan Burrows, J. G. Wright, R. C. MacDonald, J. M. Borton, W. Bathe, C. J. Boyce, Waldron Kelly and W. J. Hannagan; Assistant-Surgeons Sam. Dickson and J. K. Adams.

In September the regiment crossed over to Haslar Barracks, Gosport, under Major Powell. Forty-two men had been discharged unfit, and eight had died since landing.

In November the promotions caused by Colonel Hamilton's retirement were notified :—Major John Powell became lieut.-colonel, Captain J. H. Light major, Lieutenant H. M. Dickson captain, and Ensign Wm. Marlton lieutenant, all by purchase.

On November 21st the regiment moved into Forton Barracks. Two men had died at Haslar.

Captain Washington Carden sold out on March 25th, 1830, after twenty-five years' service. He had commanded a party of the regiment serving as marines in 1807, and the detachment which assisted in defeating the Wynaad rebels in 1813.

In July the regiment marched to Weedon and from there a detachment of two companies was sent in the following month to Northampton under Major Cramer.

On August 13th Lieutenant William Atkinson resigned the adjutancy and was succeeded by Lieutenant Ninian Armstrong.

On October 14th the regiment was concentrated at Manchester. From there Captain Mansell's company was detached to the Isle of Man.

Lieutenant C. J. Boyes died at Lymington on October 27th.

On November 19th Doctor Samuel Piper returned to the regiment *vice* Surgeon O'Reilly appointed to the Staff.

Captain Edward Neville Macready, who had been appointed on November 23rd, 1830, A.D.C. to Major-General Wilson commanding in Ceylon, ceased from January 1st, 1831, to appear among the effectives of the 30th regiment. He never joined again for duty, but to the end of his life his affection for the regiment and care for its honour were undiminished. The " Macready Fund," a legacy of £100 to the regiment, is a token of his widow's sympathy with one of her husband's strongest feelings.

The detachment in the Isle of Man was changed about this time, Captain H. M. Dixon taking the place of Captain Mansel. There was some trouble in Ashton-under-Lyne and the Headquarters and four companies marched there on February 1st, 1831 ; a fortnight later a detachment of thirty men was sent under Lieutenant Marechaux to Liverpool.

At Ashton-under-Lyne, on April 5th, 1831, new colours were presented to the regiment by Colonel Powell.

On the 18th and 19th the regiment made its first journey by railroad to Liverpool, and embarked on the latter day for Dublin, arriving on the 20th, the company from the Isle of Man arriving on the 29th. This was the first time that the regiment had been united since its return from India.

In June Major Henry Edward Robinson joined from the 48th on exchange with Major Cramer, and in July Major Henry Smith Ormond from the 45th regiment joined in succession to Major Light, who retired. On August 9th Captain C. W. Barrow exchanged with Captain and Brevet Lieut.-Colonel Edward Antony Angelo to the half-pay of the Newfoundland regiment.

Paymaster Hugh Boyd Wray died on October 13th. He was a good comrade and much liked.

In October the regiment marched to Belfast in several divisions, the last division arriving on the 15th inst. Two detachments, each of one company, were sent out, Captain Poyntz to Downpatrick, and Captain H. M. Dixon to Carrickfergus.

The regiment had no sooner settled in Belfast and out-stations than it was ordered to Enniskillen, where it was assembled on January 6th, 1832. The effective strength was 39 sergeants, 14 drummers, and 746 rank and file.

The recruiting parties were at Edinburgh, Perth and Tullamore and the proportion of recruits from Scotland was greater than it had been since the regiment crossed from there to Ireland in 1775. Of the trained soldiers two-thirds were Irish.

On reaching Enniskillen two companies were sent to Omagh, one to Sligo and one to Cavan; from Omagh a subaltern and fifteen men were detached to Lifford.

The detachment at Cavan was withdrawn on February 25th, and the company at Sligo marched to Ballyshannon on the 26th. A few days later it proceeded to Belleek. The Lifford detachment was drawn in on March 6th. On March 13th Captain Geddes with his company arrived at Donegal.

On April 8th Captain Thomas Jones died, and Lieutenant Wm. Atkinson, the former adjutant, got his company without purchase.

In July the regiment marched as follows:—

18th. Headquarters and two companies to Londonderry.
19th. Two companies to Omagh, one of them for Rutland and the other for Lifford.
20th. One company to Londonderry.
23rd. One company from Londonderry to Cardonagh and another to Buncrana.
24th. Two companies from Ballyshannon to Belleek, and the detachment from Pettigo to Headquarters.
25th. Captain Geddes' company from Donegal to Headquarters.
31st. Two subalterns, 2 sergeants, 39 rank and file marched to Clonmaney.

There were other movements during the year, including a relief of all the detachments by men from Headquarters, and at the end of December the regiment was distributed as follows:—4 companies at Enniskillen, 2 at Omagh, 1 each at Lifford, Cardonagh, Buncrana and Aughnacloy. There were also small detachments at Clonmaney and Clogher.

On July 13th, 1833, Headquarters marched from Derry to Castlebar and withdrew the northern detachments. The following fresh detachments were sent out, one company each to Ballinasloe, Westport, Ballina, Tuam and Dunmore.

On August 7th Headquarters and two companies marched to Galway. There were some more moves, but in October the regiment was given a six months' rest with Headquarters and 5 companies at Galway, 2 at Castlebar, and 1 each at Westport, Ballina and Loughrea.

Sixty recruits were at drill but there was no drill field and the barrack yard was so small that only company drill, could be practised. The recruiting parties were at Edinburgh and Omagh, but the regiment was of course taking Headquarter recruits. It was now all but complete to its establishment and was expecting to go abroad.

Before the troops were withdrawn from North America and other colonies, the proportion of foreign and home service was about eight years to four.

The establishment of the regiment was now 10 companies, 4 field officers, 10 captains, 20 subalterns, 5 staff officers, 43 sergeants, 14 drummers, 36 corporals, 703 privates.

On December 9th Lieutenant Joseph Berrige died. He was an old light company man and had served with the second battalion throughout its campaigns in the Peninsula, where he must have distinguished himself, for he was one of the first colour-sergeants appointed by Colonel Hamilton when that rank was instituted.

Lieut.-Colonel John Powell died at Cheltenham on December 31st. He was a good officer and had commanded the light company at the siege of Asseerghur. An attack of cholera which he had in India may have contributed to his early death. His death gave the regiment without purchase to Major Henry Edward Robinson and a majority to Captain John Tongue.

Colonel Powell was the seventh officer to die of those who had been present when the regiment arrived in the Isle of Wight from India in 1829.

On May 21st, 1834, the regiment marched to Fermoy, which was reached by Headquarters on June 1st. All detachments were withdrawn and new ones furnished. Headquarters and six companies were at Fermoy, and one company each at Mitchelstown, Mallow, Doneraile and Castletown Roche.

At Fermoy it was inspected previous to marching to Cork for embarkation by Major-General Thomas Arbuthnot on July 4th, 1834. It required forty-one men to complete the establishment.

It was a very handsome regiment; only fifteen privates and nine specially enlisted lads and boys were under 5 ft. 6 inches.

Owing to its having been constantly on the move and often having no space to drill in, it was uninstructed in the new drill which had been sanctioned in 1834, and by the general's advice, Lieut.-Colonel Robinson had kept it at squad drill since its arrival in Fermoy.

There had been much drunkenness and 137 courts martial in the previous six months.

General Arbuthnot took a very serious view of the amount of crime. He writes: "Having had the proceedings of courts martial before I went on parade, I expected to see anything but steadiness, but to my surprise I found the 30th regiment to be composed for the most part of a fine body of young men—steady under arms and soldierlike in their appearance, each man having a kit which both in regard to its cleanly appearance and in other respects, would do credit to any regiment." "The men are cleanly in their quarters and barrack-rooms to a marked degree." The system of interior economy General Arbuthnot declared to be "perfectly good."

The movements of the regiment in Ireland have been detailed at somewhat tedious length, in order to show how hard it was to train a battalion in those days. The 30th left Albany Barracks in July, 1829, with about 200 trained non-commissioned officers and men. From that day it had never been together except for a little under five months in Dublin, and had never occupied a barrack in which more than a company could be exercised.

The regiment now prepared for embarkation, and under the new system six companies were to be sent abroad; the other four were to form the depot under the senior major. The four-company depot could be easily formed into a second battalion if required, and it was in fact often termed the reserve battalion. The companies selected for service were No. 1 grenadiers, Nos. 2, 5, 6, 8 and 10 light. On July 19th, 1834, Nos. 6 and 8 companies embarked at the Cove of Cork on the transport *Orestes* under Major Tongue. The other officers were Captain John Proctor (8), Lieutenants C. H. Marechaux (6), and W. H. Heard, Ensigns E. J. Grant, and J. C. D'Esterre (grandfather of the Major D'Esterre of the great war of 1914–19).

On the 22nd Nos. 1, 2, 5, and 10 embarked on the *Marquis of Huntly.* The officers were Lieut.-Colonel H. E. Robinson, Captains J. G. Geddes (2), J. W. Poyntz (1), R. A. Andrews (5), Wm. Atkinson (10), Lieutenants J. M. T. Borton, R. C. Macdonald, W. A. Steele, John Moore, A. J. Barrow, A. J. H. Lumsden, Adjutant A. Macdonald, Surgeon S. A. Piper, Paymaster David Hay.

The total strength of non-commissioned officers and men was 31 sergeants, 10 drums, 23 corporals, and 444 privates. Forty-eight soldiers' wives with 74 children accompanied the regiment.

On the 29th the four companies left at Fermoy, Nos. 3, 4, 7, and 9, marched to Clonmel under Major Henry Smith Ormond. The strength was, Captains H. M. Dixon and Wm. Baxter, Lieutenants Ninian Armstrong (Paymaster), Charles Sillery (Adjutant) and E. P. Pilsworth, Ensigns Edward Lucas, S. J. L. Nicoll and the Hon. J. H. Pery, 10 sergeants, 4 drummers, 10 corporals, 206 privates with 89 women and 54 children. The only recruiting party which the depot had out was at Belfast.

The regiment disembarked at St. George's, Bermuda, on August 30th, 1834. Captain Proctor with his company was sent to Ireland Island. His subalterns were Lieutenant Wm. H. Heard and Ensign J. C. D'Esterre.

On October 1st Surgeon Samuel Ayrault Piper, M.D., returned to England on sick leave and never joined again ; he had served in the regiment for sixteen years as assistant-surgeon and four as surgeon, and if, as is said, the character of Dr. Slammer in the *Pickwick Papers* is a portrait, he must have added much to the geniality of the Mess. He became a Staff surgeon at Chatham, and it was probably during this period of his life that he met Messrs. Tupman and Jingle with the results which we all know.

He was succeeded by Surgeon Tregance, perhaps the only officer of the 30th who had been shot in action with an arrow. He had this experience in the Ceylon rebellion.

On December 12th, 1834, the regiment was inspected by Major-General Chapman, who was satisfied with the drill and more than satisfied with other things. He says " he has not witnessed such good interior economy as is established in the 30th." " The men are a fine body of men, very cleanly in their barracks and quarters." " The captains pay great attention to their companies and the number of courts martial had fallen to 35 in five months ; it was 135 at last inspection." " The good quality of the necessaries supplied and the fairness of the price was equal to anything he had ever seen." This was of great importance when necessaries were bought regimentally and not supplied from a Government store.

General Chapman's report winds up :—" The sergeants mess together, and a highly respectable Mess it is."

We have no inspection reports, but there is little doubt that as time went on, the regiment began to get slack in Bermuda. The climate had a good deal to say to it. It is very mild, but after a little while very relaxing, and the regiment stayed in the islands for seven years. In addition to this there was the absence of anything to do, and the deadly monotony of the place. Frequent prize fights were one resource; they took place on some short grass with bushes at one end behind which the officers sat and drank their wine and made their bets; they were not supposed to be seen. The greatest misfortune of all was the fact that three of the four officers at the head of the regiment were too old to keep their energy in that climate. They had rendered good service in other corps in the great war, but that had finished a quarter of a century before the regiment left Bermuda. Lieut.-Colonel Robinson was practically an invalid towards the end of his time. His two majors were not able to purchase, so he could not sell out, and the Commander-in-Chief was naturally unwilling to force deserving officers out of the service under those conditions. The right to retire on full pay after thirty years' service, which was given about this time, was a great boon to the Army. It broke up the knot of elderly men which had formed at the head of many regiments and by providing promotion without purchase benefited the poorer officers.

The regiment moved to Halifax, Nova Scotia, in November, 1841, and on arrival the men presented a very enervated appearance, but quickly picked up health in the bracing air. It was not so easy to abolish the slackness which had crept in. Perhaps it was fortunate that Lieut.-Colonel Robinson was sick, that the two majors and the two senior captains were at home at the depot or on leave, and that Captain J. W. Poyntz was in command; though he had fought in the Peninsula thirty years before, he was still a strong energetic man.

Major Tongue retired on full pay on August 5th, 1842, after thirty-eight years' service. He had commanded a company at the siege of Asseerghur. His service as a captain on full pay for twenty-two years and nine months is the longest of which we have record.

Major-General Jeremiah Dickson inspected the regiment in June, 1842, and did not find it in the best of order, but the health of the men was improving. Lieut.-Colonel Robinson was sick and was granted leave of absence to England for a twelvemonth. In General Dickson's opinion he was worn out in the service. He never rejoined but exchanged to half-pay with Lieut.-Colonel Marcus John Slade on February 10th, 1843.

Lieut.-Colonel Robinson had no war service with the regiment, but in the 48th he had served at the battles of Talavera and Busaco, the sieges of Ciudad Rodrigo and Badajoz, where he was shot through the left arm in the assault, the battles of Salamanca, Vitoria, and the Pyrenees, where he had his left leg fractured.

Brevet Major Poyntz continued to command the regiment till the arrival of the senior major, Brevet Lieut.-Colonel Henry Smith Ormond, from the depot on September 12th. Ormond was succeeded in command of the depot by the junior major, John Proctor.

While under command of Major Poyntz the regiment had marched from Halifax to St. John's, New Brunswick, in several divisions. It was united at St. John's on August 13th, 1842.

Before Headquarters left Halifax the Mayor and Corporation presented an address to Major Poyntz conveying an assurance of their respect for the soldierlike and orderly demeanour which had marked the period of the stay of the regiment in the city and "tendering to the commanding officer, the officers, non-commissioned officers and privates of the distinguished regiment their sentiments of esteem."

The regiment was now extremely healthy, but it had been necessary to invalid more than a tenth of those who landed from Bermuda. Included in the numbers of the regiment were fifty men passed on by the 8th and 37th, who had volunteered for permanent service in North America. These men passed from regiment to regiment until they were granted their discharge with pension and became settlers in the colony.

On June 6th, 1842, the depot moved from Mullingar to Galway. Its recruiting parties were at Cambridge, Sheffield, Wakefield, Chester, Glasgow and Portadown, and the recruits came in so fast that the depot was called upon to give 116 recruits to other regiments less fortunate.

In July and August the depot moved by companies to Tralee and Newcastle.

On March 9th, 1843, Lieut.-Colonel Marcus John Slade joined Headquarters at St. John's, New Brunswick, and took command. He was a man of great energy and talent and detested the slowness and apparent want of spirit which had crept into the drill not only of the 30th, but of nearly every regiment in the service during the long peace. Running drill was insisted upon and everything had to be done in double time. He devoted great attention also to training of officers and non-commissioned officers. At the summer inspection the report was very favourable. Colonel Slade's efforts had been helped by the remarkable change for the better in the health and strength of the men.

On April 7th Ensign George Frederick de Carteret died.

On October 13th Lieut.-Colonel Slade handed over the command of the regiment to Lieut.-Colonel Ormond and went home on leave.

On December 18th the service companies embarked at St. John's on H.M.S. *Resistance*, and on January 9th, 1844, they landed at Cork. The strength was :—Major (Brevet Lieut.-Colonel) Ormond; Captains J. W. Poyntz (Brevet Major), R. A. Andrews, C. Sillery, W. A. Steele; Lieutenants W. H. Heard, F. A. Edwardes, R. W. Smith, L. G. F. Broome, and J. B. Patullo; Ensigns C. D. Oliver, J. W. Keogh, and Sam. Sharpe; Paymaster R. C. Macdonald, Adjutant Alex. Macdonald, Assistant-Surgeon William Braybrooke, 26 sergeants, 9 drummers and 389 rank and file. Fifty-two men had been passed on to other corps with a view to their becoming settlers in the colony.

The depot had arrived in Cork four months before the Headquarters. The strength was Major Proctor; Captains Geddes, Nicoll, Gregory and Marechaux; Lieutenants Lumsden, Shum, O'Grady, Tongue, Bayly and Wilkinson; Ensigns Lowry, Whitmore, Rose, Butler and Gray; Assist.-Surgeon Lockwood (son of Purefoy Lockwood wounded at Quatre Bras), 437 other ranks.

Lieut.-Colonel Slade took command on January 17th. On March 1st the strength of the regiment was 861 non-commissioned officers and men, including 176 recruits at drill.

At Cork the regiment received percussion muskets in place of flint locks and new knapsacks with better adjusted straps, which did not interfere so much with the men's breathing. The great object with both the old and new pack was to "get the monkey well up on the back," as the saying went.

In July the regiment marched from Cork to Limerick and furnished the following detachments:—Killaloe, Tipperary, Newcastle and Rathkeale.

On August 16th Captain Mauleverer from the 17th regiment joined on appointment and on September 20th Captain and Brevet Major James W. Poyntz retired on full pay.

When quite a boy he had served as a volunteer with the light company at Torres Vedras, Busaco, Fuentes Onoro and Barba del Puerco. After leaving the regiment he settled in Nova Scotia, where he met his old regiment again in 1868. He had been promoted lieut.-colonel unattached on November 28th, 1852, and died in that rank at a great age.

In the winter of 1844–5 the detachments were drawn in and the regiment had 762 privates at Limerick. The recruiting parties were in Cambridgeshire and Lincolnshire. Captain William Francis Hoey joined on appointment in January from the St. Helena regiment on exchange with Marechaux.

In April, 1845, six companies were detached at Tipperary, Newcastle, Killaloe, Cahir, Rathkeale and Newport.

In May the regiment marched from Limerick to Castlebar and sent out detachments, two companies to Galway and one each to Ballinasloe, Westport, Oughterard and Loughrea.

On September 27th Lieut.-Colonel Slade exchanged to the 90th Foot with Lieut.-Colonel John Singleton, Knight of Hanover, an officer who had served with great distinction, but was now an invalid.

Colonel Slade left behind him the name of one of the most successful commanding officers the 30th had ever known. His after career in the army was equally meritorious, though he never obtained a command on service. He gained distinction as a lieut.-colonel in the Kaffir War in 1846. He was a full colonel in 1851, and inspecting field officer of recruiting in 1852, a major-general in 1857, a lieut.-general and colonel of the 50th regiment in 1862. He died in that rank in 1872. In 1857 he had been granted the Reward of £50 a year for Meritorious Service.

Lieut.-Colonel Singleton, K.H., had served with the 62nd in the Peninsula in 1811–12–13 and had been slightly wounded at the siege of the forts of Salamanca, and twice severely wounded at the battle of Salamanca. He had been granted £50 a year pension for wounds and created a Knight of Hanover for his distinguished services.

On March 2nd, 1846, two companies were placed, with other troops, in County Mayo, under command of Colonel Sir Charles O'Donnel during an election which caused some excitement. The troops were thanked by Government for their behaviour. Of the non-commissioned officers and men of the regiment, 376 were English, 147 Scottish, and 324 Irish. Nearly all were enlisted for unlimited service. The recruiting stations were Cambridge, Lincoln, Honiton, Brechin and Thornbury.

In April Headquarters and two companies were at Castlebar, five companies at Galway, and one each at Ballinasloe, Westport, and Oughterard, but before the end of the month Captain Geddes marched his company to Dublin from Oughterard as an advanced party. The regiment followed in May, but it only remained in Dublin for the drill season, and in August crossed to England, leaving two companies, and the paymaster, quartermaster and surgeon behind it in Dublin under Major Proctor.

Colonel Singleton had joined in March, but after signing the returns of April 1st he had gone on sick leave and on July 10th, 1846, Major and Brevet Lieut.-Colonel Ormond had been gazetted lieut.-colonel of the 30th, *vice* Singleton retired on full pay, and Captain Geddes had got the majority, both without purchase.

On September 1st Headquarters and three companies were at Newcastle-on-Tyne, one company at Carlisle, one at Leeds, one at Halifax, one at Bradford and two at Dublin. In October the two companies at Dublin crossed to England, one to Leeds and the other to Tynemouth. One of the Halifax companies marched in the same month to Sunderland.

In this month Surgeon Lawson was transferred to the 7th Hussars, and Captain Cavan rejoined after five years' service on the Staff in Jamaica.

In December Brevet-Major Andrews retired on full pay after thirty years' service, and Brevet-Major Gregory exchanged to half-pay with that veteran officer, Major Edward Antony Angelo, K.H., who had gone on half-pay when the regiment embarked for Bermuda. Major Angelo only came in to realize his money and sold his company to Lieutenant Henry Shum on December 26th.

By retiring on full pay Major Andrews gave a non-purchase company to the adjutant, Lieutenant Alexander Macdonald, who was succeeded as adjutant by Lieutenant Edward Augustus Whitmore; Quartermaster-Sergeant Timothy Morris was promoted ensign without purchase.

On January 1st, 1847, the distribution of the regiment was as in the previous year. The strength was 3 field officers, 10 captains, 20 subalterns, 5 staff, 47 sergeants, 15 drummers, 40 corporals, 729 privates.

In this month Captain Hoey joined the senior department of the Royal Military College. It was the forerunner of the present Staff College, and Captain Hoey was the second officer of the regiment to join. The first was the paymaster, R. C. Macdonald, who was there as a lieutenant.

Lieut.-Colonel Ormond retired on March 5th, 1847, and the junior major, James Gordon Geddes, was gazetted lieut.-colonel of the regiment, having purchased over the head of Major and Brevet Lieut.-Colonel John Proctor; Captain S. L. J. Nicoll purchased the majority. Lieutenant Robert Dring O'Grady got the company, and there being no ensign prepared to purchase, Thomas Henry Pakenham was brought in from the 59th as lieutenant by purchase. Captain Mauleverer was appointed to the light company in succession to Nicoll. On May 28th Captain H. Shum sold out and Quartermaster John Ward retired on staff pay of 8*s*. per day. He was the last of the Peninsular men and as a non-commissioned officer had served at the defence of Cadiz, in the lines of Torres Vedras, and at the battles of Sabugal and Fuentes Onoro, at the sieges of Ciudad Rodrigo and Badajoz, at the battle of Salamanca and at the action at Villa Muriel. He had the Peninsular Medal and three clasps.

In June Captain Alexander Macdonald, who had been adjutant of the regiment from 1834 to 1846, exchanged to half-pay. He subsequently became a staff officer of pensioners in New Zealand.

In July Major (Brevet Lieut.-Colonel) John Proctor sold out, Captain Cavan becoming junior major. There were several changes of quarters and in autumn the detachments were concentrated in larger bodies; at Newcastle, Headquarters and 3 companies, Carlisle 2, Tynemouth 2, and Sunderland 3.

On January 1st, 1848, the regiment was 57 under strength and had 43 recruits at drill. Recruiting was pretty brisk but so was desertion. Recruiting parties were at Cambridge, Yeovil, Longford, and Listowel.

On May 15th the regiment was ordered to be augmented. The new establishment was:—3 field officers, 10 captains, 20 subalterns, 5 staff, 57 sergeants, 21 drummers, 50 corporals, 950 privates. Recruiting parties were at Castleblaney, Roscommon, Hamilton, Hawick, Dursley, Leicester, Lincolnshire, Huntingdon, St. Ives, and Colchester.

In June the regiment travelled by rail to Manchester, detaching two companies to camp on Kersal Moor under Major Nicoll.

On August 4th the *Gazette* announced that Lieut.-Colonel Geddes had exchanged with Lieut.-Colonel Luard, who had sold out, Major Nicoll purchasing the lieut.-colonelcy and Captain Hoey the majority.

Lieut.-Colonel Samuel John Luke Nicoll was the first officer brought up in the 30th to get command since Colonel Powell died at the end of 1833, but, what was more important, he was in the full vigour of life, the long block in promotion was broken up and the field officers and captains were fit for the most arduous duties.

It was time that new men should come to the front, for the lethargy which had existed since the peace of 1815 was beginning to give way, and the nation began to take an interest in its army. From 1845 onwards there was a shower of orders which left little untouched and nearly all were, at least intended, to improve the condition of the soldier. Among the changes was the enlistment for ten years instead of for life, with the option of re-engaging to complete twenty-one years for pension, if approved.

In the autumn there was a good deal of marching of companies to and fro on account of threats of civil disturbance. Detachments visited Bury, Ashton-under-Lyne, Rochdale, and Hyde, but were always able to return in a few days without having been compelled to act.

In September the two companies from Kersal Moor marched to Stockport, where they were quartered.

At the beginning of 1849 the regiment was only seventeen short of the establishment, but there were nearly 200 recruits at drill.

In February orders were issued to reduce the establishment to 814 non-commissioned officers and men. Ninety-nine men were discharged in February and 73 in March, and the recruiting parties were drawn in. This had no sooner been done than orders were received to return to the former establishment. Recruiting parties were sent to Lincoln, Glasgow, Darlington, Durham, Clonmel and Athlone.

On June 1st Captain Heard went on half-pay. Lieutenant and Adjutant E. A. Whitmore got his company and Lieutenant Paget Bayly became adjutant.

By July 1st the regiment was only six short of the new strength and the recruiting parties were drawn in, except those at Lincoln and Darlington.

One company (Captain Sillery) was now at Birmingham and one (Captain Lowry) at Wolverhampton in addition to the two at Stockport. In October an order was issued that all barracks were to be lighted with gas, a welcome change from very poor candles.

In the beginning of 1850 the regiment was only one short of its strength and the Lincoln recruiting party was the only one out. In April the six Headquarter companies and the two at Birmingham and Wolverhampton moved to Walmer, and the two from Stockport to Canterbury; these companies (Captains Still and Rose) went on to Hythe on July 16th, and rejoined Headquarters at Walmer on October 17th. On the following day the regiment had the honour of being inspected by the Duke of Wellington who, as "Warden of the Cinque Ports," resided in Walmer Castle. Orders were now received for the regiment to prepare for foreign service and to be formed into six service and four depot companies; the order to take effect from January 1st, 1851.

The following formed the service companies :—

Lieut.-Colonel S. J. L. Nicoll, Major P. C. Cavan; Captains Charles Sillery, J. T. Mauleverer (light), J. B. Patullo, J. T. Still, C. D. Oliver, and F. H. Edwardes (grenadiers); Lieutenants A. W. Connolly, T. W. Cator, G. Le F. Dickson, F. T. Atcherley, W. R. Hepburn, J. M. Fitzpatrick, G. F. C. Pocock; Ensigns F. St. C. Hobson, J. D. Ross Lewin, W. H. Bennett, W. J. Brook and Alured Gibson; Paymaster R. C. Macdonald, Lieutenant and Adjutant Paget Bayly, Quartermaster T. Morris, Surgeon T. D'Arcy; 31 sergeants, 11 drums, 24 corporals, and 536 privates. Thirty-six women and 54 children accompanied Headquarters.

The depot (four companies) was composed as follows :—

Major Hoey; Captains E. A. Whitmore, J. Rose and T. H. Pakenham; Lieutenants John Tongue (paymaster), C. M. Green, J. O'Brien and F. Luxmore; Ensigns Mark Walker (adjutant) and G. F. La Touche; Lieutenant M. Pennefather and Ensign E. N. Falkner were on leave. The other ranks were 26 sergeants, 10 drummers, 26 corporals, 388 privates.

The service companies moved to Gosport in three divisions and on January 24th, 1851, two companies embarked on the hired ship *La Belle Alliance* and five days later the Headquarters and other four companies went on board the *Lady Flora*.

On February 1st they set sail for the Ionian Islands. Major Cavan was in command, Lieut.-Colonel Nicoll going on leave after superintending the embarkation.

At Corfu the two wings were transferred to H.M. Ships *Growler* and *Spiteful*, which landed them at Cephalonia on March 5th and 11th.

Cephalonia, one of the most southerly of the Ionian Islands, is about 32 miles long and from 5 to 12 broad. It is very mountainous—the highest point being about 5,000 feet above sea level. The chief towns are Argostoli and Luxuri. On landing, Captain Oliver's company marched to the latter town and detached Ensign Hobson with twenty-four non-commissioned officers and men to Fort George.

Lieut.-Colonel Nicoll joined and took command on April 16th.

In October a company was detached to Spartea in aid of the Civil Power, but returned in a few days without being called upon to act.

The detachment at Luxuri and Fort George was relieved every four months by one of the battalion companies.

Nothing of interest occurred during 1851 except the retirement of Captain R. W. Smith on half-pay, which brought Captain Robert Dillon into the regiment from the 97th Foot on July 25th.

The recruiting parties were at Peterborough, King's Lynn, Derby and

Bristol; the last was new ground, but the regiment enlisted many good soldiers there during the succeeding twenty years.

On January 2nd, 1852, the depot marched to Dover Castle from Walmer, and in February the establishment of the regiment was reduced to 47 sergeants, 16 drummers, 40 corporals, 810 privates. The depot sent a draft of 1 sergeant and 21 privates, under Lieutenant Mat. Pennefather, to Headquarters, and gave 121 privates to other corps.

On April 16th Lieut.-Colonel Nicoll went on leave preparatory to retiring, and on May 21st Major Philip Charles Cavan was gazetted lieut.-colonel of the regiment by purchase, Captain James Thomas Mauleverer got the majority and Lieutenant Arthur Wellesley Connolly succeeded Mauleverer as captain of the light company.

Captain Whitmore, who was on the strength of the depot, was appointed aide-de-camp to Lieut.-Colonel J. Ferguson, commanding at Malta, on May 4th.

On August 17th Paymaster Roderick Charles Macdonald died at Cephalonia. He had joined the regiment in 1827 from the 99th Foot, as a lieutenant, and in 1835 had succeeded Hay as paymaster.

On September 21st Assistant-Surgeon Augustus Purefoy Lockwood left on promotion after eleven years' service with the 30th.

On October 7th the regiment received the Horse Guards' order of September 22nd, ordering the troops to wear mourning, and conveying Her Majesty's deep grief at the loss sustained by the death of Field-Marshal the Duke of Wellington.

In January, 1853, the recruiting parties were at Derby, and in Cambridgeshire and Lincolnshire. The standard for recruits had been lowered to 5 feet 6 inches for men, and 5 feet 5½ inches for growing lads. Forty recruits were wanted to complete the battalion.

On February 24th, 1853, Headquarters and three companies left Cephalonia and landed in Gibraltar on April 5th. At the beginning of March the second division of three companies embarked on the freight ship *Poictiers* and landed at Gibraltar on April 10th.

In April Assistant-Surgeon W. J. Fyffe joined the depot in place of Assistant-Surgeon Macnamara, placed on half-pay on account of ill-health. In the same month Captain Cator exchanged to the 76th Foot, bringing into the regiment Captain and Brevet-Major Archibald Campbell, a Peninsular veteran. On May 19th the depot embarked on the steamship *Prussian Eagle*; it landed at Cork on the 22nd, whence it marched to Fermoy on the 24th. This is the first time any companies of the regiment had been on a steamship. The last time the regiment moved from Dover to Cork, in 1816, the voyage lasted thirty-three days. The depot was now armed with the Minie rifled musket which was being introduced in the army to take the place of the old smooth-bore musket. The officers with the depot were:—Major Hoey; Captains Rose, Pakenham, Dillon and Archibald Campbell; Lieutenants Tongue, Luxmore, Falkner and Bennett; Ensigns Mark Walker (adjutant), Hobbs and Macpherson, and Assistant-Surgeon Fyffe.

The recruiting parties were at Glasgow and Bristol, and in Cambridgeshire.

On October 25th Ensign Mark Walker embarked with one sergeant and thirty men on H.M.S. *Leopard*, a paddle steamer, to join the service companies. The draft disembarked at Gibraltar on November 2nd.

The regiment had now a paymaster, Mr. W. H. Fitzgerald having joined on June 25th. From September 1st Major Mauleverer was in command of the service companies, Lieut.-Colonel Cavan having gone home to sell out.

On December 16th Major Hoey was promoted lieut.-colonel and Captain J. B. Patullo major by purchase. Surgeon-Major Dowse had been appointed to the regiment *vice* D'Arcy. Dowse exchanged from the 13th Foot which was alongside the 30th in Gibraltar.

On February 2nd Captain C. D. Oliver died suddenly at Tangier.

On March 14th Lieut.-Colonel Hoey joined and took command.

The death of Captain Oliver had given Lieutenant and Adjutant Bayly his company, and Ensign Mark Walker the lieutenancy, and the latter was now appointed adjutant.

So far the movements of the regiment and changes among the officers had not been affected in any way by the anticipation of active service. The British Government persisted in assuring the world that there was no danger of war arising from the diplomatic difficulties with Russia, and the 30th was in full expectation of going on to the West Indies after a short stay at Gibraltar.

By the month of March, however, all hope of the preservation of peace was given up and the 30th was selected for active service.

CHAPTER XIII

THE CRIMEAN WAR.

1854–1856

THE following are the steps leading up to the war between Britain and Russia.

On July 2nd, 1853, the Russian army crossed the Pruth and occupied the Danubian principalities. On October 23rd the Sultan declared war against Russia.

On November 30th six Russian line-of-battle ships attacked seven Turkish frigates near Sinope and destroyed them.

On January 12th, 1854, the Czar was informed that his fleets must remain in their harbours or they would be attacked by the English and French fleets which had entered the Black Sea. Diplomatic relations with Russia were broken off on February 21st, and on March 28th England declared war. A declaration of war against Russia quickly followed from France.

On March 8th the 30th was warned for service, and the number of the service companies was ordered to be raised to eight by bringing two from the depot, where two new companies were to be formed. The strength from April 1st was to be :—Eight service companies :—3 field officers, 8 captains, 14 subalterns, 6 staff, 47 sergeants, 15 drummers, 40 corporals, 810 privates. Four depot companies :—4 captains, 8 subalterns, 20 sergeants, 8 drummers, 20 corporals, 380 privates.

To bring the service companies up to strength the 13th Light Infantry, the 39th and the Gordon Highlanders were called upon to volunteer for the 30th, a bounty of a guinea a man being offered. The 13th furnished 23 privates, the 39th 2 corporals and 39 privates, and the 92nd 2 sergeants and 100 privates. A few volunteers also joined from other corps, making a total of 2 sergeants, 4 corporals, 170 privates.

In April the General commanding at Gibraltar received orders to send the six companies at Gibraltar to the East as soon as complete, without waiting for the two summoned from the depot. The Governor and Commander-in-Chief in Gibraltar was Lieut.-General Sir R. W. Gardner, Royal Artillery, a veteran soldier who had served in the Peninsula and at Waterloo. He ordered the regiment to embark on May 1st on board the s.s. *Cambria*, and added :—" The 30th have not formed part of this garrison for so long a period as the regiments with whom they are about to be brigaded, but they have been long enough on the Rock to have gained the Governor's admiration of their order, discipline, subordination, and efficiency, and the glorious names on their colours record past days in which he had the honour to serve with them."

At half-past two p.m. on May 1st the embarkation was completed and the *Cambria* steamed out of the bay with the six companies on board.

Strength:—

OFFICERS.

Lieut.-Colonel W. F. Hoey.
Major James Thomas Mauleverer.
,, James Brodie Patullo.
Capt. (Bt.-Major) Charles Sillery.
,, Arthur Wellesley Connolly.
,, Graham Le Febre Dickson.
,, Francis Topping Atcherley.
,, Paget Bayly.
Lieut. Matthew Pennefather.
,, George Francis Coventry Pocock.
,, Charles Mingay Greene.
,, John O'Brien.
,, William James Brook.
,, Alured Gibson.
,, Augustus Henry Williamson.
Ensign John Dillon Ross Lewin.
,, Charles Jocelyn Cecil Sillery.
Paymaster William Henry Fitzgerald.
Adjutant Mark Walker.
Quartermaster Timothy Morris.
Surgeon Richard Robert Dowse.
Assist.-Surgeon William Percy Pickard Mackesy.

NON-COM. OFFICERS, DRUMMERS, RANK AND FILE.

Staff Sergeants.
Acting Sergt.-Major, Colour-Sergt. John Forbes.
Acting Quartermaster, Sergeant John Moon.
Paymaster's Clerk, John Hilder.
Armourer, William Meagher.
Drum-Major Thomas Gunning.
Hosp. Sergt. James Holmes.
Acting O.R. Clerk Thomas Jackson.
29 Sergeants.
12 Drummers.
32 Corporals.
612 Privates.

No. 1 company embarked under Lieutenant Mat. Pennefather in the absence of Captain Whitmore, who was now A.D.C. to Lieut.-General Sir George Brown, commanding the Light Division; Regimental Sergeant-Major Alexander Moncur, a soldier of twenty-two years' service, was invalided on April 30th, and Quartermaster-Sergeant John Prendergast was transferred to the Monmouth Militia. Moon was appointed quartermaster-sergeant on July 1st, *vice* Prendergast, and Sergeant Thomas Jackson became orderly room clerk. Sergeant-Major Moncur was not finally discharged till November 29th, and till then one of the colour-sergeants acted as sergeant-major.

No. 3 company, Captain Pakenham, and No. 7 company, Captain Rose, were at Cork awaiting transport to join the service companies.

The *Cambria* coaled at Malta on the 5th, and started in the afternoon of the 6th with a sailing ship, which carried some artillery, in tow. At 5.30 a.m. on the 10th she reached Gallipoli—a filthy place where some regiments landed and suffered severely from sickness. The 30th had the good luck to be sent on at once to Scutari, which it reached in the forenoon of the 11th. At two o'clock on the afternoon of the 12th it disembarked, and was quartered in Turkish barracks on the plateau above the dirty town. The troops in barracks were at first known as the "flea-bitten brigade," for obvious reasons.

The regiment was very healthy in the well-situated barracks at Scutari;

later in the war these barracks were used by the British as their base hospital and will be connected for all time with the name of Florence Nightingale.

The regiment was now in the Second Division commanded by Major-General Sir De Lacy Evans.

The 1st brigade, commanded by Brig.-General John Lysaght Pennefather, consisted of the 30th, 55th, and 95th; Captain Thackwell of the 22nd was brigade major. The 2nd, commanded by Brig.-General H. W. Adams, had the 41st, 47th, and 49th.

The 55th landed on the 19th, and on the 22nd Ensigns Wm. Young Johnston, John Simon Chandos Harcourt, and John Pennock Campbell joined from the depot. Campbell was a son of the veteran captain of No. 2 company and had followed his father to the 30th from the 97th.

Nos. 3 and 7 companies joined on the 24th from the depot, under Captains James Rose and Thomas Henry Pakenham, with Lieutenants Frederick Luxmore, Edward Newsted Falkner and Lachlan Macpherson, Ensign James Cavendish Hobbs, and Assist.-Surgeon William Johnstone Fyffe; 4 sergeants, 6 corporals, 194 privates. They had embarked at Cork on the 8th on the s.s. *Trent.*

The companies were now equalized and stood as follows:—

GRENADIERS.—Captain Dickson, Lieutenant Brook (left for Bulgaria on the assistant quartermaster-general's staff on May 27th), Lieutenant Macpherson, 113 other ranks.

No. 1. Captain Edward Augustus Whitmore (A.D.C. to Sir George Brown, commanding the light division), Lieutenant Pennefather, Ensign Campbell, 114 other ranks.

2. Captain (Brevet-Major) Archibald Campbell's company, at depot.

3. Captain Pakenham, Lieutenant Luxmore, Ensign Hobbs, 112 other ranks.

4. Captain Atcherley, Lieutenant Gibson, Ensign Ross Lewin, 113 other ranks.

5. Captain Robert Dillon's company, at depot.

6. Captain (Brevet-Major) Sillery, Lieutenant O'Brien, Ensign Sillery, 114 other ranks.

7. Captain Rose, Lieutenant Falkner, Ensigns John Charles Newcombe Stevenson (on way to join), and Johnston, 110 other ranks.

8. Captain Bayly, Lieutenant Williamson, Ensign Harcourt, 112 other ranks.

LIGHT.—Captain Connolly, Lieutenants Pocock and Green, 114 other ranks.

The total strength of the regiment in the East was 3 field officers, 8 captains, 17 subalterns, 6 staff, 39 sergeants, 15 drummers, 39 corporals, 817 privates.

On June 2nd the regiment moved into camp and on the 3rd Minie rifles were issued to it and the old smooth-bore percussion muskets which it had received in Cork in 1844 were returned to store. On the 5th the first practice with the Minie took place. Some of the men who had recently arrived from the depot must have handled it already.

On the 12th Ensign J. C. N. Stevenson and 14 other ranks joined from the depot.

The allied armies were at this time being transferred by divisions to Varna in Bulgaria, in support of the Turkish army.

On the 16th the 30th embarked on the steam troopship *Simoom*, which entered the Black Sea on the following day and reached Varna on the 18th. On the 19th the regiment disembarked and encamped a mile beyond the town, being played up to the camping ground by a French band.

On leaving Scutari the men's shell jackets and all other superfluous kit was packed in the squad bags and left in store; little of it was ever seen again.

Each man carried in the pack one shirt, one pair of boots, one pair of socks. The great-coat, blanket and smock frock were carried on the knapsack. This is the first mention of the smock frock, the best fatigue dress the army ever had. In hot weather it was a capital dress for musketry, and in cold weather it could be worn over the jacket. A clean smock frock with a black neckerchief tied in a sailor's knot made a smart enough dress for company parades.

Some sick and a few men on command were left with Lieutenant Williamson at the army depot formed at Scutari on June 8th under Brevet-Major Sillery. Most of the soldiers' wives also remained with the depot. They were granted full rations from the day they were separated from their husbands. From small beginnings this depot grew during the war into a very important command. Major Sillery became a regimental major from September 30th, but never rejoined the regiment. On September 9th, 1855, he was promoted a lieut.-colonel in the army and appointed an A.A.-General at Scutari. On the termination of the war he was given a depot battalion.

Lieutenant O'Brien was placed in permanent command of Major Sillery's company.

Hospital-Sergeant James Holmes was among the sick left behind. He never rejoined, but on recovery was employed in the base hospital. In December he was appointed purveyors' clerk and during the war rose to be a purveyor.

At Varna the troops could hear the sound of the guns at Silistria, where the Turks, helped by a few British officers, were besieged by the Russians. On the morning of June 23rd there was silence and it was believed that the place had fallen, but to the surprise of Europe it turned out that the Turks had made good their defence and that the Russians had retreated.

The allowance of bell tents was, as of old, 2 for C.O., 1 each for other F.O.'s, 1 for each captain, 1 for every two subalterns, 1 each for adjutant, quartermaster, and surgeon, 2 for the three assist.-surgeons, 1 for staff-sergeant, 1 for orderly room, 1 store, 1 for every 14 N.C.O.'s and men.

Forage was allowed for 4 private horses for the C.O., 3 for a major, 2 for a captain, 1 for a subaltern, and 1 to each of the other officers. Two public bat horses were allowed for each company and one each for medicine chest, paymaster's books, entrenching tools, armourer's tools, and one for the three tents of O.R., store and staff sergeants.

There were frequent parades at Varna, both to practise loading the bat horses and for manœuvres.

The heat during the day was very great, the thermometer sometimes reaching 108 degrees in the tents, but the nights were cold, and wet days and nights were not uncommon. The part of Bulgaria in which the 30th was serving is very beautiful. Much of it is wooded like an English park with a number of lakes, which were made full use of for bathing by all ranks, and the health of the regiment was very good, the daily average of sick being sixteen.

On July 3rd the division marched to Karagauli, and on the 5th moved on to Devna, a lovely place, where it set the example to the army of erecting arbours of boughs and covering the tents with them to lower the temperature. Some cases of cholera had appeared in the army, but so far the 30th was free from it.

On July 10th Lieutenant Pennefather left the service companies on promotion to a company at the depot. There were other promotions about this time in consequence of the augmentation ordered on May 10th. Lieutenant Pocock got his company on July 4th. Ensigns Ross Lewin and Stevenson became lieutenants on June 6th. Ensign Hobbs got a step on August 11th, and Acting Sergeant-Major John Forbes became ensign on August 10th, when the duties of sergeant-major were taken up by Colour-Sergeant John Thompson.

During the long peace, the art of supplying an army had been forgotten. The heads of the commissariat were able and energetic, but they had no experienced assistants. Pork and biscuit was the usual ration; fresh meat was sometimes issued, but was usually uneatable by the time it reached the troops, and there were no vegetables. The pay, too, was issued in sovereigns to the captains, who could not change them. The only way they had of paying their companies was to give a sovereign between a number of men, who had to divide it among themselves. Of necessity the sovereign passed to the sutler, who gave in exchange partly goods and partly tin tokens available at his shop only. Worst of all was the inadequate supply of medicines and medical comforts. In an army cantoned close to the sea, and with no enemy near, things should have been better.

The army contrived to enjoy itself, however, and its health was still very good. In the 30th the average daily sick in July was 18.

There was a great deal of cricket. General Sir Robert Hume, who served in 1854 as a lieutenant in the 55th, used to be fond of telling how, of the second division eleven in which he played, all but two were killed or wounded during the war.

Devna races were held on the 11th of the month, and it was probably from what Sir George Brown saw there that the orders to shave were rigorously enforced, and all infantry men who had begun to grow moustaches had to shave them off. Lord Raglan did not care for these things, but to Sir George a moustache was an abomination. He had a saying, which the sight of a pair of long whiskers always called forth, "Where there's hair, there's dirt; and where there's dirt, there's disease." On August 1st the regiment changed camp to Loombay and on the following day to Boerkascesse. There was now a very great change for the worse in the health of the army. Cholera was bad among the British troops and three French divisions which had been sent into the Dobrudja had been almost destroyed by it. The whole of the French and English armies were now echeloned along the road from Varna to Shumla, where the Turkish army was assembled under Omar Pasha.

On August 8th the first case of cholera appeared in the regiment and in the succeeding fortnight eight men died of that disease, after which it disappeared for a time entirely. The 30th were very lucky, for 532 men had died in the army, and in the navy it was worse, H.M.S. *Britannia* losing one man in seven of her crew.

Apart from the actual losses by death and invaliding, the strength of the soldiers was so lowered by dysentery and bowel complaints that it was doubtful if the army was fit for the expedition to the Crimea which had been planned by the Allied Governments.

On August 9th the order to shave was abolished and all branches of the service were allowed to wear moustache, whiskers and beard.

After inspections by generals commanding divisions, it was decided that the army could face a campaign and that the expedition should start for the Crimea as soon as the health of the navy would allow. All divisions now began to close in towards Varna. In most cases the men were so feeble that their packs were carried for them.

On August 28th the regiment marched to Loombay, on the 29th to Karagauli, and on the 30th to Varna. Colonel Hoey, who was seized with dysentery at Karagauli, was sent on board ship at once. Ensign Stevenson had been sent dangerously ill to Varna for embarkation ten days before, and was now on his way home.

On the recommendation of the medical authorities, a half-ration of rum was now issued to the men and the meat ration was increased by half a pound when fresh meat was obtainable. An excellent order had been published in June, directing that each man should have a cup of coffee before parade in the morning, and another in the evening, and the stoppage for groceries had been raised in consequence from 3½*d.* to 4½*d.*, but it was not always possible to provide the coffee.

In August the average daily sick in the 30th had risen to 68, chiefly from cholera and bowel complaints. The number of deaths since arrival in the East was 17. The second division was exceptionally healthy, owing to the care Sir De Lacy Evans bestowed on the camping grounds.

On August 31st the regiment embarked at Varna Bay on the s.s. *Vulcan.* The strength was as follows :—

Lieut.-Colonel Hoey (sick) ; Majors Mauleverer and Patullo ; Captains Rose, Pakenham, Connolly, Dickson, Atcherley and Bayly ; Lieutenants Green, O'Brien, Brook, Luxmore, Gibson, Williamson, Macpherson, Ross Lewin, Sillery, Hobbs ; Ensigns Harcourt, Campbell, Johnstone ; Paymaster Fitzgerald ; Lieutenant and Adjutant Mark Walker ; Quartermaster Morris ; Surgeon Dowse ; Assistant-Surgeons Mackesy, Fyffe and David Milroy ; Acting Sergeant-Major Thompson ; Quartermaster-Sergeant Moon ; Armourer Sergeant Meagher ; Drum-Major Gunning ; O.R. Clerk Jackson ; 39 sergeants, 18 drummers, 32 corporals and 657 privates.

The following were left at Varna :—Captain Pocock, left in charge of weakly men, Ensign J. Forbes, in charge of regimental transport, Paymaster's Clerk John Hilder, in charge of the regimental pay and record office. One hundred and twenty-three non-commissioned officers and men were left at Varna and Scutari, and about 35 had died in Bulgaria. All bat horses were left at Varna. Acting Sergeant-Major John Thompson had already been recommended for promotion and he became ensign three weeks later, but no official notice of his promotion was received and he continued to act as Sergt.-Major till mortally wounded at Inkerman.

While in Bulgaria the 30th had a great deal of pleasant companionship with the French soldiers, particularly with the Zouaves, from whom our buglers picked up a number of first-rate marching tunes which they played for many years after. It was then a rare thing for a British regiment to march to the bugle.

The *Vulcan* lay in Varna Bay till September 5th, when it steamed to Balchik, the general rendezvous for the British part of the expedition.

On the 7th the fleet got under way, the *Vulcan* towing two transports, one with the 55th on board and another carrying the Royal Engineers and their stores. The real object of the allies was Sebastopol, but a feint was made of attacking Odessa. For this purpose the fleet steamed north, and anchored on the night of the 9th about 50 miles from that town, and an equal distance from Cape Tarkan, the north-west point of the Crimea.

On the 11th a course was steered for the Crimea. Eupatoria was given as a rendezvous, and it was occupied by a small detachment of troops on the 13th.

At daybreak next morning the transports were in Kalamita Bay, in which Lord Raglan had selected a landing place where a strip of low land, about 200 yards in width, interposed between the sea and Lake Touzla, a piece of water about a mile in length from left to right. A ruined work on the right gave the place its name of "Old Fort," by which it was ever afterwards known in the army. The orders for the disembarkation, which were based upon those of Sir Ralph Abercrombie when he forced a landing in Egypt in 1801, had been issued on the 12th instant. All knapsacks were to be left behind, and each man was to land with fifty rounds of ammunition, a great-coat, blanket, forage cap, a pair of socks, a pair of boots, and three days' biscuit and pork. The officers carried the same as the men, with the exception of the ammunition. All were equipped with the heavy and detested shako, with white cotton cover hanging down over the ears and shoulders as a protection from the sun, a haversack on the right side and the same clumsy water keg which had been carried in the Peninsula (and it is believed in Marlborough's campaigns) on the left. Full dress was worn, epaulettes and all. Needless to say, all wore the high stiff leathern stock. Most of the officers carried revolvers.

The 30th commenced disembarking in ships' boats towed by steam launches about one o'clock in the afternoon, and by three o'clock it and the whole of the light, first and second divisions were on shore and formed about 3 miles from the landing place. The French army, under Marshal St. Arnaud, landed on our right. The men were kept under arms till nightfall, by which time a chill breeze with rain sprang up and the army passed a miserable night in bivouac. Lieut.-Colonel Hoey, though not well, was considerably better, and landed, but Lieutenants Brook and Sillery, Quartermaster Morris and sixteen men had to be left on board sick.

The 15th was a bright warm day, and the men got their clothes dried, but the cold and wet of the previous night caused some sickness, and cholera reappeared. The night of the 15th, too, was very cold though dry, and Lord Raglan, seeing that the landing of horses, artillery, and stores would take a few days longer, ordered the tents to be landed, although he had no transport for them. The 30th got theirs pitched on the night of the 16th. During the five days the army was in the neighbourhood of Old Fort, the men in their weak state felt severely the fatigue of carrying water, which had to be drawn and fetched from wells two miles off.

By the 18th the cavalry and artillery and a sufficient reserve of ammunition had been landed. The tents were, therefore, sent on board ship and the army bivouacked ready for an advance. On the 19th the division fell in at six o'clock in the morning and the army advanced southwards. The French were on the right, near the coast; next to them was the second division in

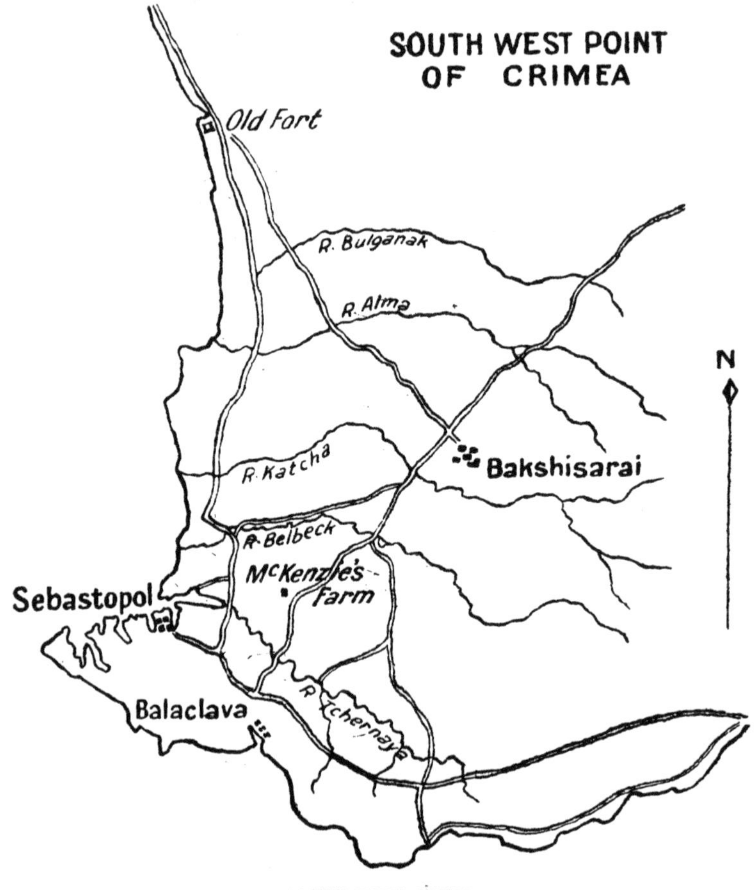

SOUTH-WEST CRIMEA.

column, and at deploying distance on its left the light division; the other divisions were in support. The 11th and 13th Hussars and a battalion of Rifles covered the front.

The sun shone brightly, the colours were uncased, the bands played, and for a time the brilliant spectacle had an exhilarating effect on all. The colours of the 30th were carried by Ensigns Harcourt and Johnston. The band was under Sergeant Horatio Bannen, afterwards quartermaster-sergeant. Drum-Major Thomas Gunning, afterwards assistant sergeant-major, led the band and drums. The 30th, as senior regiment of the 1st brigade, led the second division, in column, and was on the right, close to the French when in line.

About four o'clock in the afternoon, after a hot march, the army forded the small stream of Bulganac, where the enemy showed a strong force of cavalry and there was a slight skirmish with our cavalry and horse artillery. Pennefather's brigade formed line and bivouacked in that formation, the men lying in their accoutrements round the piled arms, a couple of hundred yards in front of the stream. The watch fires of the Russians were seen to the south. Many of all regiments had fallen out on the march; some merely from weakness, but many from cholera.

On the 20th the division fell in without sound of drum or bugle, and at 6 a.m. corrected its distance from Prince Napoleon's, the left division of the French army.

It was evident to Lord Raglan that he would meet the enemy in force during the day, and at seven o'clock arms were piled again, and the men ate their rations of pork and biscuit, breakfast and dinner in one. At 11 a.m. the advance was resumed. By this time the allied fleet had sighted the Russians on the hills south of the Alma River, and the troops could hear our men-of-war firing from time to time. At 1.30 p.m. the men of the 30th saw the enemy on the heights in their front, about a mile and a half off and beyond the river, and the division halted in order to deploy.

So far, the advance might have been that of Wellington's army in Spain forty years before, except for one marked difference. Although at Varna orders had been issued for the formation of "Flank Battalions," such as the Great Duke had used in the Peninsula, they were not employed on this day. Grenadiers and light companies, although composed, as of old, of picked men, now moved in column or line with the others as battalion companies. There was indeed a thin screen of riflemen, but the Duke's skirmishing line would have been the flank battalions, and the riflemen to boot, and both would have belonged to the divisions they covered. As a matter of fact, the Rifles bore away to their left and never covered the second division after the commencement of the action. Another doubtful change was now made from the old tactics. The second and light divisions which were to lead the attack deployed all their brigades and regiments, and neither division had any support or second line of its own, but had to depend on others—the second division on the third, and the light division on the first. As it turned out, the two front divisions had not room to deploy; and when the formation was completed, the right regiment of the light division and half of another stood in line behind the left half of Pennefather's brigade, which now formed the left of the second division, for Sir De Lacy Evans had brought Adams' brigade up on the right of Pennefather.

The Russian force under Prince Mentschikoff which proposed to dispute the passage of the Alma consisted of about 3,600 cavalry and over 35,000 infantry, in 44 battalions, with 96 guns. The British force under Lord Raglan numbered 1,000 sabres, 26,000 bayonets, and 60 guns. The French, under Marshal St. Arnaud, numbered 28,000 bayonets and 68 guns, and the 7,000 Turks with the army were under the orders of the Marshal. The French, too, were supported by the fire of the fleet. The preponderance of strength on the side of the allies was immense, but as things turned out, the real battle was between the British army of 27,000 men and 60 guns, and the right and centre of the Russian army numbering 20,000 men and 86 guns, and occupying a position selected and prepared beforehand.

The special task of the second division was to cross the river and assault the Russian centre, which was astride of the paved road from Eupatoria to Sebastopol, which follows the main pass through the hills to the south. The division had, during the advance, kept this road on its left and in the attack all except the left regiment would have to ford the river. The left regiment might use the bridge where the road crossed, if the enemy did not destroy it.

When deployed, the men of the 30th saw in front of them a gentle descent without cover, till the river was approached, but there the long straggling village of Bourliouk standing amidst vineyards fenced by low stone walls, occupied a space nearly equal in breadth to the ground covered by the brigade from right to left. Beyond the village the Alma flowed with many windings and deep pools. On the further side of the river the ground rises abruptly, the bank being 8 or 10 feet high. The hills through which the high road ascends to the south of the river are broken with many ravines and rocky knolls but have few enclosures. Their utmost height is about 400 feet. One natural feature was of great advantage to the defence; a line of small mounds about 600 yards from the river, and parallel to it, served all the purposes of an entrenchment.

The Russian force detailed to defend the great pass, and thus opposed to the second division, was four battalions of the Borodino regiment of light infantry, a part of the 6th Rifles, a battalion of sappers and sixteen field guns. The guns were placed on each side of the great road behind the line of mounds already spoken of, the sappers were at the bridge ready to destroy it and to fire the adjacent village of Bourliouk, the Rifles and two battalions of the light infantry formed a skirmishing line along the river with advanced parties in the vineyards on the northern bank and their supports in column of companies in rear. The other two battalions of the Borodino light infantry were in reserve near the guns. The general reserve of the whole Russian army, consisting of seven battalions of infantry and two batteries, was about a mile to the left rear of the batteries which the second division was about to attack.

After deploying, our line advanced, but on coming within range of the Russian artillery halted again and lay down to await the result of the advance of General Bosquet who, as arranged by Marshal St. Arnaud and Lord Raglan, was to cross the river near its mouth with 14,000 French troops, ascend the hills there, supported by the fire of the fleet, and thus turn the left of the enemy's position.

While the line was lying down our two divisional batteries of 9-pounder guns attempted to reply to the Russian fire, but the Russian guns on the hill had such a command over ours that they could throw their 6-pound shot right over our front line and our heavier guns could not reach them. The 30th had some casualties here, and Mr. Kinglake, the historian of the war, who with Lord Raglan and the Staff was immediately in rear of the regiment, praises the unconcerned behaviour of the men and the quiet way in which, when a man was struck, his comrades on either side would rise, carry him to the rear and resume their places in the ranks.

The French had met with difficulties in their advance, especially from the ruggedness of the ground which prevented their artillery from keeping up with their divisions, and after our line had been lying under fire for an hour and a half, Lord Raglan saw that nothing would come of their turning movement; in fact our allies were asking him for support. He, therefore, gave the order for his army to advance to the attack. Major Lysons of the 23rd Fusiliers, who brought the order to the second division, says in his account of the campaign, "I shall never forget the excited look of delight on every face when I repeated the order—the line will advance."

Sir De Lacy Evans had the markers out and dressed the line and then gave the word to move on.

The order of the division from right to left was, 41st, 49th, 47th, 30th, 95th, 55th; the two batteries (Turner's and Franklin's) under Lieut.-Colonel Fitzmayer were on the right. At one period of the advance the 41st and 49th were in second line, owing to want of room. As soon as the Rifles in front reached the vineyards around Bourliouk, the village sprang into flames. There were haystacks of coarse grass at every farm which burned fiercely. No man could pass through the greater part of it, but luckily some of the houses and enclosures at the east end and near the high road were untouched, and the fire must have failed in other parts near the centre, for a few men of the division managed to get through there. Sir De Lacy Evans broke up his division. He sent Major-General Adams with the 41st and 49th and Turner's battery to try for a ford to the right below the village while he with General Pennefather and the other four regiments, sought to find a way to the river, keeping the burning houses on their right. The regiments moved to their left in fours until the leading regiment, the 55th, had crossed the post road. During this movement the 95th, which was following the 55th, was by some obstacle forced out of the front line and marched in second line parallel to the 55th. This movement to the left encroached still further on the ground allotted to the light division whose right regiment, the 7th Fusiliers, found its front crossed by the 55th and 95th. While taking ground to the left, the light companies were extended to cover their own regiments.

As soon as the burning village was passed, Sir De Lacy Evans caused his four regiments to turn to their front and advance. The Russians had now the exact range and the fire was severe. The ground was too cramped for an advance in line and was intersected with fences, so, although there was no hesitation, there was a good deal of confusion as the different regiments crowded on a narrow front made their way forward, running from one enclosure to another. When far enough forward to clear the village, Sir De Lacy Evans inclined again to his right so as to reoccupy the place in the line of battle assigned to his division and to get into touch again with the 41st and 49th regiments and through them with the French army. This movement brought him to the river with his three leading regiments, the 47th, 30th and 55th, all in their proper places in line to the right of the post road. He had thus marched round the burning village; the 95th, finding its front clear by the movement of the 55th to the right, advanced up the road, but the light division advancing simultaneously with its right regiments on the same road carried the 95th along with it in its advance. The following account of the share of the 30th in Sir De Lacy Evans' movement of the four regiments round the burning village to gain the river is from the diary of Lieutenant Mark Walker, the adjutant (afterwards General Sir Mark Walker, V.C., K.C.B.):—"The men were ordered to shelter behind a wall, we, the mounted officers, sat on our horses in rear and any moment I expected one of us would be knocked over. The artillery came up behind and opened. They suffered considerably. We (here my shako was knocked off by the wind of a round shot) were then ordered to move across a small green field; in going over it many were knocked over, including Pakenham severely wounded and Luxmore killed; his servant fell with him. We then got into a vineyard on the banks of the river, which was deep and its sides steep. Here we were ordered to shelter for a little; many were wounded, many killed, some drowned. Here Dickson was wounded and my horse hit severely."

The 47th, 30th and 55th, on reaching the river, lined the bank and opened a steady fire on the Russian infantry in front and on the guns 600 yards off on each side of the post road. The light companies had been ordered during the advance to join their battalions, but the light company of the 30th under Captain Connolly found itself, when closed, not in its proper place on the left of its own battalion, but on the left of the 55th and touching their light company. Brevet-Major Rose, captain of the light company of the 55th, had just been mortally wounded and it was now commanded by Lieutenant Robert Hume, afterwards General Sir Robert Hume, K.C.B., always a warm friend to the 30th. The friendship between the regiments was so strong that they might almost be considered one corps, and in answer to Connolly's entreaty that he might be allowed to fight with the 55th, Lieutenant Hume most chivalrously told his men to take the word from Captain Connolly and the two companies advanced as one to the river. The Rifles had long since gone off to the left.

After being a short time in action at the river, Sir De Lacy Evans was strengthened by two more batteries, one from the light and one from the second division, but the enemy's guns in front were so well placed and so well served, that our people did no more than hold their own.

In a brief space of time all was altered by Lord Raglan's action and Sir De Lacy Evans and General Pennefather could advance to an assured victory.

As soon as Lord Raglan had seen the advance well begun, he had ridden to his right front, looking for some place whence he could see and control his army and at the same time be within easy reach of his colleague, Marshal St. Arnaud. He crossed the river at the ford below Bourliouk to which Sir De Lacy Evans had sent General Adams with the 41st, 49th, and Turner's battery.

Passing them, he rode up the hill and presently found himself looking into the left and rear of the guns and the troops with which Pennefather's brigade was in action. His Lordship at once sent back for Turner's battery and the two regiments to come up with all speed. Two guns were quickly in action and the third shot blew up a Russian ammunition waggon. The Russians, finding themselves enfiladed by a battery supported by infantry to their left and rear and having suffered a good deal from our rifle fire at 600 yards, limbered up and retired further up the hill, some of the pieces being dragged by the men of the Borodino regiment. Sir De Lacy Evans, an experienced soldier, at the first sign of the slackening of the Russian fire, gave the order to advance. Colonel Hoey called for the mounted officers, the colours and covering sergeants (markers); the men sprang up the bank and formed upon them and the regiment advanced in a line that was admired by all.

The line had been dressed by the centre by Major Mauleverer, his friend

Major Patullo, of course, marking the left, and before resuming their places in rear Mauleverer was heard to ask Patullo for a light for a cigar, at the same time lamenting the loss of his Irish pipe. The Russian guns at that time had not all ceased firing and the fire of their skirmishers was still heavy. When General Evans advanced, his was the only force across the river in action with the enemy, except the 7th Fusiliers. The light division had advanced very gallantly and captured what Kinglake calls the Great Redoubt, but not being supported had been driven out again. In falling back, a number of the men encountered the centre regiment of the Guards brigade advancing too late to their support and carried it away with them. The 7th Fusiliers on the left of Pennefather's brigade did not share in this retreat, but held its ground splendidly though attacked by greatly superior numbers. It was relieved by the advance of Pennefather's regiments, for the 55th, bringing up their right shoulders, fired, and by Pennefather's order prepared to charge, and the Russians gave way. The Guards halted for a moment to let their centre regiment re-form and then advanced on the left of Pennefather with Sir Colin Campbell and the Highland brigade on their left; there was some hard fighting and the Guards lost heavily, but as the Russian cavalry, 3,000 strong, did not for some reason take any part in the action, the issue was never in doubt.

The second division in its advance had been joined by Lieut.-General Sir Richard England commanding the third, its supporting division, who brought his two batteries with him, thus giving Sir De Lacy Evans command of thirty guns; the 41st and the 49th regiments with Turner's battery also came in from the right and the infantry of the third division was close in rear. The advance was bravely opposed but the fire was nothing like what it had been. Lieutenant Mark Walker, the adjutant, was hit at this time by a spent grape shot in the chest which nearly knocked him over, but he was able to keep up. The Russian infantry reserves did not wait for our men to come to close quarters, but retired from the field, driven by the fire of our artillery. The day was over and the regiments of the second division broke into column as they advanced to the top of the hill where Lord Raglan was greeted with tremendous cheers by his army.

Lord Raglan had a division of 1,000 cavalry which he was determined to keep intact, for the French had none, but it was hoped that as a great part of the French army had not been seriously engaged, Marshal St. Arnaud would pursue the Russians with a couple at least of infantry divisions, but he refused. He said his men had left their knapsacks at the river and must go back to fetch them. His real reason was that he had not landed enough ammunition and his batteries were consequently useless. There was a deeper cause still for the want of firmness which marked the French operations at that time. St. Arnaud was a dying man. Six days later his sufferings obliged him to resign and on the 29th he died on board ship. He was succeeded in command by General Canrobert.

The 30th bivouacked with the division on the hill-top where the Russian columns had been seen in the morning. The strength of the regiment at the Alma was 26 officers and 684 other ranks, and the casualties were 81.

The following are the names of those killed :—

One officer, Lieutenant Frederick Luxmore; 12 rank and file—Corporal Robert Emery (Rugeley), Privates Alexander Battie (Tipperary), Robert Bell,

Henry Chilvers (Linn), Michael Foley (Doneraile), Joseph Henshaw (Halifax), Robert Jackson (Finglass), Donald McInnes (Greenock), Thomas McNully, George Michie, Michael Reddy (Birr), John Vokes (Limerick). Died of wounds: 4 privates—John Chamberlain, Alex. Smith, Thomas Isherwood, William Luton. Wounded: 4 officers, Captains Pakenham and Dickson severely, Captain Conolly and Lieutenant Walker slightly. Two sergeants—Nicholas Day and Dominick Lydon (son of Sergeant Luke Lydon who fought in the Peninsula, and father of Regimental Sergeant-Major Lydon, distinguished in the South African War), 58 rank and file.

The total loss in the British army was 111 officers and 2,000 other ranks killed and wounded. The French loss was probably from 550 to 600. They had 3 officers killed. The Russians acknowledged a loss of 5,700, including 5 generals and 193 other officers.

There were as usual many men hit who had their wounds dressed and remained at duty; they are not included in the number of wounded. The record of their names was lost when the regimental hospital was abolished ten years after the war. Among them was Private James Mathewson, the future orderly room clerk, and father of the sergeant-major of the regiment in the Boer War. Private Mathewson was twice struck by musket balls.

In addition to the 3 officers and 16 men left sick on board ship, Captain Paget Bayly and 46 other ranks had been embarked sick during the advance. In the hurry the men were put on the nearest ship, perhaps a French one, without any proper record. If it was a cholera case the man might die and his body be thrown overboard before the ship reached any port and so all trace of him be lost.

Lieut.-Colonel Hoey was mentioned in Lord Raglan's despatch for his good service in the battle and the following appeared in divisional orders by Lieut.-General Sir De Lacy Evans:—

"The Lieut.-General has the satisfaction to publish in orders the following names of officers, non-commissioned officers and privates of the 30th regiment, reported by the commanding officer, whose gallant conduct came under his especial notice at the battle of Alma. Majors Mauleverer and Patullo; Lieut. and Adjutant Mark Walker; Colour-Sergeants John Thompson, John McLellan, William Barnes, Cornelius John Willey; Sergeants William Jameson, John McDonald, David Drake; Corporals John Johnson, Charles Dillon, John Green; Privates James Cree, Hughes, Francis McLoughlin, William Nichol, Thomas Carr, Frederick Coombes, Clarke, Charles Stickwood, Morgan, Thomas Fennell, Tyler, Thomas Curran, Richardson, Grant, John Butterworth, Byng, McDonald, Willard, David Laing, William Hale, Richard Rhodes."

Colour-Sergeant Willey mentioned above of No. 3, Captain Pakenham's company, died on the 23rd, probably of cholera which claimed hundreds of victims at this time.

On the 21st, Marshal St. Arnaud refused to carry out the original scheme of a rapid advance and an attack on the forts on the north side of Sebastopol, and the army halted till fresh plans could be made. The 21st and 22nd were spent in removing the sick and wounded and burying the dead. Lieutenant Alured Gibson was among the sick who had to be sent on board ship. On the

23rd the allied armies advanced over an open down country to the Katcha, where they bivouacked, and the following day they halted on the further side of the river Belbec. The sun's heat had been very great and the hills steep, and the march had been a very trying one. Ensign William Young Johnstone, who had been carrying the colours all day, was attacked by cholera in the bivouac and died on the morning of the 25th, and was buried at 7 a.m. before the regiment marched off. Ensign Harcourt was sent on board ship here. He and Gibson went back to the base hospital at Scutari.

It had now been determined to give up the attempt to capture Sebastopol from the north side and consequently the Allies had to abandon the west coast of the Crimea and to seek a new base on the south. The new plan involved a march past Sebastopol through a broken and wooded country. This, which was dangerous enough, Sebastopol being only 5 miles off, is known in books as the "flank march" but in the regiment as the "long day's march."

The men fell in at 7 a.m. on the 25th, and moved off about 10 a.m., towards the north front of Sebastopol. After advancing some miles, the direction was changed to the left and the army entered the woodlands. The shade was a great relief to the troops, but there were no guides and marching was by compass. The paths were at times so narrow that the men had to move in file, and it was difficult to keep brigades or even regiments together. The cavalry and Rifles forming the advanced guard got on a different track from the rest of the army, and a ludicrous incident happened. Prince Mentschikoff not doubting that we were advancing against the north front of Sebastopol, determined to quit the city and march with the field army towards Simferopol and thus maintain his communication with Russia. He started on the night of the 24th by the post road, and thus, unknown to both armies, the force we had beaten at the Alma passed across our line of march a few hours ahead of our leading troops. Lord Raglan and his staff, under the impression that they were covered by our advance guard, preceded the main body of our army and all but rode into Mentschikoff's baggage train. Both sides were equally surprised, but the Russians took to flight, leaving some waggons which carried meat as well as ammunition. When the 30th came up at 6 p.m., the men were allowed to take as much of the former as they could carry—a great piece of good luck, as no rations except biscuits were issued. This was close to McKenzie's farm, but there was no water to be had, for the troops in front had exhausted the wells. After passing McKenzie's farm the regiment began to descend the heights, and at 8.30 p.m. reached the Tchernaya river where four divisions and the cavalry were assembled near Traktir Bridge, in some sort of order, by 11.30 p.m.—when the men were ordered to bivouac. Cholera had by no means ceased.

On the 26th the rations were again short, and some men got none. The division marched about 7 a.m. towards the south, climbing the slopes of the valley of the Tchernaya and reached Balaclava, which after a slight defence surrendered. Here the army met the fleet again and the difficulty in feeding the troops ceased for a time.

On the 27th, the second and light divisions advanced to reconnoitre Sebastopol, the second division being on the left. The regiment piled arms and cooked and ate their dinner near Upton's house, while Lord Raglan and the Staff took a look at the positions. In the evening the division returned to the neighbourhood of Balaclava and bivouacked; the French and our fourth division which had followed the march of our army came in this day.

McKenzie was one of a number of Scottish agriculturists invited to Russia by the Emperor Alexander about 1820 to improve the Russian methods of farming. Mr. Upton was an English civil engineer employed in building the docks in Sebastopol.

A battalion was now formed for duty at Balaclava by taking 20 or 30 weakly men from every battalion. Two captains from each division were detailed for duty with the battalion. Captain Atcherley of the 30th was one of those furnished by the second division.

Colonel Hoey was attacked by cholera on this day and died at 5 a.m. on the 29th. He had a high ideal of what a regiment should be, and his rule was stricter than was always acceptable to some of those under him; but he gained all hearts by coming off the sick list to lead at the Alma, though so weak from dysentery as to require assistance to mount his horse. In the battle he had shown not only gallantry but judgment. His health had seemed to improve from that day, and his loss was much felt by all. He left the regiment in excellent hands, for Major Mauleverer succeeded, with Patullo as second in command.

Lieutenants Green, Falkner and Hobbs were sent on board the *Hydaspes* sick on the 29th. They were able to rejoin very soon, which was fortunate, for the regiment was now short of officers. Colonel Hoey's death had given Green his company, *vice* Brevet-Major Sillery, who became major.

On the 29th the allies, in our ancestors' phrase, "sat down" before Sebastopol, for they advanced and occupied positions on the heights above it—some of which were within gun-shot of the Russian batteries; but the siege had not yet commenced. The 30th occupied the ground to the left of Woronzoff road, where they had been on the 27th. The left of the light division was on the road.

For several days all hands were employed in bringing up siege guns and ammunition and gabions, fascines, etc.

Captain Atcherley returned from duty at Balaclava on October 2nd, and on the 4th the quartermaster, T. Morris, 3 sergeants, and 26 men rejoined from Scutari and Varna. The tents also were landed this day.

The regiment now moved two and a half miles to its right and took up the position it was to occupy during the first six months of the siege, on the extreme right of the army.

Our camps were on the upland, from which three great ravines run down to Sebastopol Harbour.

The second division camp was on Inkerman Heights, immediately behind the crest of what Mr. Kinglake calls the Home Ridge. A man standing on the crest facing north would have on his right an abrupt, almost a precipitous descent to the valley of the Tchernaya river which here runs north-west into the head of Sebastopol Harbour. To his front the ridge runs north for 1,200 yards, to a height we called Shell Hill and then turns north-west. Our pickets used Shell Hill as a point of observation. Its summit is 30 feet below the crest of the Home Ridge. On his left would be the steep sides of the Careenage Ravine, running north-west from the upland to Careenage Creek in the harbour. On the other side of Careenage Ravine was Victoria Ridge, where a brigade of the light division were encamped; to the left of this brigade was the Middle Ravine, running parallel to the Careenage, and beyond the Middle Ravine was the other brigade of the light division, which had its left on the Woronzoff road which runs down the third great ravine to the head of the Dock Yard Creek.

The first supported the second and light divisions. Its camp was less than a mile to the south of that of the second.

The position of the second division was one of considerable danger, for while called upon to use all its strength in the siege of the town to its left front, it was open to attack on its right front and on its east side, either by the Russian Field Army which now occupied the heights near McKenzie's Farm on the other side of the Tchernaya, or by troops coming from the north side of Sebastopol and marching round the head of the harbour. It was partially protected by the steepness and roughness of the ascent from the Tchernaya Valley 500 feet below, but two ravines running down into the valley gave the enemy a fairly easy access to the ridge. One, the Quarry Ravine, opens out on the ridge 800 yards in front of the second division camp. A road, fit for artillery, ascends the ravine and passes through the camp. The other ravine, called the Volovia Gorge, runs down from the north side of Shell Hill to the head of the harbour, where it meets the causeway from the north across the marshes at the mouth of the Tchernaya.

Besides the danger of attack by the Russian Field Army and by troops from the north of the harbour, the garrison of Sebastopol itself could send troops along the south side of the harbour as far as the Careenage Ravine, where they would find a road leading up to the top of the ridge. From there they could seize Shell Hill, only defended by our pickets, and their artillery would then be only 1,200 yards from the fore ridge which covered the camp of the division. All these avenues of attack were actually employed during the first two months of the siege.

The regiment had now 19 officers and 507 other ranks fit for duty. Seven officers and 227 other ranks were sick and wounded, 85 other ranks were on command. In addition to the four officers whose names have already been given as on detached duties, Lieutenant Falkner had been serving with the commissariat from the time of landing in the Crimea.

From the time the regiment encamped on Inkerman Heights till October 12th it was employed in bringing up guns, ammunition and material to the front when not on picket, a duty which it took every third day. On the 12th it was in the trenches of the "Right Attack," the nearest to its camp but still 2 miles away to its left. On the 13th the men had a night's rest, but on the 14th they took the brigade duties, furnishing 240 men for picket. The work went on increasing in severity until they were in the trenches on the nights when they were not on picket. In addition to their other labours, parties had to be employed daily in fetching water and cutting brushwood for fuel. The water supply was a mile and half off and the brushwood near the camp had all been cut, so that the fuel parties had to search the slopes up to the line of our advanced posts or beyond them. Captain Connolly of the light company had volunteered and was now acting as an engineer, and ten men of the regiment were among the sixty sharpshooters of the division who were employed in the advanced trenches in trying to keep down the fire of the place.

By the 17th our batteries were ready, those of the "Right Attack" to batter the left and enfilade the right face of the Redan, and one-half of the guns of our "Left Attack" to batter the right and enfilade the left face of that work. The remaining guns of the "Left Attack" were to play upon the Flagstaff Bastion at the head of Dockyard Creek. The French to our left were also ready to open fire and the allied fleets were to attack the sea front.

Lord Raglan had great hopes that the effect of a day's bombardment would be so to reduce the enemy's fire and demoralize his garrison that an assault would have a fair chance of success. With this object in view the infantry were kept in camp during the bombardment fully accoutred and ready to fall in at a moment's notice, but without great-coats or blankets.

Fire was opened on the morning of the 17th as soon as it was light, but the fleets did not come into action till the afternoon. The result was on the whole a failure. The fleets got decidedly the worst of it. On land the French, who were slightly inferior to the Russians in number of guns, had bad luck: a Russian shell blew up one of their magazines with fifty men and silenced the neighbouring guns. This was followed by the explosion of an ammunition waggon which disorganized the fire of another battery, and by 10.30 a.m. the Russians had so much the upper hand that the French fire ceased. The British artillerymen had not only a trifling superiority in the number and calibre of their pieces, but they were better gunners and utterly beat down the force opposing them, so that by 3 p.m. Sebastopol lay open to assault. The disasters of our allies, however, made it impossible for Lord Raglan to attack.

The bombardment was renewed on the 19th and continued till the 25th, but the Russian batteries always maintained their superiority over the French.

The Russians, however, had to confess to a loss of nearly 4,000 men from having to keep their reserves under fire in rear of the batteries in case we should assault.

During the days of bombardment the 30th had only one casualty—Private J. Byng, "slightly wounded" on the 22nd. Although reported slightly wounded he was sent to Scutari where he died in November.

There had been some changes in the regiment during the preceding weeks. Lieutenants Gibson, Hobbs and eleven recovered men had rejoined from Scutari, and Lieutenant O'Brien had been sent there sick. From Scutari he was invalided home. Captain Dickson, who had been wounded at the Alma, rejoined on the 25th in time for Little Inkerman.

The knapsacks had been landed soon after the tents and the men were glad to get them, though many had been plundered. The chako and leathern stock had disappeared and quartermasters no longer applied for those articles in their indents. The epaulettes had gone too.

Field pay of 6*d*. a day had been granted to the men and in addition working pay of 8*d*. by day, 10*d*. by night.

The ration at this time was 1½ lb. of biscuit and 1 lb. of pork or 1¼ lb. of fresh meat when procurable. Rum was issued with rather a free hand: a ration at dinner-time, another at sunset, and an extra one to men in the trenches or on picket.

The army was already in difficulties about transport. All public bat horses were given to the commissariat and officers were invited to sell their private horses for public use. Mounted officers had to send their own horses to Balaclava to bring up forage.

Since the second division had established itself on Inkerman Heights the Russian Field Army on the hills about McKenzie's Farm beyond the Tchernaya had received large reinforcements and began to show signs of life. The simple way to guard our position would have been to fortify it but we had no men to spare. Lord Raglan was determined to push on his batteries and capture Sebastopol by assault, so that when winter came his men might either be housed in the city or have sailed from the Crimea. He resolved, therefore, to adopt no precautions which would interfere with his main purpose. He had only a choice of evils, and it was not his fault that his force was so small that he had to trust to his soldiers' courage and discipline more than in most armies would be considered safe.

The first movement which the 30th were aware of on the part of the enemy was the erecting of a battery on October 20th on the heights above Old Inkerman to play upon our camp. We replied to this move of the Russians by building the Sandbag Battery for two 18-pounder guns, which played upon their battery on Sunday the 25th, while the regiment was attending Divine Service. Our artillery had such a command over that opposed to them that the Russians soon gave up the contest, and our guns having served their purpose were withdrawn, but the Sandbag Battery remained, and as it plays a great part in the story of Inkerman, its position is of importance, as is that of "the Barrier," which is near it.

Walking north along the post road which runs through the second division camp one comes, in 400 yards after passing the crest of the Home Ridge, to where the road enters the Quarry Ravine and begins to wind down to the Tchernaya. Across the mouth of the ravine our men had built a wall of loose stones: this was "the Barrier." It extended on each side of the road into the scrub oaks which grow thickly there. Standing at "the Barrier" the Sandbag Battery was 750 yards to the right front, not looking down the Quarry Ravine, but situated on a spur overlooking the Tchernaya and Old Inkerman. Slightly to the left front of a man at "the Barrier" is Shell Hill, 800 yards from the Barrier and 1,200 yards from the Home Ridge. On Home Ridge was a third small attempt at a defensive work. Along the crest from right to left, a bank had been raised to protect our gunners; there were no embrasures and throughout its length it was low enough for our field guns to fire over it. Home Ridge was our alarm post where the second division were accustomed to fall in. Our pickets held Shell Hill by day, and were confronted by Cossack videttes with supports, one of whose duties it was to prevent our patrols from getting far enough to the front to look down on the harbour and the road along its southern shore.

While the Russians had been amusing our men on Inkerman Heights by an exchange of shots, the greater part of their army had moved to its left and on the 23rd was in front of Balaclava.

On the 25th, General Liprandi attacked our force covering Balaclava harbour with 25,000 men. His operations were feeble and he was repulsed with a loss of about 600 killed and wounded. Our loss was the same, but more than half our casualties were in our weak cavalry division, both the light and heavy brigades of which distinguished themselves greatly. The 30th had no officers engaged, but 1 sergeant, 2 corporals and 36 privates received the clasp for the battle. Sir Colin Campbell in his account of the battle says that 100 invalids under Colonel Daveney were formed up on his left when he repulsed the Russian cavalry with the 93rd in line, and Captain John Hume of the 55th (brother of Sir Robert) records that the 55th detachment at Balaclava (chiefly convalescents) was on the right of the 93rd. The 30th men being from the same brigade were probably under Captain Hume, and perhaps both were under Colonel Daveney and Captain Hume has placed them on the right by mistake.

Although repulsed General Liprandi still continued to threaten Balaclava, and as a diversion in his favour and a reconnaissance of the Inkerman position, an attack was made on the 27th against the second division.

As General Liprandi had at the beginning of the action on the 25th put to flight some untrained Turkish troops and captured a standard and some guns from them, it was possible to represent the affair as a victory, and a Te Deum was celebrated in Sebastopol on the morning of the 26th to put heart into the troops selected for the attack. They fell in about noon on that day to the number of 6,000 infantry with four guns, and leaving Sebastopol by the Karabel suburb passed the head of the Careenage Creek and ascended Mount Inkerman, which they crossed, and when their leading troops had almost reached the Volovia Gorge they formed line to their right facing Shell Hill. The pickets of the 1st brigade on the right consisted of three companies of the 30th, commanded by Captain Atcherley, Lieutenant Alured Gibson and Lieutenant J. D. Ross Lewin, and one company of the 95th. They were under Major Champion of the 95th, the field officer of the day. The pickets of the 2nd brigade consisted of three companies of the 49th and one of the 41st. Major Emin of the 41st commanded them. Each company was made up to a strength of sixty rank and file.

The extreme right of the line of our advanced sentries was due east of the camp and about 600 yards from it. From there the chain ran north for about 1,500 yards along the top of the precipitous descent to the Tchernaya Valley till it reached the neighbourhood of the Sandbag Battery. Here there is a steep narrow ravine running north-east down to the Tchernaya, and then comes the Quarry Ravine with the post road winding down it. The line now turned west along Shell Hill, and its spurs, to the Careenage Ravine where the light division took up the outposts.

The right of the 49th and left of the 30th under Captain Atcherley met on Shell Hill; a company of the 30th was in support of Captain Atcherley, and the remaining company of the 30th and that of the 95th found the advanced posts and supports looking east to the Tchernaya.

The Russians, who advanced with a strong line of skirmishers supported by company columns and with a reserve in column, attacked the right company of the 49th commanded by Lieutenant Connolly on Shell Hill. The 49th made a vigorous resistance, and Major Champion, leaving his right picket in position looking down on the Tchernaya, advanced with the remainder of his force beyond the Quarry Ravine to support Lieutenant Connolly and Captain Atcherley.

The Russians, however, broke through the line, and while the 49th picket fell back to its left on its own supports, Major Champion, alarmed by a report that an enemy's column was coming up the post road in rear, moved his men back to the Barrier to close the Quarry Ravine to the Russians. Here he made a stand, and summoned his right picket to join him. The second division was now under arms and its artillery, reinforced by a battery of the first, came into action on the ridge in front of the camp, and having silenced the Russian guns, began to play upon the reserves on Shell Hill. At the same time a scratch company about sixty strong was made up from the first men of the 30th who fell in and was despatched to the Barrier under Captain Paget Bayly; the light company of the 41st was also sent to reinforce the 49th.

All the efforts of the Russians failed to move the companies of the 30th and 95th at the Barrier. Captain Atcherley was shot through the arm and disabled, and Captain Paget Bayly was shot through the face, but was able to fight on to the end. Colour-Sergeant Daniel Sullivan likewise distinguished himself by his gallantry and was slightly wounded. Corporal J. Symington's conduct was also conspicuous. The resistance of the pickets on the left was equally stubborn.

The Russian commander, Colonel Federoff, felt that he had gained no success against the pickets which would justify him in awaiting the attack of the second division, although he had brought entrenching tools with the purpose of holding Shell Hill. He gave the order to retreat, but it was already too late. Sir De Lacy Evans, keeping the 55th in reserve, had sent forward the remainder of the 30th and the 95th to the Barrier and ordered the 2nd brigade to reinforce its pickets. The Russians at first fell back in good order and their fire was well kept up, but our men were not to be denied. Federoff fell and his men broke in great disorder.

Sir De Lacy Evans in writing an account of the action to Lord Raglan, says :—

"The enemy came on at first rapidly, assisted by their guns on Mound (Shell) Hill. Our pickets, then chiefly of the 30th and 49th regiments, resisted them with a very remarkable determination and firmness. Lieutenant Connolly of the 49th greatly distinguished himself, as did Captain Bayly of the 30th, and Captain Atcherley of the same regiment. Sergeant Sullivan also at this point displayed great bravery."

At the end of his despatch the general says that the Russians turned in complete confusion and flight. "They were then literally chased by the 30th and 95th regiments over the ridges down towards the head of the bay. So eager was the pursuit that it was with difficulty Major-General Pennefather eventually effected the recall of our men.

"The conduct of the pickets excited universal admiration."

To his division Sir De Lacy Evans addressed the following order :—

"The Lieutenant-General expresses his hearty and most cordial thanks to the division he has the honour to command for their exemplary and most spirited conduct on the 26th inst. The enemy left in our hands 100 prisoners; about 130 of his dead remain within the lines of our post.

"The severest part of the fighting fell on the pickets of the 30th and 49th regiments. Impartial witnesses not belonging to the division have declared that heroic acts were performed on this occasion. The Lieutenant-General has had the happiness of hearing the conduct of the whole of the division, and of the pickets in particular, and of the Royal Artillery adverted to in terms of the warmest approval by the highest authorities of the French and English armies."

When the 30th and 95th chased the Russians headlong over Shell Hill and 2,000 yards beyond it to the rocky banks of the Careenage Ravine, they met the pickets of the 2nd brigade, which had advanced on their left. A few of the more active descended into the broken ground of the Ravine, and here Lieutenant Ross Lewin with an officer of the 41st and a party of both regiments greatly distinguished themselves.' Here, too, Sergeant Thos. Shaw and Private Andrews made the capture of the day. The ground is full of caves, quarries and lime-pits, and along with some men of the 41st, Sergeant Shaw and Andrews bailed up some Russians in a quarry and made them prisoners. The 41st men, satisfied with what they had got, marched their prisoners back to camp, but Sergeant Shaw and Andrews went on to look for something better. Presently they came on a mounted Russian officer who had been wounded. After some parley, the Russian delivered up his sword and accompanied his captors to camp. They managed to dispose of the sword, saddle and accoutrements, but had not completed the sale of the horse on November 5th, when it was killed with others in the regimental lines. For their conduct on this occasion Sergeant T. Shaw and Private John Andrews received the Sardinian War Medal at the termination of the war. Shaw was made prisoner near Balaclava in winter, but Andrews was conspicuous for gallant conduct in several other fights and received the Medal for Distinguished Conduct.

When Lieutenant Ross Lewin was bringing back his party, he met General Pennefather who shook hands with him and said they had given the Russians a good slating. He added, " The 30th behaved like gentlemen." This in the general's opinion was the highest praise he could bestow.

When the action began and before our artillery had subdued the fire of the Russian guns, there was a lively scene in the second division camp. Every hot which cleared the crest in front, pitched among the tents and all hands which could be spared from the fight were busy conveying horses, and everything movable, to a place of safety; some men lost their lives while so employed. Timothy Morris, the quartermaster, had a narrow escape, a round shot passing between his legs when he was putting the saddle on his horse. A Board assembled in November awarded him compensation for a horse killed in action.

The following were the losses of the division :—

	Killed.	Wounded.				
	Rank and File.	Officers.	Sergeants.	Drum.	Rank and File.	Total.
30th	7	2	1	0	22	32
41st	1	1	0	0	9	11
47th	2	0	0	0	7	9
49th	1	2	1	1	17	22
95th	1	0	1	0	8	10
	12	5	3	1	63	84

Sir De Lacy Evans held the 55th in reserve and they had no casualties. The following are the names of the seven privates killed in the 30th:

Charles Freeman, George Rose, Bryan Matthews, Thomas Jenkinson, Patrick Morrison, George Smith (4), Robert Hunter. The wounded were Captains Atcherley and Bayly (both severe), Colour-Sergeant Dan. Sullivan, Corporals Thomas Delaney and John Sharpe (1), and twenty privates.

Captain Atcherley was sent on board ship at once and despatched to Scutari, but Captain Bayly was kept in hospital in the Crimea and was able to rejoin for the great fight on November 5th. On the 29th General Sir De Lacy Evans met with an accident when riding and had to give up the command of the division to Major-General Pennefather and go on board ship.

On November 1st, four days before the battle of Inkerman, there were present and effective with the regiment—Lieut.-Colonel Mauleverer, Major Patullo; Captains Rose, Connolly, Dickson and Bayly; Lieutenants Greene, Gibson, Falkner, Williamson, Macpherson, Ross Lewin and J. P. Campbell; Adjutant Mark Walker, Paymaster Fitzgerald, Quartermaster Morris, Surgeon Dowse, Assist.-Surgeons Fyffe, Mackesy and Milroy, and 454 other ranks.

Ensign Hobbs and 117 other ranks were sick or wounded in the Crimea: Captain Atcherley (wounded), Lieutenants O'Brien and Harcourt (sick) were at Scutari; 161 other ranks were wounded or sick at Varna and Scutari; 56 other ranks were on command, 3 in the Crimea, 29 at Scutari, and 24 at Varna depot, of which Lieutenant Forbes was adjutant; Captain Pocock had been sent to the depot at Fermoy on promotion.

His promotion not having appeared in orders, Ensign Thompson was still acting as sergeant-major.

The second division were naturally highly elated after their brilliant action of October 26th, and many said that they had given the Russians such a lesson that the latter would think twice before they ventured on Mount Inkerman again. Lord Raglan was not so sanguine, for he knew that huge columns of men were pouring down the road through Simferopol to reinforce the Russian army.

Exactly when the blow would fall he could not judge, but he ordered the outposts of the Guards and second division to be strengthened and the duty to be most carefully performed. To enable the second division to do this he spared them some of the work in the trenches.

It was now a race for time. Lord Raglan hoped that on November 7th, he would be able after a second and a more severe bombardment to carry the defences of Sebastopol by assault, and so avoid the sufferings of a winter in the open.

He lost by two days, for the Russians were able to strike such a blow on the 5th as put an end to any idea of an assault for months. The Russian attack was directed against Mount Inkerman, and it is obvious that its success or failure depended greatly upon whether they could overwhelm the second division speedily, or whether that hard fighting division would hold on long enough for the more distant troops to come up and for the allied army to present a united front.

The Russians had, in and near Sebastopol, 115,000 men under Prince Mentschikoff.

The allies numbered 65,000, but of these only 16,000 were British infantry, upon whom the hardest fighting must fall from the nature of the position. The plan of attack was that General Soimonoff should leave Sebastopol with 19,000 men and 38 guns, and, following the route taken by Federoff on October 26th, establish himself on Mount Inkerman, where Shell Hill would give him a gun position of three-quarters of a mile from left to right.

General Pauloff with 16,000 men and 96 guns, crossing by the bridge over the marshes of the Tchernaya near the head of the harbour and by a bridge near Old Inkerman, was to join Soimonoff on the heights. General Dannenberg was then to take command of the combined force.

The Russian left of 22,000 men with 88 guns under Prince Gortschakoff was to await in the Tchernaya Valley, with its right near Old Inkerman, the result of the main attack. It was hoped that Dannenberg's 35,000 infantry and 134 guns would not only overcome the resistance of the second division without any great expenditure of time, but might follow up this success by pushing the Guards from their position and thus give Gortschakoff an opportunity of ascending the height from the Tchernaya Valley, and joining his comrades on the upland.

The men left in garrison in Sebastopol were to make a sortie to hold the French and English troops fast in the trenches and prevent their sending help to Mount Inkerman.

INKERMAN

The night had been wet and a cold mist clung to the ground, when a little before daylight on the morning of November 5th the relieving pickets of the second division marched down from the Home Ridge to their posts. This mist continued throughout the day, now thickening, now thinning, but never disappearing. It clung especially to the hollows and the men on Shell Hill and Home Ridge could see each other for long periods while the mist was so thick over the intervening low ground that it was impossible to guess what was happening except by the sound. The fact that on both sides the troops wore the grey great-coat increased the difficulty of distinguishing them.

The 30th found no men for picket on the morning of the 5th. The 1st Brigade pickets were furnished by the 55th and those of the 2nd Brigade by the 41st and 47th.

The new pickets found those whom they relieved in their usual places, but very wet and miserable, and passed through them to take up a more advanced position on the heights for the day. The relieved pickets returned to camp and the division which had, according to custom, got under arms an hour before daybreak, was dismissed and water and wood parties were sent out and on all sides men were to be seen trying to get the wet wood to light in order to cook the breakfasts.

Captain Rowland, commanding a picket of the 41st, after passing the old picket, had advanced on to Shell Hill and placed his sentries beyond the crest at about 6 a.m. He had just returned to the main body of his company and told his men they might take off their knapsacks when firing began in front. He extended and advanced, but the enemy were in enormous numbers and he fell back on the light company of his regiment which was in support. They were both forced back, but the volume of fire had warned all troops within hearing and the alarm had sounded throughout the British camps, the fatigue and working parties hastily rejoined their regiments and all fell in at their divisional alarm posts. By the time the second division had got under arms Soimonoff was able to open fire from Shell Hill with twenty-two 32-pounder howitzers and 12-pounder guns, and one of our field batteries was silenced at once.

General Pennefather pushed forward all available men "to feed the pickets," as he said. One wing of the 30th about 200 strong under Lieut.-Colonel Mauleverer he sent forward down the Post Road towards the Barrier, the other of about equal strength under Major Patullo was sent to reinforce the pickets who were retiring from Shell Hill nearly straight towards Home Ridge and who could be heard keeping up a stout fight in the mist below. They were commanded by Lieut.-Colonel Haly of the 47th.

The movements of Major Patullo's wing cannot be followed with entire certainty; it was mixed with the pickets and had no individual existence. It met the retiring pickets half-way between Shell Hill and Home Ridge, or in other words when it had advanced 600 yards.

When Captain Rowlands of the 41st with two companies of his regiment had been forced back from Shell Hill by weight of numbers, the two companies of the 47th on picket on his left had retreated likewise. All four companies had been unable to give fire as effectively as usual from the rain having choked the nipples of their rifles. After retiring some hundreds of yards, Lieut.-Colonel Haly turned and gave the order to charge. He himself rode into the midst of the enemy and cut down three men before he was unhorsed and bayoneted, but not mortally wounded. He was rescued by Captain Hugh Rowlands of the 41st, and two men of his own (the 47th) regiment, and in the history of the 47th it is recorded that "it is believed a bugler of the 30th assisted in this rescue."

Captain Rowlands and one man of the 47th received the Victoria Cross for their feat. The other man of the 47th was killed, and the bugler of the 30th cannot be traced.

It would seem from the foregoing, that Colonel Haly had been induced to turn and charge by the arrival of Major Patullo and his 200 men. Certainly Major Patullo shared in the new advance towards Shell Hill.

General Soimonoff's attack now developed. The number of his guns on Shell Hill was being constantly reinforced, and he himself advanced with six battalions against the extreme left of Home Ridge near the Careenage Ravine, while six other battalions were ordered to advance from Shell Hill, to brush aside the resistance of Major Patullo's wing and the remains of the pickets, and to make a direct attack on Home Ridge to the west of the post road. Ten thousand Russians were left in reserve.

Our men, although corps and companies were mixed together, fought with coolness and courage in an irregular line and used their rifles with great effect upon the enemy, most of whom were in deeper formation, but the Russians were in a superiority of ten to one and pushed the British back to Home Ridge, where to the right of the post road Major John Turner with three guns was waiting an opportunity to fire. Loading with grape he sent a sergeant forward to warn those of our men in front of his guns to lie down the moment they reached the rise of the hill. His order was obeyed; the retreating men threw themselves down in front of and below the guns, and Turner, having a clear field, opened fire upon the leading Russian column. At this moment the regiments formed on Home Ridge, sprang forward cheering loudly and drove the six Russian battalions back to Shell Hill and clean through the batteries there. It was impossible, however, to maintain such a forward position, chiefly from want of ammunition. From the extreme urgency of the case the men had been hurried into action in the morning as soon as they had fallen in, and no extra ammunition had been served out, and, as yet, no means had been organized for sending it to them.

Presently our force fell back again on Home Ridge. The men were wild with passion at having to retire, but they had done their work; not a single Russian followed them. Major Patullo with his wing took post to the right of the post road.

The attack of the six battalions led by General Soimonoff in person against the left of Home Ridge, near the Careenage Ravine, failed also. It was repulsed by the 47th, a wing of the 49th, and four companies each of the 77th and 88th which had been sent by the officer commanding the light division across the Ravine to assist the second division—the 77th especially distinguished itself by a charge and a prolonged advance which brought it to Shell Hill, close under the Russian guns. Some of the 30th had the honour of sharing in this charge and advance.

When General Pennefather despatched the two wings of the 30th to the front "to feed the pickets" many men detailed for special duties had not yet rejoined, but as they came in they were formed upon Home Ridge. We have an account of what happened from Corporal Colcutt, who brought back the water party of the 30th only to find that the regiment had gone to the front. The 77th in moving from the light division camp and passing the head of the ravine, came close under Home Ridge and the men on the ridge, irrespective of regiments, were ordered to join the right of the 77th and act with it. The fog was very thick close to the ground then and the men could only recognize that the 77th was British by the mounted officers, who were partly visible above the mist. On the Russian side Soimonoff alone was mounted. Up to that time the enemy had met with some success, and when the 77th came into action the combat was very sharp, but Soimonoff fell and his men broke and were pursued right up to and through their guns. The fact that of the 19,000 men who left Sebastopol under Soimonoff only 4,500 could be brought into action again after this repulse marks the severity of the handling they had received.

When Colonel Egerton halted the 77th, after his victorious advance, he ordered the men of the 30th and other regiments who had joined him back to Home Ridge.

A fresh army now took up the action on the Russian side under General Pauloff, whose attack ought to have been simultaneous with that of Soimonoff. Shortly before the fall and defeat of the latter general, the two leading regiments of Pauloff's force had reached the plateau. Of these the Borodino Light Infantry was in front of the Barrier and the other to its left, near the Sandbag Battery. They each had four battalions and a strength of 3,000 men. It will be remembered that the Borodino regiment had been defeated by Pennefather's brigade at the Alma.

Colonel Mauleverer with the wing of the 30th, 200 strong, was at that time in front of the right of Home Ridge, where he covered Major John Turner's battery. There was no other force to meet the Borodino regiment, which advanced with two battalions in front and two in second line. The Barrier lay unoccupied between the opposing forces.

As the Russians advanced Colonel Mauleverer opened fire, but so many rifles had been injured by damp and become useless that he ordered his men to cease fire and advanced to the Barrier, where the wing lay down.

When the leading battalions of the enemy were almost on him, Colonel Mauleverer gave the word and he and his officers mounted the Barrier and leapt down among the enemy, sword in hand.[1] His men followed with the

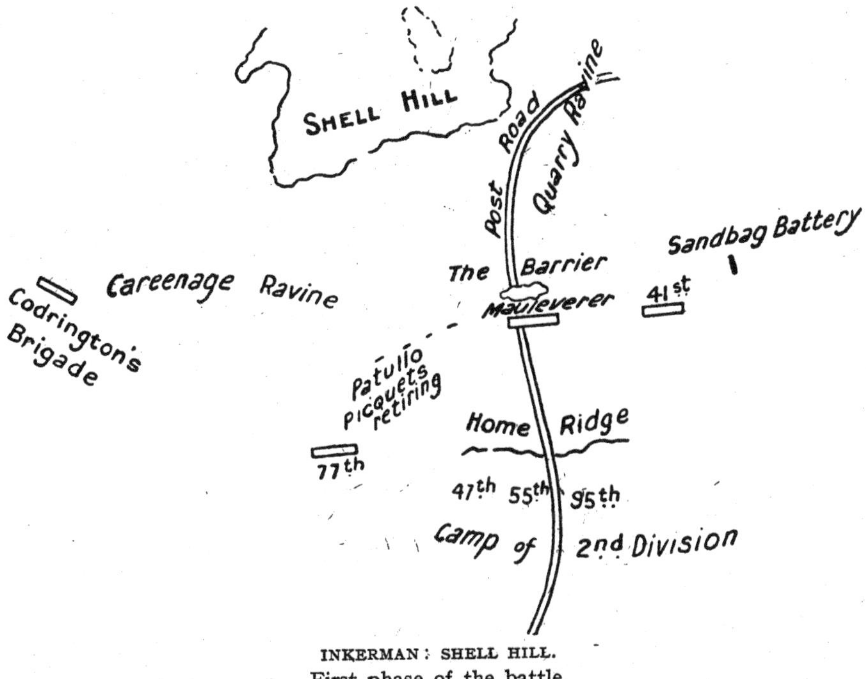

INKERMAN: SHELL HILL.
First phase of the battle.

bayonet, and, although Colonel Mauleverer and many others fell, the regiment, after a short struggle hand to hand, drove the Russians before them in a helpless mob. The two rear battalions of the Borodino regiment fell back without an effort to retrieve the fight.

To the right of the 30th near the Sandbag Battery the 41st routed the Tarontine regiment.

It was now eight o'clock and the second division with the assistance of 500 men of the 77th, and of the Connaught Rangers and a battery, had in two hours' fighting defeated 25,000 men backed by an enormously superior artillery.

Sir De Lacy Evans had arrived from on board ship, but he did not feel well enough to take the command from General Pennefather. The latter's

[1] *Note.*—To all his other qualifications as a leader of men, Colonel Mauleverer added that of extreme agility. To the end of his service he was able to vault the billiard table in the officers' mess.

voice had been heard everywhere. He was supposed to have the greatest command of strong language in the British army, and he put forth all his powers on that day, but that was merely the embroidery; the main thing was, that he was loud, cheery and confident, and that his speech and bearing were those of a man who would not give an inch. He had two horses shot under him in the course of the day.

Reinforcements were now coming up on both sides; the remaining 10,000 of Pauloff's corps had joined and General Dannenberg had now under his command 96 guns and 19,000 effective infantry.

The reinforcements on the allied side were three field batteries, 1,200 of the Guards and 2,000 of the fourth division on the ground and 1,600 French close behind.

The wing of the 30th still held its position near the Barrier, but was very much isolated. There was a gap of 500 yards between it and the 41st on its right, and in spite of General Pennefather's protests, our reinforcements as they came up inclined more and more to their right until the bulk of them were at the Sandbag Battery or beyond it.

General Dannenberg renewed the action with the 10,000 fresh men of Pauloff's corps and the 30th were the first to meet the attack. As Mr. Kinglake put it, "After a foot to foot resistance long maintained against heavy columns, they were pressed back and back until they found themselves at last behind the bank on the top of the Ridge in front of their camp." Here the wing was joined by the remainder of the regiment and Major Patullo took command in place of Mauleverer, severely wounded. When the men lay down behind the bank on the top of Home Ridge most of them fell into the deep sleep of exhaustion, regardless of the shot and shell striking the bank and screaming overhead. Probably none had broken their fast that day. When an enemy's column, which in the mists had been mistaken for one of our own, had nearly gained the summit of the ridge, the 30th were suddenly called upon and, once awakened and on their feet, our men showed all their old alacrity in charging and driving the Russians down the hill.

The battle was now drawing to a close, but the Russians launched two great attacks upon the Ridge. The 30th were then, as they had been through the day, near the post road and they were in line with the remnants of many of our regiments and in support was the French 7th Léger.

The enemy came on with great determination and actually gained a footing on the crest.

The 7th Léger was ordered up, but after firing it turned. The adjutant of the 30th wrote of this in his diary, "I thought it was all up with our position. Fortunately they rallied and fresh bodies coming up, we on the right drove them (the Russians) back with dreadful slaughter."

This was the last attack upon the position held by the second division, and upon its repulse some of our fresh troops followed up the Russians past the Barrier and down the Quarry Ravine. Nothing further could be done until the bulk of the French infantry arrived; their artillery was fast coming into action and Lord Raglan had brought up two 18-pounders on the Ridge, but the 90 Russian guns on Shell Hill held their own for a time. The leading French troops on their arrival met with an unfortunate repulse and General Canrobert, whose own gallantry was as conspicuous as his caution in risking his men, refused to do more than act on the defensive. The Russians and the British were too exhausted to attack each other, and it looked as if the battle might be drawn. It lingered on till about 10 o'clock when General Dannenberg, finding that our heavier guns were disabling his, one after the other, and that the long range rifle fire of our foremost troops was slaughtering his gunners, gave the order to retreat. There was no pursuit.

The 30th do not appear to have moved from the Home Ridge after the last repulse of the Russians.

During the battle the 30th never was more than half a mile from its own parade ground and its duties, first to keep the enemy at a distance from Home Ridge and secondly when driven back to the Ridge to defend it to the last, were so obvious that although at times mixed with other corps it was easily extricated and re-formed. The greater part of the army was less fortunate; despatched, regiment after regiment, sometimes company after company, into unknown ground it found itself fighting in the mist among ravines covered with bush and strewn with enormous boulders. Corps and companies got mixed and broke into bands, whose only aim was to fight the Russians whenever they met them. Around the Sandbag Battery and from there to the Barrier there was a succession of desperate combats. One officer of the 30th shared in this confused struggle. John Pennock Campbell had been lent to the 95th, who were short of officers, and went through the day with them. Among his comrades he had the name of being able to observe and think accurately under fire as well as of being a gallant leader.

He used to say, " It was like this—you might find yourself with a party belonging to several regiments pursuing the Russians with joyous shouts, ' They're on the run, keep them going, etc.,' presently you ran against a stronger party, perhaps a formed up body; it was now your turn to fall back, the Russians did not pursue very far or hotly, being afraid of being trapped in the mist, but your party diminished, some found their own regiments, some dispersed to collect ammunition from the pouches of the killed and wounded, some wandered and a few might go to the rear, not many, for the men were full of resolution. Presently you might come across another fellow with a party, or groups of men sometimes under non-commissioned officers, sometimes not, then you would all join and advance to have another go at the Russians, and so it went on."

Campbell came through the day unhurt, but he had the misfortune to have one of the tails of his coatee shot off, and in it was not only some of his company's money, but all his own. After a Court of Inquiry Government made good the loss of the company's money, but not the other.

As soon as the men had eaten, the pickets who had mounted that morning were sent off to complete their tour of duty and parties were sent out to search for wounded comrades and bring them in.

The losses were very heavy. The British brought 7,464 infantry, 200 cavalry and 38 guns into action and had 130 officers and 2,227 non-commissioned officers and men killed and wounded.

The French brought up a nearly equal force, but most of them did not arrive till the heat of the action was over. They had 49 officers and 880 non-commissioned officers and men hit.

The casualties in the 30th were: Killed—2 officers and 28 rank and file. Wounded—7 officers, 4 sergeants, 1 drummer, and 90 rank and file. —a total of 132 of all ranks.

The experiences of the other regiments in the division had been similar to those of the 30th, but the recollection of the surviving officers ten years later was that, owing to there being fewer men of that corps absent on duty, General Pennefather was able to send the two wings of the 30th to "feed the pickets" somewhat earlier than the other regiments.

The total casualties in the division were 721 of all ranks.

In the 30th 2 officers, 1 sergeant, 1 corporal, and 6 privates died of their wounds.

The strength of the 30th was 14 officers and 394 other ranks, a total of 408. The average strength of the other battalions of the division was slightly over 500.

The following are the names of the killed and wounded in the regiment: Killed: Captain Arthur Wellesley Connolly, Lieutenant Alured Gibson; Privates Thomas Dale, William Davis, John Elliot, Patrick English, John Featherstone, David Fletcher, William Hands, Patrick Hughes, John Hunter, John Inglis, William Manton, George MacAllister, John Mackintosh, Daniel McMurty, John Middlemas, John Mohan, William Morgan, Joseph Rodway, Henry Scott, David Sheny, Donald Simpson, Thomas Smith (6), William Stills, John Swift, Patrick Tearney, Thomas Waldock, John Williams, John Willis. Died of wounds: Lieutenant J. Dillon Ross Lewin, Ensign John Thompson. (He is shown among the colour sergeants as dangerously wounded, but his promotion was published in orders before his death on November 10th.) Colour-Sergeant William Gallaher, Corporal George Young, Privates John Allen, Thomas Behan, James Cannon, Thomas Carr and William O'Kill.

Other wounded—Lieutenant-Colonel James Thomas Mauleverer (severely) (Colonel Mauleverer had a horse killed before reaching the Barrier), Captains James Rose (severely), Graham Le Febre Dickson and Paget Bayly, Lieut. and Adjutant Mark Walker, Sergeants David Dunn, William Jameson and Patrick Lowmore, with 85 other ranks.

Of the officers killed at Inkerman, Captain Arthur Wellesley Connolly had been wounded at the Alma. He was very much liked by all ranks for his charm of manner and gallantry in action. By descent he was a Pakenham and a cousin of Captain Thomas Henry Pakenham. His father on succeeding to the Castletown estates had assumed the name of Connolly.

Lieutenant Alured Gibson was the second son of Wood Gibson, Esquire, of Bodlondeb, North Wales; he had been wounded early in the action by a musket ball and Sergeant Jameson, the pay sergeant of his company, had urged him to go to the rear to find a surgeon, but he refused to quit the field while able to keep up. A major of the regiment wrote, "I was not near him when he fell, but I hear, that though previously wounded, he was still gallantly leading his men when he received his death wound. He joined us, poor fellow, before he was quite recovered from a fever, so great was his anxiety to do his duty with the regiment. On October 26th, when we repulsed the enemy almost unaided, he behaved most gallantly, and on the 5th his conduct was equally conspicuous." (Lieutenant Gibson after the Alma had gone sick with an attack of fever, but had rejoined in time for Little Inkerman.)

Lieutenant John Dillon Ross Lewin, who was dangerously wounded in the battle and died at Balaclava on the 7th, came of a soldiers' family. His father, Major Henry Ross Lewin, of Ross Hill, County Clare, a Peninsular veteran, had been in eleven battles and sieges and two of his uncles had also distinguished themselves under the Duke of Wellington. Ross Lewin received his death wound towards the close of the day, probably in the last great charge when the most dangerous attack of the Russians was defeated. Throughout he had shown the same gallantry which had attracted Major-General Pennefather's attention in the action of October 26th.

Ensign John Thompson, who was dangerously wounded in the battle, died at Balaclava on the 10th. He fought at Inkerman as acting-sergeant-major. He was a great loss to the regiment.

Colonel Mauleverer, Captains Rose, Dickson and Paget Bayly were forced by their wounds to leave the Crimea. The three latter never rejoined. They and Captain Pakenham were given brevet-majorities for distinguished conduct. Their services in the 30th had been extremely honourable. Rose had come through Alma and Little Inkerman unhurt, Dickson had been severely wounded at the Alma, and although offered leave to England from Scutari he had preferred to return to the Crimea with his Alma wound yet unhealed. An officer of another regiment who accompanied him from Scutari wrote, " When one sees such instances of devotion to the service one ceases to wonder at the constant success of our arms." Dickson had rejoined just in time to share in the brilliant action of October 26th (Little Inkerman). Paget Bayly had been obliged to go sick soon after landing and missed the Alma ; but he rejoined when the regiment reached Balaclava and distinguished himself greatly at Little Inkerman, where he was shot in the face, a wound of which he had scarcely recovered by November 5th.

Major Patullo was given a brevet of lieut.-colonel for distinguished service in the field after Inkerman and Captain Whitmore a brevet majority. He was still on the staff of Lieut.-Colonel Sir George Brown and had a horse killed under him at Inkerman.

The Victoria Cross was awarded to Lieutenant and Adjutant Mark Walker " For daring bravery at Inkerman when to encourage his men he leaped over a wall in face of two battalions of Russian infantry, his regiment following and repulsing the foe." Mark Walker had been struck in the face by a spent ball, but was otherwise unhurt.

The question has sometimes been asked why Mark Walker received the Cross and not Mauleverer. The answer is simply that Mauleverer was in command and, therefore, not eligible. He recommended his adjutant as the one who was most conspicuous among his officers. A similar case has been narrated a few pages back of Colonel Haly, who led the pickets against a Russian column into which he rode alone and cut down three men. Two of those who followed and rescued him got the Cross on his report of their conduct, but he, the leader, could not receive it. Mauleverer's selection of Mark Walker gave pleasure to the whole regiment.

In addition to Lieutenant Mark Walker, Colonel Mauleverer recommended Corporal William Colcutt for the Victoria Cross.

After sharing in the great charge of the 77th, Colcutt with his party had returned to Home Ridge where preparations were now being made to despatch ammunition mules to supply the men fighting in front. A string of mules which had just been loaded with ammunition boxes for the 30th, was knocked over by Russian shells and the pack saddles set alight. Colcutt, at great personal risk, dashed in and beat out the flames. He then volunteered to carry ammunition himself and made many journeys to the companies in front distributing cartridges. Colonel Mauleverer's recommendation was not successful.

The death of Lieut.-Colonel E. W. Pakenham, Grenadier Guards, killed at Inkerman, had a considerable effect on the 30th, for he was Member of Parliament for County Antrim, and when he was killed at Inkerman Brevet-Major Thomas Henry Pakenham, his younger brother, succeeded him as Member for his county, and held the seat till he followed Mauleverer in command of the 30th in 1863.

In addition to the rewards bestowed upon the officers after Inkerman, the commanding officer of each regiment was ordered to recommend a sergeant for promotion to a commission. Quartermaster-Sergeant John Moon was recommended by Colonel Mauleverer and was promoted ensign to date from November 5th. Colour-Sergeant John McLellan succeeded him as quartermaster-sergeant. Sergeant William McGlade was appointed sergeant-major from November 29th. On December 24th a General Order was published conveying the Queen's high approbation of the conduct of the troops at Inkerman and announcing her intention of bestowing a silver medal upon all ranks. Her Majesty's gracious message wound up, "Let not any private soldier in the ranks think that his conduct is unheeded. The Queen thanks him, the Country honours him." Lord Raglan was promoted a field-marshal from November 5th.

On November 13th the 62nd regiment landed and joined the brigade.

On the following day the Crimea was struck by a storm of extraordinary violence and the effect of it on the British was only less disastrous than a lost battle.

The following account is from Mark Walker's diary: "A dreadful day. About 8 a.m. it commenced to blow a hurricane. My tent, bed and all were blown over, the only two left were those of Patullo and Morris, into which we all got and had considerable difficulty in keeping them up. The day was dreadfully cold and such misery we never before endured. In the middle of it Ensign Gubbins, two sergeants and 98 men joined from the depot; their first day in the Crimea was a dreadful one and the poor wretches as they stood shivering without a covering were to be pitied."

Another account says: "Imagine the bleakest common in all England, the wettest bog in all Ireland or the dreariest moor in all Scotland overhung by leaden skies black as ink, lashed by a tornado, sleet, snow, drizzle and pelting rain, a few broken stone walls and roofless huts dotting it here and there, roads turned into torrents of mud and water."

About 2 p.m. the strength of the wind lessened, but the cold increased and the snow began to lie on the hills. By nightfall the more energetic men had set to work and managed to secure the remains of their tents and to get up some cover. One of the most distressing things of the day was that the hospital marquees were the first to be blown away and the sick and wounded were exposed to the storm and no man could help them. The marquees were not replaced for months, which forced the surgeons to send off the sick and wounded to Scutari even when scarcely fit for the journey and so caused infinite suffering. What the troops endured on that day was not the worst effect of the storm. A Turkish 90-gun ship foundered with all hands, the French lost two men-of-war, and 21 transports and store ships were wrecked off Balaclava, including the *Prince*, which was driven against the perpendicular cliffs and sunk, carrying with it the stores of warm clothing sent out for the Army from England. As Field-Marshal Sir Evelyn Wood put it in his book *The Crimea* 1854 *and* 1894, " The storm was the beginning of misery so intense as to defy adequate description."

The 15th was a fine day and by night the men were partially dry and got under cover, but the tents had never been good and at Inkerman more than half were damaged. There was no reserve of tents at home, but luckily there were some in store at Malta and Lord Raglan had them brought at once, otherwise the army must have re-embarked or perished. Extra food might have strengthened the men to bear their hardships, but unluckily the supplies ran short and Lord Raglan had to knock off the extra ration he had granted a little time previously. There were no vegetables.

The principal medical officer reported on December 1st, "Health unsatisfactory from fatigue, exposure and many other privations. Bowel complaints prevalent among all, but especially among the recruits joined from England, among whom the disease degenerates into a fatal form of cholera; symptoms of scurvy have appeared."

The severe losses at the battle of Inkerman not only put an end to all idea of taking Sebastopol by main force, but brought about a reconsideration of the share which the British and French armies were to take in their joint enterprise. When the two armies landed in the Crimea they were nearly equal in force and the British were proud to take the exposed flank with all the extra labour and danger to which they were thereby subjected, but now, after having borne the brunt of the fighting at Alma and Balaclava, on October 26th, and at Inkerman, our effective strength had fallen to 14,000 men and there were no reserves at home to fill the ranks. The French, on the other hand, were continually receiving reinforcements and had now nearly 50,000 men in the Crimea, soon to be raised to over 60,000.

Marshal Canrobert, therefore, assumed the duty of guarding the Allies from attack by the Russian field army and took an increased share in the actual siege operations, but the portion still retained by the British was more than their strength justified and overwork, combined with want of shelter and insufficient food, nearly destroyed the army.

To carry out the new arrangements, two French regiments (six battalions) encamped on Inkerman Heights. Shell Hill, the scene of so many heroic deeds, was permanently included in the allied positions and fortified, and other works were thrown up overlooking the Tchernaya Valley.

On the 23rd of the month the French took over the pickets on Inkerman Heights, but the 30th and the other regiments of the second division still remained in their old camp to share if necessary in the defence of the position. Although relieved to a certain extent of outpost work, fatigues and working parties and guards of the trenches employed all their available men. The fatigues were mostly to fetch regimental stores from Balaclava. One disastrous result of the battle of Balaclava had been that we had drawn bac our lines and abandoned the Woronzoff Road to the enemy, and all supplies for the army had to pass along the direct Balaclava and Sebastopol Road, which was soon so cut up as to be almost impassable in rainy weather. This was especially severe on the troops on Inkerman heights, for the Woronzoff (Post) Road ran past the 30th camp and the distance to Balaclava by it was a couple of miles shorter than by the other. Our men were forbidden to use the Woronzoff Road, which led past the enemy's outposts into the French lines on the heights, but to the end of the siege men risked the chance of capture or punishment by using the short cut. Sergeant Thomas Shaw of the 30th was captured by the Russians while doing so in January, 1855, and remained a prisoner until the peace.

The officers present on January 1st, 1855, were Lieut.-Colonel Patullo; Captain Green; Lieutenants Falkner, Williamson, MacPherson, Stamer, Gubbins, Campbell, Austin and Walker (Adjutant); Quartermaster Morris, Surgeon Dowse and Assist.-Surgeons Milroy and Mackesy. There were in the Crimea 508 other ranks fit for duty and 57 sick. Lieutenant Hobbs had been invalided on November 16th to Scutari, where there were now eight officers and 304 other ranks sick and wounded. On recovery Hobbs was posted to a small detachment on the Bosphorus and did not rejoin in the Crimea.

Captain Dillon, Lieutenant Austin, one sergeant, and fifty privates had joined from the depot on December 2nd, but Captain Dillon had soon been obliged to go sick, and a Medical Board ordered him home on February 10th.

Although the regiment was not healthy and the recruits went sick fast, the effective strength in rank and file had risen from 409 on December 1st to 446 on January 1st, not only from the arrival of Dillon's draft of fifty men, but from the return to duty of those wounded at Inkerman. The month of December had commenced with heavy rain but turned cold towards the end. In the beginning of January the cold was intense and occasionally there were heavy falls of snow.

A mess had been established by the officers before the storm, and when Lieut.-Colonel Patullo obtained a new tent and fixed a stove in it the officers were as comfortable as was possible to men in that climate and insufficiently clad. On January 21st, Thomas Mapleson Fitzpatrick, who had left the regiment with the rank of lieutenant in 1851, rejoined as a volunteer.

There were only slight changes in the strength in the Crimea in January, though the sickness, especially among the young soldiers, was very great.

Hospital Sergeant James Holmes, who had been since the beginning of the war at Scutari, was appointed purveyor's clerk there on December 23rd, and Sergeant Francis Stanley became regimental hospital sergeant. Orderly Room Sergeant Thomas Jackson went sick to Scutari in January, and died there in February.

The paymaster clerk, John Hilder, with the office was moved from Varna to Scutari on the breaking up of the Varna depot, and Lieutenant John Forbes was ordered to rejoin the regiment.

From the time of the battle of Inkerman to the second week in January the regiment being mostly engaged in bringing up stores had no casualties from the enemy's fire, but on January 9th Private John Bowles and Private Patrick Enright were killed. Between the 22nd and 25th two men were wounded.

The weather had throughout the month been variable. Frost, rain and snowstorms succeeded each other and the men, who were on duty two nights out of three, were never dry. After returning from the trenches in the morning they were employed bringing up ammunition and material to the front. The few transport animals we had had died from overwork and want of nourishment. To march 7 miles to Balaclava and return laden through a sea of mud which came over the boots was a terrible task for men insufficiently clothed and fed and who had done a night's duty in the trenches. Fuel, which consisted of the roots of the oak or vine, was hard to get and difficult to light, and to save trouble men ate as much as they could stomach of their pork ration raw, and threw the rest away. Scurvy was very prevalent, and the sufferers were prevented by their inflamed gums from eating the hard biscuit. The rations seldom failed altogether but were frequently issued in the evening after the men for night duty had marched off to the trenches and the others were too tired to cook. Naturally there was a craving for a hot drink, but the coffee berries were green and issued unground, and it was not every one who had time or energy to bruise them with a stone in a fragment of shell and then roast and boil them in the mess tin. The men slept in their wet great-coats and blanket without removing their one pair of boots, which it would have been difficult to get on again. No part of their clothing was sound, for the people at home did not realize that clothing on a campaign has twice the wear of that in barracks and saw no reason to anticipate the usual annual issue in April. It is needless to say that the men's bodies and hair swarmed with vermin. This was at times the state of all ranks.

In spite of all, the spirit of the army was unbroken and there was a steadfast determination to win. The men even tried to conceal their sufferings from their own officers. As Sir Evelyn Wood says, " there is nothing in history grander than the enduring courage and discipline shown in the winter of 1854–5."

At the end of January the regiment had 378 effective rank and file present and 423 sick and on command. The weather in the first part of February was no better than that of January, and on the 7th Mark Walker entered in his diary " The second division is fast melting away." A few fine days from the 18th to the 20th raised the spirits of all, but the snow and cold returned on the 20th and lasted for several days. This was almost the end of one of the worst winters ever experienced by the British Army.

The hard weather had brought in a lot of ducks and some snipe in the Tchernaya Valley, and among others Major Patullo and Mark Walker went after them, though with not much success.

There was only one casualty from the enemy's fire in this month ; Lance-Corporal Charles Douglas was killed on the 3rd. A draft of one sergeant and fifty men joined on the 7th from Fermoy under Captain Robertson and Lieutenant Hill.

On the 16th the quartermaster, Timothy Morris, was ordered to proceed to Scutari to take up the duty of quartermaster of the army depot then commanded by Lieut.-Colonel Sillery. Ensign John Moon succeeded Morris as regimental quartermaster. Morris never rejoined the 30th. When the Scutari depot was broken up after the peace he was appointed to the depot battalion at Colchester commanded by Lieut.-Colonel Whitmore, late 30th regiment. He received the reward for distinguished and meritorious service on March 8th, 1867, and retired with the rank of captain on April 1st, 1870.

On the 27th of the month Lieut.-Colonel Mauleverer returned from Scutari and resumed command and on the following day General Pennefather also rejoined. Both were received with cheers by those under them. The spring had now fully declared itself and Mark Walker records that he gathered crocuses on the field of Inkerman on the 27th. The health of the regiment was very unsatisfactory in February. The daily sick rate was 98, and in spite of the arrival of the draft the effectives present had fallen to 300. At the end of January a new distribution of the regiment had been promulgated from the Horse Guards which was as follows :—

Crimea—8 Companies : Lieut.-Colonel, 1 ; Majors, 2 ; Captains, 8 ; Lieuts., 14 ; Ensigns, 6 ; total, 31.

Malta—6 Companies : Major, 1 ; Captains, 6 ; Lieuts., 8 ; Ensigns, 4 ; total, 19.

Home—2 Companies : Captains, 2 ; Lieuts., 4 ; Ensigns, 4 ; total, 10.

It was probably with a view to his organizing and commanding the reserve battalion at Malta that Brevet-Lieut.-Colonel Patullo was sent home on March 5th. In his diary Mark Walker says of him, " He is a great loss to us, a better fellow never breathed, he was always ready for his work and cheerful about it." Many years after his death, men who had served under him used to say of Patullo that it was his firmness, patience and soldier-like qualities which dragged the regiment through that awful winter in front of Sebastopol.

On the 13th Mark Walker entered in his diary, " We were inspected by General Pennefather at 2 p.m. We mustered 300 and got off very well. He certainly favoured his favourite regiment."

There was a strong liking between the general and the regiment. To others he was " Sir John " or " General," but to the 30th he was " Old Matt.," not that his name was Matthew, but Lieutenant Matthew Pennefather of the 30th being affectionately known to all as Matt., it followed naturally that the general must be " Old Matt."

The 30th were not engaged in the repulse of the sortie made by the Russians on March 22nd, and the casualties during the month were five men wounded.

Besides the loss of Lieut.-Colonel Patullo there were several changes in the regiment during the month. Captain Robertson went sick to Scutari. Lieutenant Falkner was ordered home by a Medical Board and Lieutenant Forbes rejoined the regiment on the 24th. The officers commanding companies on March 31st were :—Grenadiers, Lieutenant John Forbes ; 1, Lieutenant L. MacPherson ; 3, Lieutenant J. P. Campbell ; 4, Lieutenant A. J. Austin ; 6, Lieutenant Stamer Gubbins ; 7, Lieutenant E. Hill ; 8, Lieutenant A. H. Williamson ; Light, Captain C. M. Green.

The lieutenants were noted in Army Orders as permanently in command of companies and entitled to field and other allowances on the same scale as a captain.

Quartermaster-Sergeant John McLellan had succeeded William McGlade as sergeant-major at the end of March, and Sergeant Michael Tooner had become quartermaster-sergeant.

The colour-sergeants were, William Barnes (Scutari), James Brown,

David Dunn, William Jameson, John Lockhart (Scutari), Richard Nagle, Daniel Sullivan, and Thomas Tennent. Mr. Fitzgerald, the paymaster, after a month's overhaul of the books in the pay office at Scutari, appended to the pay list of March 31st a statement that eighteen men in the muster rolls were reported dead, but that the place, time and circumstances were unknown. No doubt these men had fallen sick, been hurriedly embarked, died on board ship and been buried at sea without the captain of the transport ever knowing their name or regiment. During the advance after landing at Old Fort when cholera was prevalent there were many cases of this sort in the army.

With the great improvement in the weather after the end of February the camps had been made much more comfortable, the standing room in the tents had been increased by digging out the interior and an ample supply of warm clothing had been received, more indeed than was necessary.

There was a slight improvement in health in March. The "daily sick" had fallen to 84½, and although no draft had been received, the effectives present had risen to 310.

The siege works were now pushed on more vigorously and the second division was called in from Inkerman Heights to the neighbourhood of Cathcart Hill so as to be nearer their work. The French were now definitely committed to the attack on the Malakoff as well as the Russian defences from the Flagstaff Bastion to the mouth of the harbour. The British were between the right and left French attacks and our main object was the capture of the Redan while supporting the attack of the French on the Malakoff on our right and on the Flagstaff Bastion on our left.

The 30th camp was moved on April 1st. On the new ground no excavation was allowed as the huts for which Lord Raglan had asked in November were said to be nearly ready to be delivered. As a matter of fact officers and men were still in tents when Sebastopol fell five months later, although a few huts were in use as hospitals.

In April both the Russians and the Allies were active in pushing forward their entrenchments and the casualties became more severe. In front of their main line of defence the Russians had established a couple of lodgments or rifle pits to annoy our working parties in the left advanced sap. On the evening of the 19th they were both taken by the light division, but on the following morning the Russians recovered one lodgment but failed in their attack against the other, which our engineers had connected with our advanced trenches. The second division came on duty in the trenches at 6 p.m. on that day, finding 800 men for the advanced parallel and 400 for the 21-gun battery. In the early part of the night the Russians attacked our men but were easily driven back. It was important to find out if they were still in the lodgment they had recovered in the morning, and after midnight Mark Walker volunteered if twenty men would join him to move out and ascertain. Colonel Mauleverer at first refused his consent, saying none of them would come back alive and he could not afford to lose his adjutant. Walker at last was allowed to go. He gave his watch to a brother officer with instructions to whom to send it, gave the bread and pork in his haversack to the men near him and called for volunteers. The number was soon made up and a volunteer working party was also assembled to follow in case the lodgment was found unoccupied. Walker and his party advanced about 3 a.m. and reached the lodgment undiscovered. They sent word back to the advanced parallel and were then joined by the working party who completely filled in and levelled the lodgment.

For this exploit Walker was mentioned by Lord Raglan in his despatch to the Secretary of State and recommended for promotion, which he soon received in the form of a company in The Buffs. He was also publicly thanked on church parade by the general on the following Sunday.

The casualties during April were :—Privates Ganley, Carey, and Walsh killed, John Desmon died of wounds and ten rank and file wounded.

On the 16th Major Whitmore rejoined the regiment from the staff and nineteen recovered men joined from Scutari. On the 23rd Francis Stanley died. He had been scarcely three months hospital sergeant. He was succeeded by Sergeant John Keegan.

In April there was a still further improvement in health. The daily sick had fallen to 59⅔, and the effectives present had risen to 326.

During the months of extreme misery no one had cared what the officers and men wore ; warmth was the one essential, but on April 28th Lord Raglan published an order saying that the winter being over, those costumes which were suitable to that season should now be laid aside and that the officers of the army would now be expected to appear in their proper uniform with black handkerchiefs round their necks. He was careful to add that, as the nights were still cold, warm clothing might be worn in the trenches.

On May 5th the second division was on duty in the trenches when the Russians managed to rush them. They were at once driven out. Stamer Gubbins on this occasion distinguished himself. He was a man of immense physical power, and disdaining the use of a sword he laid on with an enormous blackthorn he had brought from his home, till the Russians were fairly terror-stricken. Lord Raglan mentioned Stamer Gubbins in very favourable terms in his despatch to the Secretary of State. The 30th had three wounded, among whom was Lance-Corporal (Sandy) Borland, the future quartermaster.

On the 9th the division was on duty when the Russians again came on with a great show of resolution, but were stopped by the guard of the trenches when within 50 yards of the parallel. The guard who according to the Officer Commanding Royal Engineers " behaved nobly " had only one officer and fourteen men wounded, among whom was one of the 30th. On May 14th notification arrived of the promotion of Colour-Sergeant Daniel Sullivan to a commission in the 82nd for his gallant conduct at " Little Inkerman " on October 26th, 1854. From the 82nd he was transferred to the 13th Light Infantry, and after distinguishing himself during the Indian Mutiny retired from that regiment as a captain.

On the 19th Lieutenant Henry Wood and Ensigns Kerr, Neville and Sanders joined from the depot. On the 24th every man in the army got a double allowance of porter to drink Her Majesty's health on her birthday. Such a supply of a bulky thing like malt liquor shows how much the communication with Balaclava had improved ; the great cause was, of course, the railway, now almost complete, assisted by the dry weather which made the roads available for distribution from rail head. We had now full use also of both roads to Balaclava, for the French and Sardinians (who had landed in the second week of the month) advanced on May 25th across the Tchernaya and the Allies were no longer closely hemmed in by the Russian Field Army.

Although there were some cases of cholera the improvement in the physical condition of the army which had begun in April was now quite noticeable.

The casualties of the regiment in May were :—Sergeant Thomas Delaney and five rank and file wounded.

On June 1st the promotion of Lieutenant and Adjutant Mark Walker, V.C., to a company in The Buffs, who had landed at the end of April, appeared in General Orders. The loss to the regiment was very serious. The following extract from his diary shows how he felt about it :—

"I am now no longer one of the old three tens. It is with the greatest regret that I leave them. I have been with them for eight years and eight months, some of them the happiest of my life ; in bidding them farewell my sincere and heartfelt wish is that every happiness in life may attend them. I am now about to make new friends, but none I can ever make will equal those I leave behind. I have known and tried them long and always found them the same, sincere and firm."

Before he left the regiment his old friends Patullo and Atcherley arrived from England. A fresh order of March 2nd, granting a second lieut.-colonel to regiments in the Crimea, had restored Lieut.-Colonel Patullo to the service companies and Atcherley had recovered from the effects of his severe wound at "Little Inkerman."

Lieutenant John Forbes succeeded Mark Walker as adjutant.

It is worth noting that at this time the mess of the regiment was splendidly managed and it was considered a privilege to dine at it.

On June 6th and 7th a tremendous bombardment was kept up by the British and French batteries and about six o'clock in the evening of the 7th the French assaulted the Mamelon which covered the Malakoff while the British rushed the Quarries in front of the Redan. Both attacks were successful, but an attempt to push on to the main works of the Redan and Malakoff, which was made without proper support, failed badly, the number of officers hit being out of all proportion to the total casualties. Among them was Captain Matt. Pennefather, who had only been ten days in the Crimea. A fragment of shell struck him on the thigh, penetrating almost to the bone. He died at Parkhurst on November 27th, 1857.

In consequence of the return of Colonel Patullo to the Crimea Major E. A. Whitmore was sent in June to Malta to command the six reserve companies which were ordered to be formed there. After a short stay at Malta Major Whitmore, finding nothing to command, asked for and obtained leave to England on private affairs.

On the whole the Malta depot was a failure. Owing to the urgent demand for men in the Crimea, neither officers nor other ranks were allowed to stay in the island long enough to serve any useful purpose. On April 1st, 1855, there were present 6 officers and 130 other ranks, but by June 12th there were only fifteen men left. A strong draft joined the Provisional Battalion, as it was called, in August, but by October 1st the strength was reduced to Lieut. R. O. Campbell and twenty-five other ranks. By this time active operations in the Crimea had ceased, and the strength in Malta began to rise again. During their short stay in the island it is believed that all ranks received some slight instruction in musketry. All rifles had been withdrawn from the home depots and some soldiers went into action in the Crimea who did not know how to load a rifle.

On June 9th Captain Mark Walker, V.C., made the last entry in his journal with his right hand, for on the following night when on duty with The Buffs, his new regiment, in the trenches, his right elbow was shattered by the explosion of a howitzer shell close to his side. The arm was amputated above the elbow and he was sent home by a Medical Board as soon as he was able to move.

After a distinguished career he died as General Sir Mark Walker, K.C.B. His affection for the 30th remained unchanged to the end of his life.

The Allies continued the bombardment of the Russian works, especially the Redan, Malakoff, and Little Redan, till June 17th, when orders were issued for an assault on the following day. The French were to attack the Malakoff and the Little Redan and the British the Redan and Garden batteries. The second division was to assault the salient of the Redan while the light division attacked the right and the third and fourth divisions the left face, looking from our trenches.

The attack was not preceded by any artillery preparation ; the French, through a misunderstanding of signals, sent forward their columns in succession and not simultaneously, and the British force was a great deal too weak for its task. The attack failed with a loss to the British of about 1,500, to the French of 3,500 and to the Russians of 5,400 men. As the second division through a mistake did not take part in the assault a detailed description is unnecessary.

As soon as it was evident that the assault had failed our guns reopened and under cover of them we proceeded to sap closer to the Redan. On June 18th the assaulting troops had been called upon to advance 500 yards in the open after leaving the trenches.

On the 20th, Captain Pocock joined from the depot at Fermoy.

On June 26th Ensign T. Mapleson Fitzpatrick died of cholera. His commission had been restored to him on May 25th, after he had served four months as a volunteer. He did not know what fear was and on one occasion, after dining in the French camp, he accompanied his hosts as an after dinner joke in their attack on the Mamelon. Mark Walker says of him, " His high spirits and good nature endeared him to all."

On the 29th, Lord Raglan died, an irreparable loss to our army. He was succeeded by General James Simpson, the senior officer on the spot.

The regimental casualties during the month of June were one private killed, Captain Pennefather and 15 rank and file wounded.

A draft of thirty-eight rank and file without an officer arrived from Malta during the month, raising the number of effectives present at the end of the month to 366.

Throughout July the approaches to the Russian lines continued to be pushed on in spite of the enemy's fire and repeated small sorties. The *London Gazette* at this time had some announcements of interest to the regiment. Lieut.-Colonels Mauleverer and Patullo were made Companions of the Bath on July 5th.

Assistant-Surgeon Mackesy was allowed to resign his commission on June 1st, and on the same date was given a fresh commission as ensign without purchase. He had served with Headquarters throughout the war, but was now sent to the depot.

The casualties in July were :—Private Robert Dowell killed, Private Edmond Clarke and Private Isaac Bott mortally wounded, seven rank and file wounded.

On the 12th of the month Captain Stevenson had joined with a draft of nineteen men from Malta. The health of the regiment was good and on the 31st the number of effectives present was 395.

During the month Lieutenant L. MacPherson, who had not missed a day's duty since landing in the Crimea, was sent sick to Scutari on the 30th, and Ensign Neville was invalided on the 27th. He never rejoined.

On August 16th the Russian Field Army attacked the French and Sardinian covering army on the Tchernaya and was defeated with a loss of 8,000 men.

On the following day the Allies began their final bombardment with 800 guns.

The Russian and British sentries were now in shelters in front of their lines, often only a few yards apart, and as we got nearer the casualties on both sides had increased.

Those for August in the regiment were :—Privates James Dunn, Arthur Ingram, Moses Kemp, Henry Richardson and George Cann killed; Lance-Corporal Robert Tyrie and Private Alexander Still mortally wounded; eighteen rank and file wounded.

The health of the regiment was good and the number of effective privates on the 31st was 385.

On the night of August 31st–September 1st the regiment suffered a great loss in the death of the adjutant, Lieutenant John Forbes. He was on duty as adjutant of the trenches when about 12.30 a.m. the Russians made a determined sortie. At the particular point where the attack was made the supports were not sufficiently close up and the working party fell back, but upon the advance of our reserves the Russians retired. G. H. Sanders, his successor as adjutant, has left the following account of Forbes' death :—

"It was night and I was asleep in my tent when I heard a commotion outside; struck a light, and found poor Forbes borne on a stretcher, the blood dripping through the canvas. I asked, 'Who have you there?' and I heard him answer in his broad Scotch, 'It is me, Sandy; I am badly hit, and it's a' up with me, I'm thinking.' He was taken to his tent, and there, to my horror, I found his left arm literally carried away from his body (it had been hit by a round shot) and only connected by a few sinews. He died of his wounds that day."

In those days the distinctive marks of an adjutant were a steel scabbard and brass spurs. Those belonging to Lieutenant Forbes had been worn by Adjutants Whitmore, Paget Bayly and Mark Walker. At the auction of his effects after his death, the competition ran very high, as there were several newly-appointed adjutants in the Crimea all anxious to possess the insignia of their appointment. It was eventually left between Lieutenant Sanders, the newly-appointed adjutant of the 30th, and Lieutenant and Adjutant

Burke of the 49th, to settle who should possess the scabbard, which, as it was a point of honour that it should not go out of the regiment, was bought by the former. Of the officers who wore it in the Crimea, one was killed, as above related, Mark Walker, V.C., lost an arm, and Sanders a leg. While in the latter's possession it was hit during the assault on the Redan by a canister shot. It cost its fortunate possessor no less than £2, but may well be considered a bargain at that money. Few more interesting relics exist than this shattered steel scabbard, three successive owners of which lost a limb.

Lieutenant Gilbert Sanders had served for a time in the Austrian army before joining the British.

In September the casualties were:—Lieutenant and Adjutant John Forbes, mortally wounded; Drummer Walter Cleary, Privates Richard Dawson, James Cree, Thomas Hunter, and Thomas Lennon killed; Corporal John Hardy mortally wounded, and ten rank and file wounded.

THE REDAN

On the 6th a draft of 3 officers, 1 sergeant and 52 rank and file arrived from Malta. The officers were Captain Campbell, Lieutenant Moorsom and Ensign Deane.

The effective strength was now that with which the regiment attacked the Redan two days later and was as follows:—

Lieut.-Colonels James Thomas Mauleverer, C.B., and James Brodie Patullo, C.B.; Captains F. T. Atcherley, A. C. Campbell, G. F. C. Pocock, C. M. Green, A. H. Williamson, and J. C. N. Stevenson; Lieutenants H. Wood, A. J. Austin, C. J. Moorsom, H. C. Singleton, J. Fleming, M. B. Field and W. Kerr; Ensign R. G. Deane, Lieutenant and Acting Adjutant G. H. Sanders, Surgeon Francis Reynolds (attached), Assistant-Surgeons D. Milroy and J. R. Streat (attached).

Lieutenant J. P. Campbell was serving with the commissariat and Quartermaster Moon was sick. Dr. Dowse, the regimental surgeon, after a year's continuous service, had been obliged to go sick.

The staff-sergeants were Sergeant-Major John McLellan, Quartermaster-Sergeant M. Tooner, Paymaster's Clerk John McLean, Drum-Major Thomas Gunning, Hospital Sergeant John Keegan, Acting O.R. Clerk William Colcutt. Armourer Meagher had been invalided and no one had taken his place. There were six colour-sergeants, David Dunn, William Hunns, Hastings McAllister, Thomas McDonagh, Richard Nagle and Thomas Tennant.

The other ranks were 21 sergeants, 13 drummers, 24 corporals, and 421 privates, making a total of 510 of all ranks present and fit for duty in the Crimea.

The actual number who left the trenches for the assault on the Redan was:—2 F.O., 6 captains, 8 subalterns, 2 staff-sergeants (sergt.-major and drum-major), 21 sergeants, 2 drummers, 384 rank and file, 425 of all ranks.

The men left behind as camp guards, etc., were selected from the weak and untrained. There were also, as usual, a number employed by the transport and other departments.

Before describing the share which the 30th took in the assault on the

Redan, the following slight sketch of the position to be attacked and of the general course of the action is given.

By September 5th the British approaches were within about 240 yards of the Redan, the French to our right were within 25 yards of the Malakoff and still further to the right they were even nearer to the Little Redan. A closer approach on the part of the British to the Redan would have entailed a great delay and heavy loss of life, for in front of them there was an outcrop of rock which did not admit of the ordinary methods of sapping and trenching. It was, therefore, determined by the allied commanders that the assault should be delivered as soon as the fire of the place was subdued.

Although it was considered necessary to attack all along the line covering the Karabelnaia suburb, in order to occupy the enemy, the success or failure of the British attack on the Redan, however important to ourselves, would not affect the fate of Sebastopol if only the French could take and hold the Malakoff, which was the key of the position and, being a closed work, easy to hold when once occupied. The fact that the British had 240 yards of open ground to cross to reach the Redan and the French only 25 yards to the Malakoff does not give the full measure of the difference between the tasks assigned to each of the Allies.

The French trenches were provided with places of arms where large bodies of troops could be assembled and whence they could issue to the attack by suitable openings prepared for them and their supports. The British had not been able to construct more than the parallels and the communications between them, and the regiments detailed for the storming party and its supports had to file one after another into the fifth, the most advanced, parallel and from there advance to the attack in succession.

It was certain from the imperfection of our approaches and of our arrangements, that our supports would advance piecemeal and that a great part of them would be too late and that the three divisions in reserve (first, third and fourth) would take no part in the action at all. Our advanced trenches too were soon crowded with wounded. The surgeons followed their regiments through the trenches, stopping from time to time to bind up a man's wounds, but on getting near to the mouth of the advanced sap they were overwhelmed by the number of wounded and fixed their dressing stations there. Their action was according to their orders, but the result was that our advanced trenches were blocked.

The French had 100,000 effectives in the Crimea and we not a fifth of that number, which accounts for the slow progress of our approaches and their incompleteness.

The works known as the Redan and Malakoff were placed upon two rocky eminences somewhat over 1,000 yards from each other. The Malakoff was 330 feet above sea level and the Redan 306 feet. The faces forming the Redan were 70 yards long, meeting at an angle of about 65 degrees. At the salient the parapet was 17 feet above the surface of the ground, and 28 feet above the bottom of the ditch, which was 11 feet deep and 20 broad. On the faces the dimensions were a trifle smaller. The parapet was continuous with the adjoining works, but as the ground fell to the rear, those works were on a much lower level than the salient they flanked. A strong abatis had been placed 100 yards in front of the salient, but at the time of the assault it had been so shattered by our artillery fire as no longer to be of importance as an obstacle.

The Garden and Flagstaff batteries on the other side of the Woronzoff ravine swept the right face and the ground in front of the Redan with their fire as the batteries near the Malakoff did the left. A strong retrenchment had been made in rear of the Redan, and unless the enemy were driven from that the work could not be held even if captured. Our attack from the fifth parallel was 240 yards up hill, finishing with the crossing of the ditch and mounting the rampart, for which ladders had to be used. It is evident, therefore, that a halt must be made when the Redan was occupied and a fresh attack be made from there, if possible. It was certain, too, that if our troops

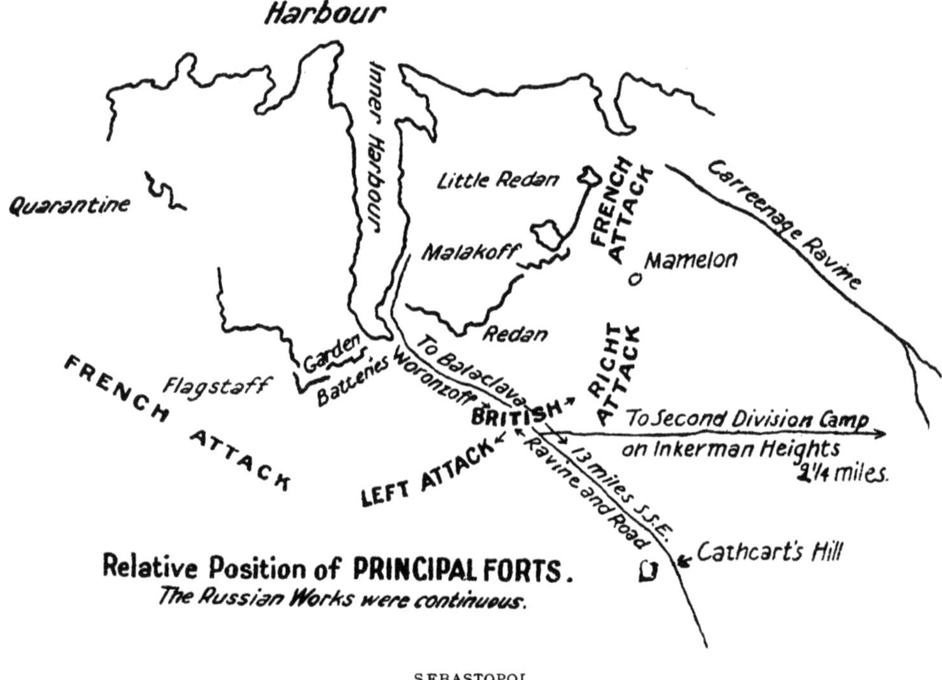

SEBASTOPOL.

in the Redan once engaged in a musketry combat with the Russians in the retrenchment, they could not be led forward and that the only course was to entrench in the Redan unless an attack on the flanks of the retrenchment was made across the open by troops from our reserves.

Even if the Redan and retrenchment had been carried they would have been difficult to hold owing to their being open to the fire of the Flagstaff and Garden batteries. Lord Raglan was strongly opposed to an assault on the Redan unless the French undertook at the same time to assault the Flagstaff bastion and, had he lived, the plan of attack would probably have been different.

On September 5th the final bombardment commenced, and on the night of the 7th orders were issued for the assault to take place at noon on the following day. It was hoped that at that hour the morning's cannonade might have driven many of the defenders to their bomb proofs and underground shelters.

The French were to rush the Malakoff, Little Redan and other defences to the right. The British were to assault the Redan as soon as a white flag displayed on the Mamelon showed that the French were in the Malakoff. Sir James Simpson decided that the second and light divisions, having borne the chief share in the labours and dangers of the siege, should have the honour of delivering the assault, the arrangements for which were left in the hands of Lieut.-General Sir Wm. Codrington, commanding the light division, assisted by Lieut.-General Sir F. Markham of the second division. Sir Wm. Codrington selected Colonel Unett to lead the stormers of the light division and General Markham named Colonel Ash Windham to lead the storming party of the second division. The two colonels tossed up for the honour of leading and the spin of the coin was in favour of the light division.

The following is the order in which the troops were disposed for the attack :—

Covering party	200 of Rifle Brigade.
Ladder party	300 of The Buffs and 97th regiment.
Stormers	500 of the light division.
	500 of the second division (41st and 62nd regiments).
Supports	750 light division (19th and 88th regiments).
	750 second ,, (30th and 55th regiments).
Working party	100 of the 90th regiment.

Red coats and forage caps to be worn, ten extra rounds of ammunition and two days' cooked rations to be carried and water bottles to be quite full.

The share of the 30th in the action was as follows: The Rouse went at 6 a.m. and at 8 o'clock the regiment paraded and marched right in front into the fourth parallel. Colours were not carried. The fifth parallel and the communications leading to it were already filled by the covering party, ladder party, stormers of the light and second division and the supports of the light division. From the fourth parallel the 30th moved gradually forward with frequent halts towards the fifth. It was followed by its old friends of the 55th and the officers of the light company of the 30th and of the grenadier company of the 55th walked between the two regiments conversing with Colonel Patullo, Sanders the adjutant and others of the Staff. Shortly before noon a halt was made and the officers of the 30th and 55th assembled round Brigadier-General Warren and chatted. Grape and round shot were flying overhead.

Exactly at noon there was a cry, "There go the French," and regardless of the fire the officers jumped on the parapet and were delighted to see the French swarming into the Malakoff, from which large bodies of Russians could be seen retiring to the Redan.

At twenty minutes after noon the white flag was displayed on the Mamelon and the officers of the 30th and 55th shook hands and ran to join their regiments. As soon as the stormers of the light and second division and the supports of the light division advanced, the 30th formed line in the fifth parallel with intervals between the companies and orders to spread out so as to embrace as much as possible of the right face of the Redan in the attack and not crowd towards the salient.

Immediately after the line was formed, the word was passed to advance and the regiment leaped the parapet. At this moment Colonel Mauleverer was struck down. As the adjutant was binding up his wounds he said faintly, "Thanks, many thanks, but tell Patullo he commands the regiment." Colonel Patullo, conspicuous from his tall handsome figure, led the regiment gallantly until he fell mortally wounded before reaching the Redan; the sergeant-major and orderly bugler fell about the same time. Captain Atcherley was now the senior officer left; four companies were without officers. The hail of grape which swept the glacis surpassed anything which had before been deemed possible and before reaching the Redan Captain Pocock fell close to the abatis, his subaltern, Kerr, was already down, Moorsom (afterwards the last lieut.-colonel of the old 30th), then quite a boy, was struck in the elbow joint when close to the counterscarp and, half-dazed by the painful wound, tried to descend into the ditch for shelter, but Atcherley, whose courtesy and kindness never failed under any circumstances, said, "No, my boy, you must not do that," and forced him to retrace his steps across the open to our trenches and probably saved his life thereby.

The first three men of the regiment into the Redan were Colour-Sergeant McAllister, Sergeant Rigney and Corporal O'Brien, closely followed by Lieutenant and Adjutant Sanders, who, when McAllister was shot through the thigh, ordered him to retire, getting for answer, "I've done nothing for old England yet," and when the adjutant tied a handkerchief round his wounded limb, and the faintness of death came over him, his one question was, "What will you say of me?" "I will say you were a good brave soldier," was the answer; but Sergeant McAllister's course was not run yet, for he survived this and four more wounds received that day, and besides being recommended for a commission he received the French war medal and an appointment in the Royal Hospital.

On arrival at the Redan, the 30th found the stormers and supports who had preceded it sheltered behind traverses or lying on the superior slope of the rampart, exchanging shots with the Russians, who had now occupied their retrenchments and whose numbers were increasing rapidly. Our regiments were mixed up and there was nothing for the scattered men of the 30th to do but to lie down where they could find room and take up the firing. They were not called on to do more. The 55th soon joined them. They also had lost their commanding officer and the brigadier had been wounded. Those men for whom there was no room took shelter on the outer slope and passed their cartridges, when required, to their comrades in front. The Russian fire was very close and deadly, and being a converging fire from a much wider front quite dominated ours. Colonel Windham, who was in command, Colonel Unett like most of the senior officers having been knocked over in the advance, sent three officers in vain for assistance and at last took the desperate resolution to quit his command and go to the rear to appeal to Sir William Codrington in person for support. He mentioned his intention to the nearest officer so that in the very probable event of his being killed his motives might not be misunderstood, but with his departure all appearance of authority ceased, for no one knew who was in command. The Russians, who had been receiving reinforcements continually and who were in overwhelming numbers, now threatened to take the offensive and our men retired, having held their position unaided for more than an hour. Their retreat was hastened by a land-slide. The 28-foot rampart at the salient gave way and 300 men were precipitated into the ditch and many hurt by their comrades' bayonets. Some were buried by the falling earth, but apparently all were extricated. The Russians followed our men when they retired, but did not quit the work. The retreat across the glacis was made under a heavy fire, but as far as the 30th were concerned, it was not so deadly as the advance.

The French managed to hold the Malakoff. Their attacks upon the open works, the Little Redan and Central Battery failed as ours did.

Among the last to leave the Redan was Lieutenant Gilbert Sanders, the acting adjutant. He was already wounded in the leg but was now struck down when near our trenches. He and others owed their lives to the gallantry of Captain Gronow Davis, Royal Artillery, who commanded the spiking party. Collecting some volunteers from his own regiment he sallied out into the "terrible fire" (this is Marshal Pellissier's expression) and conveyed as many as possible to a place of safety. The good fellows who bore off Sanders laid him down inside the advanced sap and went back into the fire for others. At this time a false alarm was raised that the Russians were advancing and all wounded who could be moved were hurried to the rear. Luckily Sanders had kept his senses and he was able to recognize and call upon Dr. Harkness of the 55th. Harkness was shocked to find the friend with whom he had advanced through the trenches in the morning, with one knee completely shattered and a bullet in the other leg. He did what he could for Sanders, who was bearing his sufferings nobly, and sent him to the rear upon a piece of a broken scaling ladder. Gronow Davis got a well-earned Victoria Cross for his gallantry upon this occasion.

The conduct of every one concerned in the attack on the Redan, from the Commander-in-Chief to the private, has been hotly canvassed. The following from the Journal of the Officer Commanding Royal Engineers is the calmest and fairest view:—

"It is very much to be regretted that the troops who with so little difficulty had forced their way into the Redan were not strongly reinforced.

"There cannot be a doubt if a large number had been sent in support the assailants could have maintained and entrenched themselves within the work. That they did not do so must not be attributed to any want of bravery on the part of the troops who so gallantly held their ground for one hour exposed to such a destructive fire as was poured on them. They retired when they found themselves without any officer of rank to command them."

The statement that our troops forced their way into the Redan with little difficulty applies only to the stormers. The Russians were to a certain extent surprised, but when the 30th left the trenches, the enemy's fire was fully developed.

British troops do not often fail, and when they do there is a good deal of wild talk about the reason. In the case of the Redan the young soldiers who joined in the Crimea received the blame for the miscarriage. It requires no argument to prove that a trained soldier is better than an untrained one, and the company officers who led at the Redan agreed that the young soldiers required a word or two where the older men followed without hesitation. The younger men, too, were not so handy and were more difficult to manage in the confusion which followed the mixing up of corps in the Redan, but that is the extent of it; there was no general hanging back. It has been said that the officers, non-commissioned officers and old soldiers attacked the Redan, and the youngsters did not leave the trenches. This, as far as the 30th is concerned, is quite unfounded. Seventeen men who had enlisted since January 1st were among the killed, which shows they were well forward and the talk of the regiment was of the cruel luck of the last draft which had far more than its share of casualties. No blame attaches to the regiments for the failure of the attack, which was simply due to want of adequate preparation and energetic direction.

The following were the losses of the 30th:—

OFFICERS:—Killed—Ensign Richard Granville Deane, son of the Rev. George Deane, of Bighton Rectory, Alresford, Hants.

Died of wounds—Lieut.-Colonel Thomas Brodie Patullo, C.B. On both sides he was descended from old East of Scotland families.

Captain John Charles Newcombe Stevenson, son of Captain John Stevenson, of the Devonshire Militia.

Lieutenant William Kerr, son of Doctor William Kerr, Northampton.

Wounded—Lieut.-Colonel James Thomas Mauleverer, C.B., Brevet-Major A. C. Campbell, Captain Pocock (arm amputated), Lieutenants A. J. Austin, C. J. Moorsom (carried his arm in a sling for life), M. B. Field, G. H. Sanders, acting adjutant (leg amputated).

OTHER RANKS:—Killed—Sergt.-Major John MacLellan, of Glasgow, Lance-Sergeant William Moore, Corporals John Ross and James Collins, Drummer Henry Corron, Privates Edward Armstrong, William Besley, John Black, John Brennan, Simon Briese, Michael Broderick, Denis Bryan, John Burke, Edmund Cantwell, Samuel Carr, James Connell, Michael Connor, William Deane, Charles Eastoe, William Fenton, James Gillain, Michael Grainger, William Hage, Henry Hillary, John Hurley, Michael Long, Wilberry Longbottom, Charles McCarthy, James Martin, Robert McDowell, Edward McGooking, George McKibbin, Patrick McQuirk, James Moore, John Munn, Michael Nowlan, Charles Palmer, John Rawlins, John Riley, William Lade, Erwin Sergisson, John Stewart, George Watts, and James White.

Died of Wounds—Corporal James Rook, Privates John Byrne, Michael Connell, John Hale, Michael Maloney, Patrick Power, William Rigsby, William Dawes Stevens, Thomas Smith and Thomas Stenson.

Wounded—Colour-Sergeants Hastings McAllister, T. McDonagh, T. Tennant, Sergeants T. Fitzgerald, J. McPhail, J. Symington, and 94 drummers, rank and file.

Total casualties—Eleven officers and 154 other ranks killed and wounded. As the number who left the trenches was 425 the loss was nearly exactly two in five.

A party went out to look for Sergeant-Major McLellan, who was very popular. His body and that of Drummer Corron, which was lying beside the sergeant-major's, were the only ones brought in that night. The 9th was devoted to bringing in the bodies of the other dead and burying them. There was no need for an armistice—the Russians had gone.

As soon as it was evident that the Malakoff could not be recaptured, they had commenced their retreat over the floating bridge to the northern side of the harbour and by morning the movement was completed; the floating bridge was then taken up and their fleet, except one vessel, scuttled and sunk. At the same time a small party left behind set fire to the town and rejoined the army in boats.

Great care was taken in the allied advance, for the town and defences were supposed to be mined. On the 10th the fortress was formally occupied and thus came to an end the siege of Sebastopol. The Crimean War practically ended with it, as was recognized by Her Majesty's Government who restricted the grant of the medal to those present with the army on or before September 10th, 1855.

The following reminiscences of Lieut.-Colonel Sir George Pocock give a vivid impression of the attack and the feelings of those engaged:—

"The next morning found us in trenches awaiting the attack; but before receiving the order to advance to the storm the grog tub appeared, and many a poor fellow had his last drink then, some got none. The Malakoff, having been stormed and taken by the French, we received the order to at once attack the Redan, and I, not wishing my company to be behind the rest of my regiment, made them move briskly on without delay, and some without their grog. My poor subaltern, Kerr, became very excited, and I had some difficulty in persuading him not to expose himself in an unnecessary manner by jumping up the parapet of the trench, as we went along to see 'what was going on.' After losing a few men, and also having some little difficulty in persuading part of my company to cross a wide gap where the parapet had been knocked away, and the shot, shell and grape were pouring through and causing great havoc, we arrived at the spot where the regiment had to leave the trenches *for the open* to storm the salient angle of the Redan between 200 and 300 yards distant. With one little passing thought of 'home' and a prayer to Almighty God for protection: 'Come on, my men,' I said, as two or three fell close to me, and away we went into the smoke and roar and excitement of the battle. I had just got through the abatis and near the ditch outside the fort, when a grape-shot went clean through me just under the left collar-bone and out through my shoulder blade, the concussion smashing my arm down to the elbow; over I went like a rabbit. At the moment I felt more stunned than hurt, and began to pick myself up, when a bullet struck me across the right wrist, tearing the flesh, but fortunately breaking no bone; this at the time was a more painful wound than the first. Bleeding freely and feeling faint and awfully thirsty, I took a good pull at the rum and water in my little barrel I had with me. My left arm being useless, I managed to lash it to my side to prevent its swinging about; that done, I began to move back to the trenches, going perhaps ten or twenty yards at a time and then resting exhausted for a drink. Thank God, I never actually fainted, and in time reached the shelter of the advanced trench. Here I met my servant, who said, when I asked him how he came to be there, that he had just brought in a wounded officer. Feeling now comparatively safe, I let him cut my jacket off and take charge of my sword, pistol, etc., etc., and this was no sooner done than two shells burst in the parapet close to me and almost buried me with dust and débris. Shortly after this had happened our men had to beat a retreat from the Redan for want of supports, and the cry being, 'The Russians are coming,' and not caring under those circumstances to remain where I was, to be perhaps bayoneted by them, I got up and crossed the open ground for about 100 yards to the other trenches in rear, and (as the Russians did not come after all) hunted for a surgeon, found one, fortunately, and in my turn got attended to and my wounds plugged and patched up *pro tem.* I was then told to go outside his bunk and look out for a stretcher. I did so, but under the circumstances was a little too particular and would not take advantage of a few very bloody ones I saw pass by. No; I waited until I saw a, comparatively speaking, clean one, which I took possession of, and gave the word, 'Home,' and a terrible journey I had; stripped to my shirt, bleeding and exhausted, my wounds painful, I did not quite care for the jolting I got on my way across the open ground to my tent; the wind was bitterly cold that day, the dust was driven about in clouds, whilst round-shot, shells, and grape were flying past too near to be pleasant. On arriving near our encampment the first person I came across was my other subaltern, Pennock Campbell, who held an appointment on the commissariat staff, and consequently was not with me in the attack on the Redan. 'Who is that?' he said to my bearers. 'It's Captain Pocock of the 30th.' 'Is he dead?' 'No,' said I, 'old fellow; no promotion for you here.' He then asked me about his father, Major Campbell (then in the 30th), and I told him that I had passed him in the trenches, slightly wounded. Well; I was taken to my tent, put on my bed, and the doctors sent for; in the meantime I tipped the men who brought me home, and then asked for a cup of tea, which refreshed me. The doctors came and examined me, and said they would call again a little later, and when they did call again they told me, after probing the wound, that my arm would probably be amputated next day. I suppose I then dropped off to sleep, for I remember nothing more until I awoke during the night in much pain, and remembering what the doctors had said about my arm coming off, was really pleased when the time came for me to be carried from my tent to the hut to be operated upon—just eighteen hours after I was wounded. There were then four of us in this hut. Next to me was my poor Subaltern Kerr, mortally wounded, but who lived for a few days. Opposite to me was Sanders, with a leg amputated above the knee, and next to him Stevenson, mortally wounded, and who died that night. My turn now came to be attended to, and the glittering row of steel knives at my side was not *very* enchanting to behold. Chloroform was then administered, under the influence of which I was not kept long, and regained my senses long before the operation was finished. Cutting through tendons, cutting and tying up the veins, cutting out of the flesh the broken bits of bone, and finally stitching up the overlapping flaps, *very nearly* obliged me to collapse or faint, but a judicious application of brandy and water soon revived me, and after having been made comfortable the doctors left me. I am naturally musical, and this accounted for the laugh the doctors and others indulged in at my expense (so they told me afterwards) during the time I was under the influence of chloroform, for I sang right merrily all the time: they were also rather amused when my arm and Sanders' leg were being taken away to feed the Crimean worms, for I called out, ' Hold hard ; take the glove off my hand and you will find a couple of rings on the little finger, give them to me.' I have one of those rings now (twenty-eight years after). Soon after the doctors had left me I began to feel rather hungry and thirsty. Nature evidently wanted support (so I thought) ; so I said to my servant, ' You be off at once and see if you can get me a mutton chop or two and a glass of porter.' He soon returned with the objects of my desire, which I soon demolished, very wrong no doubt, and so the doctors thought when they heard of it, and told me so pretty plainly, but this refreshment did me more good than harm. Being young and healthy, always looking on the bright side of things, and without fear, everything progressed favourably with me, and at the end of five or six weeks I was ordered home, and joined the P. and O. Company's steamer *Ripon* lying in Balaclava harbour."

After the fall of Sebastopol, the chiefs of the allied armies were opposed to field operations being undertaken so late in the year ; the autumn and winter were, therefore, devoted to preparing for an active campaign in spring, either in the Crimea or elsewhere. The British army was hutted before winter, the working of all the departments had improved immensely, food and clothing were ample and the strength rose rapidly through the return of men from hospital as well as by reinforcements from home. Drills, field days and ball firing were constantly practised and a great deal of road making was done, but if there was an abundance of work there was a good deal of play, as there always is in a British army. Cricket matches, races and sing-songs in the canteens at night were frequent. The improvement of communications enabled forage to be issued once more and every officer owned a nag of some sort. Leave was only too easy to obtain and those who could afford it went home, others visited Constantinople, Greece, and the Ionian Islands. There were a large number of officers available for duty. In September, after the fall of Sebastopol, the following joined the 30th from home or Malta :—

Captains R. Dillon, J. O'Brien, Lieutenants C. J. Sillery, E. St. George Smyth, N. W. Massey, S. H. Smith, J. M. Allardice, Thos. Elwyn, and T. G. Peacock and Assist.-Surgeon C. H. Tovey. The number of effective men was not so satisfactory as in the case of the officers.

On October 1st, the effective strength of non-coms. and rank and file was 387 with 32 " sick present." Those " on command " and " sick absent " numbered 214. Colonel Mauleverer managed to sign the returns, but the daily duties of commanding officer were performed by Captain Dillon till the arrival of Major Pakenham on the 7th of the month. On November 1st Colonel Mauleverer was again on the sick list on account of his wounds. Lieutenant H. Woods acted as adjutant in place of Sanders, incapacitated by wounds. Lieutenant C. J. P. Clarkson had joined on October 16th.

In November, Sir William Codrington succeeded Sir James Simpson in command of our army in the Crimea.

Bt.-Major Atcherley and Lieutenant C. J. Tolcher joined the regiment from England in this month. Although his wounds still troubled him, Colonel Mauleverer resumed command during the month and continued in command till the armistice.

On January 1st, 1856, the effective strength had risen by 45 since October through the " daily sick " falling from 72 to 16. No reinforcements had yet reached Headquarters in the Crimea, but the four reserve companies in Malta had risen to a strength of 6 subalterns, 8 sergeants, 10 drummers, 168 rank and file. The subalterns were Lieutenants Harcourt (in command), and Henry L'Estrange Herring, Ensigns R. O. Campbell, Alex. Ewens, and Charles Proby Fitzgibbon. A draft from the depot at Fermoy on January 26th under Ensign Charles Tyner raised the numbers in Malta to a total of 226, the greatest strength to which the reserve companies attained.

On February 19th, Lieutenant Herring and Ensign Fitzgibbon landed in the Crimea from Malta with a draft of 70 non-commissioned officers, rank and file.

On February 25th Sir William Codrington reviewed 46 battalions of his army, numbering 25,000 infantry, before the allied commanders-in-chief, and a huge concourse of officers of allied and neutral armies as well as of the British. The effect of the previous four months' work was at once apparent and all were impressed by the splendid bearing of the troops and the ease and exactness of their movements. What was equally important was that the army itself was satisfied and felt a renewed pride and confidence in its strength.

On March 1st an armistice between the Allies and the Russians was announced. Colonel Mauleverer was again on the sick list and Major Pakenham commanded till the arrival of Lieut.-Colonel Whitmore from Malta on the 11th. It was evident that our Government was determined to make peace, for the reserve companies there were practically abolished. Two drafts amounting to 103 non-commissioned officers, rank and file landed in the Crimea about the middle of the month and the 54 non-commissioned officers and men left in Malta were ordered to be held in readiness to join the depot at Fermoy. The officers who joined Headquarters with the drafts were Lieutenant R. O. Campbell, Ensigns C. Tyner, H. S. Harrison, and T. C. Wray. In addition there joined the regiment on appointment Paymaster E. C. Grant on the 29th and Captain John Meade on the 31st. Bt.-Major Campbell and Lieutenants Moorsom and Clarkson were now ordered home to the depot at Fermoy.

Peace followed close on the armistice and was announced to the army on April 1st. The distinguished part which the 30th had played in the Crimean War was recognized by the honours bestowed upon the regiment by Her Majesty the Queen and her allies.

Leave was granted to bear the words, " Alma," " Inkerman," " Sebastopol " on the colours.

Colonel Mauleverer, who was already a Companion of the Bath, was appointed an officer of the French Legion of Honour, and was granted a Good Service Pension of £100 a year. Lieut.-Colonel Patullo had been appointed a Companion of the Bath before his death.

The following were appointed Chevaliers of the Legion of Honour :—

Brevet-Lieut.-Colonel E. A. Whitmore, Brevet-Major Paget Bayly, Brevet-Major F. T. Atcherley, Brevet-Major C. M. Green, Lieutenant Stamer Gubbins, Sergeant-Major Richard Nagle, 3422 Private John McCormick.

The French Military Medal was awarded to Colour-Sergeants Hastings McAllister, John Richardson and Thomas McDonagh, Sergeant Owen Curran, Privates 1638 Charles Quigley, 3038 John Smith (4), 3186 Thomas McDonald, 3367 Michael Byrne, 3786 William Nicol.

Awarded the Medal for Distinguished Conduct in the Field with annuity :—Sergeant-Major J. McClellan, Sergeant-Major R. Nagle, Colour-Sergeant D. Sullivan, Corporals Charles Dillon, John Johnston, James Ollerton and Samuel Weale; Privates 1638 Charles Quigley, 2939 Thomas Fitzgerald, 2117 Henry Holmes, 2847 James Alexander, 3680 John Andrews, 3186 Thomas McDonald, 3230 Patrick Grant, 3446 Thomas Fennell, 3361 George Richardson, and 3032 John Smith. Quartermaster-Sergeant M. Tooner and Colour-Sergeant William Hunns were given the Medal for Meritorious Service with Annuity. The annuity was £20 to a sergeant-major, £15 to other sergeants, £10 to corporals, and £5 to privates.

Awarded the Turkish Order of the Medjidie :—

Fourth Class, Colonel Mauleverer; Fifth Class, Lieut.-Colonel Whitmore, Lieut.-Colonel T. H. Pakenham, Brevet-Majors James Rose, G. Le F. Dickson, P. Bayly, F. T. Atcherley, A. C. Campbell, C. M. Green, G. F. C. Pocock; Captains John O'Brien, A. H. Williamson and Lachlan MacPherson; Lieutenants J. P. Campbell and A. J. Austin, Surgeon R. R. Dowse, and Assist.-Surgeon D. Milroy.

The Sardinian War Medal was awarded to :—

Colonel Mauleverer, Majors Pakenham (Lieut.-Colonel), and R. Dillon, Lieutenant and Adjutant Gilbert Sanders, Sergeant T. S. Shaw and Private John Andrews (2).

All ranks who served in the Crimea on or before September 10th, 1855, received the British Crimean Medal, and clasps were given for Alma, Balaclava, Inkerman and Sebastopol to those who were present in those battles or in the siege. The medal was awarded to all, and not only to the survivors as had been the case with the Peninsula Medal. A silver medal was also given by the Turkish Government to all who served in the Crimea.

The armistice and the peace gave greater freedom of movement to the officers in the Crimea and many riding excursions were made into the interior, including one historic ride by Captain Brook (attached to the Quartermaster-General's Department), and two or three other staff officers along the Russian line of communications to the north, which gave much valuable information and laid bare the disorganization which prevailed. There was quail shooting to be had on the steppes beyond the Tchernaya, and many tried fishing in the river, which was now clean, but (Lieutenant) Joe Fleming always maintained that he was the only man in the army who had caught a fish properly with the fly.

On May 21st the 30th and 55th embarked on the transport *Great Britain* for Gibraltar. At Malta the remains of the four reserve companies joined the ship, 4 captains and 1 subaltern with 8 sergeants, 2 drummers and 7 rank and file for duty with Headquarters of the regiment and 52 sergeants, rank and file for transport to Ireland to join the depot.

On June 3rd the 30th and 55th disembarked at Gibraltar and the 30th under command of Lieut.-Colonel Whitmore encamped on the North Front. Its strength was :—12 companies, 2 field officers, 9 captains, 17 subalterns, 4 staff, 44 sergeants, 25 drummers, 38 corporals, 656 privates; 2 sergeants, 1 corporal and 28 privates were left in the East sick or employed.

The 30th left behind it in the Crimea many memorials to those who had fallen in battle or died of disease. No record was kept at the time of such memorials and it is doubtful if we have a complete list of them. Many were of wood and it may be that they have disappeared. The following are those of which we have knowledge :—

One on Cathcart Hill to the officers. It is a very short four-sided column and the inscriptions on it are—on the front :—" Sacred to the memory of the officers XXX Regiment who fell or died in the Crimea." On the right side the names of Lieut. Luxmore, Captain Connolly, Lieut. Gibson, Lieut. Ross Lewin, Ensign Thompson, Lieut. and Adjutant Forbes and Ensign Deane. On the left side the names of Lieut.-Colonel Patullo, Captain Stevenson, Lieut. Kerr, Ensign Johnston, Lieut.-Colonel Hoey, and Ensign Fitzpatrick. On the rear: "This monument was erected by their Brother Officers—April, 1856." Facing the Woronzoff Ravine is a monument with the following inscription :—

I.H.S.

Sacred to the memory of Sergt.-Major John MacLellan, the Non-Com. Officers and Men of the 30th Regiment, who fell in action or died of wounds or disease in the Crimea from September 14, 1854, to February 29, 1856.

There are also monuments to individuals—one to Patullo and another to Stevenson erected by their brother officers; one each to Kerr and Deane without any statement as to who erected them.

The following are of wood—one to Hospital Sergeant Stanley, another to Private W. Griffiths, a third erected by his loving brother to Private Alexander Still; one each to Corporal James Brady, Private J. Jones, and Private Mark Johnson, the last erected by H. Gore, and one to Private Thomas Egan by his comrades.

CHAPTER XIV

GIBRALTAR. HOME. CANADA. HOME. INDIA. THE END.

1856–1881

THE regiment remained for over a year in Gibraltar. Lieut.-Colonel Whitmore was in command till he left on September 6th, 1856, to take command of a depot battalion, when he was succeeded by Major Dillon.

Colonel Whitmore held many important appointments, the chief of which was that of military secretary at Headquarters. He died a lieut.-general and Knight of the Bath. To the end he kept in touch with the regiment and was never happier than when meeting old and new comrades at the regimental dinner.

In August the four depot companies moved from Fermoy to Parkhurst in the Isle of Wight, to form part of the depot battalion there.

On September 25th four companies were reduced and the establishment was fixed at 8 service and 4 depot companies with a total strength of 3 field officers, 12 captains, 24 subalterns, 6 staff, 63 sergeants, 25 drummers, and 1,000 rank and file.

In consideration of his services and sufferings Lieutenant Sanders had been allowed to retain the post of adjutant after losing his leg at the Redan. Lieutenant Herring and Lieutenant Henry Wood had acted for him, but on February 24th, 1857, Lieutenant Herring was gazetted adjutant. Her Majesty the Queen interested herself on behalf of Sanders, and he was given an appointment in the Barrack Department.

On September 3rd, 1857, the 30th and 55th embarked for home.

The 30th were on the s.s. *Jura* and the strength was 3 field officers, 5 captains, 18 subalterns, 5 staff, 743 other ranks. The *Jura* called at Portsmouth on the 8th and went on to Cork, where the regiment disembarked on September 11th, 1857. On the same day it went on to Dublin and occupied Beggar's Bush Barracks.

A week later it was sent to Belfast in consequence of some disturbance caused by a fanatic preacher, but it returned on October 20th to Dublin without having been compelled to act.

On return from service it was felt that a suitable monument should be erected to those who had fallen in action or died of disease during the Crimean War. All ranks and many friends of the regiment subscribed generously with a view to a memorial being placed in Ely Cathedral, but through a misunderstanding this plan was not carried out and with a portion of the money subscribed the present marble monument was erected in the Garrison Chapel, Arbour Hill, Dublin. The inscription runs :—

ERECTED BY THE

OFFICERS, NON-COMMISSIONED OFFICERS AND PRIVATES

OF THE

XXX (CAMBRIDGESHIRE) REGIMENT

IN MEMORY OF

Lieut.-Colonel W. F. HOEY	Died of disease
Lieut.-Colonel J. B. PATULLO	Died of wounds
Capt. A. W. CONNOLLY	Died of wounds
Capt. J. C. N. STEVENSON	Died of wounds
Lieut. F. LUXMORE	Killed in action
Lieut. A. GIBSON	Killed in action
Lieut. J. D. ROSS LEWIN	Died of wounds
Lieut. W. KERR	Died of wounds
Ensign W. J. JOHNSTON	Died of disease
Ensign T. M. FITZPATRICK	Died of disease
Ensign R. G. DEANE	Killed in action
Ensign J. THOMPSON	Died of wounds
Lieut. and Adjutant J. FORBES	Died of wounds

Also of
Sergeant-Major MACCLELLAN
and 426 Non-Commissioned Officers and Privates
who fell in action or died from Wounds or
Diseases during the war with Russia,
from 1854 until 1856.

The Colours attached to this Memorial are those carried by the Regiment during the Campaign.

The colours were afterwards transferred to Kilmainham.

On May 27th, 1859, new colours were presented to the regiment by the Countess of Eglinton, wife of the Lord-Lieutenant. The regiment with its old comrades of the 55th was drawn up in the Phoenix Park, close to the Vice-Regal Lodge. Her Excellency in making the presentation complimented Colonel Mauleverer and those present on their share in the late triumphs and dwelt upon the century and a half of successful service by the regiment. Colonel Mauleverer in the course of his reply said he was induced to refer to a successful charge of the regiment in Egypt by the presence of an old, gallant, noble and distinguished officer who had witnessed the whole transaction. The officer alluded to was Field-Marshal Lord Seaton, commanding in Ireland, who as John Colbourne, Captain-Lieutenant of the 20th, was one of those who cheered the charge of Lieut.-Colonel Lockhart and the 30th at the Green Hill, Alexandria; Lord Seaton had told the whole story to Mauleverer when riding to the Phoenix Park that morning.

On September 18th the 30th marched to the Curragh Camp.

On June 11th it embarked for the Channel Islands, Headquarters, with Nos. 2, 5, 6, 7, and 10 companies for Jersey, Nos. 1, 3, 4, 8, and 9 companies for Alderney. No. 4 afterwards proceeded to Jersey. The Alderney detachment landed on the 13th and Headquarters on the 14th.

On May 8th, 1861, Headquarters left Jersey on relief by the 55th, landed at Portsmouth on the following day and went on at once by train to Aldershot. The Alderney detachment joined on the 13th. The regiment was quartered in the South Camp.

In consequence of the civil war in the United States some regiments were sent to strengthen our forces in Canada at this time and on June 25th the 30th, which had trained from Aldershot, embarked at Liverpool on the *Great Eastern*. The great ship sailed on June 27th and dropped anchor off Quebec on July 6th. Colonel Mauleverer was in command of the troops on board belonging to the 30th, 4/60th Rifles, Royal Artillery and infantry drafts. The strength was 108 officers and 2,079 other ranks; with 418 women and children and the crew, the total reached 3,031. There were also 110 horses. No ship had up to that time embarked anything approaching such numbers.

The 30th started by river steamers for Upper Canada on July 10th and 11th, and reached Toronto on the 12th, where three companies were quartered in the Old Fort, three in the New Barracks; the remainder encamped.

The strength was Colonel Mauleverer, C.B., Major (Brevet-Lieut.-Colonel) Pakenham, M.P., Major Dillon; Captains Atcherley (Brevet-Major), Brook, Macpherson, Moorsom, Singleton, Nathaniel William Massey, Charles James Palmer Clarkson, Hyde Sergison Smith; Lieutenants Joseph Fleming, Robert Olpherts Campbell, Henry Faulkner Morewood (I. of M.), Montague Dickin Stevenson, Decimus Montagu, Fred. Harcourt Williamson, William Glascott, Pelham Thursby Pelham, Joshua E. C. C. Lindesay, William Vesey Brownlow; Ensigns Alexander Webster McKenzie, Charles Henry Garnett, James Thom, James Cooke, John Peters Charles Lowder, Henry Granville Sharpe, Thomas Brown Stewart, Alexander James Boyle, and Alexander John Macmahon; Paymaster George Fead Lamert, Adjutant Henry L'Estrange Herring, Quartermaster John Moon, Surgeon Raphael Woolman Read, Assist.-Surgeon David Milroy, M.D., 7 staff-sergeants, 40 sergeants, 21 drummers, 802 rank and file. A schoolmaster was attached to the regiment.

On March 19th, 1862, Headquarters were removed to Parliament Buildings.

On May 1st Colonel Mauleverer went on leave previous to retiring on half-pay. The whole regiment, officers and men, women and children, went to the station to see him off, and when the grenadiers raised him on their shoulders that the regiment might have a last look at its beloved colonel a good many tears were shed, and not by women only.

Colonel Mauleverer had served as a subaltern in the 17th regiment with Lord Keene's column in Beloochistan and Afghanistan in 1839, and had been present at the storming of Ghuzni and the capture of Khelat. He joined the 30th as a captain in 1844. The foregoing pages give some account of the services he rendered to the regiment and the country, but his great success as a commanding officer was due to the strong affection he inspired as much as to the universal belief that he was the best and boldest soldier in the regiment.

Colonel Mauleverer was placed on half-pay on December 19th, 1862, and died in London on October 26th, 1866.

Brevet-Lieut.-Colonel Thomas Henry Pakenham succeeded to the command of the 30th and Brevet-Major Atcherley to the vacant majority. On succeeding to the command Colonel Pakenham had to give up his seat in the House of Commons.

In 1863 a lighter and more comfortable chako of blue cloth was taken into wear.

On April 1st the establishment of the regiment was fixed at 10 service companies of 3 field officers, 10 captains, 20 subalterns, 5 staff, 48 sergeants,

21 drummers, and 650 rank and file—2 depot companies of 2 captains, 4 subalterns, 10 sergeants, 4 drummers, and 110 rank and file.

On September 23rd the 30th relieved the 1st battalion 16th regiment in Molson's College Barracks, Montreal.

In the autumn of 1864 there was an outbreak of yellow fever in Bermuda and Assist.-Surgeon Milroy was sent there for duty. He fell a victim to the fever on his first visit to the hospital and died on September 3rd. He had been through the Crimean War with the regiment and was beloved by all ranks for his care and tenderness in treating the sick and wounded. The regiment asked for and obtained Dr. James Paxton as Milroy's successor. Dr. Paxton had the naval war medal for service in the Baltic.

On March 21st, 1865, Brevet Lieut.-Colonel Atcherley retired, to the great regret of the regiment. He was a tall handsome man with a very courtly manner which was the same to all ranks and ages, and he never missed a chance of doing a kind act. He had displayed great gallantry at Little Inkerman, where he was wounded. Captain Brook bought the majority. In June Major Dillon went on half-pay, and was succeeded by Captain Macpherson.

On June 1st, 1866, news reached Montreal that a band of Fenians had crossed the Niagara River and seized Fort Erie. All regiments in the garrison were ordered to be ready to move at the shortest notice and on the following evening Colonel Pakenham was ordered to proceed with 200 men by rail to Cornwall, 40 miles above Montreal, to protect the communications by river, canal and rail between Upper and Lower Canada.

In half an hour Colonel Pakenham was ready, and on the following morning he was at Cornwall. On the afternoon of that day he was joined by the remainder of the regiment except No. 10 company which was at Chambly for musketry and was incorporated with a field force based on St. John's.

The field force assembled at Cornwall under Colonel Pakenham consisted of—the 30th, Headquarter wing of the 25th, a wing of the 47th, Grey Battery R.A., a half battery Ottawa Artillery, two very smart battalions of Upper Canadian troops, the local volunteers, and the Argenteuil Rangers (The Gentiles), physically one of the finest regiments which ever wore a red coat. They came from the lumber camps north of the Ottawa.

The duties were patrolling the St. Lawrence and finding pickets to protect the canal and railway.

Owing to the decided action of the United States Government the trouble from the Fenians quickly died away and the force was reduced. The 30th returned to Molson's Barracks, Montreal, on June 22nd. At the Cornwall Railway Station the Mayor and Town Council, with many of the citizens, and the Council of the United Counties of Stormont, Dundas and Glengarry had met the regiment and requested Colonel Pakenham to receive an address expressing the thanks of the district for the good conduct of the regiment while quartered there.

It is worth noting that at this time some of the inhabitants spoke Gaelic only and an interpreter attended the Courts.

A medal was granted thirty years later by the Canadian Government to all who served in the Fenian disturbances of 1866.

On the return of the regiment to Montreal orders were issued for each company to proceed as usual in succession to Chambly on the river Richelieu for musketry, and on completion of the course to repair to Point Levis on the southern bank of the St. Lawrence to work at the fortifications on the heights opposite Quebec. No. 7 as the best shooting company was at this time armed with the Whitworth rifle (muzzle-loader) as an experiment. The whole regiment was encamped in the autumn on the heights at Point Levis and was exceptionally healthy.

On October 15th the regiment crossed the river and occupied the Jesuit Barracks in Quebec. There had been some changes in the Staff. In August Lieutenant Edward St. George Smyth got his company and was succeeded as adjutant by Lieutenant Neil Bannatyne. Surgeon Raphael Woolman Read went home on leave in September and did not rejoin. Surgeon Alfred Hooper was appointed in his place in December.

In January, 1867, Major Charles Ashmore was appointed colonel of the regiment *vice* Lieut.-Colonel Thomas Wright, C.B. They were mere names to the regiment.

In spring Quartermaster-Sergeant James Matthewson left. He had been twice slightly wounded at the Alma and had done invaluable work in the orderly room. His son was sergeant-major of the regiment in the South African war.

While in the Jesuit Barracks the muzzle-loading rifles were withdrawn and the regiment was equipped for the first time with breechloaders. They were Enfields fitted with the Snider or snuff-box breech action.

In September the regiment relieved the first battalion Rifle Brigade in the Citadel.

On June 17th the 30th embarked on H.M.S. *Himalaya* for Halifax. The relations of the regiment with the people of all ranks in Quebec had been of the friendliest description, and previous to embarkation a resolution of the City Council was communicated to Colonel Pakenham by the City Clerk. The resolution expressed the regret of the citizens at losing the regiment, and begged Colonel Pakenham to convey to his corps the esteem in which it was held.

The strength of the regiment on board the *Himalaya* was 20 officers and 650 other ranks. On landing at Halifax on June 23rd, 1868, two companies with Headquarters occupied the Citadel; the remainder of the regiment was scattered up and down the enormous harbour to work under the engineers, for Halifax, like Quebec, was being strengthened.

A few days after landing, Lieut.-Colonel James Poyntz called upon his old regiment. He was the last representative of Arthur Poyntz, who had served as a sergeant in South Carolina in 1781, and had died as quartermaster in 1812. Fifty-eight years had elapsed since Colonel Poyntz had first been under fire as a volunteer with the light company in the Peninsula.

There were no events in the life of the regiment at Halifax except the periodical changes of the companies as their turn came for a course of musketry at Beauport, which was a unique place for musketry, for the instructor, Lieutenant (Tim) Morewood, caught a twelve-pound salmon casting a fly in the stream on the range.

On June 27th Major Lachlan Macpherson went on half-pay and Captain Richard Nagle exchanged to the Canadian Rifles. They both had served with distinction in the Crimea. Macpherson was succeeded as major by Brevet-Lieut.-Colonel Henry Prim Hutton, an old 31st man who had been severely wounded serving with the light company of that regiment in the Sikh war of 1845.

HOME SERVICE

On May 20th, 1869, the regiment embarked for home on H.M.S. *Crocodile*, and on June 1st it disembarked at Queenstown. The strength was 2 field officers, 7 captains, 9 subalterns, 4 staff-sergeants, 47 sergeants, 20 drummers, and 582 rank and file. Colonel Pakenham was in command.

On the same day Headquarters proceeded by rail to Waterford, 2 companies to Carrick, 3 to Clonmel, 1 to Dungarvan, and 1 to Duncannon Fort for musketry.

The establishment was now 3 field officers, 10 captains, 12 lieutenants, 8 ensigns, 1 paymaster, 1 adjutant, 1 quartermaster, 1 surgeon, 1 assist.-surgeon, 9 staff sergeants, 10 colour-sergeants, 30 sergeants, 21 drummers, 40 corporals, and 520 privates.

On June 5th the depot companies joined and were broken up. They had been moved from Parkhurst to Chatham at midsummer 1866.

To check the regimental returns the whole regiment was passed under the standard and the average height was found to be 5 ft. 7¾ in. If the measurement had been taken before the youths from the depot arrived the height would have been 5 ft. 8 in.

The battalion being now greatly above the strength, volunteering for regiments in India was opened with a bounty of a guinea a man.

In June the Dungarvan company was withdrawn and sent to Kilkenny. Two companies also were sent to Kilkenny from Clonmel. In November the Kilkenny detachment was withdrawn, two companies joined Headquarters and one was sent to Clonmel.

Early in March, 1870, the regiment was employed in aid of the civil power during the contested elections in Waterford and Tipperary and received the thanks of Lord Strathnairn, the commander of the forces, for their conduct while engaged on this disagreeable duty.

On March 7th, 1870, the regiment moved to Dublin by railway and was quartered in the Royal Barracks. It was now united for the first time since landing at Halifax. On March 12th the two depot companies of the 108th regiment were attached to the 30th on the abolition of the depot battalion system.

On April 1st, Lieutenant H. F. Morewood, instructor of musketry, was promoted to a company on half-pay and was succeeded as instructor by Lieutenant J. E. Goodwyn.

In May three companies were detached to Longford under Major Brook, in aid of the civil power. They returned in a few days.

On May 20th the establishment was reduced by six subalterns and sixty privates, but on August 6th the number of privates was raised by 200 to 660. On February 1st, 1871, it was altered to 560.

On March 7th, 1871, Headquarters moved from the Royal to Ship Street Barracks. Five companies were detached to Linen Hall Barracks, and two to Pigeon House Fort. On April 24th the five companies returned from Linen Hall to the Royal Barracks.

On June 9th the 30th embarked on H.M.S. *Orontes* at Kingstown, and on the 14th of the month landed in Jersey. Headquarters and three companies occupied Fort Regent, four were at St. Peter's and three on single company detachments.

On August 19th Surgeon Thomas Hooper was appointed to the Staff. He was the last regimental surgeon of the 30th. Under the new system Surgeon Thomas Teevan was attached for duty to the regiment. He had the Crimean and Turkish medals and a medal and clasp for the capture of the Taku Forts in China.

On May 4th, 1872, the number of rank and file was increased to 700.

The 30th embarked at St. Helier's on H.M.S. *Tamar* on July 7th, 1872, disembarked at Portsmouth on the 10th, reached Aldershot the same night and encamped on Cove Common.

In August the regiment marched to Salisbury Plain, where about 30,000 men were assembled for one of our earliest attempts to train the army under something like service conditions. At the termination of the manœuvres the regiment marched back to Aldershot and occupied huts in the North Camp.

On April 1st, 1873, the 30th and 59th became the First and Second Battalions of the 15th Brigade with a depot at Burnley. The establishment of the First battalion was

	Field Officers.	Caps.	Subs.	Sergts.	Drums.	Rank & File.
8 Service Companies .	3	8	13	42	16	520
2 Depot Companies. .	1	2	4	6	2	50

Two companies of the 30th proceeded to Burnley to form with the two companies of the 59th regiment, the brigade depot. The officers of the two depot companies were Captains W. H. Clarkson and D. R. Vandeleur, Lieutenants R. A. McCord and W. Kennedy. From June 15th all recruits for the 30th and 59th were ordered to receive brigade instead of regimental numbers.

In August two Crimean officers were lost to the regiment. On the 9th Captain and Brevet-Major Singleton retired. He was a great loss. On the 27th Captain and Quartermaster Michael Tooner was transferred to the newly formed brigade depot after thirty-three years' service with the 30th. He was the friend of every man in the regiment. He was succeeded by Q.M.S. Alexander Borland.

The 30th left Aldershot by train and arrived at Portsdown Hill Forts on September 2nd, 1873. Headquarters were at Widley and detachments were at Nelson, Purbrook, and Southwick Forts.

Lieutenant Fred. Clowes at this time volunteered for service with Sir Garnet Wolseley's Ashanti expedition. He commanded a company of Haussas at the action of Amoaful and subsequent occupation of Coomassie on February 5th, 1874, for which he received the medal and clasp.

On January 12th, 1874, authority was given for a gilt badge showing the Sphinx with the word "Egypt" below to be placed on the officers' forage caps.

Colonel Thomas Henry Pakenham, who had commanded the regiment for nearly twelve years, retired on half-pay on October 31st, and was succeeded by Brevet-Lieut.-Colonel Henry Prim Hutton. Colonel Pakenham, who was shortly afterwards promoted to major-general, saw no active service after leaving the regiment, but held the Portsmouth and other important commands, and after the amalgamation of the 30th and 59th the regiment had the satisfaction of seeing him return as its full colonel. He was made a Companion of the Bath.

On August 31st Lieutenant Neil Bannatyne resigned the appointment of adjutant, after holding it for eight years. He was succeeded by Lieutenant Henry Kemble.

Martini-Henry rifles were issued to the regiment in place of the Sniders during the autumn.

The 30th was encamped on Cove Common, Aldershot, in June, 1875, for the summer drills and on July 27th moved to Chester. Headquarters occupied Chester Castle with detachments of two companies at the North Fort, Liverpool, and two at Weedon. The Liverpool detachment was afterwards withdrawn and a company sent to Weedon and another to the Isle of Man.

On April 1st, 1876, the establishment was altered to 26 officers, 58 non-commissioned officers and drummers, and 820 rank and file.

Lieutenant R. F. Walker, who was attached to the Staff at Aldershot, was killed on July 8th by his horse putting its foot in a hole and turning over with him. All the officers who could be spared from Chester and Weedon, and some on leave, attended his funeral at Aldershot.

Quartermaster Alexander Borland died on May 21st, 1877. He had served throughout the Crimean War and had been wounded in the trenches. He was succeeded by Sergeant-Major Michael Ryan and Colour-Sergeant Charles Saunderson became sergeant-major.

In the spring of 1878 regimental transport was provided, and a new bayonet four and a half inches longer than the old one was issued. On April 1st the establishment of the service companies was raised to 31 officers and 1,067 other ranks. There was some fear of war in the East, and 292 reservists were sent to the regiment, so that the actual strength of the service companies on May 18th was 1,166 of all ranks.

On May 24th, Her Majesty's Birthday, the regiment paraded for the first time in a cork helmet covered with blue cloth and finished with a brass spike.

On August 1st, all danger of war being over, the reservists were sent home and the establishment lowered to 684 of all ranks.

The regiment moved on February 4th, 1879, to Dover where it occupied the Citadel. The establishment of the service companies was raised on May 1st to 25 officers, 42 non-commissioned officers, 16 drummers, and 800 rank and file.

On May 29th, Lieutenant A. G. Watson resigned the appointment of adjutant and volunteered for service in South Africa where the Zulu war was in progress. Lieutenant A. J. Wright succeeded him as adjutant and Lieutenant J. M. Piercy became instructor of musketry.

On October 31st, Lieut.-Colonel Henry Prim Hutton, having completed the regulated five years as commanding officer, was placed on half-pay, after one of the most successful periods of command in the regimental history. Brevet-Lieut.-Colonel Henry Wallace Stroud became Lieut.-Colonel of the regiment, and the vacant majority went to Brevet-Major John Pennock Campbell.

Orders were received in the autumn for the 30th to be prepared to embark for India about the 6th of the following January. Volunteering was opened for the regiment on December 24th, and on January 2nd, 1880, 101 privates from the 24th regiment and 26 from the 13th regiment joined the 30th. A reinforcement of 10 privates of other corps, liberated from prison, joined on board ship, in handcuffs; they came to no good.

In the previous six years the 30th had furnished 336 non-commissioned officers and men to the 59th in India.

The regiment left Dover for Portsmouth on January 7th, 1880, and embarked on the same afternoon on board H.M.S. *Serapis.*

The strength of the regiment embarked was 27 officers, 37 sergeants, 16 drummers, 814 rank and file.

To the list of officers given below a note has been added to show their principal services in later life.

Lieut.-Colonel Henry Wallace Stroud (command), dead before 1914.

Major C. J. Moorsom, commanded 1st East Lancashire Regiment, commanded Belfast District, commanded division in India. Dead before 1914.

Major J. P. Campbell, commanded 1st East Lancashire Regiment. Dead before 1914.

Captain F. H. Williamson, commanded 1st East Lancashire Regiment. Dead before 1914.

Captain J. Cooke retired as Lieut.-Colonel. Dead before 1914.

Captain J. E. Goodwyn, awarded the Humane Society's Medal for jumping overboard from a ship going eleven knots, in the Red Sea, to rescue a boy who had fallen overboard. Commanded 2nd East Lancashire Regiment.

Captain F. Clowes, commanded 1st East Lancashire Regiment. Died January 7th, 1918.

Captain N. Bannatyne, retired as Lieut.-Colonel.

Captain H. Kemble, retired as Lieutenant-Colonel. In the war of 1914–19 was Chief Guide of Essex Coast and charged with guidance of troops and removal of population in case of invasion.

Captain C. R. Hornby, retired as Lieut.-Colonel.

Captain H. T. P. Evans, commanded 1st East Lancashire Regiment. Dead before 1914.

Lieutenant A. G. Watson, served in Afghan war 1880–1, commanded 2nd East Lancashire Regiment. In war of 1914–19, commanded a brigade in England, and afterwards an area in France. Brigadier-General.

Lieutenant A. J. A. Wright (Adjutant), commanded 1st East Lancashire Regiment in South African War, and awarded C.B.; in war of 1914 commanded a brigade in England, and later an area in France. Brigadier-General.

Lieutenant J. M. Piercy (I. of M.). Commanded Dorsetshire Regiment in Tirah expedition.

Lieutenant J. F. Muntz.

Lieutenant B. G. Lewis, second in command 1st East Lancashire Regiment in South African War, D.S.O. In war of 1914 trained and commanded on service a brigade of New Army, till invalided and transferred to home brigade. C.B.

Lieutenant D. Carey, commanded 1st East Lancashire Regiment. Dead before 1914.

Second-Lieutenant W. G. Hamilton, commanded brigade in India. Brigadier and Adjutant-General of Army in Mesopotamia. Wounded and taken prisoner at Kut. C.B., D.S.O., C.S.I.

Second-Lieutenant F. S. Derham, served in South African War. Commanded 1st East Lancashire Regiment. In war of 1914-19 trained and commanded on service a brigade of New Army till sent home on account of age. Employed on Staff at home. Brigadier-General and C.B.

Second-Lieutenant C. A. Bray, served South African War, awarded C.M.G. and C.B. In the war of 1914 was Chief Paymaster of Army in France, promoted Major-General and awarded K.C.M.G., and created an Officer of Legion of Honour.

Second-Lieutenant C. R. M. O'Brien, served in war in South Africa. Towards the close was appointed Chief of Tribunal of Johannesburg. When war of 1914 broke out was Governor of Seychelles Islands. Later was appointed Governor of Barbados. K.C.M.G.

Second-Lieutenant C. Haynes, served in Afghan War 1880–1. When war of 1914–19 broke out was Governor of Brixton Prison.

Second-Lieutenant H. M. Browne, died in South African War.

Second-Lieutenant H. L. Gallwey (spelling altered to Galway), served in West Africa, D.S.O., Governor of St. Helena. At time of war of 1914–19 was Governor of South Australia. K.C.M.G.

Second-Lieutenant C. H. Billings.

Paymaster J. J. Morris (Hon. Captain), dead before 1914.

Quartermaster M. Ryan (Hon. Captain), dead before 1914.

Percy, the infant son of Captain Goodwyn was a passenger on the *Serapis*. He gained the D.S.O. in the South African War, and commanded the 11th Battalion of the East Lancashire Regiment in France in the great war.

At 3 p.m. on January 7th, 1880, the *Serapis* weighed anchor, and proceeded on her voyage to Bombay, via the Suez Canal. Bombay was reached on February 8th, 1880. Orders were here received for the regiment to disembark on the following day and proceed by train to the Deolali Depot. Thus after an absence of 51 years, the regiment again landed for service in the East Indies. The Headquarters reached Bareilly on February 17th, and went under canvas. The remainder of the regiment arrived the following day. Here the regiment was detained for nearly two months. The heat under canvas was frequently very great, and most trying to the constitution of young soldiers just arrived in the country from home. The germs of sickness which subsequently broke out in the regiment must, without doubt, be attributed to their long sojourn in this standing camp. Orders were received on April 7th for the Headquarters of the regiment to proceed by march route to Ranikhet. Camp was accordingly struck at midnight on April 9th, and the Headquarters, consisting of A, B, C, and G companies, proceeded to Ranikhet, after a somewhat trying march through the malarious district of the Terai, by a route never before taken by troops, and known to be most deadly to Europeans, during which the thermometer inside the tents frequently reached the temperature of 105 degrees.

Seventeen days after leaving Bareilly, the Headquarter companies marched into quarters at Ranikhet. During the march up country two men and one woman died of fever. The average number in hospital of the four companies was 72, and there were a number of deaths in the succeeding fortnight.

The remaining four companies, under command of Lieut.-Colonel Moorsom, remained in the plains, divided into three detachments, as follows:—

		Capt.	Subs.	Sgts.	Drs.	R. & F.
Shajahanpur,	D and F Comp. . .	1	2	5	4	191
Moradabad,	E ,, . .	1	1	4	2	85
Bareilly,	H ,, . .	1	1	5	1	113

On May 6th H Company, under command of Captain Hornby, rejoined Headquarters at Ranikhet.

In November, 1880, the command of the regiment became vacant, owing to the retirement on full pay of Lieut.-Colonel H. W. Stroud. It was given to Brevet-Lieut.-Colonel C. J. Moorsom. On Christmas Eve the detachments from the plains marched in, and the regiment was reunited.

Early in 1881 changes of a radical nature were foreshadowed in the newspapers, and they eventually took effect from July 1st. By the new regulations, all infantry regiments below the 25th (K.O.B.'s) lost their individuality, and the system of linking two line regiments, which had been tried in a half-hearted way for several years previously, was made permanent. Under this organization the 30th Cambridgeshire with the 59th, 2nd Nottinghamshire, were attached to the 30th Regimental District, and the Headquarters being at Burnley they became the 1st and 2nd Battalions of the East Lancashire Regiment.

Thus in common with many other famous regiments the 30th lost its individuality, its county, and even its primrose yellow facings, and in a sense it might be said to have come to an end, but it was in name only. The regiment did not die, but went on, strong in its old-established discipline and in the traditions of nearly two hundred years, to render as the First East Lancashire Regiment loyal and efficient service to King and Country in wars more deadly than those of old.

ROLL OF OFFICERS 1689–1881

Abbot, C.	1757–1762
Abercrombie, / Abercromby, J.	1724–1736
Abingdon, / Abington, A.	1702–3
Acklom, T.	1804–08
Adams, J. K.	1827–29
Adamson, G.	1807–12
Agnew, G. A.	1876–79
Airey, J. T.	1830–33
Allardice, J. McD.	1855–56
Allen, H.	1778-80
,, R.	1775–79
Alured, J.	1690
Amory, J.	1796–1803
Ancrum, Earl of	1789–90
Anderson, J. C.	1848–49
,, R.	1803
Andrews, M.	1808–27
,, R. A.	1826–29 1830-46
Angelo, E. A.	1831–34 1846
Anketell, T.	1777–88
Annesley, F.	1808
Appleton, T.	1842–56
,, W.	1803
Arbouin, M.	1800–01
Archibald, W.	1786–93
Armstrong, N.	1822–38
,, T.	1747–48
Ashbrooke, Viscount	1803–04
Ashe, W.	1816
Ashby, N.	1790–93
Ashmore, C.	1867–81
Atcherley, F. T.	1847–65
Atherly, A. P.	1840–41
Atkinson, W.	1811–34
Auchmuty, J.	1720–26
Austin, A. J.	1854–60
Aylmer, E.	1706–07
,, H.	1707–17
,, P.	1708–16
Backhouse, G. L.	1815–27
Bagshawe, S.	1740–42
Bailey, N. W.	1811–47
Baillie, A.	1809–17
,, J.	1787–93
Bainbridge, L.	1794
Baker, J.	1715–22
,, N.	1806–08
,, R.	1817–20
Baldwin, T.	1724–27
Balfour, R.	1787–91
Ball, W.	1732–46
Bamford, T. B.	1804–13
Bannatyne, N.	1864–91
Barber, A.	1755–56
Barlin, F. S.	1864–65
Barlow, B.	1821–28
,, F.	1817–18
,, J. T.	1805–19
Barrett, R.	1793–95
Barrington, T.	1689–93
Barrow, A. J.	1831–37
,, C. W.	1822–31
Barry, C. W.	1800–01
,, G. A.	1755–56
,, W.	1785–86
Bathe, J. W.	1828–33
Batley, J. C.	1822–25
Battersby, J.	1816–17
Batwell, W.	1757–72
Baxter, W.	1825–40
Bayly, H.	1746–49
,, P.	1839–55
Bayntun, C.	1798–99
Beaumont, J. P.	1803–11
,, R.	1689–92
,, R.	1692–95

Bedford, T.	1689–1709
Beere, H.	1812–15
Bell, J.	1765–75
Bendish, R.	1730–49
Bennett, M.	1742–48
,, W. H.	1849–53
Bermingham, E.	1741–42
Bernard, G.	1762–63
,, H.	1698
,, J.	1692–1703
,, M. A.	1692–1705
,, W.	1694
,, W. T. P.	1868–70
Berridge, J.	1815–33
Bertram, A.	1795–1801
Best, E.	1776–84
,, T.	1802–03
Bettesworth, P.	1718–32
Bilham, J. D.	1856–57
Billings, C. H.	1880–81
Birch, R. J. W.	1862–68
Bircham, S.	1793–1817
,, T.	1805
Bird, J.	1804
Bishop, W.	1689
Bissett, A.	1717–42
Blackall, J.	1820–26
Blakeney, J. B.	1801–03
Bland, R.	1806–08
Blunt, C.	1722–30
Boggis, E. R.	1796–99
Bolton, A.	1805
,, R.	1690–1702
Bomford, J.	1834–36
Boothby, W.	1750–60
Borland, A.	1874–77
Borton, J. M. T.	1826–36
Boswell, J.	1760–62
Bourke, T.	1702–04
Bowen, J.	1756–57
,, J.	1775–83
Boyce, O.	1856–60
Boyd, E.	1811
Boyes, C.	1815–17
,, C. J.	1828–30
Boyle, A. J.	1861–65
,, M.	1778–80
Boyton, W.	1817–18
Bradford, T.	1829–46
Braine, O. W.	1860–70
Bramley, J.	1769–78
Brathwayt, G.	1690–95
,, J.	1693–1701
Bray, C. A.	1878
Braybrooke, W.	1843–44
Brereton, R.	1770–95
,, U.	1697–1707
Brett, E.	1807–12
Brevet, D.	1724–48
Brewer, R.	1717
Bridge, T.	1747–53
Brisac, G. W. A.	1809–16
Briscoe, H. C.	1803–04
Bristowe, S. S.	1856–7
Bromley, H.	1745–6
Brook, W. J.	1849–70
Brooke, J. E.	1811–12
,, R.	1788–94
,, T.	1703–5
Broome, H.	1837–48
,, L. G. F.	1839–44
Broughton, J. D.	1791–93
Brown, E.	1800–03
,, J.	1803–07
,, J.	1703–07
Browne, C. B.	1854
,, H. M.	1878
,, L.	1782–94
,, T. M.	1779–83
Browning, H. A.	1881
Brownlow, W. V.	1859–71
Broxholme, T.	1692–95
Bruce, W. D.	1742–44
Brudenell, T.	1793–95
Brunskill, P.	1693
Brydges, J. W. H.	1814
Bulkley, P. R.	1795–1803
Bullen, J.	1814–15
Bulteel, T. H.	1790
Burdett, C. W.	1803–11
Burean, P.	1717–26
Burjaud, P.	1730–45
Burns, T. R.	1864–78
Burroughes, E.	1799
Burrowes, T. R. K.	1825–31
Burston, F.	1707–11
,, G.	1689–1713
,, G.	1713–47

Burston, T.	1704
Burton, F.	1851
Bush, A.	1740–51
Butler, C. E.	1842–46
,, C. R.	1866–70
Byng, G.	1780–81
Calder, H.	1790–91
Cameron, C.	1765–75
Campbell, A.	1795–1802
,, A.	1853–58
,, J.	1750–56
,, J.	1823–27
,, J. P.	1854
,, R. O.	1855–63
,, T	1763–93
,, W.	1811–17
Cane, W. L.	1805–11
Carden, W.	1805–30
Carey, D.	1878
Carrington, F. A.	1849
Carter, J. V.	1811–12
Casewell, J.	1702
Cassidy, T.	1863–71
Castle, W.	1711–13
Castleton, Viscount	1689–94
Cator, T. W.	1846–53
Cavan, P. C.	1834–53
Challoner, W.	1747–48
Chambers, T. W.	1803–15
Champion, C. F.	1803–09
Chapman, C.	1742–58
Charles, J. N.	1811–13
Charlewood, C. B.	1865–71
Cheape, P.	1819–26
Chester, P.	1741–42
Chisholm, D.	1805
Christian, C.	1702
Christie, J.	1745–47
Claringburn, E.	1703–06
Clark, H.	1694–1702
Clarke, B.	1806–10
,, C.	1757–62
,, P.	1812–16
,, T.	1792–99
Clarkson, C. J. P.	1855–70
,, W. H. J.	1864–78
Clayton, J.	1708–11
Cliebely, C. W.	1846–47
Clowes, F.	1863
Cobden, G. E.	1862–72
Cobham, R.	1715–16
Cochrane, J.	1718–39
,, J. G.	1828–30
Cockburn, J.	1836–43
Cockburne, P. C.	1815–16
Coghlan, G.	1742–51
Colborne, J.	1848–49
Colecroft, R.	1694–95
Collins, J.	1797–99
Collis, W. C.	1804–05
Colville, R. L.	1803–05
Compton, J.	1693
,, R.	1745–46
Congalton, G.	1793–98
Coney, E.	1691
,, F.	1689
Connell, F. J.	1855–61
Connolly, A. W.	1847–54
,, J. H.	1877–79
Conyers, B.	1755–57
Cooke, J.	1860
,, W.	1709–30
Copley, J.	1788–94
,, R.	1793–94
Corbett, A.	1705–07
,, C.	1757–59
,, R.	1759–63
Cotterell, C.	1721–30
Courtney, W. F.	1804–05
Coventry, T. W. R.	1844–50
Cox, J.	1806
,, J. H.	1780–95
Cracroft, C.	1774–82
Craig, H.	1794–96 1799–1801 1809–15
,, J. H.	1763–71
,, N.	1791–1809
,, P.	1762–68
Craigie, P. E.	1818
Cramer, H.	1803–31
Crauford, J. C.	1793–95
,, P.	1803–06
Craven, J.	1746–48
Crawford, J.	1809–11
Crosbie, E.	1805–6
Crowe, D.	1799–1804
Crowther, R. T.	1880–81

Cruckshanks, J.	1818–19
Cunningham, V.	1754–71
Currey, J.	1810–14
Cuyler, H..	1782–97
Dalrymple, J.	1817–29
Dalton, J.	1697–98
Daniel, J.	1803–05
,, R.	1809–19
D'Arcy, T.	1846–53
,, W.	1846
Darling, G.	1811–17
Davenant, C.	1831–46
Dauvergne, C.	1748–51
Davies, W. T.	1878–79
Davis, T. H.	1820
Davison, A.	1698
,, A.	1719–20
,, C.	1702–17
,, J.	1704–11
,, J.	1693
,, W.	1703–19
Dawes, C. M.	1871–79
,, L.	1708–13
,, T.	1706–15
,, W.	1703–13
,, W.	1722–38
Dawson, T.	1797–1800
Day, A.	1696–1704
Dealtry, R.	1691
Deane, C.	1815–24
,, R. G.	1855
De Berniere,	1803–04
De Carteret, G. F.	1841–43
Deedes, W.	1859–61
De Grangue, H.	1742–43
De Jersey, J.	1796–99
De La Bouchetiere, C. J.	1721–44
De la Porte, M.	1721–55
Dennie, W.	1801–02
Denistone, J.	1755–56
Derham, F. S.	1878
De Saint Just, / De Singest } P.	1698–1703
D'Esterre, J. C. E.	1832–39
Desvaux, M.	1689–90
De Vauclen, D.	1703–05
,, T.	1695–1705
Devlin, W. H.	1876–77
Dewer, H.	1801–03
Dickson, G. Le F.	1847–55
,, S.	1826–30
Digby, S.	1751–60
Dillon, J.	1806–09
,, R.	1851–65
Dixon, H. M.	1822–39
Dobson, G.	1703–08
Dolling, J. A.	1800–02
Donnelly, J.	1813–16
Dornell, W.	1702–07
Douglas, J.	1797–1803
,, R.	1795–97
,, R.	1811–16
Dowse, R. R.	1853–56
Drake, E.	1813–16
Dumas, J.	1774–81
,, P.	1751–76
,, P.	1797–1800
Dundas, T.	1757–62
Dunn, J.	1803–04
Duplex, I.	1696–98
Dussaux, J.	1719–27
Eades, M. A.	1808–12
Eager, J.	1808–13
East, H.	1804–17
Eden, H. H.	1858–78
Edmondson, J.	1834–41
Edwardes, D. J. B.	1838–44
,, F. A.	1844–53
Edwards, E.	1838–41
,, T.	1776–86
Elford, S.	1703–05
Elkington, J. G.	1813–17
Elliot, R.	1815–17
Elliott, R. C.	1809–16
Ellis, J. H.	1803–04
,, T.	1837
Elwyn, T.	1855–58
England, R.	1803–06
Erskine, A.	1761–78
,, H. D.	1854–55
Evans, H. T. P.	1870
,, J.	1811–21
Evelyn, H.	1794–95
Evered, J. G.	1826–27
Ewens, A. T.	1855–57
Fairfax, T.	1689–94
Falkner, E. N.	1850–60
Fawcett, W.	1796–1804

Feild, M. B. 1854–58
Fell, R. E. 1749–55
Fettes, A. 1803–14
Filbridge, F. 1692–1704
Finch, E. H. F. . . . 1880
Finucane, J. 1815–17
Firebrace, G. 1707–08
Fitzgerald, J. 1756–62
,, L. C. W. H. . 1834–37
,, R. . . . 1736–39
,, R. . . . 1780–82
,, R. . . . 1803–04
,, W. H. . . 1853–55
Fitzgibbon, C. P. . . 1855–58
Fitzpatrick, T. P. . . 1847–51, 1855
Fitzthomas, W. E. . . 1795–1803
Fleming, J. 1808–10
,, J. 1854–70
,, T. 1803–08
Flude, J. 1811–13
Foulke, H. 1693–96
Forbes, Viscount . . 1804–09
,, J. 1804
,, J. 1854–55
,, W. 1702–03
Forrester, A. 1716–20
,, Lord . . . 1716–17
,, J. 1716–19
Fox, S. 1801–26
Frampton, C. 1743–49
Frankland, Franklyn H. . . . 1692–1701
Fraser, R. 1813–19
Freear, A. W.. 1809–17
French, E. 1806–14
Frizell, W. B. 1814–29
Fry, W. D. 1803–05
Fullarton, G. 1755–56
,, J. 1809–21
Fullerton, R. E. . . . 1836–39
Furlong, R. T. . . . 1830–36
Fyffe, W. J. 1853–55
Gallwey, Galway, H. L. . . . 1878
Gamble, J. H. . . . 1780–81
Gardner, J. 1707–08
,, W. 1702–07
,, W. L. . . . 1794–96
Garland, J. 1807–13
Garnett, C. H. . . . 1860–73
Garnon, J. 1798–1804
Garret, P. 1782–87
Garsia, M. C. 1871–78
Garvey, J. 1808–21
Geddes, J. G. . . . 1826–48
Geekie, W. 1884–85
Gee, ... 1746
Gernon, N. 1803–04
Gervaiset, Blaise, . . 1696–98
Gibbes, J. W. . . . 1762–1778
Gibbons, J. 1718–38
Gibson, A. 1868
,, A. 1850–54
,, E. 1780–87
Gillespie, J. R . . . 1821–34
Giles, H. F. 1746–47
Gisborne, J. 1739–55
Glascott, W. 1858–70
Gladestanes, G. . . . 1780–86
Godfrey, W. 1689–95
Goldie, A. J. 1865–81
Goodwin, J. 1797–1800
Goodwyn, J. E. . . . 1863
Gordon, G. St. L. . . 1803–04
,, J. 1791–96
,, T. 1728
Gore, A. 1814–17
,, T. 1797–99
,, W. 1785–87
Gould, J. 1773–78
,, P. 1764–82
Gowan J. 1805–18
,, T. 1777–85
Grant, E. C. 1855–59
,, E. J. 1831–44
,, J. H. 1855–56
Grantham, V.. 1690–95
Gray, W. 1803–07
,, W. R. 1843–47
Gregg, E. R. 1826–29
,, J. N. 1815–26
Green, C. 1794–1804
,, C. M. 1849–66
,, J. 1690
,, T. W. 1857–64
,, W. 1778–81
Greene, W. W. H. . . 1846–49

Gregory, H. J. M.	1841–46
Grey, G.	1799–1812
,, O. W.	1804–29
Griffin, W.	1805–13
Grimes, C.	1855–57
Groves, J.	1783–90
Grylls, R. G.	1839–41
Gubbins, S.	1854–60
Gumbleton, R.	1863–65
Gunning, W. O.	1820–22
Gunsley, J. A.	1711–13
Hall, D.	1715–37
,, D.	1737
,, G. L.	1778–93
,, P.	1806–08
Hames, J.	1796–1801
Hamilton, A.	1787–1811 1811–29
,, C.	1718–25
,, G.	1796
,, R.	1789–90
,, W. G.	1878
Hammond, W.	1740–48
Hampton, C. J.	1858–59
Hancock, J. G.	1744–46
,, R.	1815–17
Harcourt, J. S. C.	1854–58
Hardyman, T.	1803–04
Hare, J.	1746–48
,, R. G.	1796–1803
Harpur, W. C.	1803–17
Harris, E.	1702–04
,, J.	1696–1706
Harrison, H. S.	1855–57
,, R.	1811–18
,, W.	1796–97
Hartley, W.	1787
Harvey, W.	1716
Harward, R.	1732–45
,, R.	1742–51
Hassall, G. R.	1857–68
Haseltine, J.	1710–13
Hawker, P. R.	1803–17
Hay, D.	1831–35
Haynes, C.	1878
Hazard, G.	1692–98
,, R.	1775–77
Head, T.	1742–47
Headley, A. W. M.	1865–69

Heard, W. H.	1827–49
Heathcote, H.	1805–06
Heaviside, R.	1808–17
Henagan, J. B.	1787–88
Henley, R.	1720–23
Hennen, J.	1807–11
Hepburn, W. R.	1847–54
Herring, D.	1720–39
,, H. L.	1855–65
,, J.	1808–11
Hervey, A.	1778–81
Hewett, C.	1747–54
Hicks, J.	1690
Hill, E. N.	1854–65
Hitchen, J.	1803–13
Hobart, J.	1713
,, R.	1778–81
Hobbs, J. C.	1852–67
Hobson, J. St. C.	1847–54
Hodges, P.	1719–44
Hodnett, J.	1778
Hoey, W. F.	1844–54
Holbrooke, F.	1801–02
Holt, R.	1805–09
Hooper, A.	1866–71
Hornby, C. R.	1864–82
Horne,	1747
Horsman, J.	1719–20
Howard, R.	1803–28
Hubbard,	1721
Huggins, J.	1815–17
Hughes, D.	1811–13
,, R.	1811–17
,, W.	1746–48
Hume, R.	1695–98
,, R.	1737
Hunter, C.	1839
Hutchinson, A.	1726–32
,, E.	1706–08
,, E.	1806–11
,, J.	1746–47
,, T.	1758–60
Hutton, H.	1801–02
,, H. P.	1868–79
I'Anson, B.	1720–24
Ibbetson, D.	1772–73
Ingle, L.	1849
Innes, A.	1709–17
Irwin, T.	1810–11

Irwine, C.	1778–81
Irwyn, J.	1707–09
Jacobs, M. E.	1803
Jackson, R.	1692–98
,, T.	1801–15
Jago, C.	1817–18
James, J.	1813–15
,, T.	1799–1803
Jauncey, J. K.	1807–08
Jefferies, C.	1719–42
,, J.	1711–20
Jennings, E.	1800–03
,, J.	1755–64
,, J.	1762–74
,, T.	1762–80
Jenoure, J.	1721–22
Jevers, / Ivers, } H. N.	1754–70
,, J. A.	1751–84
Jocelyn, G.	1732–41
,, J.	1734–39
Johnson, E.	1757–62
,, T.	1803
Johnston, W. Y.	1853–54
Jones, J.	1758–62
,, M.	1797–1818
,, T.	1806–32
Joy, E. W.	1855–57
Julian, J. C.	1868–79
Kelly, F.	1811–13
,, J.	1746–49
Kemble, H.	1865–81
Kennedy, D. H.	1819–22
,, H.	1760–62
,, J. A.	1803–07
,, W.	1870–75
Keogh, J. H.	1841–49
Kerr, W.	1854–55
Kettlewell, T..	1808–27
King, R. D.	1815
Kingsley, J. F.	1803–24
Kirkcudbright, Lord	1756–76
Knight, C.	1807–08
,, I.	1689–1707
Knuttall, G.	1754–56
Lacon, J. M.	1804–05
Lacy, E. W.	1829–30
Lahiff, T.	1797–1802
Lamert, G. F..	1859–62
Lane, F.	1795–1800
,, M. T.	1816
Lardner, C.	1815–18
La Touche, D.	1813–16
,, D. M.	1859–61
,, G. F.	1850–52
Lawson, J.	1778–79
,, S.	1844–46
Lay, J.	1748
Leach, T.	1803–14
,, W.	1692
Leader, W. N.	1867–70
Leathat, J.	1709–10
,, R.	1704–16
Lee, J.	1771–80
,, W.	1762
Leightheizer, H.	1770–76
Legg, T.	1783–86
Lestanquet, G.	1728–33
Lewes, G.	1754–65
Lewin, J.	1806–17
,, J. D. R.	1847–54
,, T.	1815–17
Lewis, B. G.	1876
,, G.	1747–65
,, H. H.	1717–27
,, L.	1761–75
Liardet, F.	1814–17
Light, J.	1808–31
Lindesay, J. E. C. C.	1858–76
,, R. S.	1844–47
Lindsay, N. J.	1800–06
,, W.	1858–62
Littleton, J.	1706–10
Livesay, J.	1690
Livingstone, H.	1790–96
Lockhart, W.	1784–1813
Lockwood, A. P.	1841–52
,, P.	1811–16
Loftus, H.	1742–68
Long, H.	1705–36
Loudoun, Earl of	1749–70
Love, H.	1797–1801
Lovell, G.	1721–30
Lowder, J. P. C.	1860–64
Lowry, C. A.	1841–50
Lucas, C.	1789–91
,, E.	1833–36
Lumsden, A. J. H.	1833–47

Lushington, W. 1753–55
Luxmore, F. 1849–54
Lynam, W. C. C. . . . 1779–81
Lynch, R. B. 1797–1828
,, J. B. 1797–99
Lyster, G. W. 1797
,, R. 1791–93
McCarthy, E. J. C. . . 1803–07
,, R. 1807–08
McCord, R. A. 1865–76
McCrohan, J. 1803
McCulloch, D. 1787–93
McDonald, A. 1834–47
,, D. 1812–17
,, R. C. . . . 1827–52
McDougald, A. 1795–96
McDougall, A. 1815–17
,, A. C. . . . 1815–20
McKenzie, A. W. . . . 1856–64
MacKenzie, J. 1746–49
,, J. B. . . . 1855–58
McKerral, W. 1754–56
Mackesy, W. P. P. . . 1851–55
McLaughlin, G. 1726
McLeod, C. R. 1818–26
MacLellan, J. (see Kirkcudbright).
McMahon, J. 1784–85
,, J. A. . . . 1861–64
McNabb, A. 1804–15
Macnamara, J. 1852–53
McNiell, M. 1857–58
McPherson, L. 1852–68
,, P. 1816–17
Macready, E. N. . . . 1814–39
Machell, R. 1808–22
Madden, G. 1811–13
Magee, H. W. 1826–31
,, W. 1806–07
Maitland, C. 1787–88
,, F. 1784–89
,, J. Lord . . . 1731–38
Major, T. 1818–19
Malbone, ... 1745
Malet, E. 1795–1813
Mallory, R. H. 1773–79
Mangin, J. 1803–08
,, S. 1806–07
Mann, W. 1820–28
Manners, R. 1799–1823
Mansel, G. 1820–32
,, H. 1830–34
,, W. 1703–05
Mansergh, D. 1796–1803
Mantell, A. W. 1803–04
Marcell, L 1742–51
Marechaux, C. H. . . . 1820–44
Margaret, P. 1718–41
,, T 1741–55
Marlton, W. 1827–30
Marolfe, B. 1692
Marshall, J. 1770–91
,, R. 1761–73
,, W. 1693
Martin, T. 1797–98
Martindale, C. De B. . 1880–81
Martyn, E. 1707–17
Mason, J. D. 1703–13
Massey, N. W. 1854–63
Masters, S. 1807–17
Masterton, C. 1782–89
Mauleverer, J T. . . . 1844–62
Maurice, T. 1744–48
Maxwell, C. 1755–94
,, C. 1794–1817
,, D. 1759–62
,, D. 1793–1803
,, G. V. 1881
,, R. 1804–06
,, R. 1815–16
,, W. 1792–93
Mayne, R. 1808–27
Meade, C. J. 1806–08
,, J. 1801–04
,, J. 1856–57
Medford, / Midford. M. 1707–17
Meggs, H. 1734–49
Meik, J. P. 1827–29
Merriden, W. 1706–09
Merrit, W. I. 1865–76
Metcalfe, N. 1778–80
Michell, W. 1702–03
Middlemore, R. 1690–95
Middleton, O. R. . . . 1857
Millar, R. 1772–75
Miller, R. 1800
Milroy, D. 1854–64

Minet, W.	1783 / 1804–14
Minnitt, J.	1747–48
Minshall, P.	1693–94
Mitford, W.	1703–04
Mohun, J.	1709–13
Moneypenny, T.	1813–17
Montagu, D.	1856–63
Montgomerie, J.	1823–29
Montgomery, A.	1756–58
,, A.	1767–73
,, I.	1778–85
,, R.	1788–1802
Moon, J.	1854–62
Moore, J.	1830–41
,, W.	1813–15
Moorsom, C. J.	1854
Morewood, H. F.	1855–70
Morris, T.	1847–55
,, J. J.	1881
Morse, C. J.	1880
Mossman, J.	1719–42
Mounsey, W. H.	1825–27
Moutray, J.	1751–62
Muller, J.	1756–63
Munroe, H,	1742–46
Muntz, J. F.	1876–85
Mure, A.	1718–40
Murray, G.	1795–96
,, R.	1795–1829
,, W.	1782–92
,, W.	1816–17
Mustemberger, G.	1689
Nagle, R.	1858–68
Napper, J.	1809–14
Nash, J.	1689
Ness, B. T.	1810–17
Neville, P. P.	1810–26
,, R. H.	1854–56
Newdigate, T.	1703–19
Newland, W. J.	1803–4
Newton, A.	1742–44
Nichol, W.	1702
Nicholl, S. J. L.	1834–52
Nicholson, B. W.	1806–17
,, W.	1766–67
Norman, W.	1706–11
Norris, W. A. D.	1870–71
North, W.	1793–94
Northey, R.	1773–78
Nunn, B.	1801–13
,, L.	1755–72
O'Brien, C. R. M.	1878
,, J.	1849–57
,, P.	1804–10
O'Flaherty, D.	1805–06
O'Grady, R. D.	1836–50
O'Halloran, T.	1811–18
O'Heighan, F.	1804–05
Oliver, C. D.	1840–54
Onebye, C.	1711–13
Onslow, R.	1719–21
O'Reilly, W. P.	1828–30
Ord, G.	1702–07
,, R.	1707–10
Orfeur, W.	1721–32
Ormond, H. S.	1831–46
,, H. S. A.	1838–39
Ormsby, O.	1734–41
Ottley, B. R.	1815–17
Ouseley, W.	1823–24
Owen, R.	1742–56
Pakenham, H.	1717–20
,, T. H.	1847–74
Pallister, / Palliser, H.	1696–1717
,, W.	1697–1717
Palmer, J.	1776–84
,, J. R.	1816–17
,, W.	1724–30
Palmes, G. St. M.	1873–75
Parke, R.	1771–73
Parry, E.	1811–15
Parslow, J.	1770–86
Parsons, T.	1732
Partridge, J.	1698
Paterson, H. A.	1797–98
Paton, J.	1818–20
Patterson, D.	1765–81
Patullo, J. B.	1840–55
Paul, A.	1742–51
Paxton, J.	1864–72
Peach, J.	1804–05
Peacock, J.	1774–75
,, T. G.	1855–58
Pearce, R.	1803–28
Peat, W.	1856–57
Pelham, P. T.	1858–70

Pennefather, M.	1848–57
,, W.	1808–17
Pennock, } B.	1697–1703
Pinnock, }	
Perrott, O. G.	1827–29
Perry, A.	1808–11
,, J. P.	1805–18
Pery, J. H.	1834–42
Phillips, E.	1689–92
Phipps, H.	1689–91
Piercy, J. M.	1870–82
Pierson, F.	1717–41
,, F.	1739–54
Pigot, R.	1803–04
Pigott, W.	1791–92
Pilcher, E. P.	1786–93
Pilkington, W. L. L.	1799
Pillsworth, E. G.	1830–40
Piper, S. A.	1806–23
	1830–34
Place, R.	1695–97
Pocock, G. F. C.	1848–56
Pogson, H. J.	1827–41
Ponsonby, F.	1717–21
Poole, J. W.	1856–58
Popple, W. P.	1755–67
Potter, N.	1698–1706
Poumies, M.	1705–13
Pountney, E.	1772–73
Powell, J.	1804–33
Pownall, T.	1702–05
,, W.	1703–09
Powys, L. W. H.	1858–60
Poyntz, A.	1796–1812
,, A.	1807
,, J.	1814–44
,, S. R.	1811–17
Pratt, J.	1811–17
Prendergast, E.	1812–15
Price, J.	1797–1801
,, T.	1746–71
Primrose, R.	1769–75
Pritchard, W.	1711–19
Proctor, J.	1826–47
,, T.	1785–89
Prosser, F.	1814–16
Pulleine, H. B.	1855–58
Purcell, J.	1718
Purdon, E	1805–10
Purvis, W.	1796–97
Quarles, E.	1707–19
Rainsford, C.	1705–21
Ralph, J.	1819–27
Ramsay, G. W.	1780–89
,, J.	1726
,, J.	1729–71
,, J.	1785–88
Ramus, H.	1818–24
Rand, H.	1745–56
Ravenhill, H.	1720–50
,, H.	1785–87
Rawlings, S.	1803–04
Read, R. W.	1856–66
,, W. T.	1769–74
Reid, C. C.	1881
Reynolds, T. V.	1795–1801
Rice, W.	1757–59
Rich, C. L. M.	1881
,, J.	1751–61
Richardson, T. G.	1804–14
Rider,	1730
Roberts, C. H.	1828–30
,, E.	1781–85
,, T.	1712–18
,, T. F.	1803–12
Robertson, C.	1794–97
,, G. D.	1803–05
,, J.	1749–55
,, W. J.	1853–57
Robinson, H. E.	1831–43
,, J.	1873–74
,, J. E.	1872–81
Robson, H.	1815–17
Roche, B.	1796–1801
,, D.	1742
,, J.	1783–84
Rochfort, W.	1771–83
Rocke, A.	1778–81
Roe, J.	1811–23
,, J.	1811–21
Rogers, A.	1691–94
,, J.	1722
,, J.	1778–81
,, R. N.	1813–1818
,, T.	1798–1801
Romaine, N.	1738–41
Romney, G. J.	1806–07
Rooke, H.	1757–62

Rooke, H.	1740–61
,, T.	1702
Roper, H.	1771–80
,, J.	1707–34
Rose, J.	1841–55
Ross, D.	1747–54
,, R.	1742–47
,, W.	1811–20
Round.	1758
Rowe, P.	1780–85
Rowley, R.	1795–98
Roy, W.	1786–90
Rugge, H.	1742–63
Rumley, C.	1817–25
,, G.	1807–11
,, J.	1809–18
Russell, J.	1785–1803
Rutherford, R.	1772–74
Ruxton, C.	1779–81
Ryan, M.	1814–17
,, M.	1877
Ryley, P. K.	1767
Salisbury, T.	1785–91
Sandall, H.	1762–63
Sanders, G. H.	1854–58
Satterthwaite, J. C.	1786–93
Saunders, H. C.	1797
,, J.	1705–07
,, J.	1690–1716
Saunderson, C.	1689–94
,, G.	1691–1706
,, T.	1689–1704
,, W.	1694–98
Schoof, M.	1824–27
Scott, R. A.	1875–78
,, V.	1710–41
,, W.	1708–29
,, W. H.	1875
Scroop, G.	1702–03
Sedgwick, T.	1695–98
Selleck, H.	1800–01
Selwyn, J.	1804–11
Serjeant, H.	1793
Sharkey, F.	1798–99
Sharman, J.	1776–78
,, W.	1719
Sharpe, J.	1695–1704
,, H. G.	1860–64
Shawe, M.	1816–17
Shawe, T.	1799
Sherard, G.	1706–09
Shewbridge, E. P.	1791–1801
Shum, H.	1835–47
Sillery, C.	1834–56
,, C. J. C.	1853–56
Sims, T	1803–04
Sinclair, D.	1804–17
,, R.	1732–40
Singleton, H. C.	1854–73
,, J.	1845–46
,, W.	1693–1703
Skerrett, J. B.	1783–84
Skipton, W.	1751–56
Skirrow, J.	1803–24
Sladden, B.	1711–19
Slade, M. J.	1843–45
Smart, H.	1726–1736
Smith,	1693–98
,, H. S.	1855–62
,, J.	1768–69
,, N.	1717–24
Smith / Smyth } R.	1784–1803
Smith, R.	1811–14
,, R. W.	1839–49
,, S.	1690–93
,, W.	1690–95
Smyth, E. St. G.	1854–68
,, T.	1849–62
Snell, W.	1746–65
Southwell, W.	1741–46
Sowle, H.	1719–21
,, R.	1717–28
Sparks, M. J.	1805–11
Spawforth, J.	1809–13
Spiers, J.	1775–76
Stacpoole, H.	1826–27
Staff, W. B.	1824–28
Stanhope, W.	1814–19
Stanwix, T.	1717
Steel, W.	1806–08
Steele, W. A.	1827–46
Steiger, / Stygar, } J. C.	1689–97
Stephens,	1811
Stephens, H.	1815
Stephenson, G.	1810–18
Steuart, J.	1814–25

Stevenson, J. C. N.	1852–55
,, M. D.	1855–70
Stewart, J.	1759
,, T. B.	1861–68
,, W.	1776
,, W.	1802–17
Stiel, W.	1749–56
Still, J. T.	1847–52
Stillingfleet, E.	1717–51
Stirling, G.	1806
Stone, T.	1741–42
Storey, A.	1705–06
Stow, R.	1703–16
,, T.	1741–47
Strong, R. H.	1829–31
Stroud, H. W.	1870–80
Stuart, C.A.	1807
,, R.	1762–77
,, W.	1737–41
Sullivan, J.	1798–1804
,, W.	1805–27
Sutherland, A.	1804
,, J. W.	1876–77
,, R.	1755–56
,, S. H.	1819–22
,, T. B. M.	1820–22
Sutton, R.	1690–98
Swain, A.	1693–95
Sye, J.	1692
Symons J.	1689–91
Taylor, G.	1803
Teevan, T.	1871–75
Tegart, E.	1794–1800
Tessier, L.	1807–08
Teulon, G.	1809–17
Thom, J.	1860–70
Thomas, E.	1756–80
,, R.	1775–76
Thomlinson, W.	1787–93
Thompson, G. W.	1822–26
,, J.	1854
,, J.	1702–15
Thorpe, W.	1833–39
Throgmorton, R.	1721
Thwaites, E.	1703–05
Timon, P.	1803–05
Tincombe, F.	1812–17
Tobin, C	1826–27
Todd, R.	1806–09
Tolcher, C. J. H.	1854–57
Tomson, J.	1822–29
Tongue, J.	1803–42
,, J.	1837–54
Tooner, M.	1862–73
Torriano, C.	1784–93
Toussaint, I.	1696
Tovey, C.	1855–58
Townrow, W.	1704–05
Tracy, W.	1721–24
Travers	1812–13
Tressider, S.	1824
Trigance, J.	1834–44
Turner, R.	1703–06
Turney, E.	1689
Tweedale, Marquis of	1846–62
Twynam, H. M.	1881
Tyner, C.	1855–60
Tyrwhit, G.	1692–98
,, N.	1697–98
Urquhart, W.	1783–94
Vachell, H.	1831–34
Vanbelle, A.	1690–93
Vandeleur, D. R.	1871–80
Vandersee, D.	1818–23
Vangensimmer, J.	1702–03
Vaughan, H.	1722–24
Vaumorel, P.	1793–1820
Vigoureux, C. A.	1813–26
Vincent, J.	1718
Voules, W. J.	1856–58
Wade, G.	1803–07
,, J.	1806–14
Waldron, F. C.	1827–34
Walker, J.	1707–11
,, M.	1846–55
,, R. F.	1865–76
Wallace, T.	1788–91
Waller, R.	1741–46
,, R. B.	1723–26
Wallop, W.	1803–04
Ward, D.	1702–16
,, J.	1694–97
,, J.	1824–47
,, R.	1748
Warner, E.	1803–04
,, T. L.	1757–62
Warren, C.	1814–20
,, W. O.	1813–17

Watson, A. G.	1870
,, B.	1805–07
,, C. S.	1803–10
Waymouth, S.	1839–41
Weddell, W.	1694–98
Wedge, R.	1811–15
Welman, G. A.	1876–77
Wemyss, D.	1720–36
Westenra, H.	1738–41
Whaley, / Walley, J.	1698–1703
Whichcott, C.	1695–1698
,, G.	1689–93
,, W.	1692
White, C. H.	1839–40
,, J. L.	1805–14
Whitehalf, J.	1693
Whithof, / Whitoft, W.	1693–98
Whiting, J.	1744–45
Whitmore, E. A.	1841–56
Wilkie, T.	1761–82
Wilkinson, T. W.	1839–48
,, W.	1781–82
,, W.	1773–1815
Williams, C.	1708–15
Williamson, A. H.	1851–59
,, C.	1797–1810
,, F. H.	1858
,, J.	1757–58
,, J.	1814–17
,, T.	1804–14
Wills, C.	1694–1701, 1705–16
,, C.	1708–11
Wilson, N.	1812–14
Willson, R.	1823–25
Wilsonn, R.	1790–91
Wiltshire, J.	1711–13
Windus, E.	1813–17
Winrow, J.	1809–18
Winter, J.	1750–82
Wiseman, J.	1764–65
Wolfe, E.	1715–18
Wolstonholme,	1692
Wood, D.	1805–10
,, H.	1855–59
,, W.	1793–95
Woodward, D.	1796–98
,, J.	1751–56
Wooldridge, T.	1793–1802
Worsley, J.	1769–72
Wray, H. B.	1806–31
,, T. C.	1855–58
Wright, A. J. A.	1870
,, J.	1742–61
,, J.	1756–69
,, J. G.	1825–38
,, T.	1862–67
,, W.	1803–08
Wynne, J.	1740
Yarborough, S.	1695–98
Young, A.	1796–1803
,, J.	1744–45
,, M.	1824–27

INDEX

Abercromby, Sir Ralph, 205–6, 209–16
Aboukir Bay, 209–10
Act of Union, 54
Addington, Prime Minister, 226–228
Agueda, 253, 259, 262, 275–6, 291
Aix-la-Chapelle, Peace of, 132
Alba de Tormes, 277, 279–80
Albany Barracks, 371, 377–8
Alcoentre, 249
Aldea da Ponte, 260
Aldershot, 446, 451–2
Alexandria, 210–12, 214, 217, 219, 221–2
Algeciras, 244
Alicante, 50–1, 54–5, 59, 66–8, 93
Almeida, 253, 256, 263, 276
Altea Bay, 49
Alten, Sir Charles, 312, 329, 331, 326, 341, 344–5, 350
Amiens, Peace of, 223
Anne, Queen, 23, 88, 94
Anson, Admiral, 130–1
Antigua, 166
Antwerp, 277–9, 300–4, 309
Appa Sahib, 367–8
Appleton, Teavil, 133–4
Arapiles or Hermanitos, 277–9
Archangel, 28
Argyll, Duke of, 154
Ariadne, transport, 232
Arlanzan, 283–4
Arms and Equipment, 3, 7, 25, 48, 98–100, 119, 136, 152–3, 171, 207, 238, 310, 384, 389, 393, 431, 449, 452
Army of Portugal, 248, 257, 264
d'Artagnan, M., 13
d'Asfeldt (*see* d'Hasfeldt)
Asseerghur, 368–70
Association, H.M.S., loss of, 61
Atcherley, Capt., 392–3, 406, 410–11, 413, 432, 436, 441, 447–8
Augusta, 159
d'Auvergne, Rev. E., 12, 13, 19

Badajoz, 257, 263–4, 266–72
Balaclava, 405, 409–10, 421, 423, 425, 428
Baldwin, Consul at Alexandria, 212
Bamford, Capt. T. B., 242, 244, 246, 272, 275, 293–4
Bandon, 181, 198–200
Bantry Bay, 200
Barba del Puerco, 256–7
Barcelona, 11, 40–1, 44–6, 49–56, 65, 67
Battles, Sieges, etc.:
  Alexandria, 211–13, 214–6
  Alicante, 50, 68
  Alma, 400
  Almanza, 54
  Annapolis, 73
  Antwerp, 300
  Asseerghur, 369
  Badajoz, 267
  Balaclava, 409
  Barcelona, 41, 49
  Bastia, 191–2
  Batavia Roads, 236
  Belle Isle, 145–7
  Burgos, 283
  Cagliari, 66
  Calvi, 194
  Cap Brun, 187
  Ciudad Rodrigo, 263
  Convention Redoubt, 191
  Denia, 68
  Eutaw Springs, 162
  Fuentes Onoro, 254–7
  Finisterre, 130
  Firth of Forth, 65
  Gibraltar, 32, 35, 108–11
  Grezzie, 236
  Gulf of Lyons, 71
  Hyères (Naval), 195
  Inkerman, 414
  Landen, 13
  Lerida, 56–9
  Little Inkerman, 410
  L'Orient, 128
  Malaga, 33
  Manantoddy, 360
  Montjuich, 41
  Namur, Defence of, 9
  Namur, Assault of, 17
  Port Mahon, 66
  Preston, 95
  Quatre Bras, 313
  Redan (Sebastopol), 433
  Sabugal, 252
  St. Cast, 141
  St. Estevan, 45
  Salamanca, 278
  Samarang Roads, 235
  Sandwich Bay, 70
  San Fiorenzo, 191
  Scheldt and Lys Lines, 12
  Sebastopol, 404–439
  Steenkirk, 10
  Sumatra Ports, 236
  Toulon, 182, 190
  Valetta, 206
  Villa Muriel, 285
  Waterloo, 325, 346
Belfast, 7, 38, 154, 382, 445
Belle Alliance, La, 326–7, 331–2, 334, 338
Belliard, General, 218
Bellona, H.M.S., 180–1
Bergen op Zoom, 8, 9, 297–9, 300
Bermuda, 382–3, 448
Berwick, Duke of, 52, 54
Bettesworth, Col., 96, 102, 106, 108, 113–4
Bevan, Col., 256–7
Bircham, Samuel, 187, 201, 207, 227, 362, 366–7
Bisset, Andrew, 96, 102, 106, 108, 116–7
Bisset's Foot (Fusiliers), 101–2, 104, 107–8
Bligh, General, 140–141
Blondin, Thomas, Mayor of Canterbury, 79–81
Blucher, Marshal, 312, 323–5, 339
Bombay, 454
Boothby, Sir Wm., 138–43
Bossu, Wood of, 313–4, 316, 321

Boufflers, De, 9, 16, 19
Bradford, Sir Thomas, 351, 377
Braine Le Comte, 320, 324, 329
Brasschaet, 298–301
Brenier, General, 256–7, 280
Brereton, Robert, 172, 184–6, 189, 192–3, 195
Brihuega, 75
Bromswell, 227
Brown, Sir George, 395, 421
Bruges, 10, 11, 13
Brunswick, Duke of, 313–4
Brussels, 11, 15, 17, 21, 302, 308–10, 324
Bulganac, 399
Burdell's Plantation, 161, 163
Burdett, Sir C., 237, 239
Buonaparte (*see* Napoleon)
Burgos, 283–4, 290
Burnley, 451
Burston, George, 2, 6, 43, 45–8, 53, 58, 59, 60, 75–6, 81, 83, 92
Byng, Admiral, 31 *et seq.*, 50, 65, 68, 101, 138

Cacadores, 247, 255, 259, 280–1
Cadiz, 31, 33, 35, 69–70, 244–6
Cairo, 209, 217–8, 222
Calder, Sir Henry, 171
Cagliari, 66, 67, 101
Calvi, 194
Cambridgeshire Regiment, 167
Cambridge Club, 371–2
Camp of Instruction, Kinsale, 156
Campbell, Major, 169, 172, 178
Campbell Pasha, 209
Campbell, John Pennock, 419, 440
Canada, 447
Cancale Bay, 139–40
Cannanore, 359–61, 363
Canrobert, General, 403, 418, 423
Canteen, 363
Canterbury, 27, 30, 38, 53, 60–2, 69, 80–8, 94, 125, 137, 142, 149–50, 233, 358, 371, 378, 388
Cape of Good Hope, 234
Carión, 284–7
Carnot, Lazare, 298, 301
Carthagena, 50
Castel Branco, 258
Castleton, Viscount, 1, 2, 14
Cateau, Le, 350
Cathcart Hill, 427, 444
Cathcart, Lord, 231–3
Celorico, 251–2, 261, 291
Cephalonia, 388
Chambly, 448–9
Chaplains, 201
Chapman, Major-General, 382
Charleston, 158, 160, 165–7
Chatham, 28, 64, 137–8, 170–1, 180, 357, 377–8
Chatham "chest," 38, 78–9
Chelmsford, 227, 305
Cherbourg, 140
Chester, 94, 123, 452
Cholera, 367, 376, 395–7, 404–6
Ciudad Rodrigo, 251, 253–4, 258–60, 263, 276, 289–91
Clausel, General, 279, 282–3
Clayton, Colonel Jasper, 109–110
Clinton, Sir Henry, 158, 165
Coa, 252, 259–62
Cobham, Richard, 93–6
Codrington, Sir William, 436, 441–2
Coimbra, 250–52
Colborne, Sir John (afterwards Lord Seaton), 221, 338, 446
Colchester, 136–7, 197–8, 305–7, 378, 387
Colcutt, William, 416, 421–2
Colours, 122, 133, 156, 228, 352, 379, 446
Colour Sergeants, 294, 363
Conyngham, Major-General, 43, 45–8
Cork, 70, 101, 133–4, 151, 200–2, 230–2, 241, 354–7, 384–5, 389, 392
Cornwall, Canada, 448
Cornwallis, Lord, 159, 164–5
Corsica, 191–5, 199
Courts Martial, 61–3, 70, 361, 363, 370, 382
Craig, Sir James, 225, 227
Crimean War, 391 *et seq.*
Cruger, Colonel, 159–61
Cunningham (*see* Conyngham)
Curragh, 230–1, 239

Dalrymple, Major, 369–70, 373, 376
Dannenberg, General, 414, 418–9
Davis, Capt. Gronow, 437
Davison, William, 107
Denia, 67–8
Depot, 123, 180, 231, 243, 296, 305–8, 353, 358, 363, 371, 378, 384–5, 388–9, 393–4, 428–30, 442, 445, 450
Devna, 395
Dominica, 168–175
Dornell, William, 23, 29, 53, 62
Douro, 263, 283
Dover, 31, 353–4, 389, 452
Doveton, Brig.-Gen., 367–8, 370
Doyle, General, 212–16, 218–21
Drill, 4, 153, 156–7, 171, 201, 306, 328
Dublin, 94, 98–9, 101, 156, 171, 229–31, 239–40, 242, 379, 381, 445, 450–1
Duelling, 370–1
Dundas, General Sir David, 171, 187, 190–1, 201, 355
Dundas, Henry, 188, 195–6, 198–9
Dunkirk, 64–5, 80
Dunkirk privateers, 148
Dunlop, Major-General, 247, 257, 260

East Lancashire Regiment, 455
Egypt, 206 *et seq.*
Elliot, Sir Gilbert, 195–6
Elphinstone, Captain Keith, 180–3, 185, 187
Elvas, 257, 263–5
Enniskillen, 4, 380
Equipment (*see* Arms)
Escorial, The, 283
Establishment, 2, 8, 29–30, 82, 89, 116–8, 132, 135–7, 139, 149, 151–4, 167–9, 175, 224, 226, 356, 364, 378, 386–9, 391, 447–8, 450–2
Estremadura, 249, 251, 254, 262–3
Eugene of Savoy, Prince, 60–1
Eutaw Springs, 161–3
Evans, Sir de Lacy, 393, 396, 399, 401–4, 411–2, 417
Exeter, 150–1

Fairfax, Thomas, 2
Faron, Mont, 182–5, 189
Federoff, Col., 411
Fermoy, 241, 381–2, 389, 425, 442, 445
Finisterre, 130
Firm Prize, 33–4
Flagstaff Bastion, 408, 427, 434
Flores de Avila, 280–1
Forbes, Adjutant, 431
Forrester, Lord, 93, 95, 96
Fort Mulgrave, 184, 186, 188–9
Fort, 96, 159–61
Fort St. George, 237
"Forty-Five." The, Summary of events, 124–5
De Fourbin Expedition, 65
Foy, General, 283
Frampton, Brig.-Gen., 122, 127, 132–3
Frederick Hendrick Fort, 300–1
Frenada, 291
French Eagle captured, 280
French expedition to Ireland, 200
French Guard at Waterloo, 334–8, 343–4

French outrages, 259
French Revolution: summary of events, 177
Fuentes Onoro, 253, 254
Friant, General, 210–12
Fuentes Guinaldo, 260–2
Fuente del Maestro, 275

GALWAY, 154, 380, 384–5
Galway, Lord, 51, 52, 54–9
Gardner, Capt., 198–9
Gardner, Lieut.-Gen. Sir R. W., 391
Garden Battery, 434
Gates, General, 159
Gemioncourt, 313–4, 316–9
Genappes, 9, 16, 323–5
George of Denmark, Prince, 61–2, 77
Gerona, 51
Gibraltar, 30, 31–9, 69, 83, 107–11, 151, 181, 206, 243–4, 389–91, 443, 445
Gibson, Lieut. Alured, 420
Gisborne, Col., 150, 154
Gitaut's Cuirassiers, 315, 318
Glasgow, 154
Gloucester, rioting at, 139
Godolphin, Lord, 71, 75
Gould, Col. Paston, 152, 159, 164–7
Graham, Sir Thomas, 203–4, 244–5, 264, 275, 296–8, 301
Grantham, 21, 25, 38
Grasse, Count de, 164, 166
Gravina, Admiral, 182, 185–6
"Great Eastern," 447
Green Hill, 212–4, 219–20
Green, Lieut.-Col., 193, 230
Greene, General, 159–60
Grezzie, 235–6
Grey, Lieut.-Col., 272, 275
Guadeloupe, 181
Guarda, 252, 261
Guard, French Imperial, 334–6, 338, 343–4
Guerillas, 254–5, 257
Gubbins, Stamer, 428
Guinaldo, 260
Guntur, 364

HAIE Sainte, La, 326–7, 329–33, 338, 340
Hal, 10, 13, 309
Halifax, Nova Scotia, 132, 383, 449
Halkett, General, 304, 312, 314–9, 329, 333–7, 342–5, 350–1
Haly, Col., 415
Hamilton, Alexander, 180, 244–7, 255–6, 259, 287, 292–7, 304, 310, 315–6, 319, 322, 350, 355, 365, 374–5, 377
Harley, 75, 82, 88
Harwich, 225, 307–8
D'Hasfeldt, Count, 46, 48, 68
Hesse Darmstadt, Prince Charles of, 36
— — — George of, 31, 33, 38, 40, 41
— — — Henry of, 49, 56–9
Hill, General Rowland, 231, 264, 275, 289, 310
Hilsea, 143, 149, 176, 180, 193, 233
Hislop, Sir Thomas, 189, 361, 370
Hoche, General Lazare, 199–200
Hodgson, General, 144–8
Hoey, Colonel, 392, 396–7, 406
Holland, Expedition to, 296–7
Holyhead, 94
Hood, Lord, 181, 183, 185, 188, 190–2, 194
Horn, Count, 11
Hotham, Admiral, 194–5
Hougoumont, 330, 339–41, 345
Howe, Admiral Lord, 139, 141
Hull, 5–7, 25–6, 94, 231, 262, 293–4
Hutchinson, General, 212–3, 216–9, 223
Hyde, Major-Gen., 176
Hyères, 31, 61, 190–1

IGUALADA, 56
Illicit whisky stills, 241
Inkerman, 413–14, 423, 426
Inkerman Heights, 406–7, 409, 423, 427
Ionian Islands, 222, 388
Ipswich, 225, 227
Irish Elections, 385, 450

JAMAICA, 166–7
James II, 1, 4
Jennings, Col. John, 146, 148–51
Jenny, transport, 232, 305, 310
Jersey, 123, 125–6, 294–6, 446, 451
Jones, Colonel, 266, 269, 271
Joseph Buonaparte, King of Spain, 282, 289
Junot, Marshal, 248–9, 252

KALAMITA BAY, 397
Kane, Colonel, 103, 107–8
Karmul, 363
Kelly's, Major, narrative, 337
Keppel, Admiral, 144–5
Kielmansegge, 326, 333–6
King's German Legion, 312, 314, 326, 334, 341
Kinsale, 1, 112, 156–7, 242

LA FAYETTE, Marquis de, 164
Lamego, 276, 291
Lanusse, General, 210, 215
Leake, Sir John, 35, 37, 49, 50, 65
Leeward Islands, 168–70
Leipsic, 296
Leith, Lieut.-General, 247, 249, 251, 267, 271, 278–9, 281
Leith Hay, 281
Le Palais, 144–6, 149
Lerida, 42–3, 48, 55–60
Lestock, Admiral, 127
Ligny, 312, 320
Light Companies first formed, 152
Lillo, 298–9, 300–1
Limerick, 354–5, 385
Lincoln, 1, 2, 9, 21, 25, 38, 52, 85, 153, 264, 387
Lisbon, 31, 33, 35, 40, 50–3, 151, 242–3, 245, 247, 249–51, 262, 293
Liverpool, 136–7, 176–7, 179, 198, 229, 379
Llerena, 264, 275
Lockhart, Wm., 166, 197, 201, 214, 220, 223, 227, 231, 236, 294, 359, 362
Locmaria, 144
Loenhout, 297–9
Londonderry, 4, 241, 380
"Long day's March," 405
Lorient, 126–9
Loudoun, John Earl of, 132–3, 137, 149, 151
Louis XIV, 1, 9, 22, 64
Louisburg, 126
Luxembourg, Marshal, 9, 11, 13, 15
Lynedoch, Lord, 302 (*see also* Graham, Sir Thomas)
Lys, 11, 15

MAADIEH Lake, 210, 215
Macao, 237
Machell, Capt., 304, 364, 371
McKenzie's Farm, 407–9
Macready, Edward Neville, 298, 346, 352, 355, 357, 367–71, 375, 378–9
Macready's "Journal," 298–304, 309–10, 319–23, 324–7, 338–46, 350–1, 353–4
Madras, 234–7, 358, 363–4, 367, 370, 376
Madrid, 54, 75, 112, 282, 290
Malaya, 33–4, 41
Malakoff, 427, 429–30, 433–5, 437, 439
Malines, 302
Malta, 204–7, 222–3, 392, 429, 441–2
Maltese Regiment, 205

Mamelon, 429, 430, 435
Manners, Robert, 225, 227–8, 374
Mareotis, Lake, 210–13, 217, 219
Margate, 233, 353
Marlborough, Duke of, 27, 29, 60, 78, 88, 100
Marmont, Marshal, 257, 259–60, 275–7, 282
Marmorice Bay, 207–9
Marshall, Capt., 173–5
Martin's Tavern, 164
Masséna, Marshal, 248–55
Matthew, General, 166, 168, 170
Mauleverer, J. T., 385, 403, 406, 415–8, 420–2, 426–7, 430, 436, 441–3, 446–7
(1) Maxwell Christopher, 155, 170–1, 178
(2) Maxwell, Christopher, 202, 208, 229, 236, 364
(1) Maxwell, David, 149, 193, 364
(2) Maxwell, David, 193, 222, 227
Memorials, 107, 373, 444–6
Medais, 261–2
Menou, General, 209, 214–6, 220–1
Mentschikoff, Prince, 399, 405, 413
Messina, 203–4
Minet, Colonel, 231, 246–7
Minorca, 66, 68, 83, 101–4, 108, 167, 202–3, 205, 219
Miquelets, 46, 50
Moira, Earl of (Lord Rawdon), 159, 362
Mondego, 250–1, 261
Montgomerie, Lt.-Gen. James, 377
Monk's Corner, 159–61
Montjuich, 40–1
Montreal, 448
Moore, Sir John, 191–3, 209–10, 216, 241
Mulgrave, Lord, 183–7
Musketry Practice, 100, 138, 171, 295, 306, 359, 370, 376, 393, 448–9

NAMUR, 9, 15–20
Napoleon, 183, 188, 191, 223, 296, 309–10, 312, 326–8, 331, 338–40
Napoleon, Prince, 399
Nava de Aver, 254–5, 257
Navas Frias, 258–9
Negro War in Dominica, 170, 173–6
Nelson, Capt. Horatio (afterwards Lord), 192, 194, 202–3, 205
Neville, Lieut. P., 245, 263, 283–4, 310, 365, 371, 374–5
New Geneva, 198, 200, 202–3
Newcastle-on-Tyne, 25, 27–8, 94, 223, 386
Newmarket, 193, 197
Ney, Marshal, 248, 251, 313, 331–2
Nicholson, Col. Francis, 73–4
Nicoll, Lieut.-Col., 388
Nivelles, 13, 320–1, 324–5
Numbers assigned to regiments, 122
Nunn, Lieut., 150

OFF reckonings, 3
Officers, Roll of, 1689–1881, 456–68
O'Hara, General, 166–7, 187–8
Old Inkerman, 409
Ollioules, Pass of, 182–4, 188
Ompteda, General, 326, 332–3
Orange, Prince of, 296, 313, 315, 324, 332–3, 342, 345
Orangeburgh, 159–60
Ord, Sir John, 169, 172
Orleans, Duke of, 55–57
Ormond, Duke of, 55, 78, 88
Ormond, Lieut.-Col., 386
Ostend, 10, 21, 308, 354
Oswald, Major-General, 285–7
Oudenarde, 11, 303
Oxford, Earl of (*see* Harley)

PACK, General, 256, 263, 279, 283, 314
Paget, Sir Edward, 283, 289
Pakenham, Sir E., 277–8, 280
Pakenham, Col. Thos. Hy., 420, 422, 447–8, 451
Palencia, 285–8
Palliser, Hugh, 18, 37, 44, 51, 54–5, 67–8
Palliser, Walter, 41, 55, 67
Paris, 350–3
Parkhurst, Isle of Wight, 445
Pas de la Marque, 182, 185
Patullo, Col. Thos. Brodie, 403, 406, 415–8, 421, 424–6, 429–30, 435–6, 442
Pauloff, General, 416–8
Pay, 2, 3, 38, 71, 72, 79, 98–9, 110, 200, 226, 240, 310, 408
Peace with United States, 167
Pellew, Sir Edward, 234–6
Penang, 235–7
Peninsula War, summary of events, 241–2
Pennefather, John Lysaght, 393, 401–3, 411–12, 415–7, 426
Perth, 154
Peterborough, Earl of, 39–43, 49
Phillipon, General, 266, 270
Picton, Sir Thomas, 252, 255, 265–6, 271, 312, 316–8, 322, 330
Pindaree War, 364 *et seq.*
Pindarries, 364, 367
Pisuerga, 284–5, 288–9
Pitt, William, the Elder, 138, 144
Plunder, 12, 33, 58, 263
Plymouth, 127, 129–30
Pocock's, Lieut.-Col. Sir George, Narrative, 439–41
Poco Velho, 255
Pointis de, Admiral, 35, 37
Poonamallee, 237, 239, 363
Port Mahon, 65–7, 101, 204
Port Royal (Nova Scotia), 73–4
Portmore, Earl of, 107, 110
Portsmouth, 8, 25–8, 34, 40, 62, 70, 73, 88, 101, 124–6, 129, 131, 138, 142, 148, 176–8, 180–1, 199, 233, 240, 305, 377–8
Poumies, Surgeon Matthew, 39, 67, 73–4, 92
Pownall, Colonel, 22, 29, 36, 39, 42
Poyntz, Arthur, 198, 234, 358
Poyntz, J. W., 262, 305, 383, 449
Pringle, Major-General, 278–9, 286–7
Promotions, Resignations and Appointments: 1689, 4–5; 1690, 6; 1691, 7; 1692, 8; 1693, 10–11; 1694–5, 14; 1696, 20; 1697, 21; 1698–1702, 23; 1703, 26; 1704, 29; 1705, 36; 1706, 44; 1707, 53; 1708, 63; 1709, 69; 1710, 72; 1711, 76; 1712, 78; 1713, 82; 1716, 95; 1717, 96; 1718, 97, 104; 1719, 104; 1720–21, 105; 1722–27, 106; 1728–32, 113; 1733–38, 114; 1739–41, 115; 1742, 116; 1743–46, 119; 1747–49, 120; 1750–54, 121, 178, 206–8, 222–5, 227, 231–2, 259, 262, 275, 293–5, 305–7, 352, 359–60, 362–5, 367, 371–90, 392–6, 424–6, 428, 430, 432, 441, 447, 449–50, 452
Provisional battalions, 292–3
Putte, 300–1

QUATRE Bras, 16, 313–7, 321–3
Quebec, 96, 126, 449
Queen, transport wrecked, 306
Queenstown, 450
Quiberon, 129
Quimperlé Bay, 128–9

RAGLAN, Lord, 395–7–9, 400–1, 405, 408–9, 413, 422–3, 430, 434
Railway in the Crimea, 428
Ramanieh, 210, 214, 217
Rampon, General, 215–6
Ramsgate, 353
Rations, 104, 144, 200, 218–9, 226, 310, 396, 408, 425
Ravenhill, Henry, 122, 124

Rawdon, Lord, 159–60
Rebellion at Wynaad, 360
Recruiting, 6, 26, 30, 52, 69, 81, 84, 94, 101, 117, 123, 142, 157, 171, 187, 193–4, 197, 225–6, 240, 371, 380, 384–5, 387–9
Recruiting Proclamation, text of, 155
Redan, 427, 429–30, 432–40
Regnier, General, 248, 252, 256
Reserve, Loss of, 28
Retreat from Burgos, 288–290
Richards, John, 68
Robinson, Lieut.-Colonel, 383
Rochefort, 27, 138
Rodney, Sir George, 165–6
Roman Camp, 211, 214–6
Rooke, Sir George, 27, 31, *et seq.*
Rooke, Hayman, 144, 147
Roseau, 168–9
Rosetta, 217–19
Ross Lewin, J. R., 392–3, 395–6, 410, 412, 420–1
Roy, General William, 169, 171
Ryswick, Peace of, 21–22

Sabugal, 252–3, 258, 261
St. Albans, 153
St. Arnaud, Marshal, 397, 399–400, 402–4
St. Briac, 141
St. Cast, 141–2
St. Clair, General, 127–8
St. Cristobal, 264, 266, 270, 275–7, 289
St. Estevan, 45 *et seq.*
St. John, Henry (Bolingbroke), 118
St. Lucia, 166–7
St. Malo, 139–41
St. Patrick's Day Celebrations, 355
St. Payo, 259
St. Servan, 140
St. Vincent, Cape, 196–7
St. Vincente, 264, 266–9
Salamanca, 254, 276, 277, 280–1, 289
Salisbury, 25, 131
Sanders, Adjutant Gilbert, 432, 435–7, 445
Sandwich, 70, 137, 354
San Fiorenzo, Gulf of, 191
Saragossa, 75
Sardinia, 65, 101
Saunderson, Sir George (*see* Viscount Castleton)
Saunderson, Thomas, 14, 23, 25–27
Scutari, 392–4–6, 425, 427
Seaton, Lord (*see* Colborne)
Sebastopol, 400, 404–7, 410, 413–4, 423, 426–7, 433, 439, 441
Secunderabad, 367–8, 371, 373–4
Seven Years' War, 137
Seville, Treaty of, 112
Shannon, Lord, 92
Shovel, Sir Cloudesley, 27, 39, 54, 61
Siborne, 318, 328, 332
Sicily, 101
Silistria, 394
Simferopol, 405, 413
Singleton, Lieut.-Colonel, 385
Sinope, 391
Slade, Lt.-Col. Marcus J., 384–5
Slavery and the slave trade, 172
Smith, Sir Sydney, 209–10
Smuggling and "owling," 123, 126
Sobral, 248–9
Soignies, 319–20
Soimonoff, General, 413–16
Souham, General, 284, 289
Soult, Marshal, 249, 251, 257, 275, 289
Southampton, 123, 131–2, 139
Spencer, Sir Brent, 217, 219–20, 254, 257
Sowle, Major Robinson, 111–13
Stair, Earl of, 123, 137
Stanhope, Lord, 40, 52, 55, 66, 69, 75, 94
Stanhope, Sir Thomas, 145
Staremberg, Count von, 65, 75
Steenkirk, 10
Storms, Nov. 23, 1703, 128
  Crimea, 422
  Leeward Isles, 169
  Off Falmouth, 301
  Marmorice, 209
Stuart, Sir Charles, 193–4, 202–4
Stuart, Col., 161–4
Sullivan Murtogh Oge, 134
Summer Camps, 138
Sutton, Capt. Richard, 18

Tarifa, 244
Tarragona, 59, 66
Tchernaya River, 406–7, 409–10, 414, 423, 428, 431
Tessé, Marshal, 37, 44, 48–9
Thomières, General, 278–9
Tordesillas, 276
Toronto, 447
Torquemada, 284
Torres, Count de las, 109–10
Torres Vedras, lines of, 248–9
Torriano, Capt., 185–7
Tortosa, 51, 65, 67
Toulon, 44, 57, 60, 61, 181–90, 192, 205
Toulouse, Count of, 33, 34
Tournai, 303
Tralee, 356
Travancore, 237–9
Trichinopoly, 238–9, 359, 375–6
Troubridge, Sir Thomas, 204, 234–5
Tullamore, 229
Turner, Major John, 415–6
Tynemouth, 5, 6, 28, 225

Uniform, 2, 25, 98–100, 117, 130, 133, 136–7, 152–4, 156, 171, 179, 198, 201, 207, 228, 242–3, 367, 394, 397, 408, 428, 447
Union transport, 297
Urquhart, Capt., 173–4
Utrecht, Peace of, 83
Uxbridge, Lord, 330

Valencia, 50, 54
Valetta, 204, 206
Valladolid, 283, 288
Valverde, 259, 265
Varna, 394, 396, 424
Vaumorel, Philip, 179, 188–9, 190–1, 209, 235, 239, 359–61, 363, 367
Vellore, 235–6
Venda do Valle, 251
Verdun, 232, 306
Victor, Marshal, 184, 188
V.C. Awards, 415, 421, 429, 437
Victualling on transports, 144
Vigoureux, Chas. Albert, 296, 314, 320, 329–30, 340–1, 371, 373–4
Villadarias, Marquis of, 35
Villa Muriel, 285–8
Villa Velha, 257–8, 275

Wager, Sir Chas., 109–11
Wakefield, 25, 231, 243, 262
Walcheren, 242
Walker, Mark, 400–1, 404, 421, 425–9, 430
Walker, Major-General, 266–71
Warships lost off Scilly Isles, 61
Washington, George, 159, 162, 164
Waterford, 100, 132, 157
Waterloo, 329–46
Waterloo Anniversary Celebrations, 355
Waterloo Casualties, 346–9
Waterloo Letters, 318
Waterloo Medals Presentation, 355

Waterloo position, 325–9
Wellesley, Sir Arthur, 243–5 (*see also* Wellington)
Wellington, Duke of, 248–50, 253–5, 259, 270–2, 275–80, 282–92, 310, 312–3, 323, 325–9, 333–4, 337–8, 343, 389, 425
Whitfield, Walter, 26, 77–8
Whitmore, Colonel, 445
Widows men, 79, 100, 117–18, 157
Wight, Isle of, 25–7, 138–40, 142, 293–4, 296, 371, 377–8, 381
Wilkinson, W., 179–80, 189, 201, 206, 219–20, 224, 230, 237–9, 359, 363–4
Willemstad, 8, 9, 297
William, Prince of Orange (*see* William III)
William III, 1, 4, 9, 11–13, 16, 23
Williams, Chas., 63–4
Wills, Colonel Chas., 20–3, 37, 39, 42, 45, 47–8, 56–7, 59–60, 63, 65, 75–6, 83–4, 89, 92–6, 112, 118
Winchester, 131–2, 149–51, 223, 296, 305
Wirtemberg, Duke of, 11–13
Wolfe, Colonel James, 129, 138
Wood, Sir Evelyn, 423, 425
Wright, Major, 237
Wynaad, 360

York, Duke of, 292–3
Yorktown, 164–5

Zeitun, 205
Ziethen, General, 323

www.ingramcontent.com/pod-product-compliance
Ingram Content Group UK Ltd.
Pitfield, Milton Keynes, MK11 3LW, UK
UKHW021839270726
14058UKWH00002B/244